Epitaxial Microstructures

SEMICONDUCTORS
AND SEMIMETALS
Volume 40

Semiconductors and Semimetals

A Treatise

Edited by R. K. Willardson
CONSULTING PHYSICIST
SPOKANE, WASHINGTON

Albert C. Beer
CONSULTING PHYSICIST
COLUMBUS, OHIO

Eicke R. Weber
DEPARTMENT OF MATERIALS SCIENCE AND MINERAL ENGINEERING
UNIVERSITY OF CALIFORNIA AT BERKELEY

Epitaxial Microstructures

SEMICONDUCTORS
AND SEMIMETALS

Volume 40

Volume Editor

ARTHUR C. GOSSARD
DEPARTMENT OF ELECTRICAL AND COMPUTER ENGINEERING, AND MATERIALS DEPARTMENT
UNIVERSITY OF CALIFORNIA AT SANTA BARBARA
SANTA BARBARA, CALIFORNIA

ACADEMIC PRESS

Boston   San Diego   New York
London   Sydney   Tokyo   Toronto

This book is printed on acid-free paper. ⊗

Copyright ©1994 by Academic Press, Inc.

All rights reserved.
No part of this publication may be reproduced or
transmitted in any form or by any means, electronic
or mechanical, including photocopy, recording, or
any information storage and retrieval system, without
permission in writing from the publisher.

ACADEMIC PRESS, INC.
A Division of Harcourt Brace & Company
525 B Street, Suite 1900, San Diego, California 92101-4495
United Kingdom edition published by
ACADEMIC PRESS LIMITED
24-28 Oval Road, London NW1 7DX

Library of Congress Cataloging-in-Publication Data

Semiconductors and semimetals.—Vol. 1—New York: Academic Press, 1966-

   v.: ill.; 24 cm

  Irregular.
  Each vol. has also a distinctive title.
  Edited by R. K. Willardson, Albert C. Beer, and Eicke R. Weber
  ISSN 0080-8784 = Semiconductors and semimetals

  1. Semiconductors—Collected works. 2. Semimetals—Collected works.
I. Willardson, Robert K. II. Beer, Albert C. III. Weber, Eicke R.
QC610.9.S48   621.3815′2—dc19   85-642319
                                                                AACR 2   MARC-S

Library of Congress   [8709]
ISBN 0-12-752140-2 (v. 40)

Printed in the United States of America
93 94 95 96 9 8 7 6 5 4 3 2 1

Contents

CONTRIBUTORS ix
PREFACE xi

Chapter 1 Delta-Doping of Semiconductors: Electronic, Optical, and Structural Properties of Materials and Devices
E. F. Schubert

I. Introduction 2
II. Crystal Growth of δ-Doped Semiconductors 3
III. Electronic Structure 6
IV. Spatial Distribution of Dopants in δ-Doped Semiconductors 27
V. Electronic Properties of δ-Doped Semiconductors 80
VI. Field-Effect Transistors 85
VII. Properties of δ-Doped Doping Superlattices 106
VIII. Optoelectronic Devices 132
IX. Concluding Remarks 145
Acknowledgments 146
References 146

Chapter 2 Wide Graded Potential Wells
A. Gossard, M. Sundaram, and P. Hopkins

I. Introduction 153
II. Epitaxial Growth of Graded Wells 158
III. Electronic Structure and Optical Properties of Undoped Wells 167
IV. Modulation-Doped Wide Parabolic Wells 176
V. Electron Transport 184
VI. Electron Excitations 195
VII. Superlattices in Parabolic Wells 207
VIII. Conclusions 212
Acknowledgments 214
References 214

Chapter 3 Direct Growth of Nanometer-Size Quantum Wire Superlattices

P. Petroff

Abstract	219
Introduction	220
I. Growth Kinetics and Interfaces in Conventional Superlattices and Quantum Wells	221
II. The Growth of Lateral Superlattices	235
III. The Serpentine Superlattice Growth	241
IV. Remaining Issues in the Epitaxy of Lateral Superlattice Type Structures: Segregation and Size Uniformity	244
V. Optical Properties of Lateral Superlattice Based Structures	247
VI. Other Lateral Superlattice Based Structures	253
VII. Another Direct Growth Method of Quantum Wire Structures	254
VIII. Conclusions	255
Acknowledgments	255
References	255

Chapter 4 Lateral Patterning of Quantum Well Heterostructures by Growth of Nonplanar Substrates

E. Kapon

I. Introduction	259
II. Lateral Patterning Mechanisms	264
III. Lateral Patterning by MBE on Nonplanar Substrates	274
IV. Lateral Patterning by OMCVD on Nonplanar Substrates	288
V. Device Applications	310
VI. Conclusions and Future Directions	327
Acknowledgments	330
References	331

Chapter 5 Optical Properties of $Ga_{1-x}In_xAs/InP$ Quantum Wells

H. Temkin, D. Gershoni, and M. Panish

I. Introduction	338
II. Computation of Energy Levels in Quantum Wells	340
III. Optical Properties of Quantum Wells and Superlattices	345
IV. Dynamic Effects	361
V. Quantum Wires and Boxes	363

VI. Electric Field Effects	376
VII. Strained Layer Superlattices	385
VIII. Thermal Stability and Impurity Induced Disordering	404
IX. Summary	415
Acknowledgments	415
References	415

| INDEX | 421 |
| CONTENTS OF VOLUMES IN THIS SERIES | 427 |

Contributors

Numbers in parentheses indicate pages on which the authors' contributions begin.

DAVID GERSHONI (337), *Department of Physics, Technion, Haifa, Israel.*

ARTHUR C. GOSSARD (153), *Department of Electrical and Computer Engineering, and Materials Department, University of California, Santa Barbara.*

PETER F. HOPKINS (153), *Department of Electrical and Computer Engineering, and Materials Department, University of California, Santa Barbara.*

ELI KAPON (259), *Bell Communications Research Laboratories, Red Bank, New Jersey.*

MORTON B. PANISH (337), *AT&T Bell Laboratories, Murray Hill, New Jersey.*

PIERRE M. PETROFF (219), *Materials Department, University of California, Santa Barbara.*

E. FRED SCHUBERT (1), *AT&T Bell Laboratories, Murray Hill, New Jersey.*

MANI SUNDARAM (153), *Department of Electrical and Computer Engineering, and Materials Department, University of California, Santa Barbara.*

HENRYK TEMKIN (337), *Electrical Engineering Department, Colorado State University, Fort Collins.*

Preface

This volume focuses on the development of new semiconductor microsctructures that are directly fabricated by epitaxial growth. In the past, the concept of artificial quantum wells in semiconductors has been extremely fruitful and has led to discoveries of fundamental new scientific phenomena as well as to useful new practical applications. After initial development of carefully lattice-matched quantum well materials, the quantum structures field expanded to include strained-layer structures, magnetic semiconductors, and silicon- and germanium-based quantum well structures, as described in earlier volumes of this series. The goal of the present volume is to identify the principal new trends in epitaxial microstructures and to give a view of the extended limits of possibilities that these structures now make available. The possibilities include both the unique physical phenomena found in the structures and the extended range of physical structures that are becoming attainable. The work described here pushes the limits of epitaxial growth to the single-monolayer as well as the sub-single-monolayer deposition level. It concerns both newly achievable physical structures and the development of heterojunctions in hybrid systems of arsenide and phosphide compound semiconductors. Some of the research described uses lateral atomic motion on growth surfaces during the epitaxial growth process to produce lateral structures on a finer scale than could be obtained by any currently available lithographic techniques. It is remarkable that the same atomic motion that is so successful in producing exquisitely smooth surfaces and producing graded structures with highly accurate composition profiles can also be used on non-planar surfaces to define tightly controlled lateral structures with the highly organized atom arrangements described here. The atom-incorporated engineering that is described here literally arranges and stores atoms at special places on a growing crystal surface and thus puts thermodynamics to work in atomic structure engineering, both for smooth surfaces, for graded compositional-profile structures, and for approaches to wires of nanometer-scale dimension.

Chapter 1 describes the phenomenon of "delta-doping," a process in which dopant atoms are incorporated into a semiconductor within single atomic planes. Just as the engineering of the shape of the confinement potential in a quantum structure is crucial in determining the functionality of the quantum structure, the placement of dopant atoms is of comparable importance in determining properties of the structure. The ability to control the position of the dopant atoms within an atomic plane gives nearly ultimate flexibility in the use of dopants in semiconductors. Some of the benefits that accrue from this capability are 1) the ability to reduce autocompensation and increase doping efficiency by tailoring growth conditions during a growth pause; 2) the possibility of creating precisely shaped potential barriers as large as the semiconductor bandgap; and 3) the increase in flexibility with which remote doping and modulation doping of heterostructures can be performed. The technique has application to both optical and electronic semiconductor devices. Furthermore, it allows an excellent opportunity to examine the diffusion and segregation of dopants at high resolution. It thus both broadens the range of design tools for semiconductor synthesis and gives fundamental information on crystal growth.

Chapter 2 presents the new phenomena that are accessible in quantum wells with graded composition profiles and the techniques that are needed to grow and characterize these materials. Whereas the largest amount of previous research on quantum wells has been focused on narrow quantum wells with readily observable quantum effects and strong quantum confinement, there are several motivations to create specially tailored graded quantum wells with weak confinement in the growth direction and with the possibility of easily induced motion in this direction normal to the quantum well layers. The parabolically shaped wide well is especially important because when electrons are introduced into such a well by doping of adjacent barrier layers, the electrons will spread out to screen the parabolic potential with a charge distribution that is very nearly uniform. This creates a high mobility electron gas in the well with nearly perfect three-dimensional behavior. Both the electrical and optical properties of the electrons in such a well are interesting. They display striking magnetotransport behavior as well as a rich variety of strongly resonant dynamic behavior associated with the motion of the electrons in such a potential. These materials form solid-state electron resonators with much higher quality factors than could be achieved in conventional uniformly doped semiconductors. The synthesis of the wide graded wells is demanding both in tems of purity and in terms of compositional control. The technique of digital alloy growth is considered in some detail in the chapter as a means of close control of compositional profiles in these graded materials as well as in graded structures in general, where a variety of device applications appear.

Chapter 3 presents an approach to making lateral arrays of nanometer-scale wire-like structures. It is based on the periodicity of atomic-height steps that separate the atomic planes of a crystal surface whose orientation is slightly different from a principal crystal axis. This approach to quantum wires attempts to surpass the size limits of lithographic technologies by eliminating lithography completely in the fabrication of epitaxial quantum wires. Instead, it uses the actual atom motion that occurs in the step-flow mode of crystal growth to position atoms of different species over a finer length scale than previous techniques could allow. In combination with transmission electron microscopy, this epitaxial fabrication technique gives an unprecedented view of the fundamental atomic processes of crystal growth and an opportunity to confirm

PREFACE

xiii

our theoretical knowledge of these processes. The quantum states in these directly grown superlattices can be probed by the optical properties of the materials. This chapter also describes a novel concept known as the "serpentine superlattice" that is intended to reduce the non-uniformity and imperfect order in the materials.

Chapter 4 presents another promising technique for achieving lateral definition in quantum structures finer than can be produced by lithography. This employs epitaxial growth on intentionally non-planar surfaces, particularly surfaces in which grooves have been fabricated before growth. Here again, the kinetics of the atomic motion on the growth surface is crucial. The greater tendency of an atom to incorporate itself near the bottom of a groove, where there are more neighbors to which it can bond, is exploited in this technique to concentrate atoms into a fine wire-like structure which can subsequently be overgrown with a cladding layer. The tendency of epitaxial growth to be slower on certain principal crystal planes is also exploited in this fabrication technique and leads to remarkably well defined wire faces and even to the possibility of forming wires above each other in a single groove. The technology can be executed either with molecular beam epitaxy or with organ-metallic vapor phase epitaxy crystal growth. High-performance lasers can be fabricated from the materials and benefit from interfaces that can be free of the contaminants that are commonly associated with other fabrication technologies and with exposed surfaces. The cross-sectional electron microscope images of the structures grown by the technique are remarkably informative with regard to the microscopic growth mechanisms and thus provide an added bonus to the studies that are described here.

The performance and phenomena that can be observed in quantum structures are increasingly being enhanced by the growth of compound semiconductors in which both the anion species and cation species are changed on passing between adjacent layers. This is more difficult technically than growth with a single anion species, but it is of particular technological importance in the gallium indium arsenide/indium phosphide materials system, where the quantum well structures described in Chapter 5 have been formed. When the gallium indium arsenide composition is engineered to be approximately lattice matched, high-performance quantum wells with light effective masses for electrons and with low recombination rates at surfaces are of great benefit for laser operation in technologically useful wavelength ranges and for the formation of quantum wires and dots. This chapter presents the structural and optical properties of these gallium indium arsenide/indium phosphide quantum wells, wires and dots. Both strained and unstrained structures are considered. The materials described are largely grown by the technique of gas-source molecular beam epitaxy, which is particularly useful for handling the alternating anion species that are needed in the fabrication, most especially the phosphorus.

It is hoped that the material presented in this volume will make it possible for the reader to assess the likely future impact of the several new technologies that are described. It will probably be apparent that the direct device application of quantum wires and dots will be technologically much more difficult than applications of more conventional quantum wells have been. But it should also be apparent that wire and dot technologies and the insights that they are giving into the details of growth and processing will be useful tools for realization of several generations of future electronic devices. Applications of the delta doping and arsenide/phosphide technologies are already clear, and applications of graded heterostructure growth are also already in place. But it is hoped that the reader will also appreciate the possible implications of the wide graded well systems for further fundamental studies of electronic properties in three dimensional systems and for the applications of strongly resonant solid state electron cavities.

Arthur C. Gossard

CHAPTER 1

Delta-Doping of Semiconductors: Electronic, Optical, and Structural Properties of Materials and Devices

E. F. Schubert

AT&T BELL LABORATORIES
MURRAY HILL, NEW JERSEY

I. INTRODUCTION	2
II. CRYSTAL GROWTH OF δ-DOPED SEMICONDUCTORS	3
III. ELECTRONIC STRUCTURE	6
1. *The V-Shaped Quantum Well*	6
2. *Analytical Solutions of the V Shaped Quantum Well*	9
3. *Self-Consistent Solutions of the V-Shaped Quantum Well*	18
4. *Superlattices with Periodic V-Shaped Potentials (Sawtooth Superlattices)*	22
IV. SPATIAL DISTRIBUTION OF DOPANTS IN δ-DOPED SEMICONDUCTORS	27
5. *Spatial Localization of Dopants*	27
6. *Diffusion of Dopants*	51
7. *Segregation of Dopants*	60
8. *Coulomb Correlation Effects*	70
9. *Saturation of Free Carrier Concentration*	77
V. ELECTRONIC PROPERTIES OF δ-DOPED SEMICONDUCTORS	80
10. *Electronic Properties of δ-Doped Homostructures*	80
11. *Electronic Properties of Selectively δ-Doped Heterostructures*	85
VI. FIELD-EFFECT TRANSISTORS	91
12. *Homostructure Metal–Semiconductor Field-Effect Transistors*	91
13. *Selectively δ-Doped Heterostructure FETs*	101
VII. PROPERTIES OF δ-DOPED SUPERLATTICES	106
14. *Quantum-Confined Absorption*	109
15. *Photoluminescence Spectroscopy*	113
16. *Transmission Spectroscopy*	122
17. *Perpendicular Transport*	126
VIII. OPTOELECTRONIC DEVICES	132
18. *Light-Emitting Diodes*	132
19. *Doping Superlattice Lasers*	135
20. *Modulators*	143
IX. CONCLUDING REMARKS	145
ACKNOWLEDGMENTS	146
REFERENCES	146

I. Introduction

The miniaturization of the spatial dimensions of semiconductors is motivated by increased speed, reduced power consumption, and higher functional density of semiconductor devices and integrated circuits. Among the limits of scaling-down of semiconductor devices are those set by materials science, growth, and processing. In addition, more fundamental physical limits, which are almost exclusively in the quantum regime, represent boundaries of the scaling process. It is a vital, important part of present and future semiconductor research to advance materials science in order to reduce the spatial dimensions of semiconductor structures and, at the same time, to realize and understand the physical mechanisms that impose fundamental limits on further scaling.

This chapter is devoted to the electronic, optical, and structural properties of doping distributions in semiconductors that are scaled down in one dimension to their ultimate spatial limit. This limit is reached, if the dopants are confined to a single or few monolayers of the semiconductor lattice. The thickness of the doped region is comparable to the lattice constant, i.e., only few Angstroms thick. The doping distribution is then narrower than other length scales, most important the free carrier de Broglie wavelength. Such narrow doping profiles can be mathematically described by Dirac's delta function. Semiconductor with such dopant distributions will be referred to as δ-doped semiconductors.

Experimentally, δ-doped semiconductors are obtained by crystal growth interruption and subsequent evaporation of dopants on the nongrowing epitaxial crystal surface. After the growth suspension, which can typically last seconds up to several minutes, the epitaxial growth is resumed. The first report of such a growth–interrupted dopant deposition–is by Bass (1979), who found that a strong surface adsorption of Si to the nongrowing GaAs surface results in sharp doping spikes. The width of the doping spikes is not explicitly mentioned in this publication; however, a doping profile shown in the publication reveals a width of approximately 250 Å. Adsorption of dopants to the nongrowing crystal surface may also be the origin of high dopant concentrations of the substrate–epilayer interface found earlier by DiLorenzo (1971).

The versatility of growth-interrupted dopant deposition was realized by Wood et al. (1980), who mentioned that complex doping profiles can be achieved by "atomic-plane" or δ-doping. They further found that Ge-doping of GaAs leads to reduced autocompensation. However, the publication does not investigate possible diffusion of dopants from the atomic plane. Clear indications of diffusion in δ-doped semiconductors were first found by Lee et al. (1985). They concluded that diffusion over 126 Å occurred during crystal growth.

It was not until 1988 that Schubert et al. (1988d) showed that highly spatially confined Si-doping profiles in GaAs can be achieved at substrate temperatures $\leqslant 550°C$ during growth. They concluded that Si-dopants are spatially confined to a layer whose thickness is comparable to the lattice constant. They excluded diffusion over more than two lattice constants. A high degree of spatial localization was also found for Si in $Al_xGa_{1-x}As$ (Schubert et al. 1989f) and Be in GaAs (Schubert et al. 1990b). The techniques used to assess the degree of spatial localization were the capacitance–voltage profiling technique and secondary ion mass spectroscopy measurements. Both techniques were shown to be in agreement. The result of strong spatial localization of dopants in δ-doped GaAs and $Al_xGa_{1-x}As$ and the requirement of low substrate temperatures was confirmed by subsequent work (Santos et al. 1988, Webb 1989).

Novel, improved semiconductors devices were fabricated from δ-doped structures. The δ-doped metal-semiconductor field-effect transistor (MESFET) (Schubert et al. 1985e, 1986c) has the advantages of narrow channel-to-gate distance, high transconductance, and an enhanced breakdown voltage as compared to the homogeneously doped MESFET. Application of δ-doping in selectively doped $Al_xGa_{1-x}As$/GaAs heterostructures (Schubert et al. 1987d) results in high electron densities exceeding $1 \times 10^{12} cm^{-2}$. At low temperatures the population of the ground-state subband and of the first excited subband were observed for the first time in the $Al_xGa_{1-x}As$/GaAs heterostructure system. High-transconductance selectively δ-doped heterostructure transistors exhibit superior characteristics over conventionally doped transistors (Schubert et al. 1987d). Other interesting applications of the δ-doping technique include the planar-doped barrier diode (Malik et al. 1980a, 1980b), mixer diodes (Malik and Dixon, 1982), unipolar light-emitting superlattice structures (Schubert et al. 1987a), and negative differential conductance oscillators (Baillargeon et al. 1989).

The optical characteristics of doping superlattices (n-i-p-i doping schemes) are greatly improved by using the δ-doping technique. Quantum-confined optical transitions were observed in absorption (Schubert et al. 1988e) and photoluminescence experiments (Schubert et al. 1989b). The improved doping superlattice structure, which has sawtooth-shaped band-edge potentials, allowed us to fabricate the first doping superlattice light-emitting diodes, lasers, and tunable lasers (Schubert et al. 1985a, 1985b, 1989g).

II. Crystal Growth of δ-Doped Semiconductor

The key procedure to grow δ-doped III–V compound semiconductors is the suspension of epitaxial crystal growth, the evaporation of dopants on the nongrowing crystal surface, and the subsequent resumption of the epitaxial growth. As an example, we consider the growth of epitaxial GaAs by molecular beam epitaxy (MBE). The components of the MBE system within the ultra-high vacuum chamber are schematically shown in Fig. 1(a) and include the resistively heated effusion cells of Knudsen-type geometrical dimensions, the mechanical shutters, and the rotating GaAs substrate. The growing surface is monitored by reflection high-energy electron diffraction (RHEED).

Epitaxial growth of GaAs is interrupted by closing the group-III element

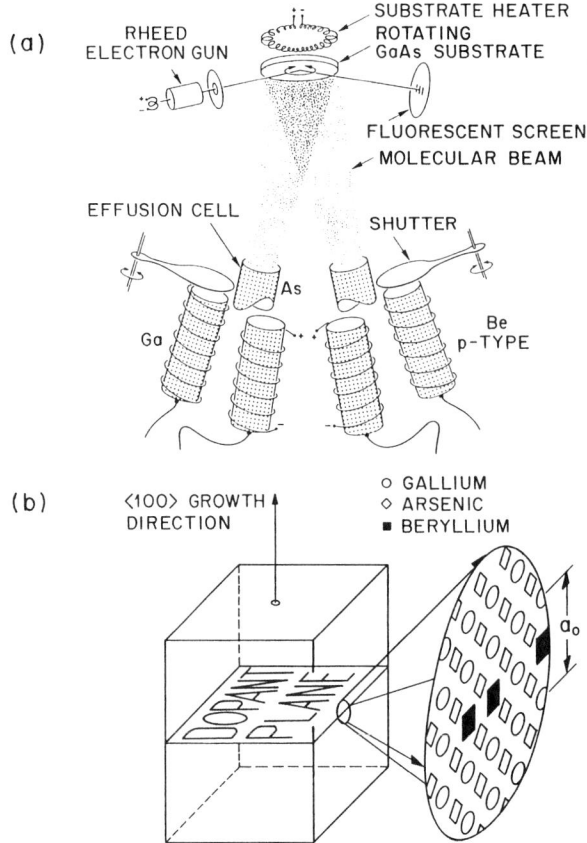

FIG. 1. (a) Schematic illustration of a molecular-beam epitaxy system comprising effusion cells with thermo-couples and heating coils, shutters and a rotating substrate with heater. Also shown is a reflection high-energy electron diffraction gun with fluorescent screen. (b) Schematic illustration of a semiconductor grown in $\langle 100 \rangle$ direction containing Be impurities in a single Ga-plane of the GaAs zincblende structure.

(Ga) shutter. The group-V element (As) effusion cell is kept open to provide an As-stabilized surface. Below the congruent sublimation temperature of 630°C for GaAs, Ga and As reevaporate from crystalline GaAs at the same rate. Note that published data on the congruent sublimation temperature of GaAs scatters between 590°C and 640°C. Autocompensation may be reduced by δ-doping an As-stabilized GaAs surface. Since most popular dopant species occupy cation sites, the anion-stabilized surface makes cation sites readily available. Typically, the growth temperature is kept between 400°C and 630°C for the growth of δ-doped GaAs, i.e., below the congruent sublimation temperature. It is well-known that the sticking coefficient of As on an As-rich GaAs surface is 0 at these growth temperatures (see for example Joyce, 1985), i.e., the surface stoichiometry is maintained.

Upon termination of epitaxial crystal growth, the dopant effusion cell is opened and dopants are evaporated on the nongrowing semiconductor surface. It was first realized that silicon (Si) and silane (SiH_4) are strongly adsorbed (chemisorbed) by the GaAs surface (Bass 1979) and exhibit no noticeable reevaporation. Dopants thus accumulate on the crystal surface during the deposition. The evaporation of dopants on a As-stabilized surface was shown to result in reduced autocompensation for Ge-doping (Wood et al. 1980), since virtually all As-sites are occupied and Ge is incorporated predominantly on Ga-sites. Similar observations were made for Si in GaAs (Schubert et al. 1987b) and for Si in $Al_xGa_{1-x}As$ (English et al. 1987).

The two-dimensional (2D) density of donors or acceptors, $N_{D,A}^{2D}$, deposited during growth interruption is calculated from the growth suspension time, τ, and the dopant cell temperature T. The 2D dopant density is thus given by

$$N_{D,A}^{2D} = N_{D,A} v_g \tau \tag{1}$$

where $N_{D,A}$ is the three-dimensional dopant density at the cell temperature T and a growth rate v_g. For example, a time $\tau = 18$ sec is required to deposit 10^{12} dopants per cm^2 at a dopant cell temperature, which results in a 3D concentration of $2 \times 10^{18} cm^{-3}$ and at a growth rate of $1 \mu m/hr$. It is advisable to choose a high dopant cell temperature and keep the growth interruption time minimal to reduce the incorporation of unwanted impurities.

After dopant deposition, the regular crystal growth is resumed. In the absence of diffusion or segregation processes, dopants are confined to a single atomic monolayer in the host semiconductor. Figure 1(b) shows dopants (e.g., Beryllium) localized in a single Ga-lattice plane of the GaAs lattice. Certainly, the confinement of dopants to a single monolayer is an ideal situation, which can be achieved only if no diffusion and segregation occur.

The surface reconstruction changes during growth interruption and during dopant deposition. During growth below the congruent sublimation temperature at Ga/As$_4$ flux ratios close to 1, a (1×1) surface reconstruction on GaAs (001) is found by reflection high-energy electron diffraction. For Ga/As$_4$ flux ratios <1, a (2×4) surface reconstruction is found. When the Ga-flux is terminated the surface remains As-stabilized the surface reconstruction remains a (2×4) As-rich reconstruction. After prolonged deposition of Si on GaAs, a change from the (2×4) to the (1×3) surface reconstruction occurs (Schubert et al. 1986b).

Hitherto, δ-doping was done in a wide range of materials, including III–V semiconductors, but also II–VI compounds (deMiguel et al. 1988) and Si (Headrick et al. 1989a, 1989b, Eisele 1989). The technique can thus be applied to a wide variety of dopants and materials. Various dopants were employed in δ-doped compound semiconductors including Si, Be, C, Ge, and Zn in GaAs (Schubert et al. 1985d, 1985e, Malik et al. 1988, Wood et al. 1980, Hobson et al. 1989). For group-IV semiconductors B, Ga, and Sb were employed (Headrick et al. 1989a, 1989b, van Gorkum et al. 1987).

III. Electronic Structure

1. THE V-SHAPED QUANTUM WELL

Two different concepts allow one to generate a V-shaped quantum well in a semiconductor; namely, by means of a sheet dopant change or by means of linear compositional grading of the mole fraction of an alloy semiconductor, e.g., Al$_x$Ga$_{1-x}$As (Levine et al. 1982, Capasso et al. 1983). Given the nature of this chapter we will concentrate on the former method. A sheet of positive dopant ions in the xy-plane at $z = z_0$ is shown in Fig. 2. The doping profile in the z-direction can be described by the Gaussian function

$$N_D(z) = N_D^{2D} \frac{1}{\sigma\sqrt{2\pi}} \exp\left[-\frac{1}{2}\left(\frac{z-z_0}{\sigma}\right)^2\right] \tag{2}$$

where N_D^{2D} is the two-dimensional dopant density (per cm^2) and σ is the standard deviation of the Gaussian function. (The standard deviation is related to the diffusion coefficient according to $\sigma = \sqrt{2L_D} = \sqrt{2D\tau}$, where D is the diffusion coefficient and τ is the diffusion time). The thickness of the doped region in the z-direction is approximately two standard deviations; i.e., 2σ.

1. DELTA-DOPING OF SEMICONDUCTORS

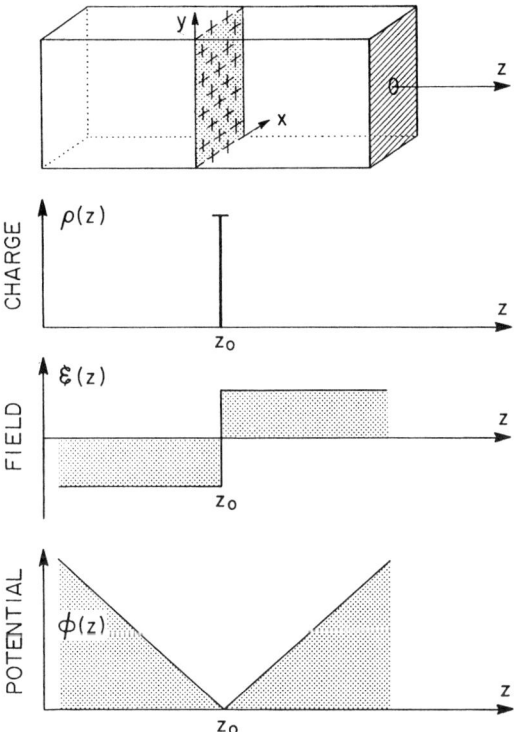

FIG. 2. Delta-function–like charge distribution and resulting step-function–like electric field and V-shaped potential distribution. Potential, field, and charge density are related by Poisson's equation.

If the standard deviation is sufficiently small, the doping profile is represented by Dirac's delta function (Schubert *et al.* 1985d)

$$N_D(z) = N_D^{2D} \delta(z - z_0) \tag{3}$$

The δ-function is a convenient and powerful function to describe such spatially localized doping profiles. Mathematically, the δ-function can be understood as a Gaussian function in the limit that σ approaches 0. However, how small a standard deviation of the Gaussian function is required *in a physical context* to be representable by the δ-function? As will be seen in the next section of this chapter, the subband structure, i.e., electronic structure, is not influenced by the magnitude of the standard deviation σ, as long as σ is much smaller than the spatial extent of the ground-state wave function of electrons or holes in the δ-doped layer. For example, the spatial extent (in the z-direction) of the ground-state electron wave function in GaAs is 50–100 Å for typical doping densities. Thus, if the thickness of the doped region is ≤ 25 Å, i.e., much smaller than the ground-state wave function for electrons, the doping profile is adequately described by the δ-function. However, if the thickness of the doped region is, e.g., 100 Å, the doping profile cannot be considered as δ-function–like. We therefore define δ-doped semiconductors as semiconductors that contain a doping profile much narrower (at least a factor of 2) than the corresponding ground-state wave function.

The electric field and the potential of a δ-function–like dopant charge distribution is obtained by integration of Poisson's equation

$$\frac{d^2\phi}{dz^2} = -\frac{d\mathbf{E}}{dz} = -\frac{eN_D(z)}{\varepsilon} \tag{4}$$

where e is the elementary charge and $\varepsilon = \varepsilon_r \varepsilon_0$ is the permittivity of the semiconductor. The electric field is obtained by integration as

$$\mathbf{E}(z) = \begin{cases} -\dfrac{1}{2}\dfrac{eN_D^{2D}}{\varepsilon} & \text{for } z < z_0 \\[6pt] +\dfrac{1}{2}\dfrac{eN_D^{2D}}{\varepsilon} & \text{for } z > z_0 \end{cases} \tag{5a}$$

or

$$\mathbf{E}(z) = -\frac{1}{2}\frac{eN_D^{2D}}{\varepsilon} + \frac{eN_D^{2D}}{\varepsilon}\sigma(z - z_0) \quad \text{for all } z \tag{5b}$$

where $\sigma(z - z_0)$ is the step function. The integration constant is chosen in Eq. (5) to obtain a symmetric field, i.e., $|\mathbf{E}(z < z_0)| = |\mathbf{E}(z > z_0)|$, as shown in Fig. 2. The second integration of Poisson's equation yields the potential

$$\phi(z) = \begin{cases} -\dfrac{N_D^{2D}}{2\varepsilon}(z - z_0) & \text{for } z < z_0 \\[6pt] \dfrac{eN_D^{2D}}{2\varepsilon}(z - z_0) & \text{for } z > z_0 \end{cases} \tag{6a}$$

or

$$\phi(z) = -\frac{eN_D^{2D}}{2\varepsilon}(z - z_0) + \frac{eN_D^{2D}}{\varepsilon}(z - z_0)\sigma(z - z_0) \quad \text{for all } z \tag{6b}$$

The potential well given by this equation is V-shaped and is illustrated in Fig. 2. Equation (6) is actually the negative potential, i.e., the potential for electrons.

It was pointed out that the singularity of the δ-function is physically problematic (Schubert et al. 1986b): According to Eq. (3), N_D tends to infinity for $z \to z_0$, i.e., $N_D(z = z_0) \to \infty$. Of course, the doping density can not exceed the host lattice site concentration (e.g., 2.2139×10^{22} cation sites per cm^3 in GaAs) in any real semiconductor. It is useful to recall that mostly the *integral* properties of the doping distribution are relevant. For example, the potential and the electric field are (according to Poisson's equation) *integral* properties of the δ-function. Therefore, the singularity of the δ-function does not represent an obstacle in analytic calculations of the V-shaped potential well.

2. Analytical Solutions of the V-Shaped Quantum Well

The spatial and energetic distribution of free carriers in a V-shaped potential well are calculated in this section using exclusively *analytical* methods. The slopes of the V-shaped potential well (i.e., the field) are $> 10^4$ V/cm for typical doping densities. The spatial width of the V-shaped potential well then becomes comparable to the de Broglie wavelength of electrons and holes. Size quantization (Schrieffer 1957) occurs, and quantum effects must be taken into account via the Schrödinger equation. The V-shaped potential well can be termed a *quantum* well.

The V-shaped quantum well problem will be solved (i) exactly and analytically using Airy functions, (ii) with the variational method, (iii) the WKB method, and (iv) in a zero-order approximation by matching the carrier de Broglie wavelength to the width of the V-shaped well.

The calculation of the eigenstate energies and wave functions are carried out in the effective mass approximation; the one-dimensional V-shaped potential then leads to textbook-type Hamiltonian operators. The wave functions in such a one-dimensional quantum-well potential can be obtained by the product method

$$\psi_{n,k_x,k_y}^{c,v}(x, y, z) = e^{ik_x x} e^{ik_y y} \psi_n^{c,v}(z) u^{c,v}(x, y, z) \tag{7}$$

where $u^{c,v}(x, y, z)$ is the periodic Bloch-type eigenfunction close to the zone center of the conduction (c) or valence (v) band (where the effective mass approximation is valid). The following calculations are restricted to conduction-band-type Bloch functions, $u^c(x, y, z)$, and to those of the valence-band, $u^v(x, y, z)$. The terms $e^{ik_x x}$ and $e^{ik_y y}$ in Eq. (7) indicate the free movement of carriers in the x- and y-directions. The function $\psi_n^{c,v}(z)$ is the eigenfunction of the nth bound state of the Hamiltonian

$$H = -\frac{\hbar^2}{-2m}\frac{\partial^2}{\partial z^2} + eV(z) \tag{8}$$

where $V(z)$ is the potential of either the conduction or the valence band edge. The function $\psi_n^{c,v}(z)$ is varying slowly on the length scale of the lattice constant ($a_0 \cong 5\,\text{Å}$), while the Bloch function is periodic in the lattice constant. Therefore, $\psi_n^{c,v}(z)$ is called the *envelope function*. The following calculations will be carried out within the framework of the envelope-function approximation (Bastard 1981). For simplicity we consider one band only and omit the superscripts c, v.

a. Exact Solution of the V-Shaped Quantum Well

The three lowest quantized energies and the corresponding wave functions are schematically sketched in Fig. 3. The subscript n of the eigenstate coincides with the number of nodes of the wave function. The ground-state wave function is therefore denoted as ψ_0 (no nodes).

The Schrödinger equation can be solved exactly and analytically, if the potential is piecewise linear. The V-shaped potential consists of two linear pieces, e.g.,

$$V(z) = |\mathbf{E}|z \qquad (z \geq 0) \tag{9a}$$
$$V(z) = -|\mathbf{E}|z \qquad (z \leq 0) \tag{9b}$$

where the absolute value of the field is given by Eq. (5). The solution of

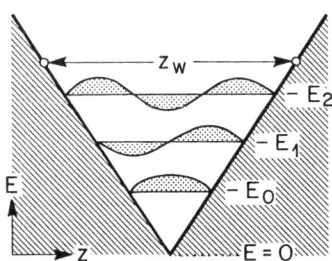

FIG. 3. Schematic illustration of quantized charge distribution in a V-shaped potential well. The quantum number n represents the number of nodes of a wave function. The lowest state is thus referred to as the $n = 0$ state.

Schrödinger's equation in a linear potential is a linear combination of Airy functions (Abramowitz and Stegun 1972). The subsequent calculation closely follows the solution of the V-shaped well using Airy functions reported earlier (Schubert et al. 1988e).

The Schrödinger equation

$$-\frac{\hbar^2}{2m^*}\frac{\partial^2}{\partial z^2}\psi_n(z) - (E - eE|z|)\psi_n(z) = 0 \tag{10}$$

is transformed into the Airy differential equation using the substitution

$$\alpha = -\left[\frac{2m^*}{(e\hbar E)^2}\right]^{1/3}(E - eEz) \tag{11}$$

Using differential calculus

$$\frac{\partial^2}{\partial z^2}\psi_n(z) = \left(\frac{\partial \alpha}{\partial z}\right)^2 \frac{\partial^2}{\partial \alpha^2}\psi_n(\alpha) \tag{12}$$

one obtains the Airy differential equation

$$\frac{\partial^2}{\partial \alpha^2}\psi_n(\alpha) - \alpha\psi_n(\alpha) = 0 \tag{13}$$

For convenience, we introduce the following notation

$$\beta = (2m^* eE/\hbar^2)^{1/3} \tag{14a}$$

$$\gamma_n = E_n/(eE) \tag{14b}$$

and consequently

$$\alpha = \beta(z - \gamma_n) \tag{14c}$$

The general solution of the Airy differential equation is then given by

$$\psi_n(z) = c_1 \text{Ai}[\beta(z - \gamma_n)] + c_2 \text{Bi}[\beta(z - \gamma_n)] \tag{15}$$

where c_1 and c_2 are constants and Ai and Bi are the Airy functions. The eigenstate energies can be obtained from the general solution of Eq. (15) by using the customary boundary conditions of continuity of $\psi_n(z)$ and $\psi'_n(z)$ at $z = 0$.

However, it turns out that the solutions are more conveniently written as eigenfunctions of even and odd symmetry. In the even states we have $\psi_n(z) = \psi_n(-z)$, and

$$\psi_n(z) = \begin{cases} c_1 \text{Ai}[\beta(z - \gamma_n)] + c_2 \text{Bi}[\beta(z - \gamma_n)] & (z > 0) \quad (16a) \\ c_1 \text{Ai}[\beta(-z - \gamma_n)] + c_2 \text{Bi}[\beta(-z - \gamma_n)] & (z < 0) \quad (16b) \end{cases}$$

In the odd states we have $\psi_n(z) = -\psi_n(-z)$, and

$$\psi_n(z) = \begin{cases} c_1 \text{Ai}[\beta(z - \gamma_n)] + c_2 \text{Bi}[\beta(z - \gamma_n)] & (z > 0) \quad (16c) \\ -c_1 \text{Ai}[\beta(-z - \gamma_n)] - c_2 \text{Bi}[\beta(-z - \gamma_n)] & (z < 0) \quad (16d) \end{cases}$$

Appropriate boundary conditions for even states are

$$\lim_{z \to 0} \frac{\partial}{\partial z} \psi_n(z) = \lim_{z \to -0} \frac{\partial}{\partial z} \psi_n(z) \tag{17a}$$

and for odd states are

$$\lim_{z \to 0} \psi_n(z) = 0 \tag{17b}$$

yields $c_2 = 0$ and $c_1 = 0$ for even and odd symmetry states, respectively. With an asymptotic expansion for the zeros of the Airy function and its derivative one obtains the eigenstate energies of the V-shaped quantum well according to

$$E_n = \left[\frac{3\pi}{4}\left(n + \frac{1}{2}\right)\right]^{2/3} \left(\frac{\hbar^2 e^2 \mathbf{E}^2}{2m^*}\right)^{1/3} \quad n = 0, 1, 2, \ldots \tag{18}$$

To obtain the exact (nonasymptotic) eigenstate energies, the term $(n + \frac{1}{2})$ in Eq. (18) is replaced by 0.437, 1.517, 2.484, 3.508, 4.491, and 5.506 for $n = 0, 1, 2, 3, 4$, and 5, respectively.

Even though the Airy functions provide the exact solutions of the V-shaped potential well, they are quite tedious to work with. Typically, computers are required to obtain the wave functions of the quantum well. An excellent balance between accuracy and simplicity is the variational solution of the quantum well, which is considered in the next section.

b. Variational Solutions of the V-Shaped Quantum Well

The variational method yields eigenstate energies and wave functions of good accuracy. The usefulness of the variational approach for quantum wells was first demonstrated for the triangular well of a Si/SiO$_2$ inversion layer (Fang and Howard 1966). The variational method was applied to the V-shaped quantum well by Schubert et al. (1989b) and yields wave functions of simple analytic form.

The following trial function for the ground state is of even symmetry (see Fig. 3), and is exponentially decaying for large absolute values of z:

$$\psi_0(z) = \begin{cases} A_0(1 + \alpha_0 z)\, e^{-\alpha_0 z} & (z \geq 0) \quad (19a) \\ A_0(1 - \alpha_0 z)\, e^{\alpha_0 z} & (z < 0) \quad (19b) \end{cases}$$

The constant A_0 is determined by the normalization condition $\langle \psi | \psi \rangle = 1$ to be $A_0^2 = 2\alpha_0/5$. The trial parameter α_0 determines the exponential decay of the wave function in the classically forbidden region, i.e., beyond the classical turning points. The expectation value of the ground state energy is given by

$$\langle E_0 \rangle = \langle \psi | H | \psi \rangle$$
$$= \frac{1}{5}\left(\frac{\hbar^2}{2m^*} \alpha_0^2 + \frac{9}{2} \frac{e\mathbf{E}}{\alpha_0} \right) \quad (20)$$

Minimization of $\langle E_0 \rangle$ with respect to the trial parameter α_0 yields

$$\alpha_0 = \left(\frac{9}{4} e\mathbf{E}\, \frac{2m^*}{\hbar^2} \right)^{1/3} \quad (21)$$

Insertion of this result into Eq. (19) yields the normalized ground-state wave function. Insertion into Eq. (20) yields the ground-state energy according to

$$E_0 = \frac{3}{10}\left(\frac{9^2}{2} \right)^{1/3} \left(\frac{e^2 \hbar^2 \mathbf{E}^2}{2m^*} \right)^{1/3} \quad (22)$$

Comparison of this variational result with the mathematically exact result of Eq. (18) yields that both methods agree to within 1%.

The calculation for the wave function and energy of the first excited state is performed analogously. The trial parameter α_1 is used in the ansatz

$$\psi_1(z) = \begin{cases} A_1 z e^{-\alpha_1 z} & (z \geq 0) \quad (23a) \\ A_1 z e^{\alpha_1 z} & (z < 0) \quad (23b) \end{cases}$$

The normalization condition yields $A_1^2 = 2\alpha_1^3$. The expectation value for the energy is given by

$$\langle E_1 \rangle = \frac{\hbar^2}{2m^*} \alpha_1^2 + \frac{3}{2} \frac{e\mathrm{E}}{\alpha_1}. \quad (24)$$

The minimization of the energy expectation value yields the trial parameter

$$\alpha_1 = \left(\frac{3}{4} e\mathrm{E} \frac{2m^*}{\hbar^2} \right)^{1/3} \quad (25)$$

and finally the eigenstate energy of the first excited state

$$E_1 = \frac{3}{2} \left(\frac{9}{2} \right)^{1/3} \left(\frac{e^2 \hbar^2 \mathrm{E}^2}{2m^*} \right)^{1/3} \quad (26)$$

The requirements for the second excited state wave function are (i) even symmetry, (ii) two nodes, and (iii) exponential decay of the wave-function amplitude beyond the classical turning points. The following trial function satisfies these requirements

$$\psi_2(z) = \begin{cases} A_2(\alpha_2^2 z^2 - 1)(1 + \alpha_2 z) e^{-\alpha_2 z} & (z \geq 0) \quad (27a) \\ A_2(\alpha_2^2 z^2 - 1)(1 - \alpha_2 z) e^{\alpha_2 z} & (z < 0) \quad (27b) \end{cases}$$

where the constant $A_2^2 = \frac{4}{9}\alpha_2/7$ is obtained from the normalization condition. The calculation of the energy expectation value is quite tedious. It is obtained as

$$\langle E_2 \rangle = \frac{1}{7} \left(3\alpha_2^2 \frac{\hbar^2}{2m^*} + \frac{47}{2} \frac{e\mathrm{E}}{\alpha_2} \right) \quad (28)$$

The minimum of $\langle E_2 \rangle$ is obtained for

$$\alpha_2 = \left(\frac{47}{12} \right)^{1/3} \left(e\mathrm{E} \frac{2m^*}{\hbar^2} \right)^{1/3} \quad (29)$$

which results in the eigenstate energy of the second excited state

$$E_2 = \frac{9}{7}\left(\frac{47}{12}\right)^{2/3}\left(\frac{e^2\hbar^2\mathbf{E}^2}{2m^*}\right)^{1/3} \tag{30}$$

The deviation of this variational result from the exact solution is <2%.

It is worthwhile to point out that the V-shaped well and the triangular well share some of the eigenstate energies and wave functions. The so-called triangular well has one infinitely steep wall and a second wall of slope $\mathbf{E}$. Such wells are realized in selectively doped $Al_xGa_{1-x}As/GaAs$ heterostructures and in Si/SiO_2 inversion layers. Both types of quantum wells are depicted in Fig. 4. As evident from this figure, the first excited state of the V-shaped well coincides with the ground state of the triangular well. Furthermore, the third excited state of the V-shaped well coincides with the first excited state of the triangular well.

c. WKB-Solutions of the V-shaped Well

The Wentzel–Kramers–Brillouin (WKB) approximation is a further method that provides good results for the eigenstate energies in a V-shaped quantum well. If the potential in the vicinity to the two classical turning points is linearly dependent on z (which is valid for the V-shaped potential well) then the WKB approximation for bound states can be written as

$$\int_a^b k_n(z)\,dz = \left(n + \frac{1}{4} + \frac{1}{4}\right)\pi \tag{31}$$

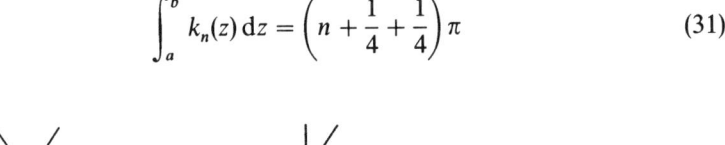

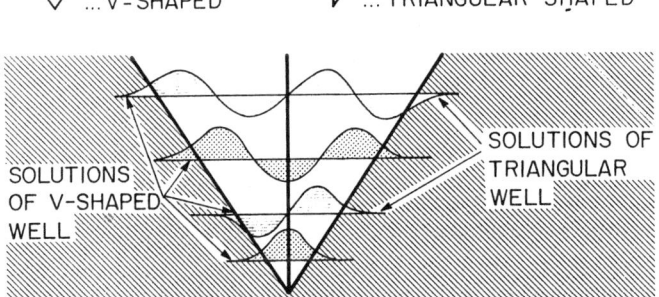

FIG. 4. Similarity of the V-shaped and the triangular potential wells. The $n = 1, 3, 5, \ldots$ solution of the V-shaped well coincides with the $n = 0, 1, 2, \ldots$ state of the triangular well.

where $z = a$ and $z = b$ are the classical turning points and the two 1/4 terms are the phase corrections at both classical turning points. (A phase factor $\pi/4$ applies for a linearly dependent potential. If the potential is discontinuous, such as in a square-shaped quantum well, the corresponding phase-factor is 0.) The wave vector $k_n(z)$ is given by

$$k_n(z)^2 = \frac{2m^*}{\hbar^2}(E_n - e\mathbf{E}z) \qquad (z > 0) \qquad (32)$$

and is real and imaginary in the well and beyond the classical turning points, respectively. Insertion of Eq. (32) into Eq. (31) and subsequent integration yields

$$E_n = \left[\frac{3\pi}{4}\left(n + \frac{1}{2}\right)\right]^{2/3}\left(\frac{e^2\hbar^2\mathbf{E}^2}{2m^*}\right)^{1/3} \qquad (33)$$

This result is identical to the exact Airy function solution with asymptotic expansions of the zero points of the Airy function. The unnormalized wave functions in the WKB approximations are given in the classically forbidden region by

$$\psi_n(z) = \frac{1}{\sqrt{\kappa_n}}\exp\left(-\int_z^a \kappa_n \, dz\right) \qquad (z < a) \qquad (34a)$$

where $\kappa_n(z)$ is given by

$$\kappa_n^2(z) = \frac{2m^*}{\hbar^2}(-e\mathbf{E}z - E_n) \qquad (z < a) \qquad (34b)$$

The wave function of Eq. (34) is also valid for $z > b$ by changing the integration variables. The wave function in the well is given by

$$\psi_n(z) = \frac{2}{\sqrt{k}}\cos\left(\int_a^z k \, dz - \frac{\pi}{4}\right) \qquad a \leqslant z \leqslant b \qquad (35)$$

The evaluation of the wave functions again requires numerical calculations. For practical use it is therefore convenient to use the previously mentioned variational wave functions.

d. Solutions of the V-Shaped Quantum Well by Zero-Order Approximation

The following method is a convenient, simple way to estimate eigenstate energies and wave functions in a quantum well. The method, originally introduced for triangular wells (Schubert *et al.* 1985c), assumes a *constant* wave vector in the well (i.e., constant de Broglie wavelength) and neglects any tunneling processes; that is, the amplitude of the wave function is 0 at the classical turning points. Such wave functions are shown in Fig. 5. The method can be understood as a further simplification of the WKB method: Whereas the WKB method takes into account small changes of the wave vector k in a spatially varying potential, the wave vector is constant in this zero-order approximation. The de Broglie wavelength of the nth state is given by

$$\lambda_{dB} = 2\pi\hbar/\sqrt{2m^*E_n} \tag{36}$$

Since the spatial width between the two classical turning points of the V-shaped well is given by (see Fig. 5 for illustration)

$$z_n = 2\frac{E_n}{e\mathbf{E}} \tag{37}$$

the wavelength is matched to the width of the quantum well according to the condition

$$\frac{1}{2}(n+1)\lambda_{dB} = z_n \tag{38}$$

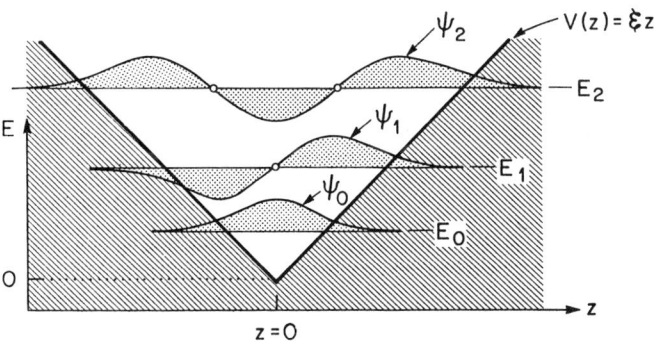

FIG. 5. Simple solution of the V-shaped well problem using sinusoidal wave functions and neglecting tunneling effects.

Elimination of z_n and λ_{dB} from Eqs. (36)–(38) yields the eigenstate energies of the V-shaped quantum well according to

$$E_n = \left[\frac{\pi}{2}(n+1)\right]^{2/3} \left(\frac{e^2\hbar^2\mathbf{E}^2}{2m^*}\right)^{1/3} \tag{39}$$

The spatial extent of the nth wave function is given by

$$z_n = [\sqrt{2}\pi(n+1)]^{2/3} \left(\frac{\hbar^2}{2m^*e\mathbf{E}}\right)^{1/3} \tag{40}$$

The relative errors for the eigenstate energies of the $n = 0$ and $n = 1$ states are 20% and 7%, respectively. Thus, considering the simplicity of this zero-order approximation, the results obtained are quite good.

3. Self-Consistent Solutions of the V-Shaped Quantum Well

In the previous sections the energetic subband structure of a V-shaped potential well was calculated without taking into account the charge of *free carriers* in the potential well. The total charge of free carriers equals the total charge of ionized dopants in a δ-doped semiconductor (the n^- or p^--type background concentration, which always occurs in a realistic semiconductor, is neglected). Therefore, the δ-doped carrier system is *neutral*, i.e., the electric field $\mathbf{E} \to 0$ for distances sufficiently far from the doped region.

A *self-consistent* calculation is required, if the charge of free carriers is taken into account. Such a self-consistent calculation satisfies simultaneously Poisson's and Schrödinger's equation. Poisson's equation allows one to calculate the potential as a function of the spatial coordinates x, y, z from a given charge distribution (both, ionized dopant change and free carrier charge). On the other hand, Schrödinger's equation enables one to calculate the charge distribution in a given potential. The calculation is termed *self-consistent* if both equations are solved in a recursive, numerical calculation, provided the solutions converge.

Self-consistent calculations are greatly simplified, if the charge distribution is assumed to depend on one spatial coordinate only. Such one-dimensional self-consistent calculations of the energy-space structure of semiconductor quantum systems (see, e.g., Ando, 1982a, 1982b, Stern and Das Sarma 1984, Vinter 1984) were initially carried out mostly for selectively doped $Al_xGa_{1-x}As/GaAs$ heterostructures. The solution of Poisson's equation is straightforward; the accurate solution of Schrödinger's equation is more elaborate. Therefore, several groups used approximate solutions of the Schrödinger equation, including the variational solution (Stern 1983, Fang and Howard 1966) and the WKB approximation (Ando, 1985) within the framework of a one-dimensional self-consistent calculation.

The band diagram obtained from a totally self consistent calculation is shown in Fig. 6 for a δ-doped GaAs sample. The two-dimensional doping density is $5 \times 10^{12}\,\text{cm}^{-2}$ located at $z = 599\text{–}601\,\text{Å}$. The doping change is assumed to be distributed homogeneously over $2\,\text{Å}$. A low temperature of $T = 4K$ was assumed for the calculation. Figure 6 reveals that four subbands are populated at the chosen doping density; the actual electron

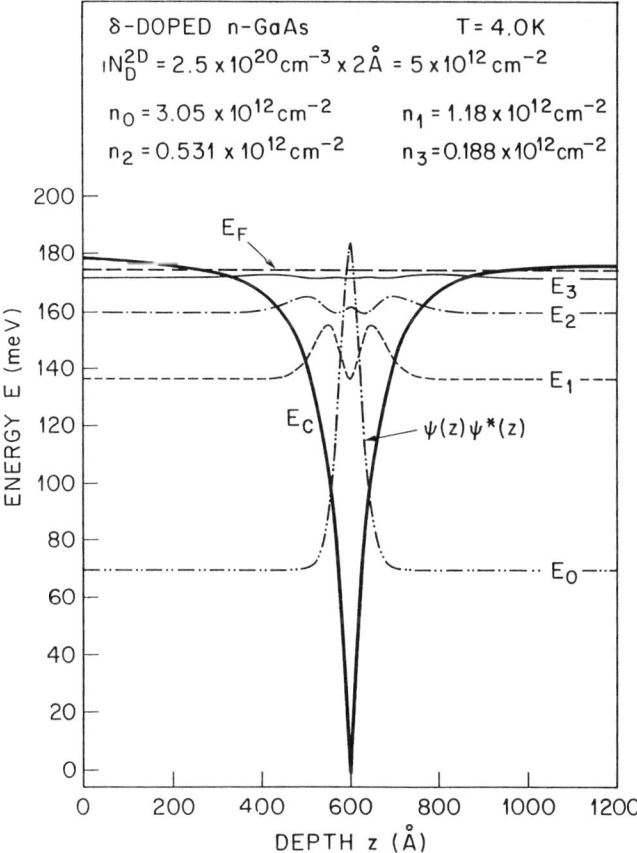

FIG. 6. Charge distribution in a δ-doped potential well obtained from the self-consistent solution of Poisson's and Schrödinger's equations. At a doping density of $5 \times 10^{12}\,\text{cm}^{-2}$, four eigenstates are populated.

densities $\psi_n(z)\psi_n^*(z)$ are also illustrated in Fig. 6. The electron densities in each subband are as follows $n_0 = 3.05 \times 10^{12}\,\text{cm}^{-2}$, $n_1 = 1.18 \times 10^{12}\,\text{cm}^{-2}$, $n_2 = 0.531 \times 10^{12}\,\text{cm}^{-2}$, and $n_3 = 0.188 \times 10^{12}\,\text{cm}^{-2}$.

Application of a voltage to the semiconductor, e.g., via a Schottky contact, results in charge accumulation or charge depletion of the quantized electron system. The subband structure of a δ-doped GaAs layer located 400 Å below a metal-semiconductor Schottky contact is shown in Fig. 7 for a temperature of 300K and a doping density of $N_D^{2D} = 2.5 \times 10^{12}\,\text{cm}^{-2}$. The Schottky barrier height is assumed to be 1.0 eV. A forward bias of $+0.75$ V and a reverse bias of -0.45 V is assumed for Fig. 7(a) and (b), respectively. In the accumulation case (positive bias) a total of two subbands are below the Fermi energy, E_F. For depletion (negative bias) only the ground-state subband remains below the Fermi energy. However, for both bias conditions several subbands are thermally populated.

From the calculations under different bias conditions the total change of the two-dimensional electron-gas density can be calculated as a function of the bias voltage. The charge-voltage relationship in turn can be used to establish the capacitance-voltage relationship (C versus V curve), since the differential capacitance at any given voltage is defined as $C = dQ/dV$.

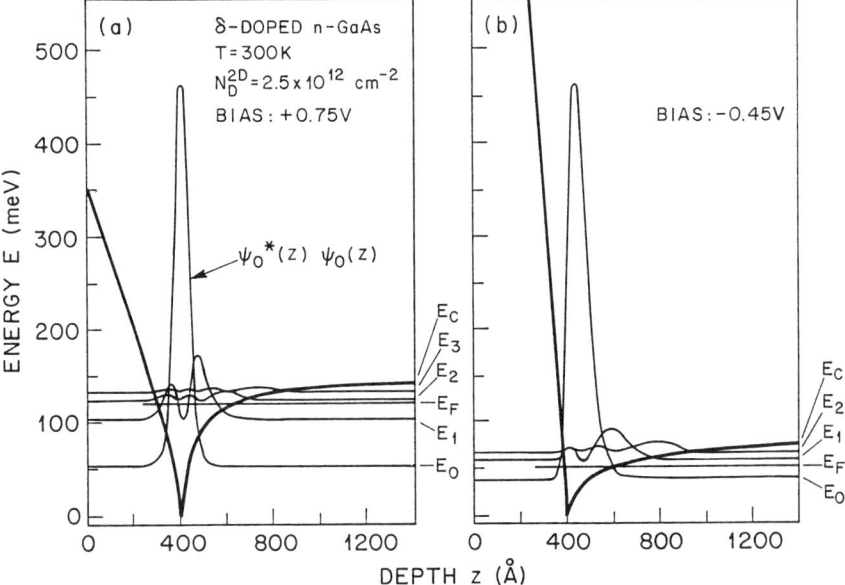

FIG. 7. Self-consistently calculated charge distribution of δ-doped GaAs at two voltages applied to a Schottky contact of $V = -0.45$ V and $V = 0.75$ V.

The discreteness of charge has not been taken into account in the preceding self-consistent calculation. Instead, the charge was assumed to be distributed homogeneously in the dopant plane. In the following we will show that this assumption leads to some uncertainties in the calculation. The coherence length of the wave function in the lateral direction (i.e., the direction of the dopant plane) allows us to estimate the dopant-density fluctuations within the doped plane, assuming random distribution of dopants. The phase coherence length in lateral direction can be estimated from the inelastic scattering time

$$\tau_{inel} = \mu m^*/e \qquad (41)$$

Certainly, any scattering event, elastic *and* inelastic, changes the phase of the electron; since the elastic scattering time is difficult to estimate, only the inelastic time is estimated in Eq. (41). The Fermi velocity

$$\frac{1}{2}m^* v_F^2 = E_F - E_0 \qquad (42)$$

allows one to estimate the mean free path between two inelastic scattering events according to

$$l_{inel} = v_F \tau_{inel} \qquad (43)$$

If we consider electrons in GaAs of approximate density $N_D^{2D} = 1 \times 10^{13}\,\text{cm}^{-2}$, a mobility of $\mu = 2000\,\text{cm}^2/\text{Vs}$, and a Fermi energy of $E_F - E_0 = 100\,\text{meV}$, the mean free path is estimated to be $l_{inel} = 550\,\text{Å}$. Within a squared area determined by the inelastic mean free path, there are an average of 300 dopant atoms. This number fluctuates depending on location, since the distribution of dopants can be assumed to be random. The average fluctuation is $\sqrt{300} = 17$ doping atoms. The relative variation in doping atoms therefore is $\pm \sqrt{N}/N = 0.06$. Because the energy of the lowest state is proportional to $(N_D^{2D})^{2/3}$ (see, for example Eq. (22), the energies are uncertain by approximately $\pm 4\%$. At lower densities (e.g., $N_D^{2D} = 1 \times 10^{12}\,\text{cm}^{-2}$) the density fluctuations of doping atom becomes more and more significant. The uncertainty can even exceed $\pm 10\%$.

The relatively large uncertainties in the subband structure of δ-doped semiconductors are due to the fact that dopant atoms are a rather infrequent event. Only several 10 to several 100 dopant atoms are within the area defined by the coherence length of the electron wave function. In compositional quantum wells (e.g., $Al_xGa_{1-x}As/GaAs$) such uncertainties are exceedingly small ($\ll 1\%$), due to negligible random potential fluctuations.

Such potential fluctuations in turn are negligibly small because the potential is defined by a large number of Al and Ga atoms. Consequently, the Poisson statistical $\sqrt{N}/N$ fluctuations are negligible in compositional quantum wells.

Exchange effects, correlation effects, and an energy-dependent effective mass due to a nonparabolic dispersion relation were taken into account in previous self-consistent calculations on selectively doped $Al_xGa_{1-x}As/GaAs$ heterostructures (see, e.g., Ando 1982a, 1982b, Stern and Das Sarma 1984).

Exchange, correlation, and nonparabolicity effects are adequate also for δ-doped structures. However, their influence is typically smaller as compared to the uncertainties due to random dopant distribution. Thus, it does not appear useful to consider minor effects such as exchange and correlation and, at the same time, ignore effects due to random dopant distribution.

Random potential fluctuations could be taken into account by truly three-dimensional self-consistent calculations, in which a random placement of dopants is realized. Unfortunately, such three-dimensional calculations are quite elaborate and were hitherto not carried out.

4. Superlattices with Periodic V-Shaped Potentials (Sawtooth Superlattice)

In a superlattice, quantum-mechanical coupling occurs between adjacent wells. Such a coupling does not occur in single, isolated quantum wells (considered in previous sections) and in wells separated by large barriers. In this section we consider a periodic potential, which is described by linear potentials, i.e., potentials that depend linearly on the spatial coordinate. The Kronig–Penney model is used to calculate the wave functions and the dispersion relation in the superlattice.

Superlattices with a periodic V-shaped potential can be generated by alternating n-type and p-type δ-doping in a semiconductor. Such a doping superlattice structure is depleted of free carriers, because electrons recombine immediately with holes. The band diagram of such a sawtooth doping superlattice is shown in Fig. 8. If the donor and acceptor sheets have the same density and are equidistant, a *symmetric* sawtooth structure is obtained (Schubert *et al.* 1985d). If a donor sheet is separated from the two adjacent acceptor sheets by two unequal distances, an *asymmetric* sawtooth structure is obtained (Glass *et al.* 1989), as shown in Fig. 8(a). Note that conduction and valence bands are *parallel* in the sawtooth doping superlattice.

Sawtooth-shaped band edges can be also generated in compositional superlattices, e.g., by a gradual change of the alloy composition of ternary

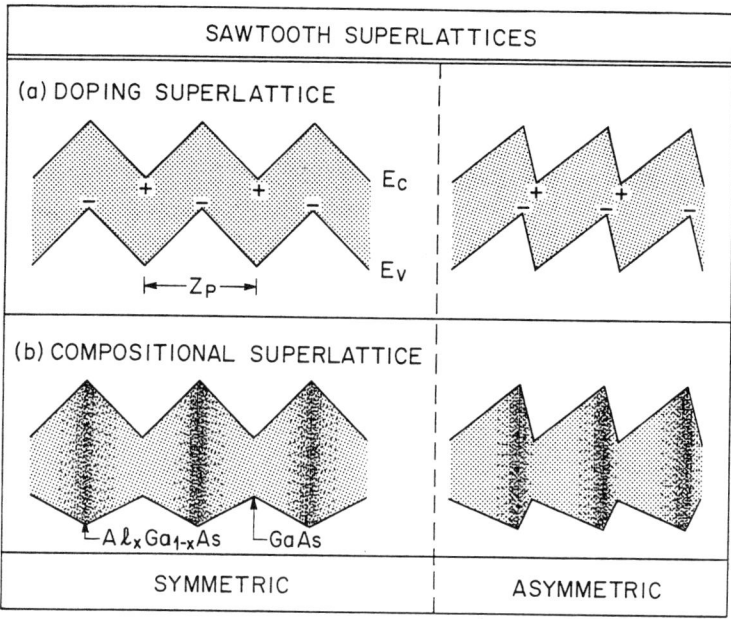

FIG. 8. Symmetric and asymmetric sawtooth superlattices. In doping superlattices the conduction and valence band edges are parallel. In compositional superlattices the band edges are antiparallel.

alloy semiconductor (Allyn et al. 1980). Such compositionally graded $Al_xGa_{1-x}As$ (Levine et al. 1982) was used in high-speed, graded-gap phototransistors (Capasso et al. 1983). Periodic grading of the alloy composition results in a compositional sawtooth superlattice potential, which can be either symmetric or asymmetric, as shown in Fig. 8(b). Note that the conduction and valence band edges are *antiparallel*, in contrast to the sawtooth doping superlattices. Capasso et al. (1983) reported transient electrical polarization in such compositional sawtooth superlattices. The polarization was induced by photoexcited carriers and is due to the lack of inversion symmetry in asymmetric sawtooth superlattices.

The calculation of the superlattice dispersion relation and the eigenstate energies are carried out in the effective mass approximation. The sawtooth-shaped periodic potential is given by

$$V(z) = \begin{cases} (z - mz_p)E & \\ (-z - mz_p)E & \end{cases} \quad \text{for } m = 0, \pm 1, \pm 2, \pm 3, \ldots \quad \begin{matrix}(44a)\\(44b)\end{matrix}$$

where z_p is the superlattice period. The electric field, E, and the amplitude of the superlattice potential, V_{zz}, are related by $V_{zz} = Ez_p/2$. The wave function in this periodic one-dimensional potential is given by the product ansatz of

$$\psi^{c,v}_{n\mathbf{k},k_y}(x, y, z) = e^{ik_x x} e^{ik_y y} \psi^{c,v}_n(z) u^{c,v}(x, y, z) \tag{45}$$

where $e^{ik_x x}$ and $e^{ik_y y}$ are the plane waves in the lateral directions (x- and y-directions), $\psi^{c,v}_n(z)$ is the envelope function of the conduction (c) or valence (v) band, and $u^{c,v}(x, y, z)$ is the Bloch function of the host material. For simplicity, the superscripts c and v, which refer to conduction and valence band, will be omitted. Note that the envelope function $\psi_n(z)$ depends on z only and describes the quantized motion in z-direction. The subscript n refers to the nth quantized state of the superlattice.

The solution of the envelope function in a linear periodic potential are Airy functions (Abramowitz and Stegun 1972). Within the one-electron model (i.e., neglecting exchange and correlation corrections), the wave function for the left half of the mth quantum well can be written as

$$\psi_n(z) = c_1 \text{Ai}\{\beta[-(z - mz_p) - \gamma_n]\} + c_2 \text{Bi}\{\beta[-(z - mz_p) - \gamma_n]\} \tag{46a}$$

and similarly for the right half of the well

$$\psi_n(z) = c_3 \text{Ai}\{\beta[z - mz_p) - \gamma_n]\} + c_4 \text{Bi}\{\beta[(z - mz_p) - \gamma_n]\} \tag{46b}$$

where

$$\beta = (2m^* e E/\hbar^2)^{1/3} \tag{47a}$$

$$\gamma_n = E_n/(eE) \tag{47b}$$

The wave function in an adjacent quantum well has an additional phase factor $e^{ik_z z_p}$, where k_z is the superlattice momentum, i.e., the momentum perpendicular to the dopant planes.

The wave function $\psi_n(z)$ and its derivative $d\psi_n(z)/dz$ must be continuous at the boundaries ($z = 0$, $z = \pm z_p/2,\ldots$). If these two conditions are applied to the wave functions of Eq. (46), and the periodicity of the wave functions is taken into account, one obtains four equations for $c_1, c_2, c_3,$ and c_4. The system of four equations can be solved if the determinant of the system vanishes. This condition leads to a relationship between the energy and the perpendicular momentum. The dispersion relation can be written as

$$\cos\left(\frac{k_z z_p}{2}\right) = \pi^2 \{[\mathrm{Ai}'(z_w)\mathrm{Bi}'(z_b) - \mathrm{Ai}'(z_b)\mathrm{Bi}'(z_w)]$$
$$\times [\mathrm{Ai}(z_b)\mathrm{Bi}(z_w) - \mathrm{Ai}(z_w)$$
$$+ [\mathrm{Ai}'(z_b)\mathrm{Bi}'(z_w) - \mathrm{Ai}'(z_w)\mathrm{Bi}'(z_b)]$$
$$\times [\mathrm{Ai}(z_w)\mathrm{Bi}'(z_b) - \mathrm{Ai}'(z_b)\mathrm{Bi}(z_w)]\} \quad (48)$$

where z_w refers to values of z in the center of the wells $z_w = m z_p$ and z_b refers to values of z in the center of the barriers $z_b = (m + \frac{1}{2})z_p$; the primes represent the derivatives of the Airy function with respect to the entire argument (not to z). The dispersion relation can be evaluated numerically and is shown for electrons in a GaAs sawtooth doping structure in Fig. 9. The parameters used for the calculation are a two-dimensional doping density of $N^{2D} = 10^{13}\,\mathrm{cm}^{-2}$ and a period of $z_p = 150\,\mathrm{\AA}$. From these parameters the electric field $\mathbf{E} = eN^{2D}/2\varepsilon = 9 \times 10^5\,\mathrm{V/cm}$; the amplitude of the superlattice potential modulation is given by $V_{zz} = \mathbf{E}z_p/2 = 675\,\mathrm{meV}$. The dispersion relation for the lowest subband is flat, as depicted in Fig. 9, indicating very little coupling between adjacent potential wells of the superlattice. In contrast, the second excited state, $n = 2$, has a finite dispersion for perpendicular motion. With increasing energy the barrier becomes narrower and tunnelling between wells becomes a relevant process.

The broadening of the subband into a broader miniband allows one to estimate the efficiency of vertical transport in doping superlattices. The Kronig–Penney model allows one to calculate the effective mass along the z-direction, i.e., perpendicular to the doping planes. Using the definition of the effective mass $m_z^* = \hbar^2 [d^2 E_z(k_z)/dk_z^2]^{-1}$ and the linearized dispersion relation from the Kronig–Penney model one obtains the effective dispersion mass for motion in perpendicular direction

$$m_{z,n}^* = \frac{\hbar^2}{z_p^2 \Delta E_n} \quad (49)$$

where $2\Delta E_n$ is the energetic width of the nth miniband. For the lowest state ΔE_0 is very small and consequently the mass $m_{z,0}^*$ is extremely heavy. In contrast for $n = 1$, the miniband width is $2\Delta E_n \cong 10\,\mathrm{meV}$, and Eq. (49) yields a mass $m_{z,2}^*/m_0 \cong 0.07$, which has the same order of magnitude as the regular electron mass in GaAs of $m^*/m_0 \cong 0.067$.

The calculation of the dispersion relation was also performed for heavy and light holes in GaAs for the same period and doping density (Ulrich et al. 1989). The broadening of subbands into minibands was found negligible for heavy holes. The light hole dispersion was found to be similar to the electron dispersion illustrated in Fig. 9.

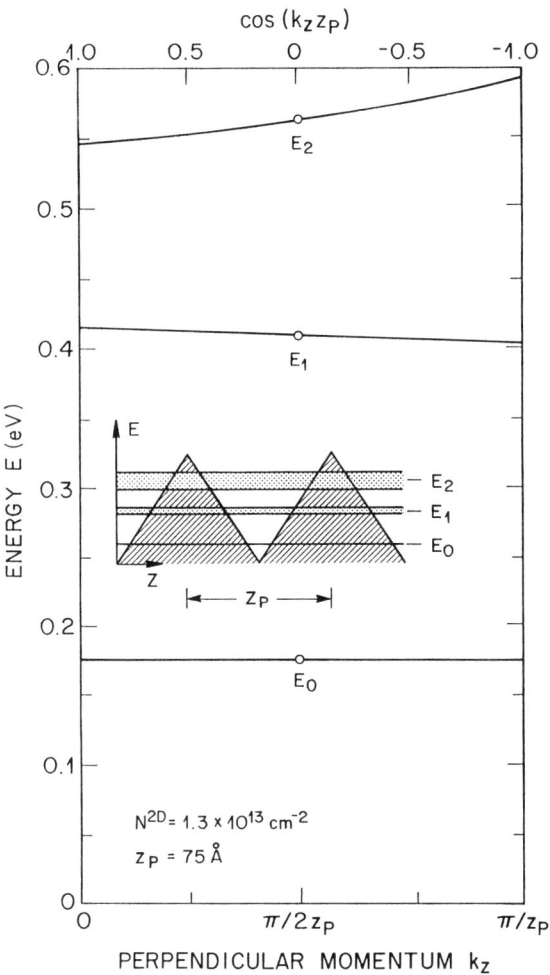

FIG. 9. Calculated dispersion relation of a sawtooth doping superlattice. The miniband width increases with the quantum number. The flat dispersion of the $n = 0$ state indicates the lack of coupling between states of adjacent quantum wells.

IV. Spatial Distribution of Dopants in δ-Doped Semiconductors

In this section, the spatial localization of dopants in a δ-doped semiconductor is investigated. What is the actual spatial width of the dopant distribution in a δ-doped semiconductor? How does the width depend on growth conditions? Is dopant diffusion relevant, and what are the actual diffusion coefficients? Does surface segregation and migration of dopants toward the growing crystal surface, occur? What is the driving force of such dopant segregation? How do the aforementioned effects depend on dopant concentration? Are there electrostatic Coulomb correlation effects at high concentrations? Those are some of the intriguing questions of basic importance for δ-doped semiconductors. The following sections are an attempt to answer those questions.

The experimental methods used to assess the spatial distribution of dopants are (i) capacitance–voltage profiling and (ii) secondary ion mass spectroscopy. Under optimized conditions, the resolution of both techniques is $<20\,\text{Å}$ and therefore well suited for this purpose.

Results are summarized on a number of dopants and host materials. The dopants include Si, Be, C, and Zn. The compound semiconductor materials include GaAs and $Al_xGa_{1-x}As$. Most of the materials are grown by molecular-beam epitaxy and gas-source MBE. Zn-doping is used for GaAs grown by organo-metallic vapour-phase epitaxy (OMVPE).

5. Spatial Localization of Dopants

a. Definition of δ-Doping

The degree of spatial localization of dopants in δ-doped semiconductors is of fundamental importance. Only if dopants can be spatially confined to one or a few monolayers of a semiconductor, are δ-doped semiconductors feasible. For the sake of clarity, the two extreme dopant profiles are illustrated in Fig. 10: the δ-doped quantum well and the homogeneously doped quantum well. The homogeneously doped well can be thought to emerge from the δ-doped quantum well by spreading out the dopants, i.e., increasing the width dz, as shown in Fig. 10.

The spreading of dopants has significant consequences. The subband structure as well as the shape of the potential well changes. In the δ-doped semiconductor, for example, the subband energies are proportional to $n^{2/3}$, where n is the subband index. In the homogeneously doped well, the subband energies are proportional to n, in analogy to the harmonic oscillator potential. Therefore the question arises as to the maximum

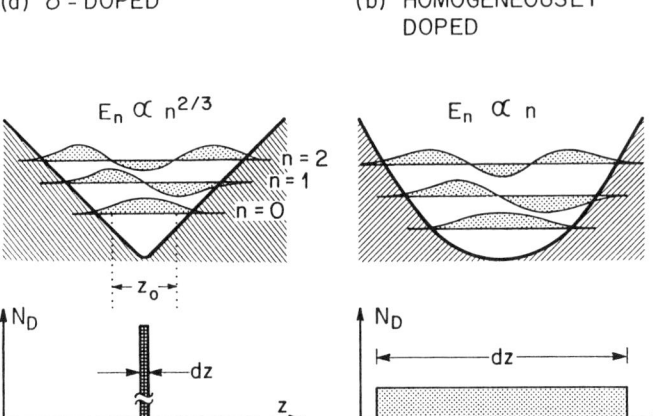

FIG. 10. Comparison of δ-doped and homogeneously doped potential wells. The subband energies are proportional to $n^{2/3}$ and to n for the δ-doped and the homogeneously doped wells respectively.

spreading of dopants for which the semiconductor can be considered to be δ-doped.

A doping profile can be considered δ-function-like as long as the spread of dopants is much smaller than the spatial extent of the ground-state wave function. As illustrated in Fig. 10, the width of the dopant profile, dz, is much smaller than the spatial extent of the ground-state wave function, z_0. The width dz may be 5, 10, or 100 times narrower than the spatial extent of the lowest wave function, without any significant changes in the electronic subband structure. The absence of significant changes in the subband structure for any dz as long as $dz \ll z_0$ is intuitively quite obvious. It can be easily verified by any of the calculations mentioned in the previous section, e.g., the variational calculation.

We define a δ-doped semiconductor as a structure that contains dopants spatially confined to a narrow region, whose thickness is small as compared to the ground-state wave function of the corresponding free carrier system.

This definition allows one to estimate an upper limit for the width of the doping distribution in a δ-doped semiconductor. To estimate this upper limit, the spatial extent of the ground-state wave function is required. It is given by

$$z_0 = 2(\langle z^2 \rangle - \langle z \rangle^2)^{1/2} \tag{50}$$

where $\langle z \rangle$ is the position expectation value. Using the variational calculation described in the previous section, z_0 can be readily evaluated. One obtains

$$z_0 = 2\sqrt{\frac{7}{5}}\left(\frac{4}{9}\right)^{1/3}\left(\frac{\hbar^2}{eE2m^*}\right)^{1/3} \tag{51a}$$

Using $\mathbf{E} = eN^{2D}/2\varepsilon$, the width of the ground-state wave function can be expressed as a function of the doping density:

$$z_0 = 2\sqrt{\frac{7}{5}}\left(\frac{4}{9}\right)^{1/3}\left(\frac{\varepsilon\hbar^2}{e^2 N^{2D}m^*}\right)^{1/3} \tag{51b}$$

The spatial extent of the lowest subband can also be estimated by the zero-order approximation, which yielded Eq. (40) in the previous section. With $\mathbf{E} = eN^{2D}/2\varepsilon$ one obtains

$$z_0 = (\sqrt{2}\pi)^{2/3}\left(\frac{\varepsilon\hbar^2}{e^2 N^{2D}m^*}\right)^{1/3} \tag{51c}$$

The width of the ground-state subband obtained from Eqs. (51b) and (51c) agree within $\cong 30\%$. As an example we choose a doping concentration of $N_D^{2D} = 5 \times 10^{12} \, \text{cm}^{-2}$ and obtain for the spatial extent of the lowest electron wave function in GaAs ($m^*/m_0 = 0.067$) a value of $z_0^e \cong 50\,\text{Å}$. Consequently, all dopant distributions with a width $\ll 50\,\text{Å}$ are equivalent to a δ-function-like dopant distribution. It is not relevant if Si dopants are localized in a single Ga monolayer ($dz < 1\,\text{Å}$) or if they are distributed over $10\,\text{Å}$. In a second example we choose p-type, δ-doped in GaAs with the acceptor concentration $N_A^{2D} = 5 \times 10^{12} \, \text{cm}^{-2}$ and obtain for the spatial extent of the heavy-hole ground-state wave function $z_0^{hh} \cong 25\,\text{Å}$. Consequently, all acceptor distributions with a width $\ll 25\,\text{Å}$ are equivalent to a δ-function-like distribution. As a general guideline, dopant distributions with a full width at half-maximum of less than or equal to $15\,\text{Å}$ are equivalent to δ-function-like dopant distributions. This practical rule holds for most n-type and p-type III–V compound semiconductors.

b. Theory of Capacitance–Voltage (CV) Profiling in Quantum Systems

The capacitance–voltage (CV) profiling technique is suited to spatially resolve doping profiles in semiconductors. The technique is typically performed on p^+n-, pn^+-junctions or on reversely biased Schottky contacts. The technique was originally developed for homogeneously doped semiconductors and is described in standard textbooks (see, for example, Sze 1980). The resolution of CV profiles on homogeneously nondegenerately doped semiconductors was shown to be limited by the Debye–Hückel screening length (Kennedy *et al.* 1968, Johnson and Panousis 1975), which depends on the temperature T and the free electron concentration according to

$$L_D = [\varepsilon kT/(e^2 n)]^{1/2} \tag{52}$$

For *degenerately* doped semiconductors the thermal energy $\tfrac{3}{2}kT$ is replaced by the Fermi energy $(E_F - E_c)$ and the resolution of CV profiles is thus determined by the Thomas–Fermi screening length

$$L_{TF} = \left[\frac{2}{3}\varepsilon(E_F - E_c)/(e^2 n)\right]^{1/2} \tag{53a}$$

as pointed out by Schubert and Ploog (1986e).

In the following we will show that the basic theory of CV profiling does not apply to semiconductors with quantum-confined carriers. We will establish the theoretical framework of CV profiles on quantum-mechanical systems and show that the resolution of the profiles is given by the spatial extent of the ground-state wave function.

Capacitance–voltage measurements can be performed by means of a Schottky contact on an *n*-type semiconductor as shown in Fig. 11(a). The semiconductor is depleted of free carriers (depletion region) below the Schottky contact, yielding a charge profile $\rho(z, V)$ as shown in Fig. 11(b). By increasing the external bias from V to $(V + dV)$ the depletion region will increase from a width W to $W + dW$. One can show that the measured CV concentration, N_{CV}, coincides with the real doping concentration, N_D. The CV concentration is given by the well-known equation

$$N_{CV} = \frac{C^3}{e\varepsilon} \frac{dV}{dC} \tag{54}$$

where C is the capacitance per unit area, e is the elementary charge, and ε is the permittivity of the semiconductor. The CV concentration given by Eq. (54) occurs at a depth

$$z = \varepsilon/C \tag{55}$$

Thus a CV profile (N_{CV} versus z) can be inferred from the CV measurement.

In a quantum-mechanical system, the depletion approximation (which

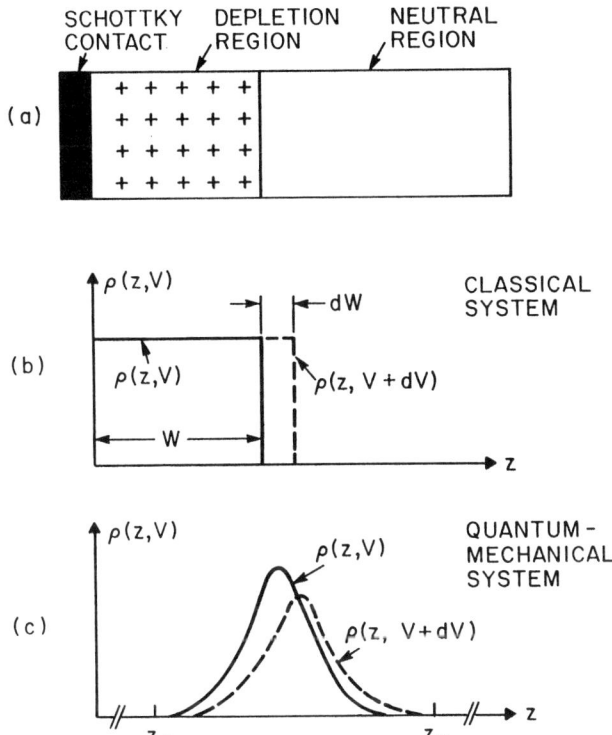

FIG. 11. Illustration of the differences in charge distribution of a classical electrostatic system and a quantum-mechanical system. Upon application of a voltage increment to the Schottky contact (a) the depletion region increases in width by an increment dW. In a quantum-mechanical system, the external perturbation results in a charge redistribution. The depletion approximation is not applicable in the quantum regime.

assumes that the semiconductor is totally depleted slice by slice for each voltage increment), which is one of the fundamental assumptions for the CV proifiling theory, fails to be valid. Instead, the entire wave function of the carrier system will be perturbed as a response to the external voltage change, as illustrated in Fig. 11(c). Upon a pertubation, the wave function may change its expectation value, $\langle z \rangle$, amplitude, and shape.

Assume a charge distribution $e\psi^*(z, V)\psi(z, V) = \rho(z, V)$ and $\rho(z, V + dV)$ at a bias of V and $V + dV$, respectively. The change of charge density is then given by

$$d\rho(z, V) = \rho(z, V) - \rho(z, V + dV) \tag{56}$$

The total change of charge per unit area is given by

$$dQ = \int_{-\infty}^{\infty} d\rho(z, V) dz \tag{57}$$

An equivalent change of sheet-charge density $d\rho_{equ}$ at $\bar{z}$ is expressed by

$$d\rho_{equ}(z, V) = \int_{-\infty}^{\infty} d\rho(z, V) dz\, \delta(z - \bar{z}) \tag{58}$$

where $\delta(z - \bar{z})$ is the delta function and

$$\bar{z}(V) = \int_{-\infty}^{\infty} z\, d\rho(z, V) dz \Big/ \int_{-\infty}^{\infty} d\rho(z, V) dz \tag{59}$$

The proof of the equivalence of the two charge distributions of (56) and (58) is straightforward and can be obtained by insertion of both distributions in Poisson's equation and by subsequent twofold integration. The value of $\bar{z}(V)$ is the *centroid of charge*. For sufficiently small voltage changes dV one can show that $\bar{z} = \langle z \rangle$; i.e., the centroid of the charge change is identical to the position expectation value. Thus, under these restricting conditions, the capacitance represents the quantum-mechanical position expectation value $C \cong \varepsilon/\langle z \rangle$. Using this interpretation, the capacitance technique is a powerful method to determine the position expectation value of a quantum-mechanical system.

However, two additional contributions are made to the capacitance of a quantum-confined two-dimensional electron gas, due to a variation of the subband energy and the Fermi level. The two additional terms were first pointed out for Silicon MOS-inversion layers (Stern 1974) and other quantum well systems (Luryi 1988, Ullrich et al. 1988). The total capacitance measured in a small-signal CV measurement is then given by

$$C = \left(\frac{dV}{dQ}\right)^{-1} = \left[\frac{dV_{el}}{dQ} + \frac{1}{e}\frac{d(E_F - E_0)}{dQ} + \frac{1}{e}\frac{dE_0}{dQ}\right]^{-1} \tag{60}$$

where dV_{el} is the pure electrostatic potential change due to the change of charge density; dV_{el} can be inferred from Poisson's equation. Following the terminology used by Beltram et al. (1988), we call the second term of Eq. (60) the *quantum capacitance* and the third term the *Stark capacitance*. The quantum capacitance is easily evaluated for a two-dimensional electron system from the Fermi energy

$$E_F - E_0 = (kT \ln e^{n_{2DEG}^{(kT \, DOS2D)}} - 1) \qquad (61)$$

where $DOS^{2D} = m^*/\pi z^2$ is the two-dimensional density of states. The Stark capacitance is a second-order effect for potential wells with inversion symmetry. Both, the quantum and the Stark capacitances are exceedingly small for the geometries used in our studies and can therefore be neglected. Thus, the measured capacitances are, as a good approximation, given by

$$C \cong \varepsilon/\langle z \rangle \qquad (62)$$

where $\langle z \rangle$ is again the position expectation value of the electron gas.

CV *profiles*, which are obtained from CV *measurements* using Eqs. (54) and (55), do conserve the integral carrier density; i.e.,

$$\int_{-\infty}^{\infty} N_{CV}(z)dz = \int_{-\infty}^{\infty} n(z)dz \qquad (63)$$

where $n(z)$ is the true physical concentration, and $N_{CV}(z)$ is the concentration of the CV profile. Equation (63) can be proven easily by using the definition of the CV profiles given in Eq. (54) and (55). Equation (63) holds for any carrier distribution, $n(z)$, including quantum-mechanical distributions. Thus, integration of the CV profile allows us to determine the concentration of free carriers. We will make use of this characteristic later. Furthermore, Kroemer et al. (1980) showed that the centroid of CV and true concentrations do coincide approximately; that is,

$$\frac{\int z N_{CV}(z)dz}{\int N_{CV}(z)dz} \cong \frac{\int z n(z)dz}{\int n(z)dz} \qquad (64)$$

Thus, even though the CV profile has no immediate physical meaning in a quantum-mechanical system, the integral and the centroid of the CV profile do have a definite physical meaning.

The resolution of the CV profiling technique in a quantum-mechanical semiconductor system was hitherto not analyzed. Assume an ideal δ-doped semiconductor in which all dopants are localized in a single monolayer, i.e., the width of the dopant distribution ($dz \ll 1$ Å) is negligible. Nevertheless, the width of the CV profile on such a structure will be several tens of Angstroms. What determines this width of the measured CV profile? In the following, it will be shown that the resolution of the CV technique is limited by the spatial extent of the ground-state wave function. As previously mentioned, the depth scale of the CV profile is derived from the inverse capacitance, which in turn is given by the expectation value of the electron system. The spatial resolution of the CV profiles is therefore determined by the change of the position expectation value upon perturbation by the external voltage.

For δ-doped semiconductors the majority of carriers are in the ground state (e.g., 60% for $N_D^{2D} = 5 \times 10^{12}\,\text{cm}^{-2}$, see Fig. 6). The fraction of carriers in the lowest subband increases even further for negative biases used during the CV measurement due to depletion of the potential well. We will therefore consider the ground-state wave function only and neglect excited states. The ground-state wave function is shown in Fig. 12 for two biases. The expectation value of the unperturbed, symmetric V-shaped potential well coincides with the dopant plane. Upon perturbation by the voltage increase from V to $V + dV$ the expectation value of the perturbed wave function moves away from the dopant plane. It is intuitively quite clear that for any perturbation

$$V \ll \frac{eN_D^{2D}}{\varepsilon} d,$$

where d is the distance between the applied voltage and the dopant plane, the change in expectation value is smaller or comparable to the spatial extent of the unperturbed wave function. It is therefore suggestive that the

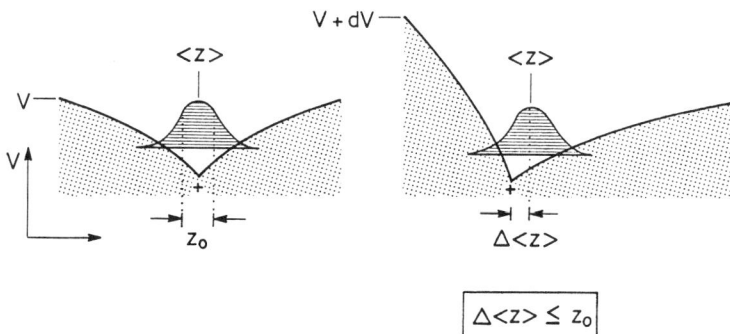

FIG. 12. Schematic illustration of the change in position-expectation value, $\langle z \rangle$, upon an external perturbation by an electric field. It is shown in the text that for typical perturbation the change in position-expectation value, $\Delta\langle z \rangle$, is smaller than the spatial extent of the lowest subband, z_0, i.e., $\Delta\langle z \rangle \lesssim z_0$.

resolution of the CV profile on a quantum-mechanical system is given by the spatial extent of the ground-state wave function. Furthermore, it is obvious that the resolution of CV profiles of quantum-mechanical systems is not given by the Debye or Thomas–Fermi screening length. In the following the resolution of CV profiles will be calculated. The calculation consists of two steps. First, we calculate the spatial extent of the ground-state wave function in a δ-doped semiconductor using a variational wave function. In the second step, the change of the expectation value is calculated using second-order perturbation theory.

The spatial extent of a wave function can be obtained from the position uncertainty of the electron system using a variational wave function. The position uncertainty of a quantized system is defined as

$$\sigma_z = \sqrt{\langle z^2 \rangle - \langle z \rangle^2} \tag{65}$$

For a Gaussian wave function the position uncertainty coincides with the standard deviation of the Gaussian function. The spatial extent of such a wave function is approximately twice the position uncertainty (for the example of a Gaussian wave function the full width at half-maximum is actually $2.35 \times \sigma_z$. Since we are interested only in an estimate, the factor of $2.35 \cong 2$ is sufficiently accurate.) The spatial extent of the lowest wave function is then given by

$$z_0 = 2\sigma_{z_0} = 2\sqrt{\langle z^2 \rangle - \langle z \rangle^2} \tag{66}$$

The spatial extent z_0 can be evaluated using variational wave functions given in Eqs. (19) and (21). One obtains

$$z_0^2 = (2\sigma_{z_0})^2$$

where

$$\psi_0(z) = \sqrt{\frac{2\alpha_0}{5}} (1 + \alpha_0 z) e^{-\alpha_0 z} \tag{67b}$$

$$\alpha_0 = \left(\frac{9}{4} eE \frac{2m^*}{\hbar^2}\right)^{1/3} \tag{67c}$$

Evaluation of the integral yields

$$z_0 = 2\sqrt{\frac{7}{5}} \left(\frac{9}{4} eE \frac{2m^*}{\hbar^2}\right)^{-1/3} \tag{68a}$$

or

$$z_0 = 2\sqrt{\frac{7}{5}} \left(\frac{4}{9}\right)^{1/3} \left(\frac{\varepsilon \hbar^2}{e^2 N^{2D} m^*}\right)^{1/3} \quad (68b)$$

which is a convenient analytic expression for the spatial extent of the ground-state subband.

Next, the change in expectation value upon application of an external bias is calculated. Assume an electron system in a V-shaped quantum well. Upon perturbation by an external field (i.e., applied voltage) the position expectation value, $\langle z \rangle$, is perturbed. First-order perturbation theory does not change $\langle z \rangle$ due to the inversion symmetry of the well. A second-order perturbation calculation is therefore required to estimate the change in $\langle z \rangle$.

The change of the position expectation value of the lowest wave function is obtained from second-order perturbation theory according to

$$\langle z \rangle_0 = \langle z \rangle_0^0 + \sum_{n \neq 0} \frac{\langle \psi_0 | z | \psi_n \rangle^2}{E_n - E_0} e\mathbf{E} \quad (69)$$

where $\langle z \rangle_0^0$ is the unperturbed expectation value of the lowest subband. If the origin of the coordinate system is at the dopant plane, then $\langle z \rangle_0^0 = 0$ due to the inversion symmetry of the V-shaped well. Using the variational wave functions introduced earlier, Eq. (69) can be evaluated. The first term in the sum of Eq. (69) is obtained as

$$\langle \psi_0 | \hbar | \psi_n \rangle \cong \sigma_{z_0} \quad (70a)$$

Inclusion of further terms of the sum results in

$$\sum_{n \neq 0} \langle \psi_0 | z | \psi_n \rangle \leqslant 2\sigma_{z_0} \quad (70b)$$

The application of a bias results in free carrier depletion as well as the perturbation of the position expectation value. Choosing the magnitude of the perturbing field as $\mathbf{E} = e N_D^{2D}/2\varepsilon$ allows one to determine the change in the position expectation value according to Eq. (69). One obtains

$$\langle z \rangle_0 \cong 2\sigma_{z_0} \cong z_0 \quad (71)$$

This is a quite important result, since it describes the resolution of CV profiles in a semiconductor with a size-quantized carrier system.

$$\text{Resolution of CV technique} = 2\sigma_{z_0} = z_0 \quad (72)$$

i.e., the resolution of the CV profiling technique on semiconductors with a quantized carrier system is given by the spatial extent of the ground-state wave function. It is obvious that the Debye screening length does *not* apply to semiconductors with quantum confinement.

Comparison of this theoretical result with experimental measurements yields excellent agreement. As an example we calculate (using Eq. (68)) the resolution width of CV profiles for *n*-type GaAs with $N_D^{2D} = 7.5 \times 10^{12}$ cm^{-2} and obtain $z_0 = 40$ Å in agreement with experimental results, which will be discussed in the next section (see also Figs. 16 and 17 and Schubert et al. 1988d). As a second example we calculate the resolution for *p*-type GaAs with $N_A^{2D} = 5 \times 10^{12}$ cm^{-2} and obtain $z_0 = 24$ Å again in agreement with experimental results (see Fig. 19).

Note that the effective mass does not enter the classical Debye screening length (Eq. (52), while the effective mass does enter the quantum-mechanical interpretation of the CV profile resolution (Eqs. (68) and (72). Such a dependence on the carrier mass is in fact observed experimentally (much narrower profiles are observed for *p*-type GaAs than for *n*-type GaAs due to the heavier hole mass).

c. Localization of Si in GaAs

The CV technique was used by a number of groups to study spatial localization in δ-doped semiconductors (Wood et al. 1980, Schubert and Ploog 1985e, 1986e, Sasa et al. 1985, Kobayashi et al. 1986, Ishikawa et al. 1987, Schubert et al. 1988, Ullrich et al. 1988). The layer sequence of a typical sample used for the CV measurements is shown in Fig. 13 and consists of a highly doped substrate, an undoped buffer layer, the dopant sheet, and the undoped top layer.

The GaAs:Si samples used for the CV study were grown by gas-source molecular-beam epitaxy (Vacuum Generator V80 system) and by conventional solid-source molecular-beam epitaxy (Varian Gen II system). Both systems provide epitaxial GaAs of excellent quality as concluded from optical and transport characterization experiments. The growth temperature for Si-doped GaAs was kept in the range 500°C to 550°C to avoid excessive diffusion and segregation.

The thickness of the top layer was chosen between 300 Å and 1000 Å. Eutectic AuGe/Ni/Au contacts alloyed at 420°C for 30 sec were used as ohmic substrate contact. Circular Ti/Au (diameter 500 μm) Schottky contacts were evaporated through a shadow mask onto the epitaxial layer. The band diagram of the structure is shown in the lower part of Fig. 13. The current-voltage characteristics of the Schottky contacts were measured with an Hewlett-Packard 4145B Parameter Analyzer. Excellent diode character-

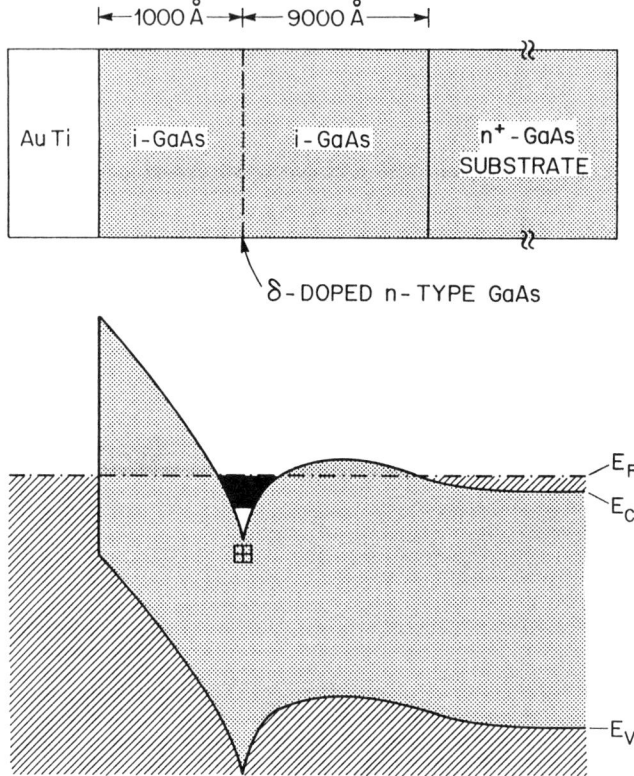

FIG. 13. Schematic layer sequence (top) of a δ-doped n-type semiconductor used for the CV profiling and SIMS technique. The band diagram is shown in the lower part of the figure.

istics with typical ideality factor of $n = 1.03$ were obtained. The CV measurements were performed with a Hewlett-Packard 4194A Impedance/Gain-Phase Analyzer. The phase angle of the measured small-signal admittance was close to $90°$ (always $>87°$) indicating the capacitive character of the Schottky diodes. The frequency used for the CV measurements was $f = 1 \, \text{MHz}$. The CV curves were measured at $T = 300 \text{K}$. Additional measurements at the low temperature of $T = 77 \text{K}$ yielded very similar CV curves and CV profiles. Further experimental details were described by Ullrich et al. (1988).

Schrödinger's and Poisson's equation are solved self-consistently for δ-doped n-type GaAs using the Hartree approximation. Such computer programs allow one to calculate the subband structure for an arbitrary doping profile (Schubert et al. 1988d). The background concentration is assumed to be n-type with concentration $10^{14}\,\mathrm{cm^{-3}}$. A parabolic dispersion relation is used for the present results. The inclusion of nonparabolicity, however, does not change the results significantly. Many-body effects are not taken into account. They are not significant for high concentrations, because the average kinetic energy exceeds the average interaction energy (Ando et al. 1982).

To obtain a CV profile, a set of 15–20 configurations with various bias conditions are solved self-consistently. The capacitance of the system at a given bias is then determined according to $C = dQ/dV$, and the CV profiles are finally obtained by using Eqs. (54) and (55).

CV profiles were calculated self-consistently for concentrations ranging from $1 \times 10^{12}\,\mathrm{cm^{-2}}$ to $7.5 \times 10^{12}\,\mathrm{cm^{-2}}$. For one set of calculations we assume a localization of donors within $dx = 2\,\text{Å}$, i.e., no diffusion. For another set of calculations we assume a spread of donors on a length of $dz = 50\,\text{Å}$, i.e., after diffusion. For simplicity we did not assume a Gaussian, but a rectangular, top-hat dopant distribution. In Fig. 14 two CV profiles are shown for $T = 300\,\mathrm{K}$ and a dopant concentration of $N_D^{2D} = 7.5 \times 10^{12}\,\mathrm{cm^{-2}}$. The profile obtained shows a narrow width of 42 Å and a

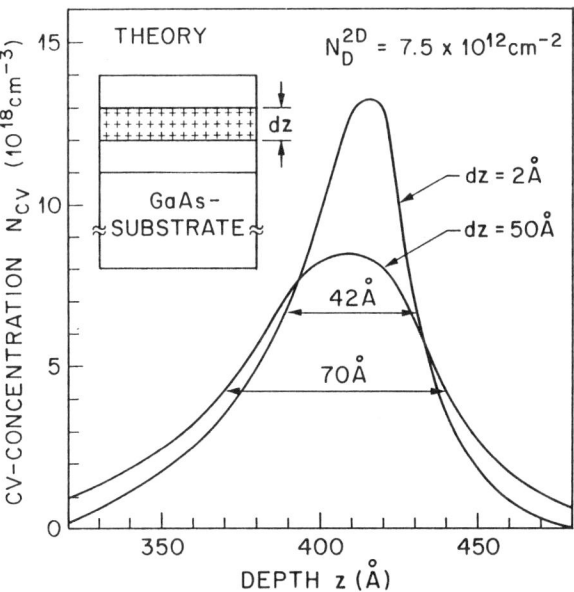

FIG. 14. Self-consistently calculated CV profiles ($T = 300\,\mathrm{K}$) for δ-doped GaAs assuming the absence of significant diffusion ($dz = 2\,\text{Å}$) and assuming diffusion over $dz = 50\,\text{Å}$. Significant changes in CV profile width and concentration are found upon dopant diffusion.

peak concentration of $1.32 \times 10^{19}\,\text{cm}^{-3}$ for the nondiffused sample. The full width at half-maximum increases drastically to a width of 70 Å when we assume diffusion over $dx = 50\,\text{Å}$. Furthermore, the peak concentration drops to a value of $8.5 \times 10^{18}\,\text{cm}^{-3}$. From the drastic increase of the CV profile width and the clear decrease of the CV peak concentration, it becomes obvious that CV profiling is an ideal tool to study the degree of localization of dopants in δ-doped semiconductors.

An experimental CV profile obtained on a δ-doped n-type GaAs sample is shown in Fig. 15, which shows the CV concentration versus depth. The profile is shown on a (a) linear and (b) logarithmic ordinate scale. The abscissa of Fig. 15 corresponds to the total thickness of the epitaxial layer of 1 μm. We observe an extremely sharp peak at a depth of 388 Å, which is very close to the anticipated 400 Å. The profile has a full width at half maximum of ca. 40 Å and a peak concentration of $1.2 \times 10^{19}\,\text{cm}^{-3}$, which is higher than any other reported value. The narrow CV profiles already strongly suggested that diffusion does not play a significant role in δ-doped GaAs. The profile shown is a typical profile; very similar CV profiles are obtained on most of the Schottky contacts on the same wafer. Typical deviations of the peak concentration and of the profile width are $\pm 5\%$.

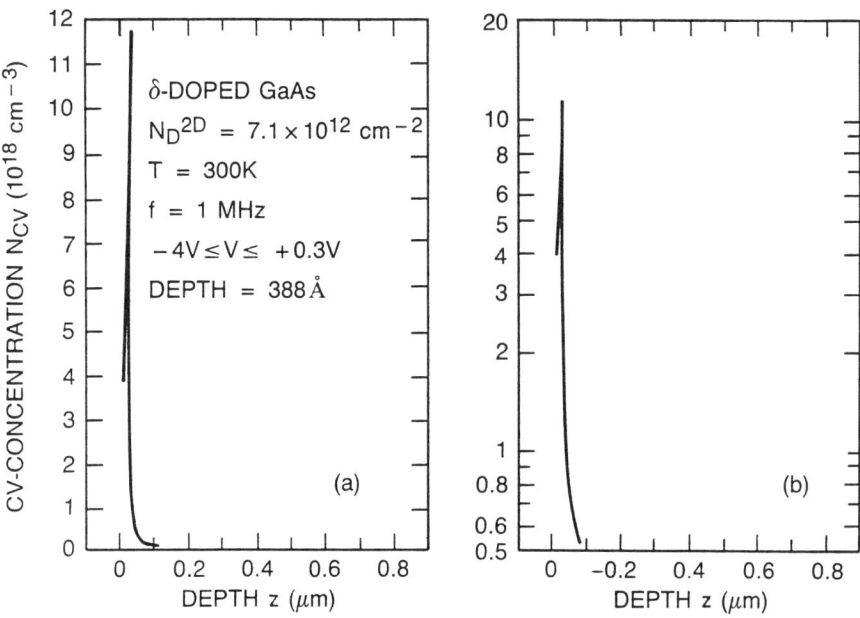

FIG. 15. Experimental CV profiles of δ-doped GaAs on (a) a linear and (b) on a logarithmic ordinate scale. The abscissa of 1 μm coincides with the entire thickness of the epitaxial layer.

The CV profile of the same sample is illustrated on a magnified abscissa scale in Fig. 16. The free carrier concentration and the doping concentration can be obtained from the CV profile by integration of the profile. As illustrated in Fig. 16, the free carrier concentration of a δ-doped sample at zero bias amounts to $n_{2DEG} = 5.4 \times 10^{12}$ cm^{-2}. In addition to *free* carriers, some carriers originating from donors in the semiconductor transfer to surface states. There are three ways to determine the carrier concentration trapped in surface states: (i) by integration of the CV profile for positive voltages (i.e., integrating the nondashed area of Fig. 9), (ii) with a self-consistent calculation, and (iii) using the depletion approximation formula $N_{ss} \cong \varepsilon \phi_B / e z_d$, where ϕ_B is the Schottky barrier height of the metal–semiconductor contact and z_d is the distance between the dopant plane and

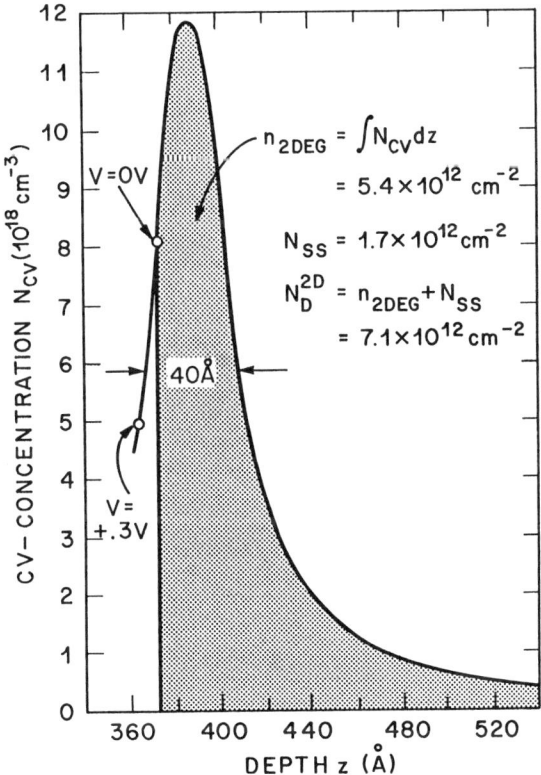

FIG. 16. Experimental CV profile of a Si δ-doped GaAs sample. Integration of the profile (shaded area) yields a free-carrier concentration of 5.4×10^{12} cm^{-2} and a total doping concentration of 7.1×10^{12} cm^{-2}.

the metal–semiconductor interface. All three methods yield approximately $N_{ss} = 1.7 \times 10^{12} \, \text{cm}^{-2}$. The total doping concentration is consequently $N_D^{2D} = n_{2DEG} + N_{ss} = 7.1 \times 10^{12} \, \text{cm}^{-2}$. The intended dopant concentration for this specific sample was $7.5 \times 10^{12} \, \text{cm}^{-2}$. If diffusion plays a role, then it should first become obvious at high doping concentrations such as the one shown here.

Next, we compare our experimental CV profiles with profiles calculated self-consistently. Namely, we compare the theoretical and experimental full width at half-maximum of the CV profiles. In Fig. 17 theoretical and experimental widths of CV profiles are plotted as a function of doping concentrations. The two solid curves of Fig. 17 are the theoretical full width at half-maximum of CV profiles calculated self-consistently at $T = 300 \, \text{K}$. The lower solid line is valid for no diffusion, i.e., $dz = 2 \, \text{Å}$. The upper solid line is valid for a diffusion of $dz = 50 \, \text{Å}$. The experimental points were

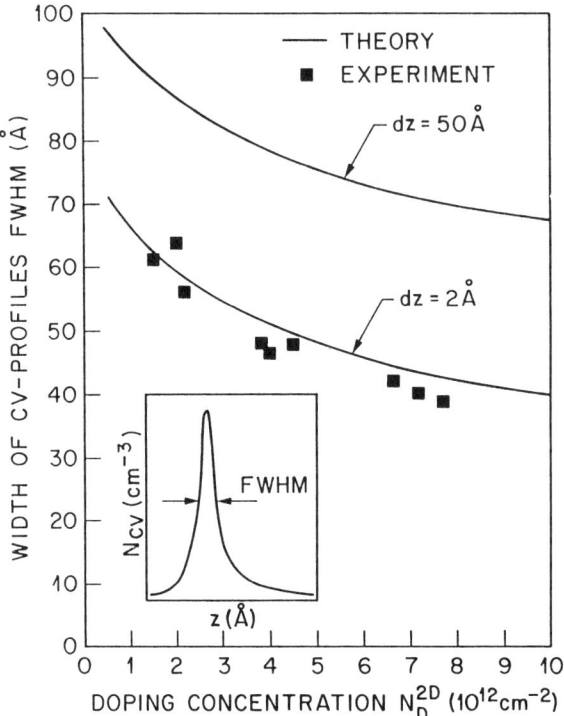

Fig. 17. Comparison of theoretical (solid lines) and experimental (squares) full widths at half-maxima of CV profiles on δ-doped GaAs. Good agreement is obtained only if diffusion is assumed to be negligible.

obtained from a number of samples with different doping concentrations. Comparison of the theoretical curves and experimental data shows agreement only if we assume a strong localization of dopants on a very short length scale. The comparison between experimental and theoretical CV profile widths proves that no significant diffusion or segregation has occurred in the samples. Even if we include both experimental and theoretical uncertainties, diffusion and segregation over more than two lattice constants ($\cong 11$ Å) is not likely. Thus, spatial localization of Si dopants on the length scale of the lattice constant is possible. Since the spatial extent of dopants is much smaller than the ground-state wave function, such dopant distributions are δ-function-like.

Shortly after it was concluded that diffusion of Si in δ-doped GaAs is limited to less than or equal to two lattice constants (Schubert et al. 1988d), an independent study of migration in Si δ-doped GaAs was reported by Beall et al. (1988). They conducted secondary ion mass spectroscopy on Si δ-doped GaAs grown at substrate temperatures between 370°C and 600°C. They concluded that Si-atoms may be confined to a single Ga-plane at substrate temperatures below 500°C. At substrate temperatures between 500°C and 550°C, the dopants were found not to be confined to a single lattice plane. In this temperature range the redistribution of Si was attributed to diffusion of dopants. At temperatures > 550°C segregation of Si, i.e., preferential migration of Si in the direction of the growth front, was found to be the dominant redistribution process. In a subsequent study Beall et al. (1989) confirmed that segregation dominates at elevated substrate temperatures and found their experiments in agreement with kinematically limited surface segregation of Si, which was originally developed to describe the incorporation of Sn in GaAs (Harris et al. 1984).

In the same year Santos et al. (1988) investigated the effect of substrate temperature on migration of Si in δ-doped GaAs by magneto-resistance measurements. Their results indicated that there is negligible spread of dopants in structures grown at a substrate temperature $T_s \lesssim 530$°C. At higher substrate temperatures they found measurable spread which increases with T_s. For $T_s = 640$°C, the Si spread was determined to be $\cong 220$ Å. They concluded that the dominant mechanism for the spreading of Si for $T_s > 600$°C is the migration of Si to satisfy the solid solubility limit.

Auger electron spectroscopy was employed by Webb (1989) to characterize silicon segregation in δ-doped GaAs. He showed that Si segregation is manifested at higher temperatures by the absence of the expected attenuation of the Auger signal. Si segregation was measurable at temperatures above 520°C.

Lanzillotto et al. (1989) studied the distribution of Si in δ-doped GaAs and $Al_{0.25}Ga_{0.75}As$ by the SIMS technique. They found for GaAs growth temperatures < 580°C that the width of the Si-SIMS profile is determined by the resolution function of the measuring technique. At growth temperatures of 580°C to 640°C the spread of Si increased to 80 Å. They found more spread of Si in $Al_{0.25}Ga_{0.75}As$ as compared to Si in GaAs.

In another study, Zrenner et al. (1988a, 1988b) argued that segregation over 195 Å occurs in their δ-doped GaAs. They proposed that a spread of 110–200 Å of Si-dopants in Si-doped GaAs occurs at a concentration of $N_D^{2D} = 1 \times 10^{13} \, cm^{-2}$ and scales linearly with concentration. The authors claimed that δ-doping is not feasible. The results of Zrenner et al. clearly contradict the results described earlier, which yielded spatial localization of dopants on the length scale of the lattice constant. The findings of Zrenner et al. may apply to the samples used for their study, whose growth conditions were not published. However, their claim, that δ-function-like doping profiles are not feasible, has no general validity. Furthermore, it was pointed out (Schubert et al. 1989a) that the analysis of Zrenner et al. is erroneous: The authors calculate the Fourier transform of the *derivative* of the magneto resistance instead of the magneto resistance that results in a systematic error (e.g., 9% for the $n = 4$ state). The growth and analysis of δ-doped samples require sophisticated control of numerous parameters, which might not have been achieved.

d. Localization of Be in GaAs

In contrast to Si δ-doped GaAs, much less is known for Be δ-doped GaAs and the feasibility to spatially confine Be to layers, whose thickness is comparable to the lattice constant. Secondary ion mass spectrometry (SIMS) measurements and capacitance–voltage profiling were employed to study the spatial localization of Be in GaAs (Schubert et al. 1990b).

The samples studied are grown by molecular-beam epitaxy (Varian Gen II) on semi-insulating and heavily Zn doped (001) GaAs substrates at growth temperatures of 500, 580, and 660°C. The GaAs growth rate is 0.9 μm/hr. During growth interruption the Be effusion cell temperature was adjusted to obtain a flux of $4 \times 10^{11} \, cm^{-2}/sec$.

The SIMS doping profile of a sample containing three δ-doped spikes at 500 Å, 1000 Å, and 1500 Å below the GaAs surface is shown in Fig. 18. The concentration of each Be-layer is $N_{Be}^{2D} = 4 \times 10^{12} \, cm^{-2}$ and the growth temperature was 500°C. The SIMS profile reveals three clearly resolved peaks at the anticipated depths of 500 Å, 1000 Å, and 1500 Å and exhibits a full-width half-maximum of 37 Å for the shallowest of the doped layers. The SIMS profile clearly indicates the strong spatial localization of Be in δ-doped GaAs grown at low substrate temperatures.

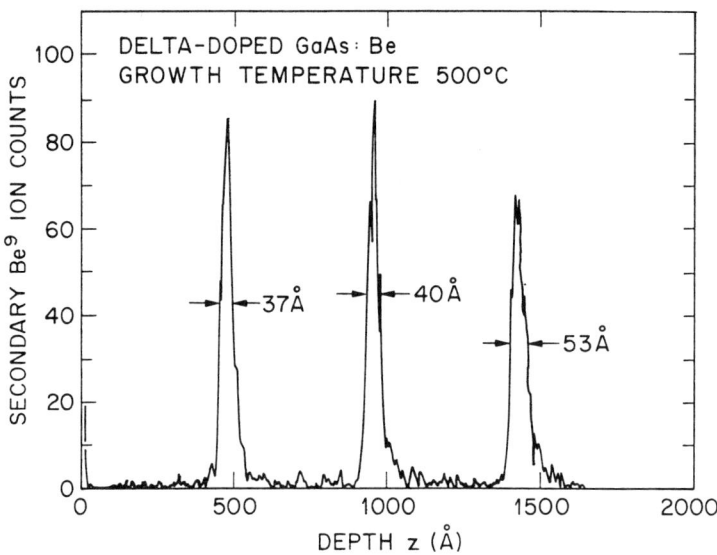

FIG. 18. Beryllium SIMS profile of a δ-doped GaAs sample with three doping spikes at 500, 1000, and 1500 Å below the epitaxial crystal surface.

The resolution of SIMS profiles is limited by roughening of the sputtered crater and the knock-on effect, also called the *mixing effect*. The roughening of the sputtered crater is due to the statistical nature of the sputtering process and increases with sputtering time, which is illustrated by the increasing width of the SIMS peaks in Fig. 18. The second limitation of the SIMS resolution is given by the knock-on effect, where sputtering primary ions transfer their momentum elastically to Be dopants and this causes Be atoms to be implanted deeper into the crystal. To minimize the knock-on effect a low acceleration potential of 3 kV is used in our measurements.

Taking into account both broadening mechanisms, we estimate the total true broadening of the Be dopant profile to be less than 20 Å for the shallowest doping spikes in Fig. 18. Thus, δ-function like doping profiles can be obtained in Be-doped GaAs.

CV profiles on samples with a single δ-doped layer below the surface confirm the SIMS results. A CV profile measured at room temperature on a sample grown at $T_s = 500°C$ is shown in Fig. 19. The profile has a full width at half-maximum of 20 Å and is the narrowest CV profile reported so far in GaAs.

As calculated previously, the resolution function of the CV measurement is given by the spatial extent of the ground-state wave function, which is

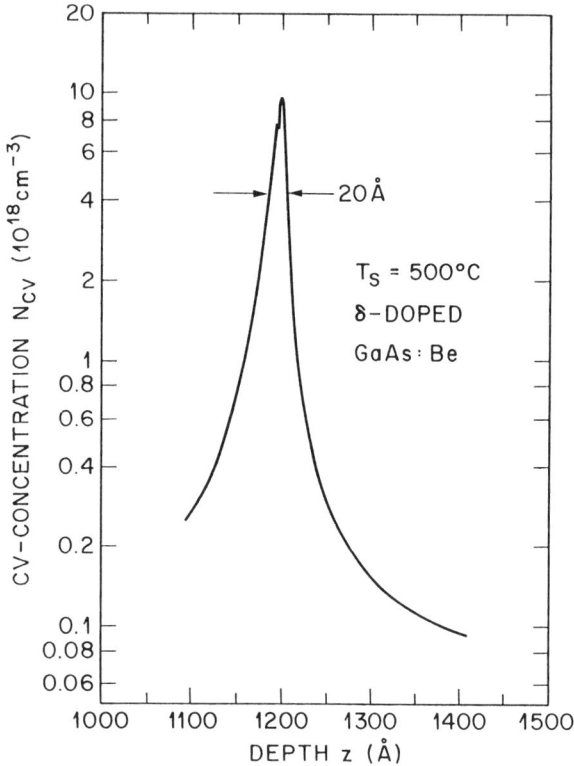

FIG. 19. Capacitance–voltage profile on a Be δ-doped GaAs sample grown at 500°C by molecular-beam epitaxy.

given in Eq. (68). With $N_A^{2D} \cong 5 \times 10^{12}\,\text{cm}^{-2}$ one obtains $z_0 = 24\,\text{Å}$ in excellent agreement with the experiment. Therefore, a broadening of less than 15 Å of the Be dopant profile is concluded from the CV profiles in good agreement with the SIMS measurements.

Elevated substrate temperatures of 580°C and 660°C are used to evaluate spatial spreading of dopants as a function of temperature. Such elevated temperatures result in stronger diffusion and, as will be discussed later, also enhanced segregation. Both, CV profiles and SIMS profiles are used to monitor the spread of dopants at elevated substrate temperatures. CV profiles of three samples grown at $T_s = 500°\text{C}$, 580°C, and 660°C are shown in Fig. 20. The profiles indicate that δ-function-like Be profiles can be obtained at 500°C, whereas elevated temperatures $T_s > 500°\text{C}$ result in noticeable spreading of Be ions. At $T_s = 660°\text{C}$ the CV profile broadens to 85 Å.

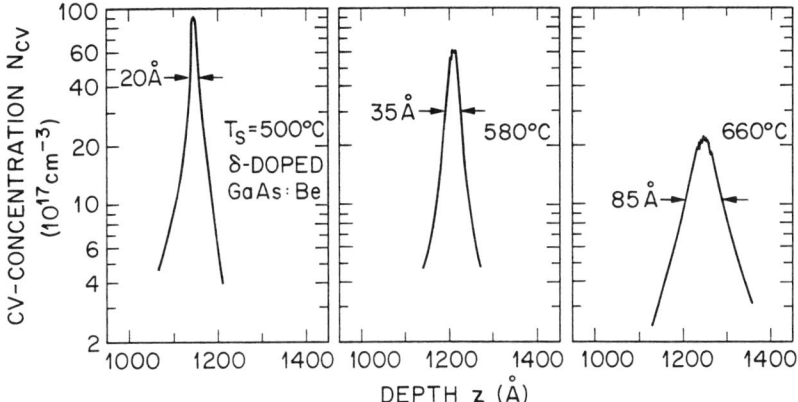

FIG. 20. Capacitance–voltage profiles on Be δ-doped GaAs samples grown at 500°C, 580°C and 660°C by molecular-beam epitaxy. A broadening of the CV profiles is observed at elevated substrate temperatures.

The SIMS profiles shown in Fig. 21 are from the same set of samples and confirm the CV profiles. An increasing spread of dopant ions becomes evident at elevated substrate temperatures. The individual spreading mechanisms such as diffusion and segregation will be discussed in subsequent sections.

The profiles in Figs. 20 and 21, which show the same three samples, allow us to compare the two profiling techniques. Both techniques agree to a fairly high degree, especially at elevated substrate temperatures. At small dopant spreads, the CV technique yields narrower profiles, with superior resolution. As the spread of dopants increases at substrate temperatures of 660°C, the resolution of CV profiles and SIMS profiles is comparable. This can be explained by the fact that the limits of the SIMS resolution are concentration independent. In contrast, since the CV profile resolution is concentration dependent (see Eqs. (68) and (72), the widths of CV profiles are expected to track the SIMS profile widths at increased spread (i.e., decreased concentration).

e. Localization of Si in $Al_xGa_{1-x}As$

The spatial localization of dopants in $Al_xGa_{1-x}As$ is important for many heterostructures based on the $GaAs/Al_xGa_{1-x}As$ material system, including selectively doped heterostructures and quantum well structures for high electron mobility. In such structures it is essential to place dopants at some given distance (e.g., 50 Å) from the hetero interface. Redistribution of the

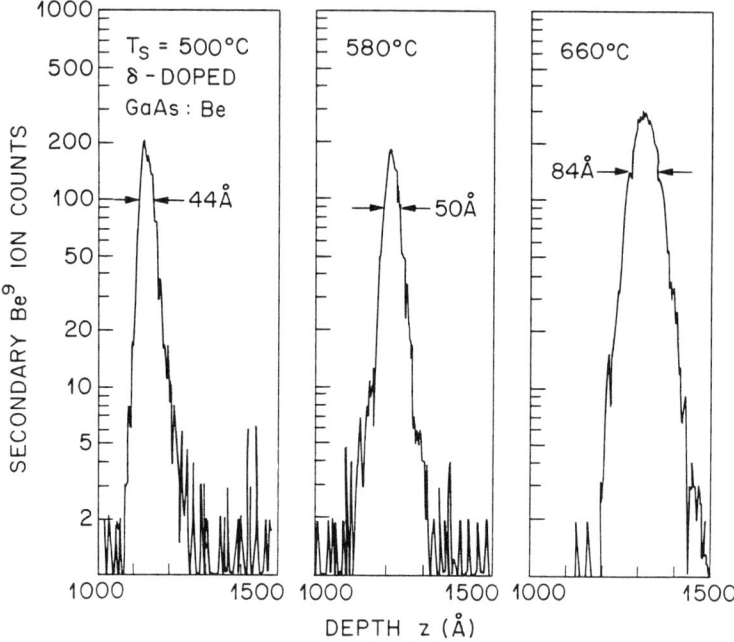

FIG. 21. Secondary ion mass spectrometry profiles on three Be δ-doped GaAs samples grown at 500°C, 580°C, and 660°C. The SIMS profiles broaden systematically at elevated substrate temperatures.

dopants can result in a degradation of the transport properties of the structure, especially if the dopants migrate into the nominally undoped GaAs layer.

Without regarding the nature of the broadening mechanism (e.g., diffusion or segregation), the spatial localization of dopants is investigated as a function of the substrate temperature during growth by molecular-beam epitaxy. For this purpose, epitaxial $Al_xGa_{1-x}As$ was grown at the substrate temperatures 500°C, 600°C, and 700°C and capacitance–voltage profiling was used to study the spatial localization of dopants (Schubert et al. 1989f).

The CV profiles of the $Al_xGa_{1-x}As$ epitaxial layers doped with Si at nominally 1000 Å below the crystal surface are shown in Fig. 22. At a substrate temperature of 500°C a narrow CV spectrum is observed at room temperature. The peak of the CV profile occurs at a depth of 990 Å in agreement with the experimentally anticipated value. The full width at half-maximum of the peak is 51 Å. The width of the CV profile is determined mostly by the resolution of the technique. For a Si dopant density of $4 \times 10^{12} \, cm^{-2}$, the resolution of the CV profile is estimated using Eq. (68)

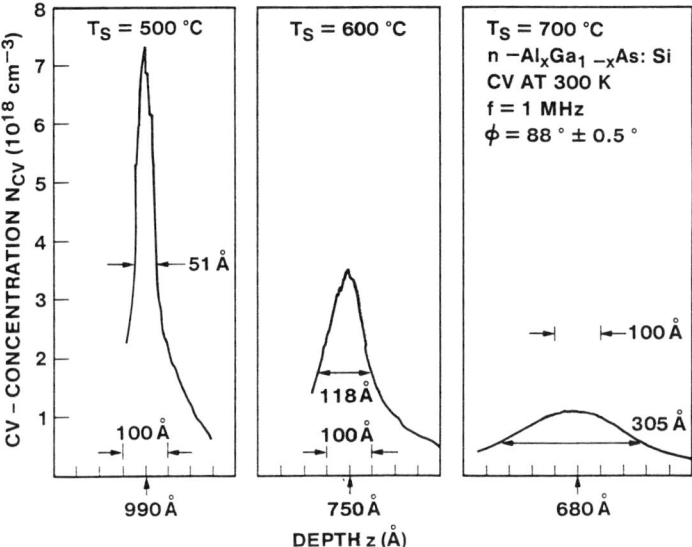

FIG. 22. Capacitance–voltage profiles of Si δ-doped $Al_xGa_{1-x}As$ grown at 500°C, 600°C, and 700°C by molecular-beam epitaxy. Low substrate temperatures are required to spatially localize Si-dopants in $Al_xGa_{1-x}As$.

to be 49 Å. The spatial width of the true Si-dopant distribution is estimated (by deconvolution) to be $\leqslant 15$ Å. Thus, δ-function-like doping distributions can be achieved in Si-doped $Al_xGa_{1-x}As$ at low growth temperatures.

At elevated growth temperatures of 600°C and 700°C, significant broadening of the CV profiles is obtained. As shown in Fig. 22, the full width at half-maximum of the CV profiles increases to 118 Å and 305 Å for substrate temperatures of 600°C and 700°C, respectively. The broadening is considerably stronger than for Si or Be in GaAs, as discussed previously. The relatively strong broadening of Si-dopants profiles in $Al_xGa_{1-x}As$ may be due to the fact that the slightly mismatched alloy semiconductor is inherently more strained than the binary GaAs. The stronger broadening may also be due to the fact that Si-atoms can form the well-known DX-centers in $Al_xGa_{1-x}As$, which may diffuse more readily.

It was found at the beginning of the 1980s, that $Al_xGa_{1-x}As$ requires elevated substrate temperatures of 660–700°C to achieve high optical quantum efficiency (see, for example, Tsang and waminathan 1981). At that time and in later years it was generally believed, that the growth of high-quality $Al_xGa_{1-x}As$ requires high substrate temperatures. However, an improvement in source purity as well as further improvement of the ultra-high vacuum in the MBE system may extend the range of tolerable growth temperatures. In addition, the use of lower growth rates could improve the crystalline quality. High-quality $Al_xGa_{1-x}As$ can in fact be obtained at a growth temperature of 500°C in such an improved MBE-system (Schubert et al. 1989f). The low-temperature photoluminescence spectrum of an $Al_{0.34}Ga_{0.66}As$ sample grown at 500°C is shown in Fig. 23. The spectrum of the undoped $Al_{0.34}Ga_{0.66}As$ epitaxial layer has indeed the typical features of high-quality material, such as strong bound exciton intensity, clearly resolved transitions, and narrow photoluminescence line widths. The excitonic line width of 6.9 meV is in agreement with the theoretical line width of excitons broadened by the random-alloy nature of $Al_xGa_{1-x}As$ (Schubert et al. 1984). Thus, $Al_xGa_{1-x}As$ of good crystalline perfection can be grown by MBE at low growth temperatures of 500°C,

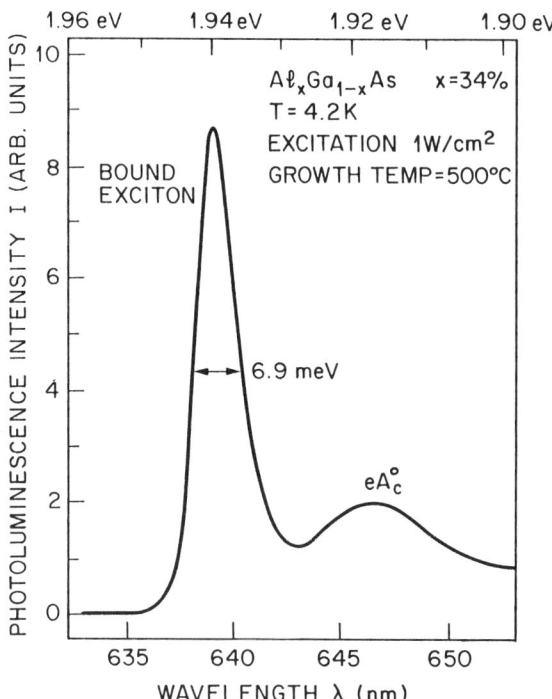

FIG. 23. Low-temperature photoluminescence spectrum of $Al_xGa_{1-x}As$ grown at 500°C by molecular-beam epitaxy. The luminescence spectrum has the features of high-quality epitaxial material despite the relatively low substrate temperature.

which is the required temperature range for the spatial localization of Si dopants δ-doped in $Al_xGa_{1-x}As$.

6. DIFFUSION OF DOPANTS

Diffusion of dopants, i.e., the random Brownian motion of impurities in the semiconductor lattice, is one of the major mechanisms that result in a redistribution of dopants during and after their deposition on the nongrowing crystal surface. The knowledge of the diffusion coefficients and their temperature dependence is therefore essential for the selection of growth and processing temperatures and times for δ-doped semiconductor structures and devices. A second redistribution mechanism, namely, segregation of dopants, is discussed in the next section.

The diffusion data presented in this section were obtained by a novel, very sensitive method that employs the CV profiling technique on δ-doped semiconductors subjected to postgrowth rapid thermal annealing (Schubert et al. 1988c, 1989f, 1990b). The samples used for the diffusion study were grown at low substrate temperatures of typically 500°C to minimize diffusion and segregation during epitaxial growth. The availability of a sensitive method to determine diffusion coefficients is especially important for III–V semiconductors. Postgrowth annealing at high temperatures (e.g., 1000°C) can be done for short times only (e.g., 5 sec) due to the degradation of the crystal surface caused by the formation of group III and group V vacancies. The method must therefore be sensitive enough to assess the diffusion occurring during such a short time. Postgrowth annealing at low temperatures (e.g., 600°C) for times on the order of 5–600 sec also results in relatively small amounts of diffusion. A sensitive method is therefore required to cover the entire temperature range of interest, which spans from 500°C to 1000°C.

The experimental procedure of the technique is as follows. Epitaxial semiconductors containing a δ-doped layer are grown at low substrate temperatures of 500°C by molecular-beam epitaxy. The typical layer sequence and band diagram is shown in Fig. 13. After growth the sample is cleaved into small pieces ($5 \times 10 \, mm^2$), which are subjected to rapid thermal annealing in an AG-Associates 410 processing unit. The annealing temperature is measured by a thermocouple. The absolute temperature is controlled to $\pm 10°C$. The heating and cooling rates are 220°C/sec and 80°C/sec, respectively. Proximity annealing is used to reduce crystalline degradation at high temperatures. Typical annealing times are 5 sec; longer times up to 600 sec are used at low annealing temperatures, if required by a low diffusion coefficient. Subsequently, ohmic substrate contacts are evaporated and alloyed. AuGe/Ni/Au (2000 Å/500 Å/2000 Å) and AuBe (2000 Å) crystallization is used for n-type and p-type substrates, respectively. Finally, circular Ti/Au (500 Å/1500Å) are evaporated through a shadow mask. The CV curves are measured and converted into profiles with a Hewlett-Packard 4194A impedance gain-phase analyzer. The phase-angle during the measurement is typically $88° \pm 1°$ at a frequency of 1 MHz.

The theory of Fickian dopant diffusion with an initially Dirac-delta-function-like distribution is straightforward. The initial dopant profile is given by

$$N_D(z) = N_D^{2D} \delta(z - z_0) \tag{73}$$

where N_D^{2D} is the dopant density per unit area and z_0 is the position of the dopants on the z-axis. A Gaussian distribution of dopants is obtained after diffusion

$$N_D(z) = N_D^{2D}(4\pi D\tau)^{-1/2} \exp\left[\frac{-(z-z_0)^2}{4D\tau}\right] \tag{74}$$

This Gaussian distribution represents the solution of the one-dimensional Fickian diffusion equation (for reviews, see, e.g., Casey 1973 or Tuck 1988). The diffusion time is denoted as τ and D is the so-called diffusion coefficient, which depends exponentially on temperature according to

$$D = D_0 \exp(-E_a/kT) \tag{75}$$

where E_a is the activation energy of the diffusion process, k is Boltzmann's constant, and T is the absolute temperature.

Frequently, a *diffusion length*, L_D, is used and defined according to

$$L_D = \sqrt{D\tau} \tag{76a}$$

Note that the standard deviation of the Gaussian distribution is related to the diffusion length according to

$$\sigma = \sqrt{2D\tau} = \sqrt{2}L_D \tag{76b}$$

It is further necessary to differentiate between the full width at half-maximum and the standard deviation of a Gaussian probability distribution. The full width at half-maximum is approximately 2.35 times the standard deviation.

The CV profiling technique measures the free carrier capacitance and is thus a free carrier effect. The CV technique therefore offers the unique possibility to measure the integrated carrier density (see Eq. (63)) before and after the annealing process. It turns out that the carrier concentration does not change significantly during the annealing process. (A change of ≤10% was detected for Si in GaAs, see Schubert *et al.* 1988c.) Therefore, dopants can be assumed to occupy substitutional sites before and after the annealing process.

It is necessary to relate the *CV profile* to the *dopant distribution* to obtain the dopant diffusion length. Self-consistent calculations and an analytic approximation were employed to obtain the dopant diffusion length from the measured CV profiles (Schubert *et al.* 1988c, 1989f, 1990b). In the first method, CV profiles are calculated numerically by solving Schrödinger's and Poisson's equations self-consistently for Gaussian or top-hat distributions with different diffusion lengths. The comparison of measured and calculated profiles allows one to extract the diffusion length associated with the measurement. In the second, analytic method the diffusion lengths and coefficients are obtained by attributing a diffusion broadening, σ_D, and a resolution-function (intrinsic) broadening, σ_i of the CV profiles. The total broadening of the CV profile is then given by

$$\sigma = (\sigma_i^2 + \sigma_D^2)^{1/2} \qquad (77)$$

The resolution-function broadening, σ_i, is due to the finite CV profile width, even if no diffusion has occurred at all. The resolution-function or intrinsic broadening can be determined either experimentally from the nonannealed sample or theoretically using Eqs. (68) and (72). The total broadening σ in Eq. (77) is obtained experimentally from the CV profiles on the annealed samples. (Note that the full width of the CV profile equals about 2.35σ.) The diffusion broadening, σ_D, can then be evaluated, and the diffusion coefficient is given by

$$D = \sigma_D^2/2\tau \qquad (78)$$

In the following the experimental results and the diffusion coefficients of Si in GaAs and $Al_xGa_{1-x}As$, and Be and C in GaAs will be summarized.

The CV profile obtained from a δ-doped GaAs sample measured at 300K is shown in Fig. 24. The CV profile peak occurs at a depth of 1123 Å, which is in good agreement with the depth of 1000 Å anticipated during crystal growth. The full width at half-maximum of the profile is 30 Å. The samples used for the Si-diffusion studies were grown by chemical-beam epitaxy (Tsang 1987, 1989). The slightly higher *p*-type background concentration of this growth method (Chiu *et al.* 1987, 1988) leads to a stronger confinement of carriers and explains the narrow 30 Å width of the CV profile. The free

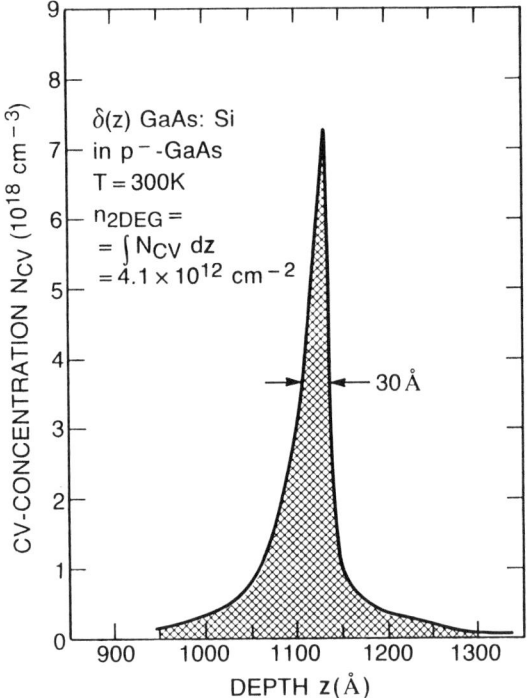

FIG. 24. Capacitance–voltage profile of a Si δ-doped GaAs sample grown by chemical-beam epitaxy. Integration of the profile yields a free carrier density of $n_{2\text{DEG}} = 4.1 \times 10^{12}\,\text{cm}^{-2}$.

carrier concentration can be obtained by integration of the CV profile, as indicated by the shaded area in Fig. 24. The integral yields a density of carriers in the two-dimensional electron gas of

$$n_{2\text{DEG}} = \int N_{\text{CV}}\,dz = 4.1 \times 10^{12}\,\text{cm}^{-2} \tag{79}$$

in good agreement with the anticipated value. After annealing samples at temperatures ranging from 600 to 1000°C for 5 sec, the shape of the CV profiles changes drastically, as shown in Fig. 25. The full widths at half-maximum increase from 30 Å for the asgrown sample up to 137 Å for the sample annealed at 1000°C. Furthermore, the maximum CV concentration drops from $\geqslant 7 \times 10^{18}\,\text{cm}^{-3}$ for the as-grown sample to $\leqslant 3 \times 10^{18}\,\text{cm}^{-3}$ for the sample annealed at 1000°C. The free carrier concentration (the integral of the CV profile) decreases by about 10% from

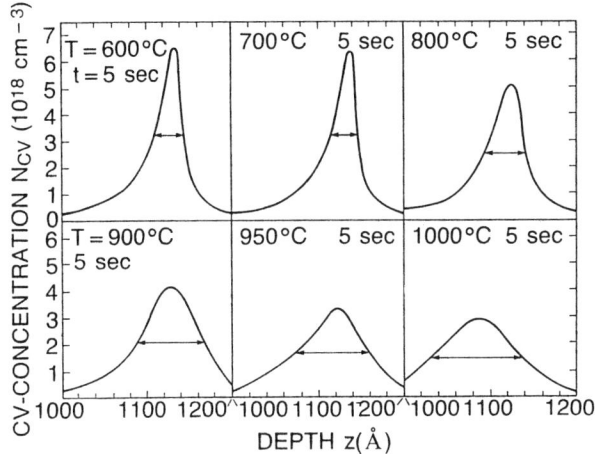

FIG. 25. Capacitance–voltage profiles of different pieces of a Si δ-doped GaAs sample annealed at temperatures of $600°\text{C} \leqslant T \leqslant 1000°\text{C}$ for 5 sec.

$4.1 \times 10^{12}\,\text{cm}^{-2}$ to $3.7 \times 10^{12}\,\text{cm}^{-2}$. This minor reduction suggests that some Si impurities might have diffused on As vacancies, i.e., acceptor sites.

The diffusion coefficients, which are determined from the diffusion lengths according to $D = \sigma_D^2/2\tau$, are shown versus reciprocal temperature in Fig. 26. The activation energy of the diffusion process determined from the Arrhenius plot is $E_a = 2.45\,\text{eV}$. The extrapolated diffusion coefficient for $T \to \infty$ is $D_0 = 4 \times 10^{-4}\,\text{cm}^2/\text{sec}$. The diffusion coefficient found in this study is extremely small compared to other impurities in III–V semiconductors (Casey 1973). Previously, the diffusion coefficient for atomic Si-diffusion was determined for a rather small temperature range, e.g., for 800–900°C. Our data are in agreement with data compiled by Greiner and Gibbons (1985). They summarized published diffusion coefficients on Si-ion implanted GaAs that range from 1×10^{-15} to $2 \times 10^{-14}\,\text{cm}^2/\text{s}$ at 850°C. In earlier experiments, Greiner and Gibbons (1984) studied Si-pair diffusion in GaAs. They presumed that paired Si atoms have a higher diffusivity than single Si atoms due to conservation of charge and vacancy type during substitutional diffusion. Omura et al. (1986) determined the temperature dependence of the diffusion coefficient of Si pairs in GaAs under excess As pressure and obtained results comparable to those of Greiner and Gibbons. The relatively low diffusion coefficient of Si in GaAs shown in Fig. 26 is in agreement with the presumption of Greiner and Gibbons (1984), who expected a larger diffusion coefficient for Si complexes (pairs) in GaAs.

The CV profiles of Si δ-doped $\text{Al}_{0.30}\text{Ga}_{0.70}\text{As}$ that was thermally an-

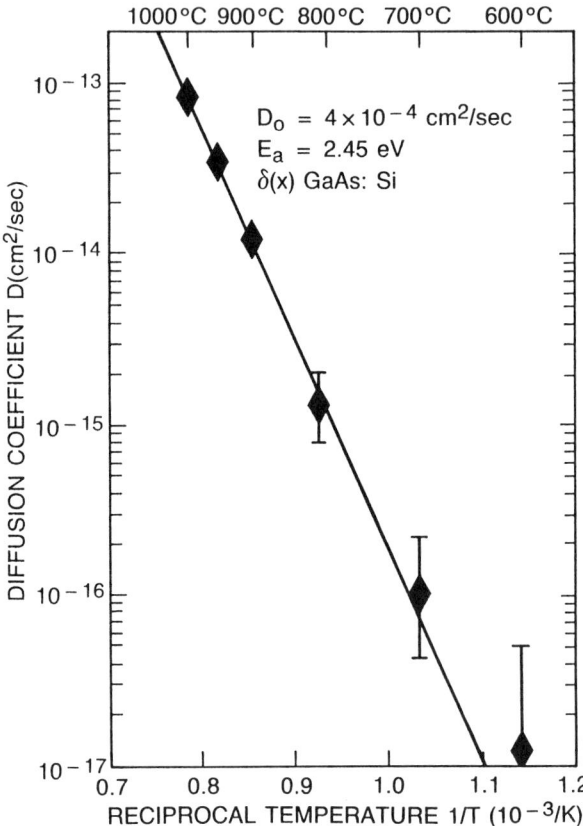

FIG. 26. Diffusion coefficient for Si in GaAs versus reciprocal temperature.

nealed at different temperatures for 5 sec are depicted in Fig. 27. A systematic decrease in the peak CV concentration as well as an increase in width is observed with increasing annealing temperature. The profile width increases to 410 Å after annealed at 1000°C for 5 sec. The increase in CV profile width is much more pronounced for $Al_xGa_{1-x}As$ than GaAs. The diffusion coefficient of Si in $Al_{0.30}Ga_{0.70}As$ is shown in Fig. 28. In addition, Fig. 28 shows some data points obtained by growth at elevated substrate temperatures of 580, 600, 660, and 700°C. The diffusion coefficients of the as-grown samples appear to be somewhat lower than the postgrowth annealed samples. This fact may be due to a dependence of the diffusion coefficient on the growth temperature. $Al_xGa_{1-x}As$ grown at 500°C has a slightly larger diffusion coefficient than the material grown at $T > 580°C$.

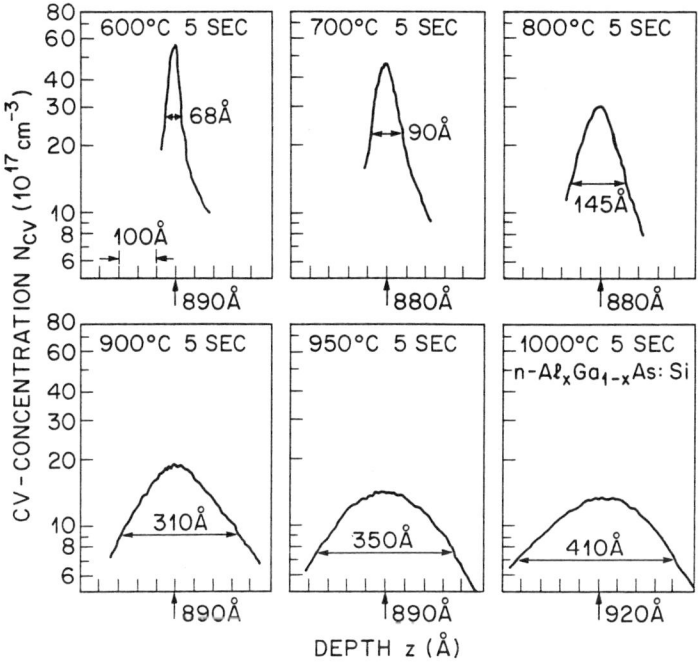

FIG. 27. Capacitance–voltage profiles of different pieces of a Si δ-doped $Al_xGa_{1-x}As$ sample annealed at temperatures $600°C \leq T \leq 1000°C$ for 5 sec.

The higher temperatures are known to be optimal for the growth of the alloy semiconductor. The average diffusion coefficient of Si in $Al_{0.30}Ga_{0.70}As$ is best described by a D_0 of 4×10^{-8} cm^2/sec and an activation energy of 1.3 eV.

Silicon redistribution in MBE-grown GaAs and $Al_xGa_{1-x}As$ during rapid thermal processing was also reported by Tatsuta et al. (1985). They also found a larger diffusion coefficient of Si in $Al_xGa_{1-x}As$ than GaAs and used the results for the optimization of selectively doped $Al_xGa_{1-x}As$/GaAs heterostructures.

Beryllium is known to diffuse more rapidly than Si. However, there is a lack of diffusion data for Be in GaAs, especially in the low-temperature regime of 600–800°C. Be diffusion in GaAs was studied by preparing buried Be-doped layers, grown by MBE and subjecting the samples to annealing at 900°C (Ilegems 1977, McLevige et al. 1978, and Tuck 1988). The diffusion of Be was monitored using both secondary ion mass spectrometry and electrical measurements. Values of 10^{-13} to 10^{-14} cm^2/sec were reported at 900°C for Be concentration of about 10^{19} cm^{-3}.

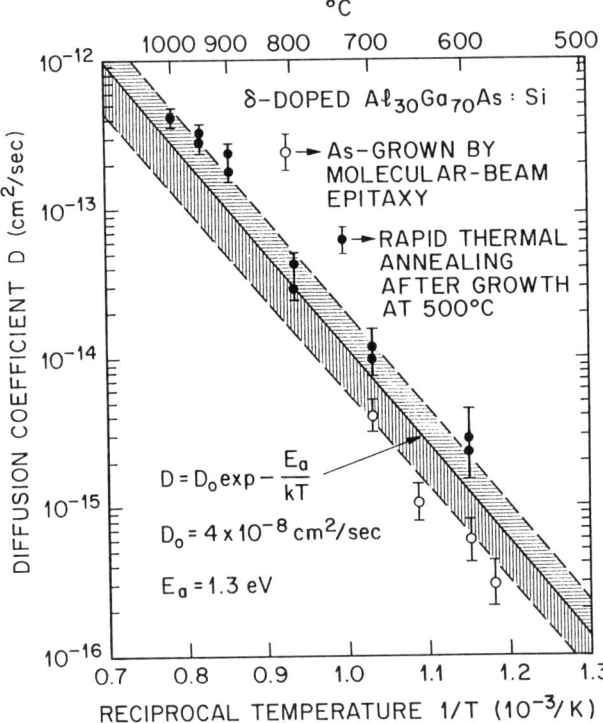

FIG. 28. Diffusion coefficient for Si in $Al_xGa_{1-x}As$ versus reciprocal temperature.

The CV profiles of Be δ-doped GaAs annealed at different temperatures are shown in Fig. 29. The CV profile width of the unannealed sample (not shown in Fig. 29) is 20 Å. As expected, a reduced peak concentration and an increased width is observed at higher annealing temperatures. The broadening of the CV profiles for Be in GaAs is more pronounced than for Si in GaAs but less pronounced than for Si in $Al_xGa_{1-x}As$.

The temperature-dependent diffusion coefficient of Be in GaAs is shown in an Arrhenius plot in Fig. 30. The data are described by a D_0 of 2×10^{-5} cm²/sec and an activation energy of $E_a = 1.95$ eV. The data agree with the single-temperature data of Ilegems (1977) and McLevige et al. (1978), who found $D = 10^{-13}$ to 10^{-14} cm²/sec at 900°C.

Carbon diffuses extremely slowly in GaAs. Diffusion times of 25 sec at 700°C to 950°C are used to obtain sufficient broadening of the CV profiles. The temperature-dependent diffusion coefficient for C-acceptors in GaAs is shown in Fig. 31. In the measured temperature range the diffusion coefficient

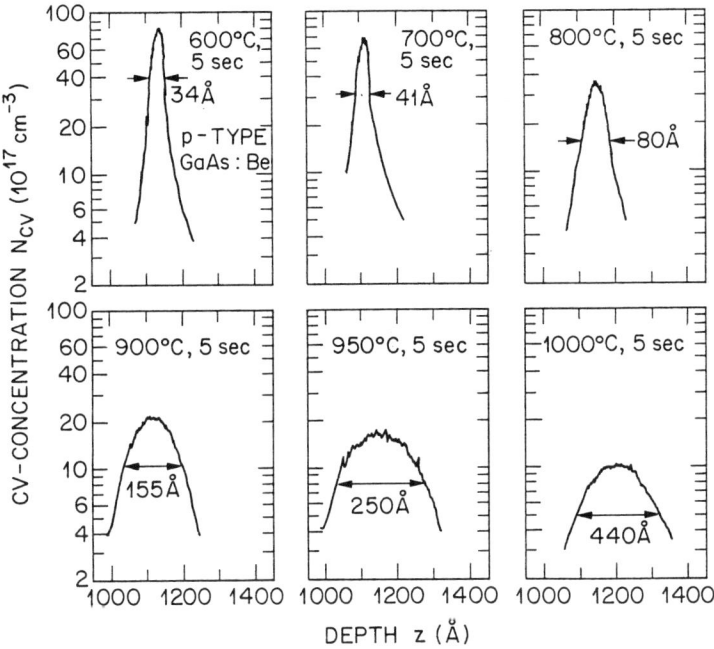

FIG. 29. Capacitance–voltage profiles of different pieces of a Be δ-doped GaAs sample annealed at temperatures $600°C \leqslant T \leqslant 1000°C$ for 5 sec.

of C is more than one order of magnitude smaller than the diffusion coefficient of Be in GaAs. A diffusion coefficient of 6×10^{-15} cm^2/sec at 900°C was reported by Saito et al. (1988) and is slightly larger than the result shown in Fig. 31. A diffusion coefficient of 2×10^{-16} cm^2/sec at 800°C was reported by Kobayashi et al. (1987) and is slightly lower than the value shown in Fig. 31. Abernathy et al. (1989) reported $D \leqslant 10^{-16}$ cm^2/sec for GaAs:C grown by MOMBE.

The knowledge of the diffusion coefficient allows one to estimate the significance of diffusion during crystal growth. As examples, we consider Si and Be in GaAs. At a temperature of 500°C the diffusion coefficients can be estimated by extrapolation of the high-temperature data. One obtains $D \cong 4 \times 10^{-20}$ cm^2/sec and $D \cong 4 \times 10^{-18}$ cm^2/sec for Si and Be, respectively. Using a growth time of 1000 sec, the diffusion length $\sigma_D = \sqrt{2D\tau}$ is calculated to be $\cong 1$ Å and $\cong 9$ Å for Si and Be, respectively. Thus, the diffusion length of Si and Be is on the order of the lattice constant. Such slightly diffused profiles are nevertheless δ-function-like, since their spatial extent is much smaller than the ground-state wave function.

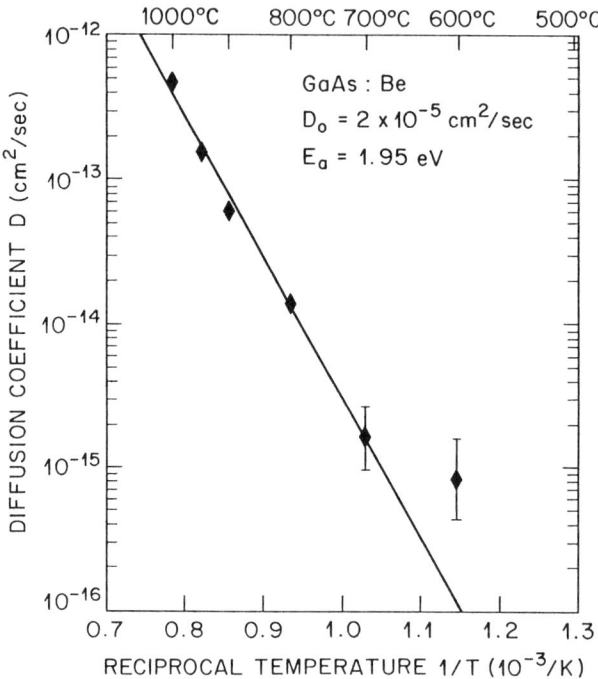

FIG. 30. Diffusion coefficient for Be in GaAs versus reciprocal temperature.

7. SEGREGATION OF DOPANTS

Segregation of dopants toward the crystal surface during crystal growth is an additional broadening mechanism that becomes prominent at elevated growth temperatures. Although diffusion, as discussed previously, broadens the dopant distribution symmetrically toward the substrate and toward the growing crystal surface, the term *segregation of dopants* is used to describe the migration of dopants predominantly toward the surface of the growing crystal, i.e., along the growth direction.

Surface segregation of Sn was previously observed during GaAs crystal growth by molecular-beam epitaxy (Cho 1975). The observation of Sn segregation was confirmed by Wood and Joyce (1978). They found that Sn is incorporated in the GaAs lattice after a steady-state surface population of Sn is formed. Evidence for Si-migration predominantly toward the surface in δ-doped GaAs was found by Beall *et al.* (1988) using secondary ion mass spectrometry. They found segregation to be important at growth temperature exceeding 550°C at Si densities of 1×10^{13} cm^{-2}. Their data appeared

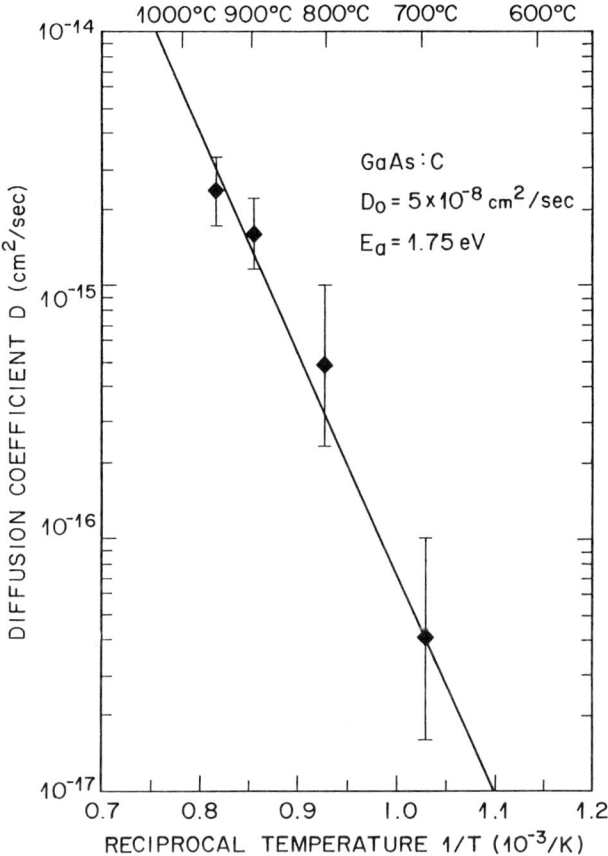

FIG. 31. Diffusion coefficient for C in GaAs versus reciprocal temperature.

to fit a model of kinetically limited surface segregation proposed for Sn in GaAs (Harris *et al.* 1984). Surface segregation was also found for Si in GaAs by Chiu *et al.* (1988) as well as Santos *et al.* (1988). The latter group proposed that segregation occurs due to the solid solubility limit of Si in GaAs. Using SIMS measurements Schubert *et al.* (1989d) showed that even stronger segregation occur for Si in Al_xGa_{1-x} and Be in GaAs than for Si in GaAs. They showed that background doping could reduce segregation significantly and proposed Fermi-level pinning at the growing crystal surface causes the segregation.

The SIMS technique allows one to measure the dopant concentration directly as a function of depth. Thus, the technique is well suited to resolve asymmetries in the dopant distribution. The SIMS profiles of δ-doped

$Al_{0.30}Ga_{0.70}As$ grown at 500°C, 580°C, and 660°C by molecular-beam epitaxy are shown in Fig. 32. The Si density is $2 \times 10^{12}\,cm^{-2}$ and the dopant sheet is located nominally 1000 Å below the crystal surface. The SIMS measurements are carried out on a PHI 6000 and an Atomica instrument using Cs^+ and O_2^+ ions for sputtering. A low acceleration energy of 3 keV is used to minimize the well-known knock-on effect.

At the low growth temperature of 500°C, a sharp Si-peak is observed as illustrated in Fig. 32. The leading and the trailing slopes of the profiles are evaluated to be 35 Å/decade and 80 Å/decade, respectively. Such slopes characterize the depth interval in which the secondary ion counts decrease (exponentially) by one order of magnitude (one decade). At a growth temperature of 500°C the trailing slope is less steep than the leading slope

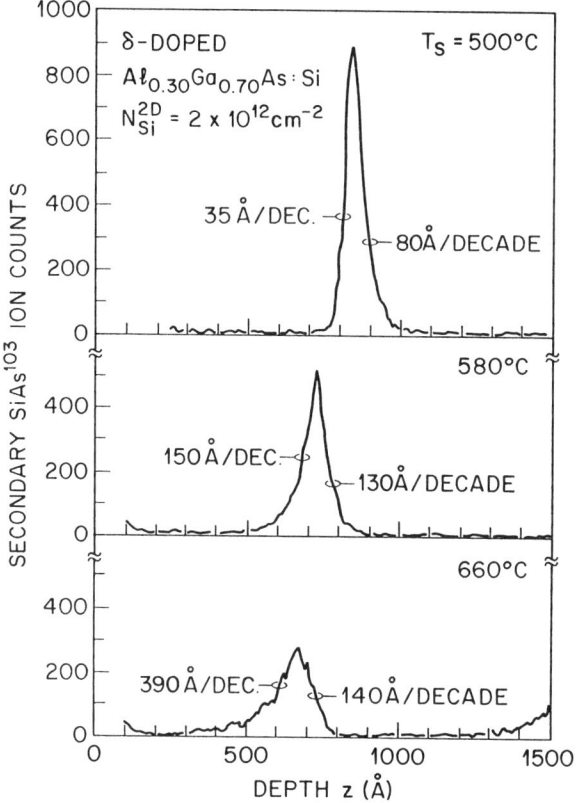

FIG. 32. Secondary ion mass spectrometry profiles of Si δ-doped $Al_xGa_{1-x}As$ grown at 500°C, 580°C, and 660°C by molecular-beam epitaxy. At elevated substrate temperatures, segregation, i.e., migration of dopants predominantly along the growth direction, is observed.

due to the knock-on effect: Impinging primary sputtering ions knock onto the Si-dopants and partially transfer their momentum to the Si-atoms. Thus, the Si-dopants are "implanted" deeper into the crystal and result in a SIMS profile slightly skewed toward the substrate side.

As the growth temperature is increased to 580°C and 660°C, the profiles exhibit significant asymmetry. Surface segregation of dopants is evident from the profiles, especially at $T_s = 660°C$. The leading slope increases from 35 Å/decade to 390 Å/decade indicating the *surface segregation* of dopants. The trailing slope increases from 80 Å/decade to 140 Å/decade and indicates diffusion of dopants, which is expected to be symmetric with respect to both sides of the structure.

In the following we will show that the observed segregation can be conclusively explained by Fermi-level pinning of the semiconductor–vacuum interface during crystal growth by molecular-beam epitaxy. The Fermi-level pinning results in a dipole consisting of electrons localized in surface states and positive donor ions close to the semiconductor surface. Drift of the donor ions in the dipole fields results in dopant redistribution predominantly toward the semiconductor surface. The interaction between surface states and donor ions can be screened and reversed by background doping.

Fermi-level pinning at an interface of a semiconductor and a second medium, such as a metal, insulator, or vacuum, is one of the classic phenomena in semiconductor physics. The Fermi level at the interface is pinned at any energy $e\phi_B$ below the bottom of the conduction band. Bardeen (1947) attributed Fermi-level pinning to interface states that are of donor and acceptor types and are energetically located around the middle of the fundamental gap in most semiconductors. In an earlier model, Schottky (1938, 1940) and Mott (1938) proposed that the energy $e\phi_B$ of a metal–semiconductor contact equals the difference of the work function of the metal and the electron affinity of the semiconductor. The physical origin of Bardeen's interface or surface states was proposed to be due to metal-induced gap states (Heine 1965) or localized states at the semiconductor surface or interface. Such localized states could be due to atomic steps of the surface (Huijser et al. 1977), defects (Spicer et al. 1975, 1980), or surface reconstruction (Ihm et al. 1983, Chiang et al. 1983). Different pinning energies were found for different crystal orientations, e.g., the $\langle 100 \rangle$ and $\langle 110 \rangle$ GaAs orientations (Kahng 1964). Chiang et al. (1983) showed by angle-resolved photoemission of GaAs (100) surfaces, that the Fermi level is pinned at 0.55 ± 0.1 eV above the valence-band maximum for all reconstructed surfaces studied by the group. They proposed that defect states associated with slight nonstoichiometry, which is quite common for As-stabilized III–V semiconductor surfaces, pin the Fermi level. Fermi-level pinning was found to be insensitive to temperature in the entire temperature range $20°C \leqslant T \leqslant 500°C$ (Nanichi and Pearson 1969). Such an insensitivity of the surface Fermi level to temperature can be explained by the fact that the thermal energy kT is always much smaller than other relevant energies. Therefore, we will assume that the Fermi level is pinned during crystal growth, and we will focus on the consequences of such pinning.

Fermi-level pinning at the semiconductor surface causes the doped layer to be depleted of the free carriers and a localization of electrons in surface states. If the resulting dipole field is the driving force toward surface segregation, then this segregation process can be reduced by screening the dipole field. Figure 33 shows two SIMS profiles in which a p-type Be background doping is included. The concentration of Be is chosen to be $N_A = 4 \times 10^{18}\,\mathrm{cm}^{-3}$ to compensate for the Si dopants within 50 Å. Figure 33 reveals that the segregation length is drastically reduced from 150 Å/dec. to 80 Å/dec. at a growth temperature of 580°C. The decrease of the segregation length is attributed to the screening of the surface dipole caused by Fermi-level pinning. The trailing slope of the SIMS profile remains

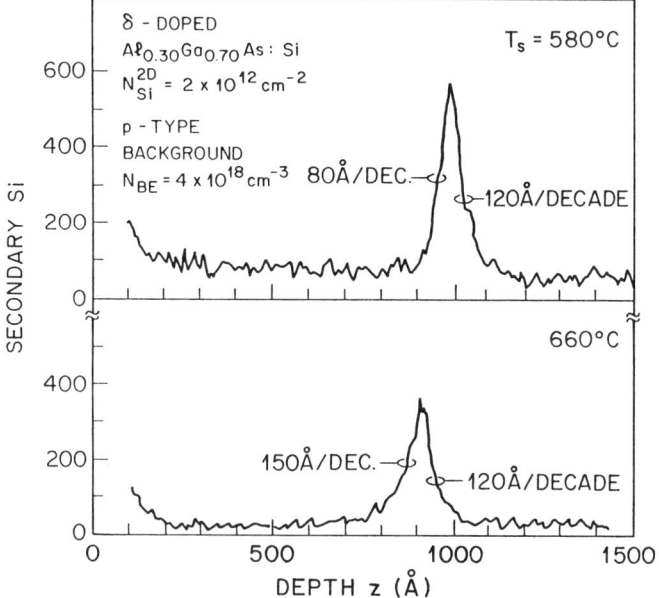

FIG. 33. Secondary ion mass spectrometry profiles of Si δ-doped $Al_xGa_{1-x}As$ grown at 580°C and 660°C, which contains intentional p-type background doping. A reduction of segregation is observed upon p-type background doping.

constant at 120 Å/dec., indicating that diffusion and the "knock-on" effect are not influenced by background doping. The SIMS profile for the sample grown at $T = 660°C$ shows the same qualitative trend as the sample grown at the intermediate temperature. The surface segregation length is reduced and the SIMS profile has a more symmetric shape, indicating that diffusion dominates the dopant redistribution. The results show that the electronic property of Fermi-level pinning causes significant (structural) redistribution of dopants during growth and that this interaction can be screened by appropriate background doping.

The observed effect is even more striking for Be surface segregation in GaAs, as shown in Fig. 34(a)–(c). At low growth temperatures (see Fig. 34(a)), the shape of the SIMS profile is dominated by the knock-on effect. At elevated growth temperatures (see Fig. 34(c)) surface segregation of Be becomes evident. The Be profile for Si background doping is shown in Fig. 34(d). The Be profile is strongly skewed toward the substrate side: Due to reversal of the surface electric field, Be impurities are driven away from the surface in direction of the substrate. The very asymmetric shape of the Be profile of Fig. 34(d) cannot be explained by the knock-on effect.

Previous theories on surface segregation suggested that a lower surface energy and a higher volatility of dopants are driving forces toward surface segregation. A classical thermodynamic calculation (Overbury *et al.* 1975, Beall *et al.* 1989) suggested that surface migration of Si in GaAs is unlikely. Panish (1973) showed that Sn incorporation in GaAs grown by liquid phase epitaxy (LPE) is controlled by equilibrium between the liquid and the crystal surface. The incorporation was therefore concluded to be influenced by the position of the Fermi level at the growing surface.

In the following we develop a model for this surface segregation process, without any fitting parameters and show that the experimenal data is in agreement with theory. We will first neglect any diffusion effects and consider only segregation. The band diagram of a semiconductor containing a δ-function-like doping profile is shown in Fig. 35(a). Figure 35(b) shows that inclusion of appropriate background doping results in a *reversal of the surface electric field*, which represents a driving force for impurities toward the substrate.

The surface of the semiconductor is assumed to be moving along the z-direction with a velocity v_s. For a two-dimensional doping density of N_D^{2D} the electric field of the dipole is given by

$$\mathbf{E} = eN_D^{2D}/\varepsilon \qquad (80a)$$

$$\mathbf{E} = \phi_B/(z_s - z_d) \qquad (80b)$$

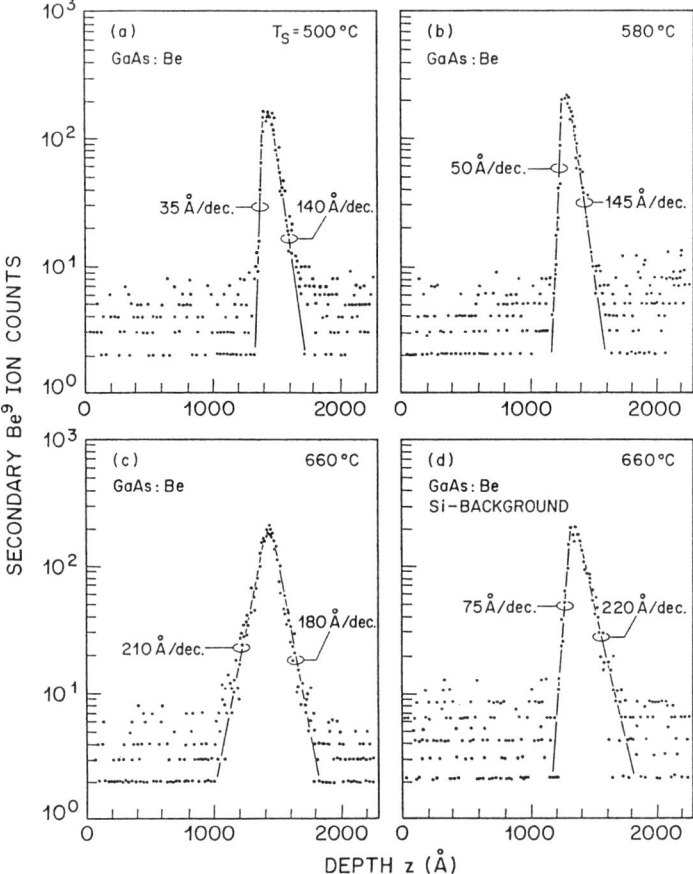

FIG. 34. Secondary ion mass spectrometry profile of Be δ-doped GaAs grown by molecular-beam epitaxy at (a) 500°C, (b) 580°C, and (c) 660°C. Inclusion of Si background doping (c) reduces the surface segregation.

where e is the elementary charge, ε is the permittivity of the semiconductor, and $z_s = v_s t$ and z_d are the position of the surface and the doped layer, respectively. Equation (80a) is valid if the doped layer is depleted of all free carriers, i.e., $z_s - z_d \leqslant \phi_B \varepsilon / e N_D^{2D}$, whereas Eq. (80b) is valid if the doped layer is partly depleted of free carriers, i.e., $z_s - z_d \geqslant \phi_B \varepsilon / e N_D^{2D}$. The segregation velocity of dopants in the electric field is given by

$$v_d = \frac{dz_d}{dt} = \mu \mathbf{E} \qquad (81)$$

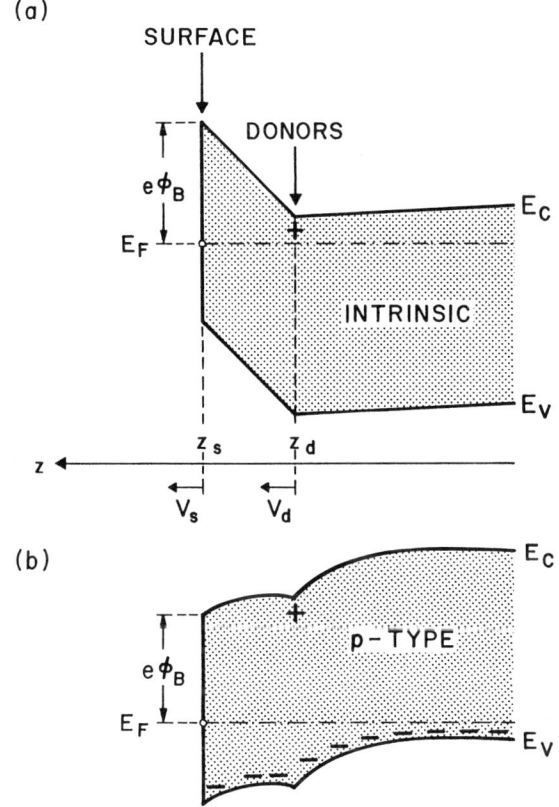

FIG. 35. Band diagram of a (a) growing n-type δ-doped semiconductor. The surface moves at a velocity v_s. Dopants segregate at a velocity v_d. (b) Inclusion of p-type background doping reverses the electric field at the semiconductor surface.

where μ is the doping ion mobility which can be obtained from the diffusion coefficient of Si in $Al_xGa_{1-x}As$ and the Einstein-relation $\mu = De/kT$.

For small distances between the doped layer and the surface, the dipole field is given by Eq. (80a), and the segregation velocity of the dopants is given by

$$v_d = \frac{De}{kT} \frac{eN_D^{2D}}{\varepsilon} \qquad (82)$$

At 500°C the diffusion coefficient of Si in $Al_xGa_{1-x}As$ is $D = 2.5 \times 10^{-16}$ cm²/sec and Eq. (82) yields $v_d = 0.1$ Å/sec for the segregation velocity, which is rather small. However, at $T = 660°C$ one obtains $D = 6 \times 10^{-15}$ cm^2/sec, where the segregation velocity is 2 Å/sec, which is comparable in magnitude to the growth rate (v_g) of 1.29 μm/hr = 3.6 Å/sec for $Al_{0.30}Ga_{0.70}As$ (0.9 μm/hr for GaAs). We see that the segregation velocity has the same order of magnitude as the growth rate. Thus, significant segregation is expected to occur at $T = 660°C$.

To obtain a single differential equation for small $(z_s - z_d)$ and large $(z_s - z_d)$ the electric field given in Eqs. (80a) and (80b) is approximated by

$$\mathbf{E} = \left[\left(\frac{e}{\varepsilon} N_D^{2D} \right)^{-1} + \left(\frac{\phi_B}{z_s - z_d} \right)^{-1} \right]^{-1} \quad (83)$$

This field approaches the exact field of Eqs. (80a) and (80b) for $z_s - z_d \ll \phi_B \varepsilon / N_D^{2D}$ and $z_s - z_d \gg \phi_B \varepsilon / N_D^{2D}$, respectively. The field represents a lower limit on the true field in the intermediate range. The differential equation then becomes

$$\frac{dz_d}{dt} = \frac{De}{kT} \left(\frac{\varepsilon}{eN_D^{2D}} + \frac{z_s - z_d}{\phi_B} \right)^{-1} \quad (84)$$

with $z_s = v_s t$. (Note that diffusion is neglected in Eq. (84).)

The segregation length is estimated by solving the nonlinear differential Eq. (84). The solution of this equation, i.e., segregation length versus time, is illustrated in Fig. 36 for $Al_{0.30}Ga_{0.70}As$ grown at the three temperatures of 500°C, 580°C, and 660°C for a dopant density of 2×10^{12} cm^{-2}. Specifically, for a 1000 Å-thick top layer ($t = 280$ sec) the segregation length for 500, 580, and 660°C is 12, 67, and 293 Å, in good qualitative agreement with the experimental results.

The calculations allow us to estimate the relative importance of the diffusion and the segregation process. The diffusion length is known to equal $\cong \sqrt{Dt}$. According to Eq. (84), the segregation length is proportional to Dt (where we assume that the field term is independent of t). Since D depends exponentially on temperature, diffusion ($\sqrt{Dt}$) dominates at low temperatures, whereas segregation (Dt) will dominate at higher temperatures; this trend is clearly confirmed by our experiments.

The preceding calculation, although giving a very good explanation of the physical process causing surface segregation, is unrealistic in two respects. First, because diffusion of impurities was neglected, the impurity profile remains δ-function-like. The second insufficiency of the calculation is the omission of screening. At the growth temperature of 660°C the concentratration of thermally excited, intrinsic carriers reaches a value of

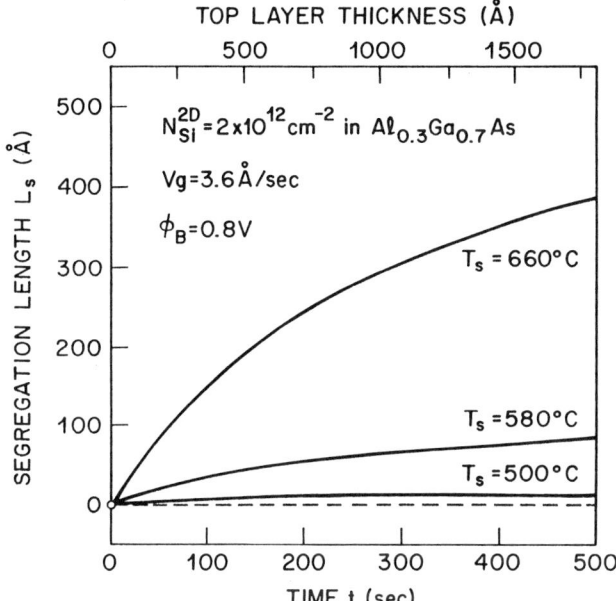

FIG. 36. Calculated Si-segregation in $Al_xGa_{1-x}As$ versus growth time. The electrostatic dipole interaction between ionized dopants and charged surface states is the driving force for the segregation process.

$n_i \cong 10^{16}$ cm^{-3}. This concentration corresponds to a Debye screening length of $\cong 550$ Å.

The understanding of the segregation mechanism opens up new ways to either make use of the mechanism or to avoid impurity segregation in III–V semiconductors. The possible uses include the controlled field-driven redistribution of dopants close to the surface. On the other hand, possibilities to reduce the segregation include (i) high-intensity illumination of the growing surface to increase the free carrier density and the screening, (ii) growth on different surface orientations such as the (110) plane on which the Fermi-level pinning may not occur (all surface Ga and As bonds are saturated on the (110) plane), and (iii) growth at low temperatures at which segregation is less pronounced.

Our findings on Fermi-level-pinning induced segregation also explain a number of observations reported previously: Tejwani et al. (1988) found that the abruptness of Be dopant distribution improves upon Si-background doping in GaAs, which can be explained by the Fermi-level-pinning model. Nichols et al. (1980) reported high concentrations of impurities close to the surface in GaAs, which can be consistently identified as a result of Fermi-level-pinning induced segregation.

8. Coulomb Correlation Effects

The spatial distribution of dopants in semiconductors is frequently assumed to be random, i.e., following Poisson statistics (see, for example, Shockley 1961). Shklovskii and Efros (1984) pointed out that correlated impurity distributions occur in compensated semiconductors. In this section it is shown that Coulomb correlation effects become important at high dopant concentrations and may result in a deviation of the dopant distribution from the purely random distribution. Experimental results on highly Be δ-doped GaAs indicate that correlation effects are indeed important at high concentrations.

The SIMS profiles of six Be δ-doped GaAs samples of nominal concentrations $2 \times 10^{12} \, \text{cm}^{-2}$ to $4 \times 10^{14} \, \text{cm}^{-2}$ are shown in Fig. 37. (The Hall carrier concentrations of the same samples are about a factor of 1.5 higher than the nominal concentration and will be shown in Fig. 42.) At Be densities $< 10^{14} \, \text{cm}^{-2}$ good spatial confinement of the Be dopants is evident from the narrow width (typically 50 Å) of the SIMS profiles. At increased densities of $\geqslant 10^{14} \, \text{cm}^{-2}$ a dramatic spreading of Be becomes obvious. At the highest concentrations, Be dopants are spread over several 100 Å along the growth axis. Such doping profiles can not be considered to be δ-function-like.

It was previously suggested that Be has a concentration dependent diffusion coefficient with D increasing rapidly, as the Be concentration exceeds $10^{19} \, \text{cm}^{-3}$. Such a concentration dependence can be explained by an increasing concentration of defects such as Be interstitials, which facilitate diffusion processes. A second explanation is enhanced Be-pair formation at high Be concentrations. Be pairs are expected to diffuse more easily as compared to atomic Be. However, we find no indication of an increased defect concentration or the formation of compensated Be pairs. Such defects or Be pairs would reduce the free carrier concentration as well as the mobility. A reduction of either free carrier concentration or mobility is not indicated by the transport (Hall) results (see Fig. 42 and 43). We therefore propose that repulsive Coulomb interaction of ionized Be dopant acts as a driving force to spread dopants out of the thin, highly doped sheet. At low concentrations, the mean dopant-to-dopant separation is large, and Coulomb interaction is insignificant. (At $N_A^{2D} = 5 \times 10^{12} \, \text{cm}^{-2}$ the mean distance between Be dopants is approximately $d = (N_A^{2D})^{-1/2} \cong 50 \, \text{Å}$.) At

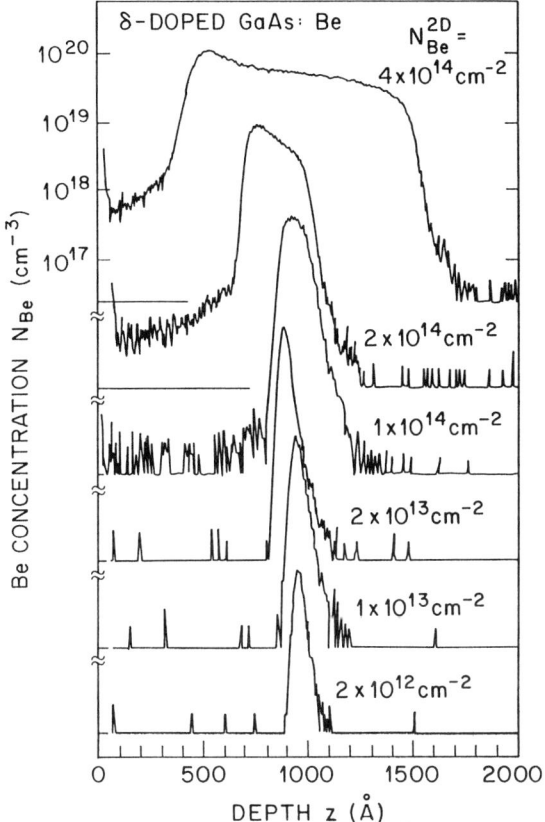

FIG. 37. Secondary ion mass spectrometry profiles of Be δ-doped GaAs for different Be densities. At densities $<10^{14}$ cm^{-2}, δ-function-like doping profiles can be achieved. At higher densities significant broadening of the Be profile occurs.

higher Be concentrations the mean Be separation could be as small as $\frac{1}{2}\sqrt{2}a_0 \cong 4$ Å; i.e., the distance of two cation sites. Such small separations lead to significant Coulomb interaction. The strong repulsive Coulomb interaction is supported by the following calculation, in which the drift of two closely separated ionized dopants due to Coulomb repulsion is estimated. If two Be dopants occupy two adjacent cation sites their separation is $s_0 = \sqrt{2}a_0/2$. Both dopants drift with a velocity

$$v = \frac{dr}{dt} = \mu \mathbf{E} \tag{85}$$

in opposite directions due to Coulomb repulsion. The drift mobility of the dopant, μ, is estimated from the Einstein relation

$$\mu = D \frac{e}{kT} \qquad (86)$$

and the Coulomb electric field, **E**, is given by

$$\mathbf{E} = \frac{e}{4\pi\varepsilon r^2} \qquad (87)$$

Insertion of Eqs. (86) and (87) into Eq. (85) and taking into account that *both* dopants drift at velocity v yields the differential equation

$$\frac{1}{2} \frac{dr}{dt} = D \frac{e}{kT} \frac{e}{4\pi\varepsilon r^2} \qquad (88)$$

which can be solved by separation of variables. One obtains

$$r(t) = \left(6 \frac{De}{kT} \frac{e}{4\pi\varepsilon} t - r_0^3 \right)^{1/3} \qquad (89)$$

which is the distance of the two dopants as a function of time. To estimate the distance of two dopants due to drift occurring at low growth temperatures, we choose the parameters $T = 500°C$, $r_0 = 4\,\text{Å}$, $D = 10^{-17}\,\text{cm}^2/\text{sec}$, and $t = 300\,\text{sec}$. The time t corresponds to the experimental growth time used for the 1000 Å-thick layer on top of the δ-doped layer. As a result one obtains for the separation $r(300\,\text{sec}) \cong 14\,\text{Å}$. That is, closely spaced dopants drift 14 Å in opposing directions due to Coulombic repulsion.

Certainly, conventional diffusion, i.e., the *random* movement of dopants, occurs simultaneously. The pure diffusion length can be estimated to be $L_D = \sqrt{D\tau} \cong 5\,\text{Å}$ for the same diffusion coefficient and time. Thus, it is obvious that drift rather than diffusion dominates at the extremely high doping concentration used in this tudy. We point out that the occurrence of drift can be misinterpreted as an enhancement of the diffusion coefficient at high doping concentrations.

The occurrence of drift also represents a fundamental limit for the highest Be-doping concentration achievable in GaAs. Assuming that the average distance between dopants is at least 14 Å, as inferred from the drift calculation, the highest achievable three-dimensional concentration would be at the most $N_A = d^{-3} = 3.6 \times 10^{20}\,\text{cm}^{-3}$. A concentration of the same order of magnitude is indeed found experimentally, as indicated by the SIMS profiles of Fig. 37.

These considerations raise the question, how can the highest Be concentration in GaAs be increased? The answer to this question is straightforward. If Coulomb repulsion effects, and the resulting drift, is the limiting factor in getting high Be concentrations, then a further lowering of the growth temperature to, e.g., 400°C reduces the diffusion coefficient and, via the Einstein-relation, also the drift mobility. Concentrations of 2.3 × 10^{20} cm^{-3} in highly Be-doped GaAs were achieved during growth at temperatures of 400°C. (Schubert et al. 1990b). Figure 38 shows the SIMS profiles of two GaAs samples grown at 500°C and 400°C. The peak Be concentration increases by a factor of 2 upon lowering of the growth

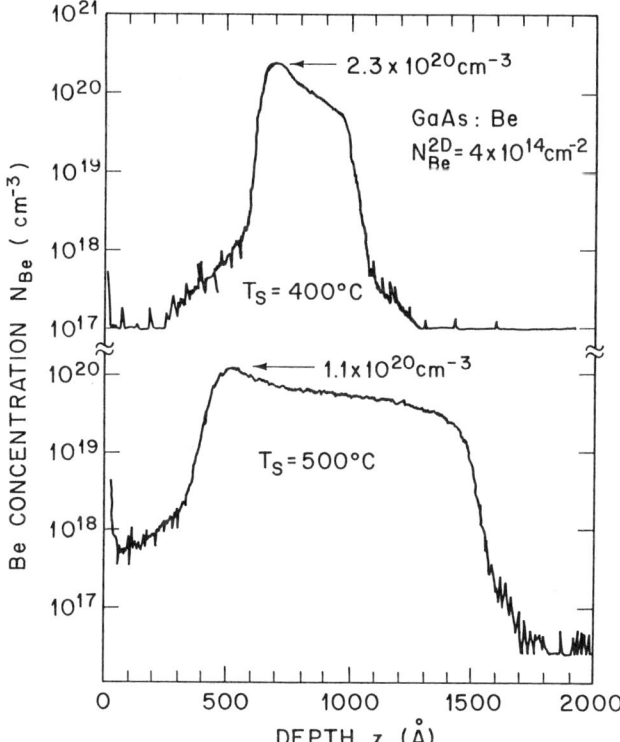

FIG. 38. Secondary ion mass spectrometry profile of Be δ-doped GaAs grown at (top) 400°C and (bottom) 500°C. The lower growth temperature results in reduced broadening and a higher Be peak concentration.

temperature from 500°C to 400°C. Similar observations were made by Hamm *et al.* (1989) for Be in $Ga_{0.47}In_{0.53}As$.

Next, the dopant distribution, influenced by Coulomb correlation effects, is calculated. The calculation follows the same concept used previously by Shockley (1961). To facilitate the calculation of the potential energy in semiconductors, we assume that dopants occupy the sites of a simple cubic lattice, but their concentration varies spatially, as indicated in Fig. 39. Areas of high ionized-dopant concentration also represent areas of high potential energy. Areas of lower concentration have a smaller electrostatic energy. If dopants were randomly distributed, and their average concentration were N_A, then the probability of A dopants being in a volume V is given by the Poisson probability distribution

$$p(A) = \frac{N^A}{A!} e^{-N} \qquad (90)$$

where $N = N_D V$ is the average number of dopants within the volume V. For large N the Poisson distribution can be approximated by a Gaussian distribution, which facilitates the following calculation. The Gaussian dis-

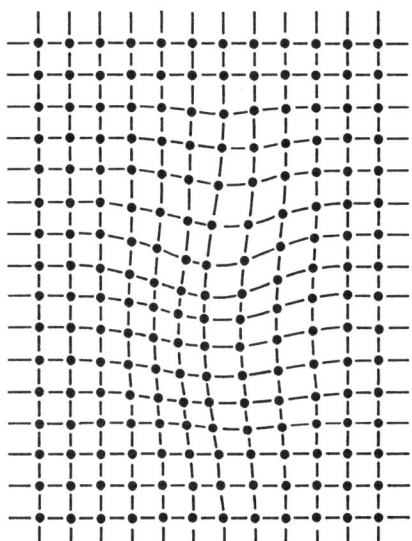

FIG. 39. Semi-random distribution of dopants that are assumed to occupy sites on a simple square lattice. Density variations are assumed to occur.

tribution then has the variance $\sigma^2 = N$ and the expectation value N:

$$p(A) = \frac{1}{\sqrt{2\pi N}} \exp\left[-\frac{1}{2}\left(\frac{A-N}{\sqrt{N}}\right)^2\right] \quad (91)$$

This distribution is shown in Fig. 40 by the solid curve for $N = 100$. The exponential term of the Gaussian distribution corresponds to the entropy term, $\exp(TS/kT) = \exp(S/K)$ in a calculation based on statistical mechanics. The potential energy of the repulsive dopants of charge e is given by

$$E = \frac{1}{2} \sum_{i \neq j} \frac{e^2}{4\pi\varepsilon r_{ij}} \quad (92)$$

This energy is reminiscent of the Madelung energy in ionic crystals. The

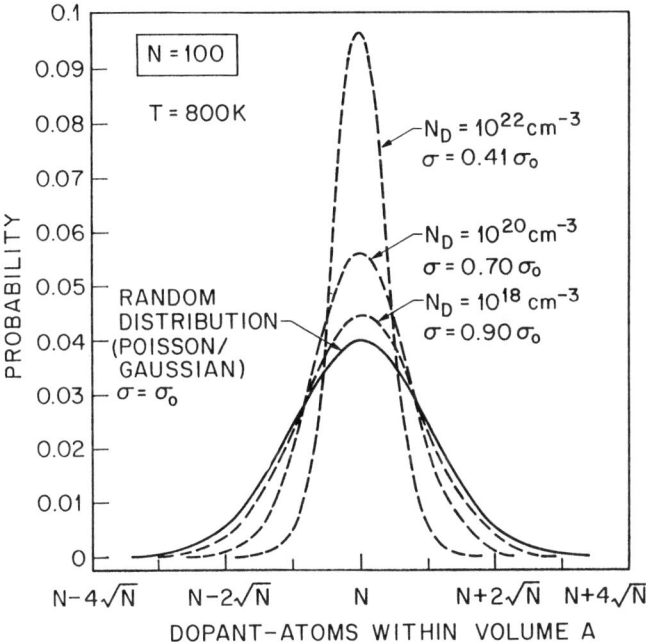

FIG. 40. Probability distribution of dopants in a semiconductor of volume A. The random dopant distribution (solid line) is valid for small concentrations. At high concentrations Coulomb repulsion effects result in a correlated nonrandom dopant distribution, i.e., the width of the distribution, σ, decreases at high concentrations.

potential energy is minimized if the dopants are ordered in a *face-centered cubic lattice*. However, it is very unlikely that such a minimum-energy configuration will be achieved. For simplicity, we assume that dopants are rather in a *simple cubic lattice* with the distance between dopants being $d = N_A^{-1/3}$. The volume of an ordered impurity lattice contains N ions, but different volumes may contain $N \pm \sqrt{N}$ ions for a random dopant distribution. The mean distance between impurities then changes by

$$\sigma_d = \pm \frac{1}{3} N^{-5/6} V^{1/3} \tag{93}$$

The introduction of the dimensionless parameter

$$\lambda = (A - N)/\sqrt{N} \tag{94}$$

allows one to change continuously from the ordered ($\lambda = 0$) to the random ($\lambda = 1$) distribution of dopant ions. The relative change of distance between dopants is then given by

$$\frac{\sigma_d}{d} = \pm \frac{1}{3} \lambda^2 N^{-1/2} \tag{95}$$

If a dopant atom is displaced by σ_d in the directions of the three Cartesian coordinates the mean increase of potential electrostatic energy (only next neighbors considered) is given by

$$\begin{aligned} \sigma_{E,1} &= 3 \left\{ \frac{1}{2} \left[\frac{e^2}{4\pi\varepsilon(d + \sigma_d)} + \frac{e^2}{4\pi\varepsilon(d - \sigma_d)} \right] - \frac{e^2}{4\pi\varepsilon d} \right\} \\ &= 3 \frac{e^2}{4\pi\varepsilon} \frac{1}{d} \left(\frac{\sigma_d}{d} \right)^2 \end{aligned} \tag{96a}$$

Thus, the mean separation between dopants, d, and its standard deviation, σ_d, are related to the electrostatic energy of a dopant ion. Within a volume V, the random distribution has an energy higher than the ordered distribution. The additional energy due to the randomness of dopants is given by

$$\sigma_{E,N} = \frac{e^2}{12\pi\varepsilon} N_A^{1/3} \lambda^2 \tag{96b}$$

Under conditions close to thermal equilibrium, the energies corresponding to different doping distributions can be assumed to be distributed according to the Boltzmann distribution

$$p(A) \propto \exp -\frac{\sigma_E}{kT} \quad (97)$$

The effect of the Boltzmann tail is to reduce the probability of configurations, which are very dense (i.e., configurations of high electrostatic energy). Combination of the Boltzmann distribution with the entropy factor of the Gaussian distribution yields after renormalization

$$p(A) = \frac{1}{\sqrt{2\pi N}} \sqrt{1 + \frac{e^2 N_A^{1/3}}{6\pi\varepsilon kT}} \exp\left[-\frac{(A-N)^2}{2N}\left(1 + \frac{e^2 N_A^{1/3}}{6\pi\varepsilon kT}\right)\right] \quad (98)$$

As indicated in Fig. 40, the standard deviation of the narrowest distribution is ($\sigma = 0.41\,\sigma_0$) less than half of the standard deviation of the purely random distribution (σ_0). Thus a significant change of the distribution occurs at high doping concentrations due to Coulomb correlation effects. The change in distribution could result in a modification of transport properties or other properties related to the nature of the dopant distribution.

Screening effects were neglected in the above calculation. They are of minor influence, since the Thomas–Fermi screening length exceeds the interparticle distance at high doping concentrations.

9. SATURATION OF FREE CARRIER CONCENTRATION

The free carrier concentration in a δ-doped semiconductor is expected to coincide with the dopant concentration. The intentional dopant density is estimated from the effusion cell temperature and the time the semiconductor is exposed to the dopant flux. The free carrier concentration can be measured, e.g., by a Hall measurement or a capacitance–voltage measurement.

Considering the electron gas as a quantized multisubband system, in which carriers of different mobilities occupy different subbands, the Hall carrier densities are given by the following relation, if the Hall factor is assumed to be unity

$$n_H = \frac{(\sum_n n_n \mu_n)^2}{\sum_n n_n \mu_n^2} \quad (99a)$$

where n_n and μ_n are the densities and mobilities in the nth subband. The Hall mobility is given by

$$\mu_H = \frac{\sum_n n_n \mu_n^2}{\sum_n n_n \mu_n} \tag{99b}$$

Equations (99) were first derived by Petritz (1985) for ambipolar semiconductors. Evaluation of Eqs. (99) yields that the Hall concentration coincides with the free carrier concentration, i.e., $n_H = \sum_n n_n$, if the mobilities in all subbands are identical, i.e., $\mu_0 = \mu_1 = \mu_2 = \ldots$. At room temperature, where phonon scattering affects the mobility in each subband in a similar manner, the Hall density is a good approximation for the true carrier density. If the mobilities in the different subbands assume different values, the Hall densities are still a *lower limit* to the true free carrier density, i.e., $n_H \lesssim \sum_n n_n$.

The CV method yields the exact number of carriers by integration of the CV profile. Both the Hall and CV techniques were shown to agree well for Si-doped GaAs (Schubert et al. 1990a).

The free-carrier Hall density versus the dopant density is shown in Fig. 41 (Schubert et al. 1986c) for δ-doped n-type GaAs:Si and p-type GaAs:Be. In addition, the Hall mobilities are shown. The solid line indicates the coincidence of the free carrier density with the dopant concentration. For Si δ-doped GaAs, the free carrier density follows the dopant concentration in the expected linear fashion, i.e., the room-temperature Hall densities coincide with the dopant density. The 77K densities are slightly lower. However, at dopant densities $N_D^{2D} \geq 1 \times 10^{13} \, \text{cm}^{-2}$, a saturation of the free carrier concentration becomes obvious. In the saturation region, the free carrier density increases sublinearly with dopant density. The exact density at which the onset of saturation occurs is most likely dependent on the growth conditions. Gillman et al. (1988) found that the saturation occurs at a higher density of $2.5 \times 10^{13} \, \text{cm}^{-2}$.

Two mechanisms were shown to cause saturation of the free carrier concentration in Si-doped GaAs; namely, autocompensation of Si and the population of DX-center. Autocompensation of highly Si-doped GaAs was shown to occur by several groups (see e.g. Chai et al. 1981). Si-Si pairs associated with autocompensation were identified by local-mode spectroscopy (Spitzer et al. 1969). A more recent study by Maguire et al. (1987) showed that a number of compensation mechanisms occur, including Si_{As}, Si_{Ga}-Si_{As}, and Si-complexes contribute to the saturation of the free carrier density. A second mechanism of saturation in highly Si-doped GaAs was proposed to be the DX-center (Theis et al. 1988). The DX-center, which acts as a deep donor in $Al_xGa_{1-x}As$, is resonant with the conduction band in GaAs. At high doping densities the center becomes populated and thus reduces the free carrier concentration.

Beryllium δ-doped GaAs does not exhibit any saturation of the free carrier density. Densities up to $10^{13}\,cm^{-2}$ are shown in Fig. 41. Higher densities, up to $6 \times 10^{14}\,cm^{-2}$, are shown in Fig. 42. The hole mobilities of the samples are shown in Fig. 43 for $T = 60K$ and room-temperature. A saturation of the free hole density is not observed. However, at very high concentrations significant spread of Be occurs, as shown in Fig. 37. The

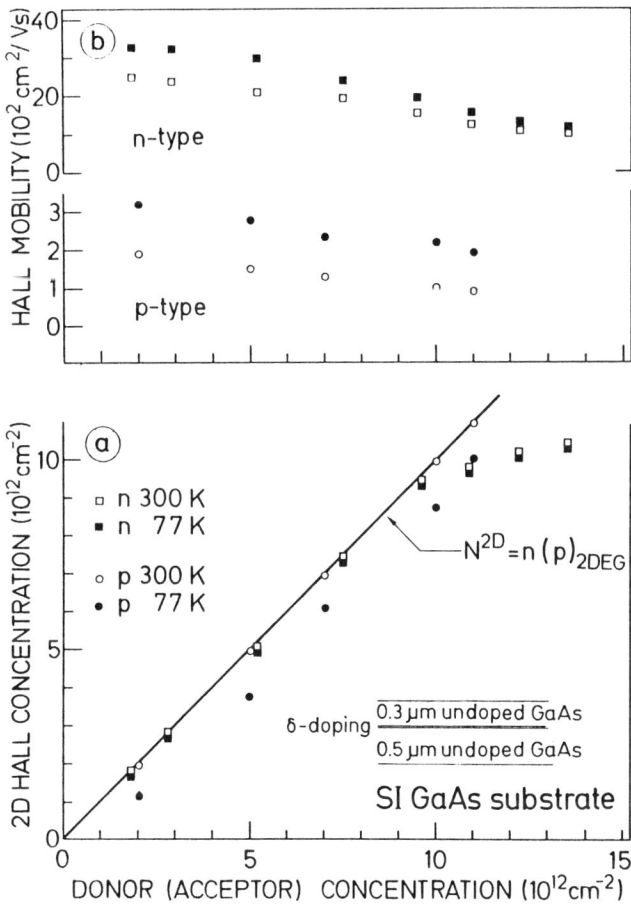

FIG. 41. Mobility (top) and free carrier Hall concentration (bottom) of Si and Be δ-doped GaAs versus dopant concentration.

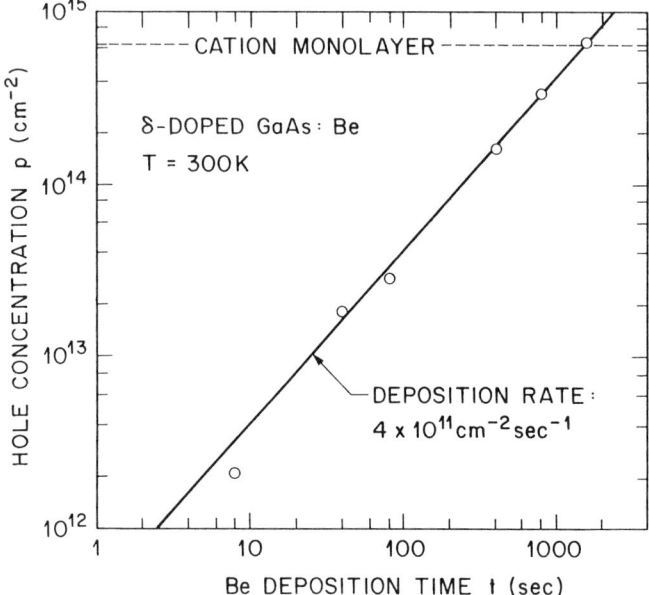

FIG. 42. Free hole concentration versus Be deposition time in δ-doped GaAs. No saturation of the hole concentration is observed even at the highest Be concentration.

samples with high doping density do not have δ-function-like doping profiles. Only samples with $p = N_A^{2D} < 10^{14}$ cm^{-2} exhibit spatial localization of dopants on the length scale of the lattice constant.

V. Electronic Properties of δ- Doped Semiconductors

10. ELECTRONIC PROPERTIES OF δ-DOPED HOMOSTRUCTURES

How are the transport properties of δ-doped semiconductors distinguished from structures doped in the conventional, i.e., homogeneous, fashion? To answer this question, it is useful to differentiate between two transitions between homogeneous doping and δ-doping. Both transitions are illustrated in Fig. 44 and characterized by a *constant 3D concentration* and a *constant 2D density*.

Consider first the scaling of a doping profile in which the 2D density is kept constant. Obviously, the carrier mobility decreases, as the doped layer is reduced in thickness. More interestingly, since the concentration of

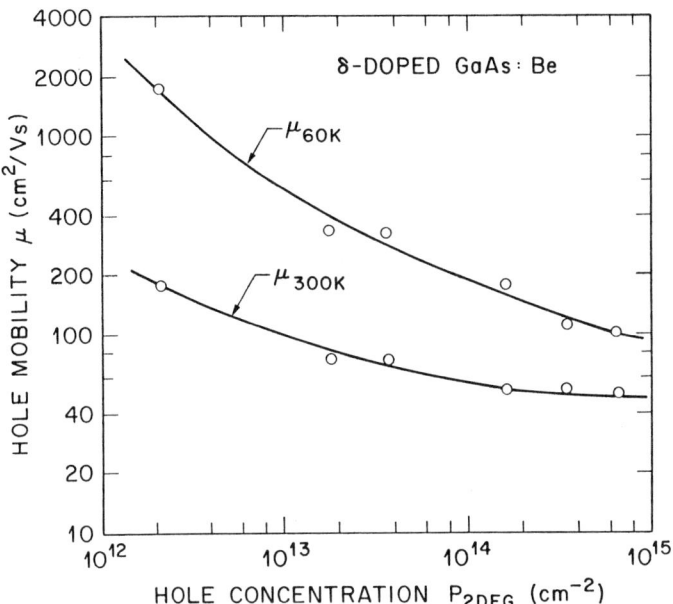

FIG. 43. Hall mobility of Be δ-doped GaAs versus Be dopant density at $T = 60K$ and $T = 300K$.

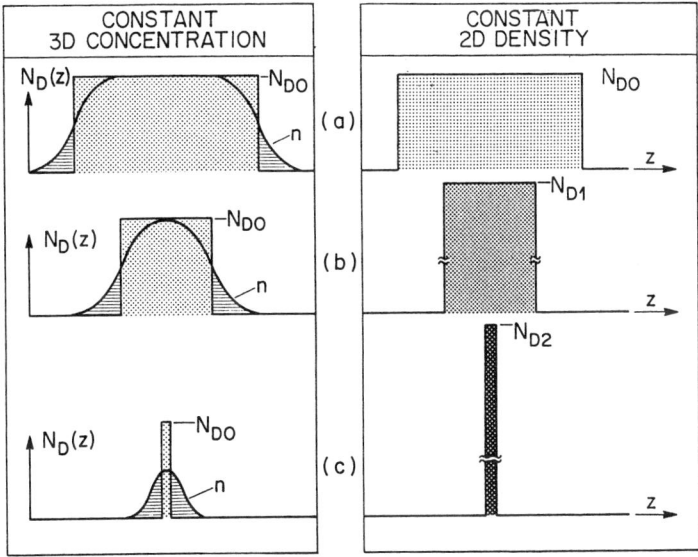

FIG. 44. Schematic illustration of constant concentration scaling (left) and constant two-dimensional density scaling (right) of a doping profile. For δ-function-like doping profiles the free carriers have a larger spatial extent than the dopants.

dopants increases for narrower dopant distributions, the Mott transition occurs at some concentration. Thus, freeze-out effects are less pronounced or even absent beyond the insulator–metal transition.

Considering the scaling with constant 3D concentration of dopants, the changes in transport properties are more intriguing. An increasing spatial separation between ionized impurities and free carriers occurs when the dopant distribution becomes more δ-function-like, as indicated in Fig. 44. A reduction of ionized impurity scattering results (Schubert et al. 1987b).

Experimental electron mobilities of δ-doped GaAs at $T = 300$K are shown in Fig. 45 as a function of the equivalent 3D doping concentration. This equivalent concentration is inferred from the 2D density according to $N_D = (N_D^{2D})^{3/2}$. The experimental points are compared with theoretical electron mobilities of n-type GaAs obtained from the Hilsum relation (Hilsum, 1974) shown by the solid line.

At low doping concentrations the mobilities of 3D and 2D doped GaAs coincide. However, at high doping concentrations, the mobility of 2D doped GaAs exceeds significantly the mobility of homogeneously doped material (see Fig. 45 for illustration). At a concentration of $N_D^{3D} = 3.5 \times 10^{19}\,\text{cm}^{-3}$ the mobility enhancement μ^{2D}/μ^{3D} is more than a factor of 4. The enhancement is shown in the inset to Fig. 45. The distinguishing feature is the *increasing enhancement* of the electron mobility *at high doping concentrations*.

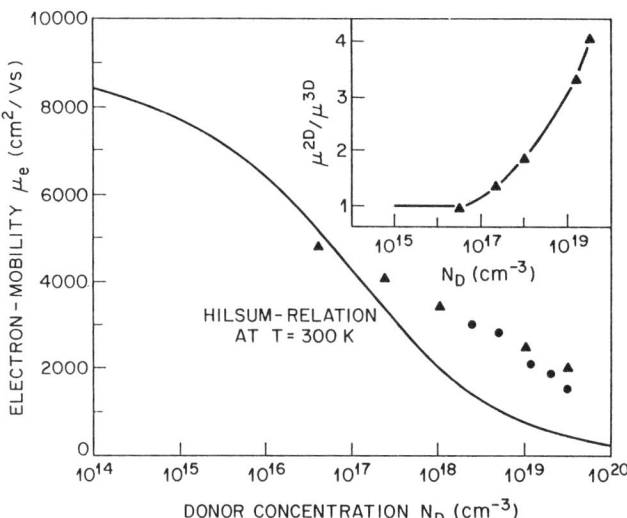

FIG. 45. Electron mobility versus doping concentration in homogeneously doped (solid line) and δ-doped GaAs. Significantly higher mobilities are found in δ-doped GaAs. The mobility enhancement is shown in the inset.

Temperature-dependent Hall measurements were performd to clarify the origins of the mobility enhancement. The electron mobility of a δ-doped GaAs sample ($n_{2DEG} = 1.5 \times 10^{13}$ cm^{-2}) is shown in Fig. 46 for temperatures 4.2K $\leqslant T \leqslant$ 300K. The electron mobility depends weakly on temperature in the entire temperature range, indicating a reduced role of impurity scattering. Ionized-impurity scattering in conventionally, *nondegenerately* 3D-doped samples depends on the sample temperature, T (or better; kinetic energy), according to (Brooks–Herring)

$$\mu_{II} = \frac{128(2\pi)^{1/2}\varepsilon^2(kT)^{3/2}}{(m^*)^{1/2}N_D^{3D}e^3} \left[\ln \frac{24m^*\varepsilon(kT)^2}{ne^2\hbar^2} \right]^{-1} \quad (100)$$

where ε, k, e, n, and $\hbar$ are the permittivity, Boltzmann's constant, effective mass, elementary charge, electron concentration, and Planck's constant divided by 2π, respectively. If impurity scattering were important, a dependence of $\mu \propto T^{3/2}$ would be expected at low temperatures. Ionized-impurity scattering in *degenerately* doped semiconductors is beset with a number of difficulties (Chattopadhyay and Queisser 1981) and cannot be expressed in closed form. A temperature dependence of $\mu \propto T^{0.29}$ is found in low temperatures for the δ-doped sample shown in Fig. 46. The weak $T^{0.29}$ dependence is significantly smaller than the $T^{1.5}$-dependence inferred from Eq. (100).

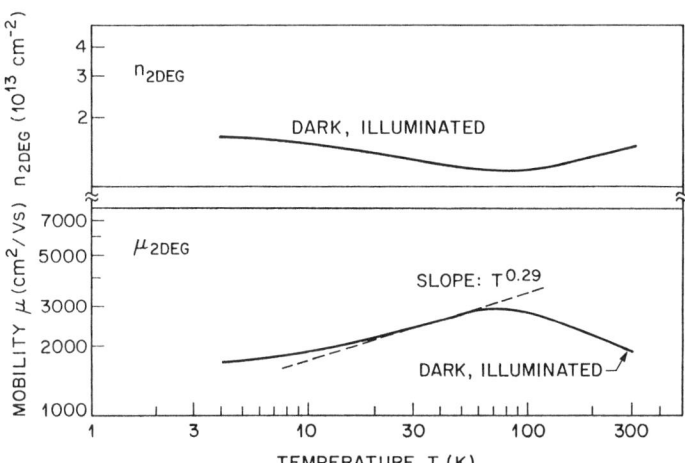

FIG. 46. Electron density and electron mobility in highly Si δ-doped GaAs versus temperature.

Three theoretical origins will be discussed for the mobility enhancement: (i) high degeneracy of the Fermi gas, (ii) spatial electron–donor separation in the $n = 1, 3, \ldots$ subbands, and (iii) screening of impurity charges. The 2DEG of δ-doped epitaxial layers has a high degree of degeneracy (i.e., high kinetic energy parallel to the donor plane) amounting to several 100 meV. Carriers energetically close to the Fermi surface, which are in turn most likely to incur an elastic or inelastic scattering event, have a kinetic energy of $E_F - E_n$, where E_n is the eigenstate energy. The kinetic energy exceeds the thermal energy kT considerably. Consequently, the temperature-dependent elastic impurity scattering is reduced as long as the kinetic energy term, $E_F - E_n$, dominates; that is, at high concentration of the 2DEG. The latter conditions is fulfilled in the entire temperature range presently investigated. The experimental results shown in Fig. 46 confirm the reduced temperature dependence.

It is worthwhile to note that the eigenstate energy itself is partly potential and partly kinetic (in perpendicular direction to the well plane). The kinetic part can be calculated from

$$\langle E_{\text{kin}} \rangle = E_n - \langle \psi(x) | E_c(x) | \psi(x) \rangle \tag{101}$$

and is usually a considerable fraction of the total eigenstate energy. The kinetic part of the *eigenstate* energy does not, however, enter the calculation of ionized impurity scattering in Eq. (100).

Spatial separation of free electrons from their parent ionized-donor impurities occur in the odd-labeled ($n = 1, 3, \ldots$) subbands. Carriers in the $n = 1$ state therefore have zero probability of being in the donor plane, resulting in a mobility enhancement of electrons in odd-numbered eigenstates. This electron–donor separation in real space in δ-doped structures is reminiscent of the carrier–impurity separation in selectively doped heterostructures.

A third origin for the enhanced mobility is due to screening of impurity charges by the high concentration of the 2DEG. The result of linear screening theory is a weak $N_D^{1/6}$ dependence of the screening radius on the electron concentration for a *degenerate* electron gas.

A mobility enhancement in δ-doped GaAs was also found by Koenraad *et al.* (1989) by cyclotron resonance. They attributed the enhanced mobility to the small overlap between the ionized donors and the electron-wave function, especially for states with high subband index.

Further electron-mobility data of Si δ-doped GaAs is shown in Fig. 47 for different Si densities. The plot reveals that the mobility is less temperature dependent for high densities. This trend is expected at high densities due to a larger degeneracy, i.e., higher kinetic energy, of the electron system.

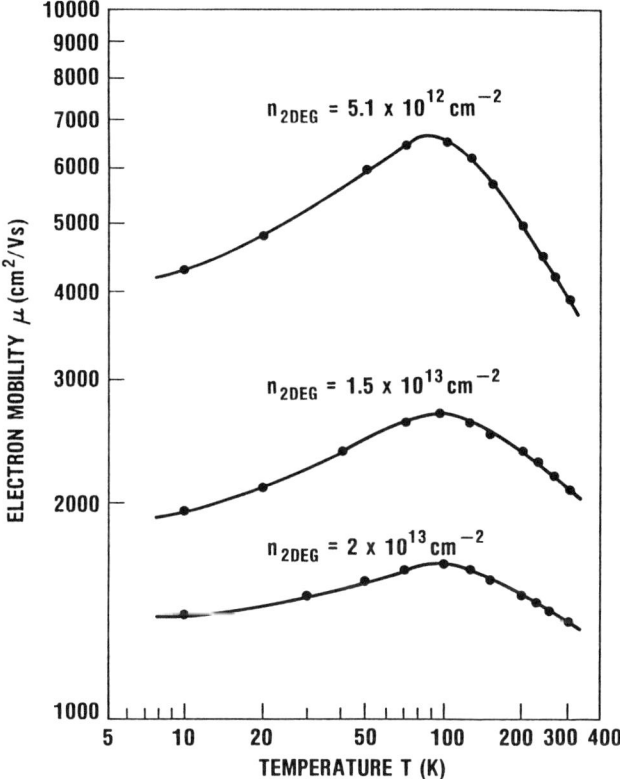

FIG. 47. Electron mobility of δ-doped GaAs of different densities.

11. ELECTRONIC PROPERTIES OF SELECTIVELY δ-DOPED HETEROSTRUCTURES

Selectively doped $Al_xGa_{1-x}As/GaAs$ heterostructures are an area of semiconductor physics that had a great impact on the direction of semiconductor science and technology during the 1980s. Such selectively doped heterostructures (Störmer *et al.* 1978) are characterized by a spatial separation of ionized impurities and free carriers and a resulting reduction in ionized-impurity scattering. Reviews that emphasize the transport properties (Störmer 1983), growth by MBE (Gossard 1985), and transistor applications (Solomon and Morkov 1984, Dingle *et al.* 1985, Abe *et al.* 1986) have become available.

The properties of selectively doped heterostructures are significantly improved by the δ-doping technique. The density of the two-dimensional electron gas is enhanced and can assume values exceeding $1 \times 10^{12}\,\text{cm}^{-2}$ in selectively δ-doped $\text{Al}_x\text{Ga}_{1-x}\text{As/GaAs}$ heterostructures (Schubert et al. 1987d, Cunningham et al. 1988). Furthermore, using δ-function-like doping profiles in the wide-gap $\text{Al}_x\text{Ga}_{1-x}\text{As}$ represents an optimized structure with minimized potential fluctuations, i.e., maximized ionized-impurity mobility (Schubert et al. 1989e). Thus, employment of the δ-doping technique enhances the density and the mobility of the two-dimensional electron gas (2DEG). The enhancement in electron density is advantageous for transistors and will be discussed in Sect. 13. In the following, the optimization in electron mobility due to δ-doping will be analyzed.

The conduction-band diagrams of two selectively doped heterostructures with different doping configurations are shown in Fig. 48. The doped region has a thickness of z_d and its centroid is at a distance of z_c from the heterointerface. The doping configurations shown in Fig. 48 are represented by the doping profile

$$N_D(z) = N_D^{2D}/z_d \{\sigma[z - (z_c - \tfrac{1}{2}z_d)] - \sigma[z - (z_c + \tfrac{1}{2}z_d)]\}, \qquad (102)$$

where the thickness z_d is a free parameter, $N_D^{2D}/z_d = N_D$ is the three-dimensional (3D) doping concentration in the doped layer, and $\sigma(z)$ is the step function. One can easily verify that the doping profiles given in Eq. (102) result in the same electric displacement and potential at the heterointerface for different z_d. Thus, Eq. (102) provides *different doping configurations* but *identical free carrier densities*: $N_D^{2D} = n_{2\text{DEG}} = \text{const}$. Note that as z_d increases the undoped $\text{Al}_x\text{Ga}_{1-x}\text{As}$ spacer layer $(z_c - z_d/2)$ decreases, and simultaneously the 3D doping concentration N_D^{3D}/z_d is reduced. The trend of the mobility is, therefore, not obvious as z_d is changed. Present mobility calculations do not take into account the randomness of the impurity distribution. Furthermore, in the mobility calculations the total impurity potential is taken as the relevant scattering potential. However, it would be more appropriate to use only the smaller amplitude of the potential *fluctuation* as the scattering potential. An analytic approximation for the dependence of the electron mobility on the design parameters of the heterostructure has been proposed (Price 1981, 1982). However, subsequent work of the author showed that the exact functional dependences are yet to be determined (Price 1984).

Next, we determine the magnitude of potential fluctuations at the interface due to random impurity distribution within the doped layer. First, we calculate potential fluctuations analytically using unscreened Coulomb potentials. Subsequently, we calculate potential fluctuations numerically using screened Coulomb potentials. As shown in Fig. 49 the volume element dV_i contains a mean total number of dopant atoms of $N = N_D dV$. The

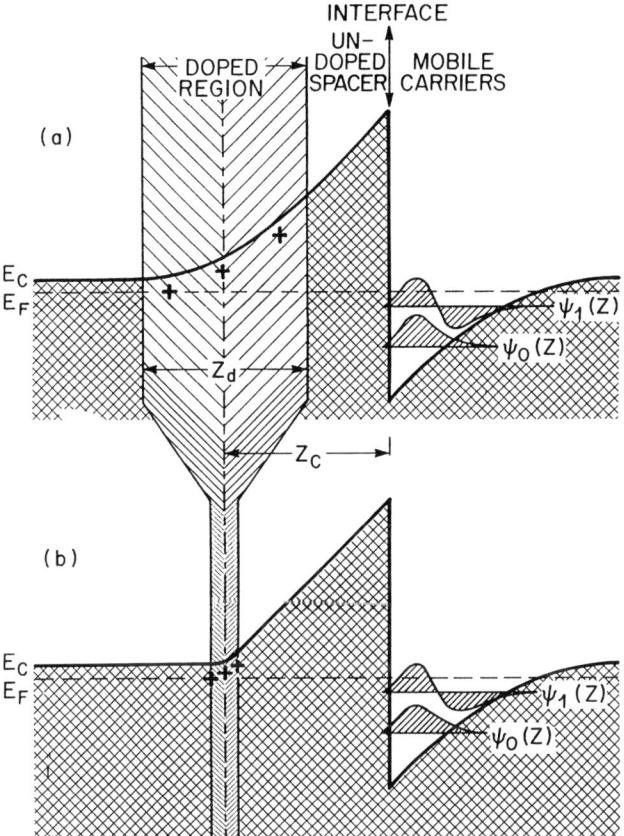

FIG. 48. Band diagram of a selectively doped heterostructure in which the thickness of the doped region is varied. The total two-dimensional density of dopants as well as the free carrier density remain constant.

number of atoms, N, varies from unit volume to unit volume due to statistical dopant distribution. The charge-density fluctuation causes a potential fluctuation at the plane of the heterointerface. The statistical variance of the potential fluctuation at the interface caused by one unit volume located at (r, α, z) can be obtained using Poisson statistics (r, α, and z are the coordinates of a cylindrical coordinate system). Using unscreened Coulomb potentials for the impurities yields for the variance of the potential fluctuation

$$d\sigma_\phi^2 = (e/4\pi\varepsilon)^2 N_D \, dr \, r d\alpha \, dz (r^2 + z^2)^{-1} \qquad (103)$$

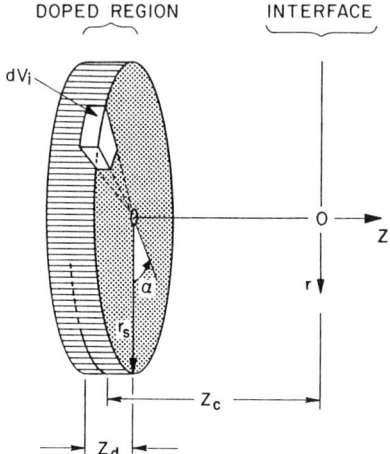

Fig. 49. Schematic illustration of dopant distribution used for the calculation of potential fluctuations at the heterointerface due to dopant density fluctuations in the doped region.

where e is the elementary charge, $dr\,r\,d\alpha\,dz$ is the unit volume dV_i in cylindrical coordinates as shown in Fig. 49, and ε is the permittivity of the semiconductor. The total mean potential fluctuation at the interface is obtained by integration over the entire doped layer

$$\sigma_\phi^2 = \int_z \int_r \int_\alpha d\sigma_\phi^2 \tag{104}$$

We restrict our calculation to charge fluctuations occurring within a screening radius r_s. One obtains for $z_c \ll r_s$

$$\sigma_\phi^2 = 2\pi \frac{N_D^{2D}}{z_d} \left(\frac{e}{4\pi\varepsilon}\right)^2 \left[\left(z_c - \frac{z_d}{2}\right)\left(\ln\frac{z_c - z_d/2}{r_s} - 1\right) \right. \\ \left. - \left(z_c + \frac{z_d}{2}\right)\left(\ln\frac{z_c + z_d/2}{r_s} - 1\right)\right] \tag{105}$$

and for $z_c \gg r_s$

$$\sigma_\phi^2 = \pi N_D^{2D} (e/4\pi\varepsilon)^2 r_s^2 /(z_c^2 - \tfrac{1}{4}z_d^2). \tag{106}$$

We now minimize the variance of the potential fluctuation with respect to the thickness of the doped layer. Calculating $\partial\sigma_\phi^2/\partial z_d = 0$ for Eqs. (105) and

(106) yields that σ_ϕ^2 is minimized, if $z_d \to 0$; that is, if the doped layer thickness approaches 0. The doping profile is then given by the δ function:

$$N_D(z) = N_D^{2D}\delta(z + z_c) \tag{107}$$

The minimum standard deviation of the potential is for $z_c \ll r_s$ given by

$$\sigma_\phi = (e/4\pi\varepsilon)\{2\pi N_D^{2D}[1 - \ln(z_c/r_s)]\}^{1/2} \tag{108}$$

and for $z_c \gg r_s$

$$\sigma_\phi = (e/4\pi\varepsilon)(\pi N_D^{2D})^{1/2} r_s/z_c \tag{109}$$

where the two-dimensional (2D) screening radius is given by (Ando et al. 1982)

$$r_s = (2\varepsilon/e^2)(\pi\hbar^2/m^*) \cong 52\,\text{Å} \tag{110}$$

where $\hbar$ is Planck's constant divided by 2π and m^* is the effective electron mass. As an example, for $z_c = 500\,\text{Å}$ and $N_D^{2D} = 1 \times 10^{12}\,\text{cm}^{-2}$ one obtains from Eq. (109) a mean potential fluctuation of $\sigma_\phi \cong 2\,\text{mV}$.

The employment of screened Coulomb potentials rather than unscreened potentials refines this calculation. The Coulomb potential of an impurity located at (r, z) screened by electrons that are assumed to be in the plane $z = 0$ is given by (Ando et al. 1982)

$$\phi(r, z) = \int_0^\infty q A(q) J_0(qr) dq, \tag{111}$$

where J_0 is the Bessel function of zero order and

$$A(q) = \frac{e}{4\pi\varepsilon} \frac{e^{-qz}}{q + 1/r_s}. \tag{112}$$

For large r, the Coulomb potential screened by a 2D sheet of electrons follows the proportionality

$$\phi(r, z) \propto 1/r^3 \tag{113}$$

and decays more rapidly than the unscreened Coulomb potential. The screened Coulomb potential is obtained by a numerical integration of Eq. (111) and then inserted in Eq. 103. The minimization of potential fluctuations can be still achieved, if the doped layer thickness approaches zero.

The minimization of potential fluctuations is equivalent to a maximization of remote ionized-impurity mobility, since the scattering matrix element is proportional to the magnitude of the scattering potential. Thus, the electron mobility and the magnitude of potential fluctuations obey a monotonical relationship, which is valid even in the presence of screening.

Experimental Hall mobilities μ (left ordinate) and carrier concentrations n_{2DEG} (right ordinate) of the samples are shown in Fig. 50. The thickness of the doped layer is changed as illustrated in the inset. The electron mobilities at 4.2K range between $860\,000\,\text{cm}^2/\text{V sec}$ for $z_d = 800\,\text{Å}$ and $4.53 \times 10^6\,\text{cm}^2/\text{V sec}$ for $z_d \to 0\,\text{Å}$. Mobilities are measured in the dark after illumination. Electron mobilities clearly increase as z_d decreases as shown in Fig. 50 with the highest mobility being obtained as z_d approaches zero. The mobility follows the qualitative trend predicted by the theoretical model. At small thicknesses of the doped layer, only minor changes of the mobility are expected: Eq. (106) shows that the change of the scattering potential becomes negligible as z_d approaches zero (i.e., $d\sigma_\phi/dz_d = 0$ for $z_d \to 0$). We indeed observe such a weak dependence of the mobility on z_d as $z_d \to 0$.

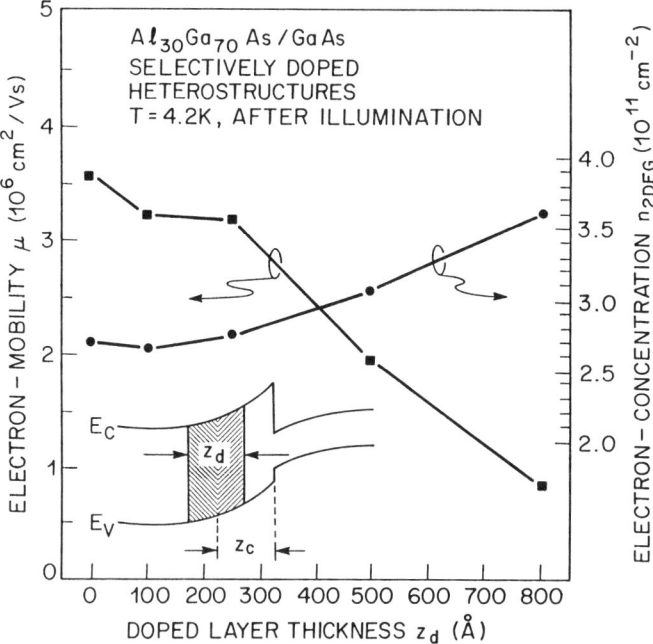

FIG. 50. Electron mobility and two-dimensional electron concentration of selectively doped heterostructures with different thicknesses of the doped layer. The highest mobilities are obtained when the thickness of the doped layer approaches zero.

Figure 50 further illustrates the carrier density as a function of the doped layer thickness. An increase of $n_{2\text{DEG}}$ is observed as z_d increases, which is not expected if only shallow donors are taken into account. A probable cause for this increase is the coexistence of shallow and deep donors in the $\text{Al}_x\text{Ga}_{1-x}\text{As}$. As z_d increases, more deep donors are elevated above the Fermi level in the doped region close to the interface as illustrated in Fig. 48. Consequently, the deep donors nearest the interface become ionized and the mobile carrier density is enhanced. This concentration enhancement is expected for carriers in the dark. It is apparently maintained even after illumination in samples with large z_d, due to photoionization of deep donors closer to the interface.

The historical development of electron mobilities in selectively doped $\text{Al}_x\text{Ga}_{1-x}\text{As}/\text{GaAs}$ heterostructures is shown in Fig. 51 (Pfeiffer *et al.* 1989). The plot includes data of homogeneously doped *n*-type GaAs (lowest trace), the 5×10^6 cm^2/Vsec mobility reported by English *et al.* (1987), and the $>10^7$ cm^2/Vsec mobility reported by Pfeiffer *et al.* (1989).

VI. Field-Effect Transistors

The δ-doping technique represents the ultimate limit for scaling of doping profiles. Such scaling of the dimensions of a semiconductor structure is of importance for semiconductor devices. As the spatial, lateral, and vertical dimensions of devices shrink, the switching speed and the power consumption of the devices decrease. It is thus of fundamental interest to investigate field-effect transistors with doping profiles scaled to their ultimate limit.

In this section, the properties of field-effect transistors (FETs) that include δ-function-like doping profiles are investigated. Homostructure as well as heterostructure FETs will be considered.

12. Homostructure Metal–Semiconductor Field-Effect Transistors

A field-effect transistor that contains a δ-function-like doping profile is shown in Fig. 52. The doped layer is sandwiched between nominally undoped GaAs, which is epitaxially grown on GaAs. The δ-doped FET (Schubert and Ploog 1985e) shown in Fig. 52 has a number of advantages over conventional, homogeneously doped metal–semiconductor field-effect transistors (MESFETs), including short gate-to-channel distance, high transconductance, high densities of the 2DEG, large breakdown voltage, and reduced short channel effects. In addition, such devices could be

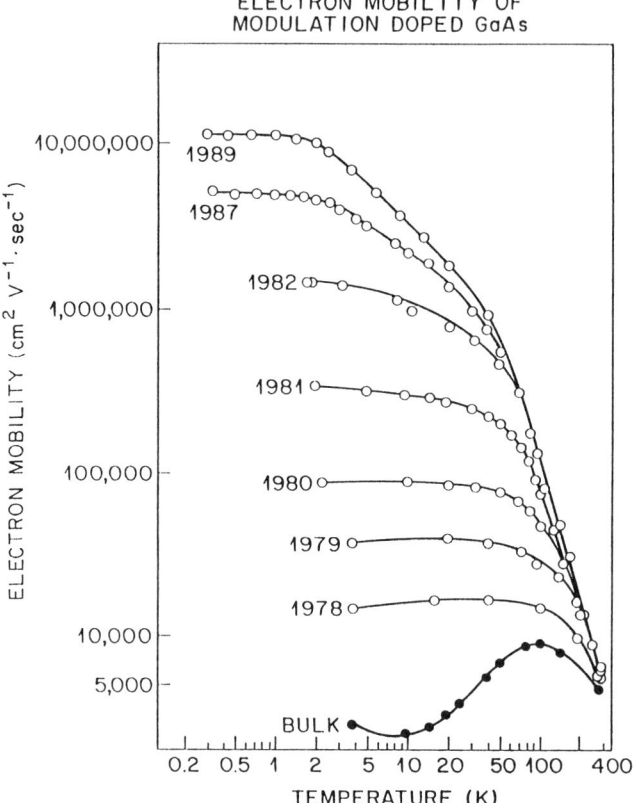

FIG. 51. Historical development of electron-mobilities in selectively doped heterostructures as a function of temperature. The two top curves were obtained using δ-doped heterostructures (from Pfeiffer et al. 1989, with permission of the authors).

advantageous for high-power applications (Wood et al. 1979). The intrinsic transconductance of a FET can be given in analytic form according to

$$g_m^* = \frac{1}{L_G}(e\mu W_G n_{2DEG})\left[1 + \left(\frac{e\mu n_{2DEG}d}{ev_s L_G}\right)^2\right]^{-1/2} \quad (114a)$$

where W_G, L_G, and d are the width and length of the gate, and the distance between the gate and the 2DEG, respectively. The saturation velocity is designated as v_s. For short gate lengths $L_G \to 0$ the equation reduces to

$$g_m^* = ev_s W_G/d \quad (114b)$$

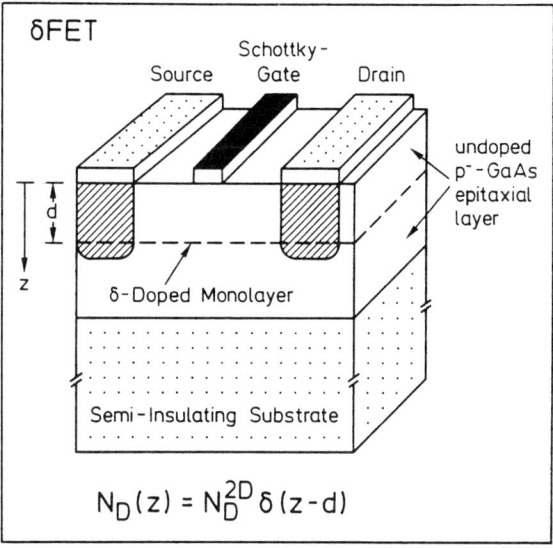

FIG. 52. Delta-doped GaAs metal–semiconductor field-effect transistor. The conductive channel in a Si δ-doped layer located at a distance d below the Schottky gate.

which can also be derived from the well-known saturation-velocity model. Equation (114a) is based on a two-region model for the velocity–field characteristic and a velocity saturation at the drain end of the gate (Hower and Bechtel 1973, Delagebeaudeuf and Linh 1982). The equation suggests that a short gate-to-channel distance d is desirable and yields a high transconductance. An intrinsic transconductance exceeding 500 mS/mm was estimated for the δ-doped FET (Schubert et al. 1986c).

A high carrier density of the 2DEG is desirable as well to achieve a high transconductance, as concluded from Eq. (114). Since high carrier densities are easily achievable by the δ-doping technique, a high transconductance is feasible. Note that the concentrations of the 2DEG can exceed the densities achievable in selectively doped heterostructure transistors.

A further advantage of the δ-doped FET is the large gate breakdown voltage. It was shown (Schubert et al. 1985e, 1986c) that the maximum electric field under the gate is *smaller* for the δ-doped FET than the conventional, homogeneously doped FET. A large breakdown voltage is consequently observed in δ-FETs.

Finally, short-channel effects are less pronounced in δ-doped FETs. Short-channel effects are prominent if the gate length becomes smaller than the gate-to-channel distance. Short-channel effects manifest themselves as a lack of pinch-off and a large output conductance in the saturation regime.

The δ-doped FET with its inherently small gate-to-channel distance has thus advantageous properties in the short gate-length domain. Little short-channel effects are expected for the δ-doped FET for gate lengths down to 1000 Å.

The current-voltage characteristic of the first δ-doped FET is shown in Fig. 53. The characteristic shows excellent pinch-off and saturation regime. The device has a gate length of 0.5 μm and a source-drain separation of 5.0 μm. The measured transconductance is 75 mS/mm. Transconductances of 270 mS/mm were reported in a more recent study by Ishibashi et al. (1988) on a δ-doped FET.

The δ-doping technique was also used to fabricate nonalloyed ohmic contacts to a semiconductor (Schubert et al. 1986b). Such nonalloyed ohmic contacts consist of one or more highly δ-doped sheets in proximity to the semiconductor surface. Since the contacts require no alloying, their surface morphology is very smooth and no "balling-up" occurs. The band diagram of such an ohmic contact, shown in Fig. 54, consists of the metal and the semiconductor doped at a distance z_0 from the metal-semiconductor interface. For sufficiently high doping concentrations and sufficiently small distances z_D, the barrier of thickness t (which depends on the applied voltage) is so small that tunneling through the surface barrier becomes the dominant transport mechanism. Contact resistances $<10^{-8}\,\Omega\,\text{cm}^2$ were

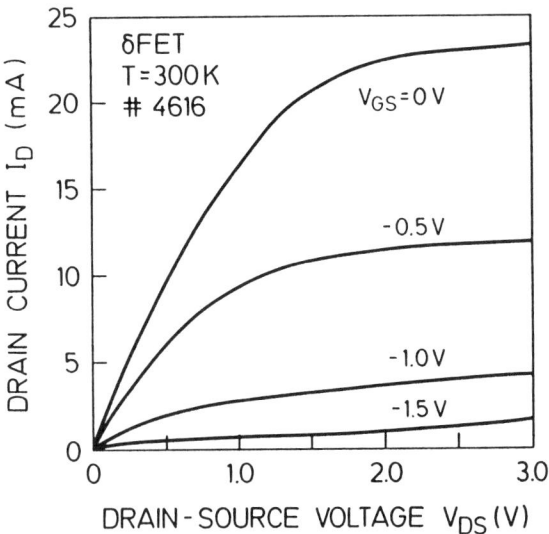

FIG. 53. Output characteristic of the δ-doped MESFET at $T = 300$K. The transconductance of the device is 75 mS/mm.

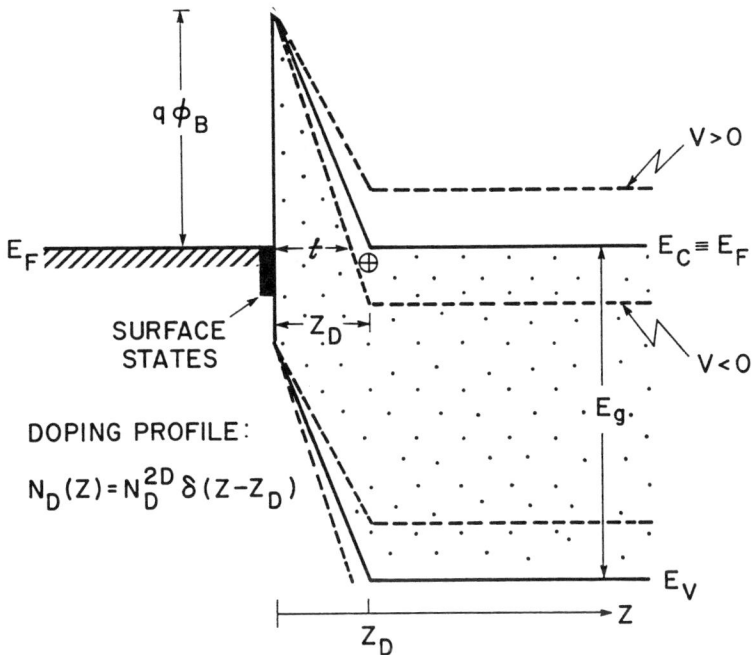

FIG. 54. Band diagram of a δ-doped nonalloyed ohmic contact. The tunneling thickness t is small (~ 20 Å), which makes tunneling the dominant transport mechanism across the Schottky barrier.

calculated for small distances between the doped layer and the metal–semiconductor interface.

Experimental I–V characteristics of the nonalloyed ohmic contact is shown in Fig. 55. The I–V curve is a straight line and shows no S or N shape, as is sometimes observed in nonideal ohmic contacts. The evaluation of the specific contact resistance yields a typical value of $6.3 \times 10^{-6}\,\Omega\,\text{cm}^2$ as shown in the lower part of Fig. 55. Contact resistances in the low $10^{-6}\,\Omega\,\text{cm}^2$ range can be obtained.

The nonalloyed ohmic contacts were used in self-aligned FETs, which were fabricated using a two-mask process (Schubert et al. 1986a). The FET structure, shown schematically in Fig. 56, consists of several δ-doped regions including the channel region and several highly δ-doped sheets for nonalloyed ohmic contacts. Additional doped layers are included to reduce the series resistance between the ohmic contact and the channel.

The processing steps of the two-mask process are as follows. Titanium (500 Å) and gold (1500 Å) are deposited in an evaporator immediately after

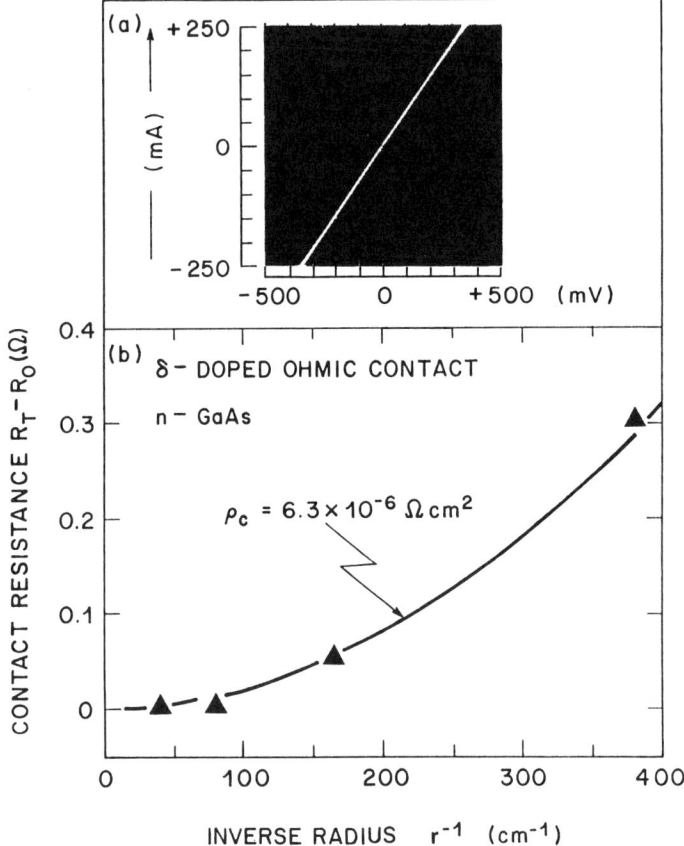

FIG. 55. (a) Current–voltage characteristic of the nonalloyed δ-doped ohmic contact to n-type GaAs. Evaluation of the specific contact resistance of a typical δ-doped contact reveals a contact resistance of $6.3 \times 10^{-6}\,\Omega\,\text{cm}^2$. The lowest contact resistances measured are in the $1-2 \times 10^{-6}\,\Omega\,\text{cm}^2$ range.

the sample is removed from the ultra-high vacuum MBE system. The first photolithographic step defines the active mesas, by etching (i) the Au, (ii) the Ti, and (iii) the GaAs. The height of the GaAs mesa is 0.1 μm. The second photolithographic step defines both the source and drain contacts, as well as the gate electrode. The Ti/Au contact is etched through the photoresist gate window to separate the mesa into source and drain contacts. Thereupon, the gate recess follows with wet chemical etching and an etching depth of 300–500 Å. Finally, the gate metallization (Ti/Au) is formed by a lift-off process. The self-alignment of Schottky gate and ohmic electrodes by one

δ-Doped GaAs field-effect transistor

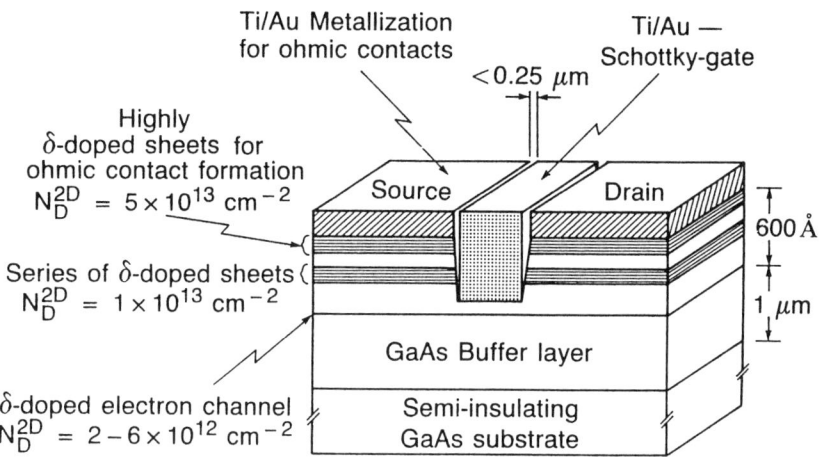

FIG. 56. Schematic illustration of the self-aligned MESFET using the δ-doping technique for the contacts as well as the electron channel.

single mask is made possible by elimination of the thermal annealing of the ohmic contact. Thus, a separation of ohmic and Schottky contacts of $<0.25\,\mu m$ can be achieved with this two-mask process. A scanning electron micrograph of the source, gate, and drain contacts is shown in Fig. 57. The length of the gate is $1.0\,\mu m$. The micrograph reveals the smooth morphology of the ohmic contacts as well as the small distance between source and gate as well as gate and drain.

The energy-band diagram below the gate of a depletion-mode δ-doped FET is shown in Fig. 58 for both zero bias (a) and negative pinch-off voltage (b). The donor plane of density N_D^{2D} is located at a distance z_D away from the Schottky barrier of height ϕ_B. One part of the electrons originating from the donors occupies surface states at the metal–semiconductor interface, and the remaining part forms the 2DEG. The threshold voltage of a δ-doped FET can be determined to be

$$V_{PO} = \phi_B - (z_D/\varepsilon)eN_D^{2D} \tag{115a}$$

where the quantum size effect is neglected and the condition of validity is limited to long gate lengths. Equation (115a) shows that for an appropriate choice of the distance z_D and doping concentration N_D^{2D}, the pinch-off

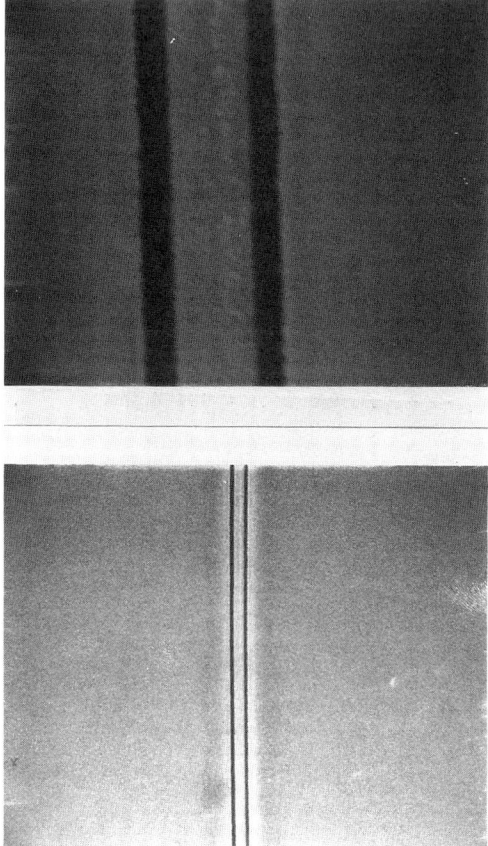

FIG. 57. Source, gate, and drain contacts of the self-aligned δ-doped MESFET. The scanning electron micrograph reveals a very smooth morphology of the nonalloyed ohmic contacts as well as a small separation between source, gate, and drain.

voltage can be < 0 or ≳ 0, corresponding to the depletion and enhancement-type FET operation. In a *conventional, homogeneously doped* metal–semiconductor FET, the pinch-off voltage depends quadratically on the channel depth z according to

$$V_{\text{PO}} = \phi_B - [z^2/(2\varepsilon)]eN_D \qquad (115b)$$

Clearly, this quadratic dependence makes the threshold voltage control more sensitive to the etching depth than the linear dependence of the

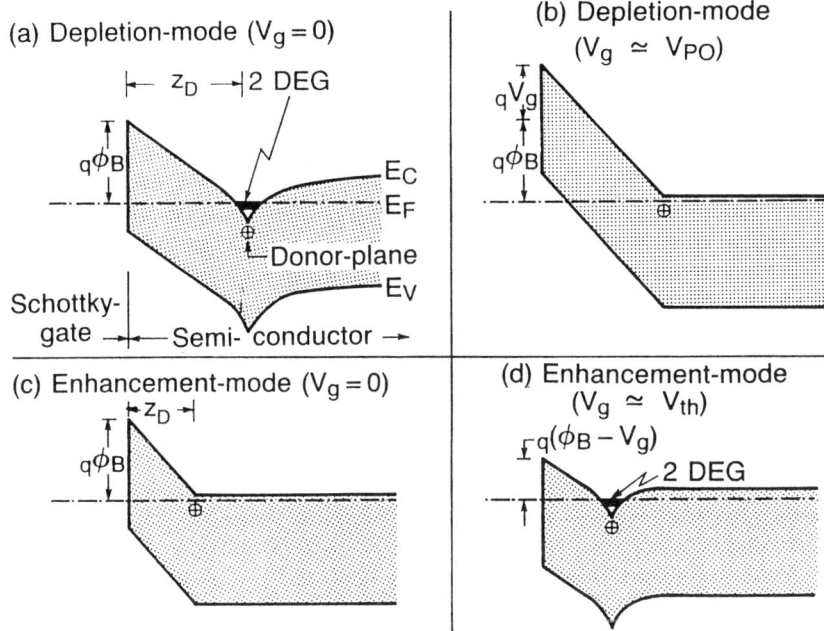

FIG. 58. Schematic band diagram of the δ-doped depletion-mode MESFET at (a) zero gate voltage and (b) at pinch-off, as well as of the enhancement-mode MESFET at (c) zero bias and (d) at positive gate voltage.

δ-doped FET inferred from Eq. (115a). Therefore, the threshold-voltage control in a δ-doped FET is superior to conventional MESFETs.

The enhancement-mode operation of a δ-doped FET is shown schematically in Figs. 58(c) and 58(d) for the off-state (zero bias) and on-state (positive gate voltage), respectively. Enhancement-type operation of the δ-doped FET can be achieved with either a lower doping concentration or a reduced distance z_D as indicated in Fig. 58(c). Practically, we determine the pinch-off voltage during the recess etch by simultaneously monitoring the source–drain current and stopping the etching at an appropriate drain current. Both types of FETs have a gate recess of *less* than 600 Å, which is relatively shallow and results in improved controllability of the gate recess.

Figure 59 shows experimental current–voltage (I–V) characteristics of an enhancement-mode FET, including the source–gate Schottky characteristic and the drain–source I–V traces ($L_g = 1.3\,\mu$m, $W_g = 150\,\mu$m, $V_{th} = -0.3$ V). The turn-on voltages of the gate diodes are typically +0.8 V. The output characteristic shows excellent pinch-off and a small output conductance in

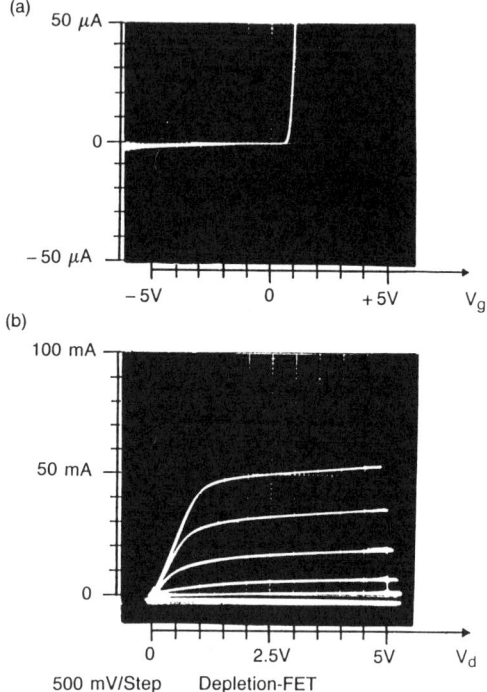

FIG. 59. (a) Gate–source and (b) drain–source current–voltage characteristics of a δ-doped depletion-mode MESFET with a gate length of 1.3 μm, a width of 150 μm, and a transconductance of 240 mS/mm. The gate voltage is +0.5 V for the top drain current trace.

the saturation regime. The transconductance of the FET shown in Fig. 59 is 153 mS/mm.

In addition to enhancement-mode FETs, depletion-mode δ-doped FETs were fabricated with gate lengths of 1.3 μm (nominal length of 1 μm). Experimental results of the source–gate I–V characteristics and of the output characteristic are shown in Figs. 60(a) and 60(b), respectively. The depletion-mode FETs have excellent pinch-off characteristics and a small output conductance in the saturated regime. The extrinsic transconductance of the δ-doped FET is 240 mS/mm. This transconductance is among the highest ever reported for *homostructure* GaAs FETs employing the same device geometry. Both depletion-mode and enhancement-mode FETs have a relatively high gate breakdown voltage of $V_b < -5$ V, typically -10 to -15 V. The origins of the high breakdown voltage are the *linear* potential drop between gate and channel and the absence of doped material between gate and channel.

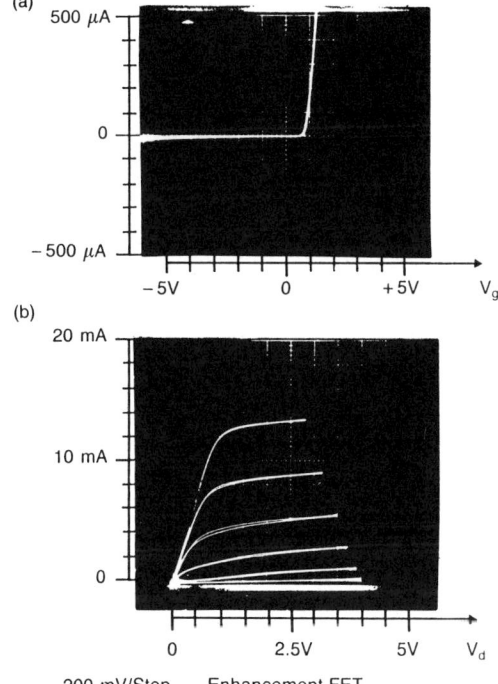

FIG. 60. (a) Gate–source and (b) drain–source current–voltage characteristics of a δ-doped enhancement-mode MESFET with gate length of 1.3 μm, and width of 150 μm. The gate voltage is +0.7 V for the top drain current trace.

13. SELECTIVITY δ-DOPED HETEROSTRUCTURE FETs

Selectivity doped heterostructure transistors (SDHTs) take advantage of the enhanced electron mobility in such a heterostructure and the spatially well-defined two-dimensional electron gas (2DEG). An extensive review on SDHTs was given by Pei and Shah (1989). Today, SDHTs, also called HEMTs, TEGFETs or HFETs, are used in high-speed integrated circuits and for low-noise microwave applications.

An important figure of merit of SDHTs is the density of the 2DEG at the heterointerface. The density is determined by the inherent properties of the heterostructure, such as the conduction band discontinuity, the spacer thickness, or the doping concentration. Densities are typically limited to $\lesssim \times 10^{12}\,\mathrm{cm}^{-2}$ in the $Al_xGa_{1-x}As/GaAs$ system. However, higher densities are desirable to improve the transconductance of the transistor.

Employment of the δ-doping technique allows one to enhance the density in selectively doped heterostructures over values achievable in conventional, homogeneously doped SDHTs.

The energy-band diagrams of the selectively δ-doped heterostructure (SΔDH) and the conventional SDH are shown in Figs. 61(a) and 61(b), respectively. In the SΔDH, all donor impurities are localized in a plane at a distance W_s from the $Al_xGa_{1-x}As$/GaAs interface. The donor localization results in a V-shaped quantum well with a lowest subband energy of E_0^δ. In the conventional SDH, the $Al_xGa_{1-x}As$ conduction band has a parabolic

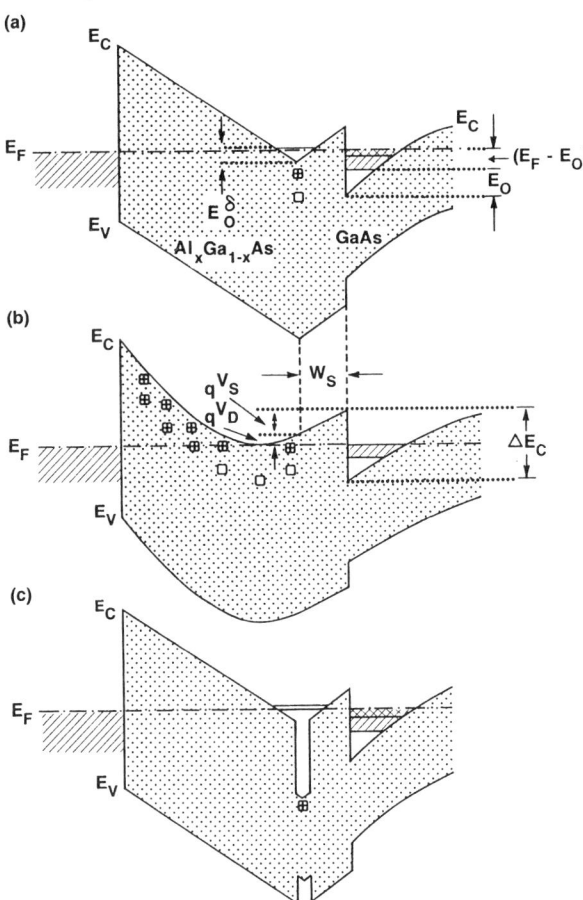

FIG. 61. Schematic energy band diagram of a (a) selectively δ-doped heterostructure and a (b) homogeneously doped heterostructure. A heterostructure (c) that is δ-doped in a GaAs quantum well avoids persistent photoconductivity effects.

shape (Fig. 61(b)) with negligible size quantization. For the selectively δ-doped heterostructures (SΔDH) we can write the energy balance:

$$\Sigma E = 0 = -E_0 - (E_F - E_0) + \Delta E_c - e\mathbf{E}W_s + E_0^\delta + (E_F - E_0^\delta) \quad (116a)$$

where E_0 is the lowest subband energy of the 2DEG, $(E_F - E_0)$ is its degeneracy, ΔE_c is the conduction band discontinuity, E_0^δ is the lowest eigenstate energy in the V-shaped quantum well of the $Al_xGa_{1-x}As$, $(E_F - E_0^\delta) = 0$ (no mobile carriers in the $Al_xGa_{1-x}As$), and $\mathbf{E}$ is the electric field within the spacer. In Eq. (116a) we assume that neither of the two quantum wells (the V-shaped quantum well in the $Al_xGa_{1-x}As$ and the triangular well at the semiconductor interface) perturbs the eigenstate energy of the other corresponding quantum well. For the homogeneously selectively doped heterostructure (SDH) we can write the following sum of energies:

$$\Sigma E = 0 = -E_0 - (E_F - E_0) + \Delta E_c - e\mathbf{E}W_s - eV_D \quad (116b)$$

where V_D is the potential drop within the depletion region, as shown in Fig. 61(b). Comparison of Eqs. (116a) and (116b) yields two differences. First, in the SΔDH, the eigenstate energy in the V-shaped quantum well of the $Al_xGa_{1-x}As$, E_0^δ, adds up to the barrier height, ΔE_c. Therefore, we can understand the sum $(E_0^\delta + \Delta E_c)$ as an "effective conduction band discontinuity" which is *enhanced* as compared to the conventional SDH. Second, the potential drop in the depletion region (Eq. (116b)), $-qV_D$, does *not* enter Eq. (116a). The depletion width approaches zero thickness due to the localization of donor impurities in the δ-doped $Al_xGa_{1-x}As$. The SΔDH consequently has two advantages: effective discontinuity enhancement due to size quantization in the $Al_xGa_{1-x}As$, and absence of depletion-region potential drop due to localization of donor impurities in the δ-doped $Al_xGa_{1-x}As$. Both characteristics will result in the desired increase of the density of the 2DEG. The increase in concentration was previously calculated in greater detail (Schubert *et al.* 1987d) and will not be repeated here.

An interesting structure is shown in Fig. 61(c), which uses a GaAs quantum well that is δ-doped. Previously, a superlattice donor layer was shown to improve the low-temperature characteristics of SDHTs (Baba *et al.* 1983, 1984, Tu *et al.* 1986). If the GaAs quantum well is thin enough (<10 Å), the eigenstate energy in the $Al_xGa_{1-x}As$ is not purturbed significantly. Such a structure would maintain the advantages of the SΔDH and, in addition, would reduce the problem of persistent photoconductivity associated with the deep donor in the $Al_xGa_{1-x}As$ by spatially separating donors from the $Al_xGa_{1-x}As$.

A high-density 2DEG is indeed observed in the selectively δ-doped heterostructures. The magnetoresistance of a SΔDH is shown in Fig. 62(a). Shubnikov–de Haas oscillations with two distinct periods are observed and indicate the first observation of two subbands being occupied in an $Al_xGa_{1-x}As/GaAs$ heterostructure. We attribute the two oscillations to the lowest and first excited subbands of the 2DEG. The density within the two subbands are evaluated by plotting the Landau quantum numbers of the

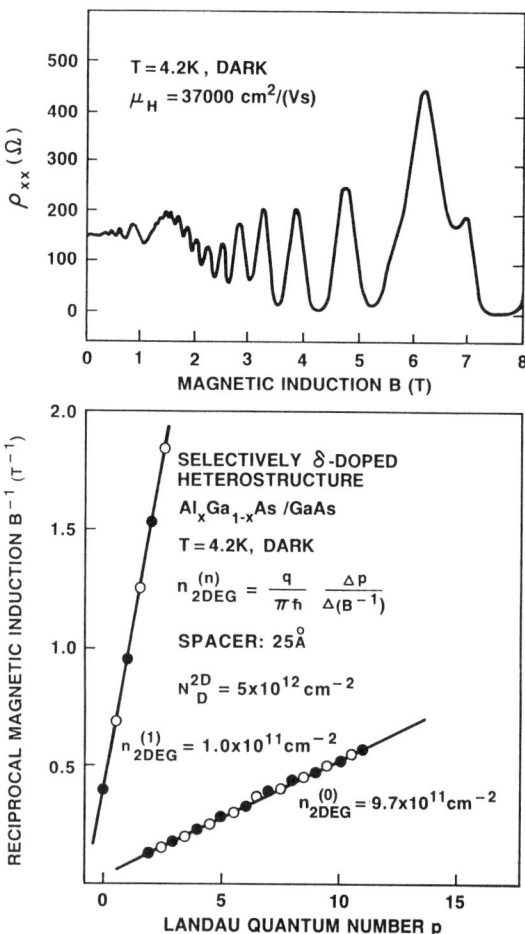

FIG. 62. (a) Low-temperature magnetoresistance of a selectively δ-doped heterostructure. (b) Evaluation of the two periods of the Shubnikov–de Haas oscillations yields a density of 9.7×10^{11} cm^{-2} and 1.0×10^{11} cm^{-2} for the lowest and the first excited subband, respectively.

minima (solid circles) and maxima (open circles) versus reciprocal magnetic induction, as shown in Fig. 62(b). The slope of this plot yields the concentrations of 9.7×10^{11} cm^{-2} and 1×10^{11} cm^{-2} for the lowest and first excited subbands, respectively. The total density is then $n_{2DEG} = 1.07 \times 10^{12}$ cm^2 at 300 mK. The corresponding mobility is $\mu = 37000$ cm^2/V sec. At room temperature a density of $n_{2DEG} = 1.7 \times 10^{12}$ cm^{-2} and a mobility of 8900 cm^2/V sec have been obtained from Hall measurements on the same sample.

Conventional selectively doped heterostructure transistors have typical concentrations of $n_{2DEG} < 1 \times 10^{12}$ cm^{-2}. The high 2DEG concentration that can be obtained in the selectively δ-doped heterostructures is favorable for field-effect transistor performance. In Fig. 63 we show the output characteristics of two depletion-mode SΔDHTs. The SΔDHTs have low *on* resistance ($R_{ON} = 1.83\,\Omega$ mm), excellent saturation characteristics, low differential output conductance in the saturation regime, and good pinch-off characteristics. A very high transconductance of up to $g_m = 360$ mS/mm is obtained from the SΔDHT. A transconductance of 320–360 mS/mm is measured in a considerable number of SΔDHTs on the same wafer. The processed wafers have good homogeneity and yield. The contact resistance is measured to be $R_{ci} = 0.07\,\Omega$ mm. At a low temperature of $T = 77$K a transconductance of $g_m = 420$ mS/mm is obtained. The lower part of Fig. 63 shows the gate–source current–voltage characteristic. A large breakdown voltage of $V = -6$ V is measured in the reverse direction.

The density of the 2DEG is increased further by using a quantum-well structure, instead of the single-interface structure (Chen et al. 1987, Kuo et al. 1988). The quantum well structure can be doped at both interfaces as shown in Fig. 64. Since carriers transfer from the doped region underneath and on top the quantum well, the density of the 2DEG is approximately twice as large as compared to the single-interface structure.

The experimental output current–voltage characteristic of a selectively δ-doped quantum well transistor is shown in Fig. 65. The transistor has a gate-length of 1.2 μm and a gate width of 150 μm. The maximum experimental transconductance of the device is 267 mS/mm. The device operates in the depletion mode and the top current trace is for zero gate bias. The gate voltage steps are 200 mV. A slightly higher transconductance (285 mS/mm) is measured at positive gate voltage of $+0.4$ V. The transconductance of the quantum well transistor has not yet achieved comparable values of the single-interface transistor, which may be due to well-known difficulties of the first interface of the quantum well. Charge carriers at the first interface are usually of lower mobility as compared to carriers at the second (top) interface, even though both interfaces of the quantum well structure are nominally identical.

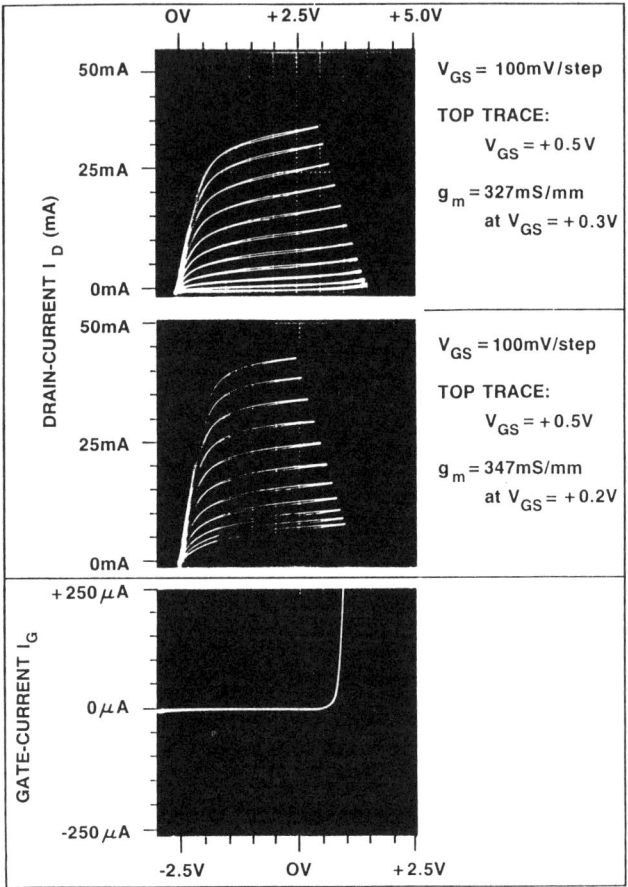

FIG. 63. Drain–current versus drain–source voltage of a selectively δ-doped heterostructure transistor with a transconductance of up to 347 mS/mm. A gate–source current–voltage characteristic is shown in the lower part.

VII. Properties of δ-Doped Superlattices

The initial proposal for superlattices did not only comprise compositional superlattices, e.g., of the type $Al_xGa_{1-x}As/GaAs$, but also doping superlattices (Esaki and Tsu 1970). Such doping superlattices are a periodic sequence of alternating n-type and p-type doped regions in a semiconductor. A periodic potential results in a growth direction due to positive and negative donor and acceptor charges, respectively. The periodic potential or

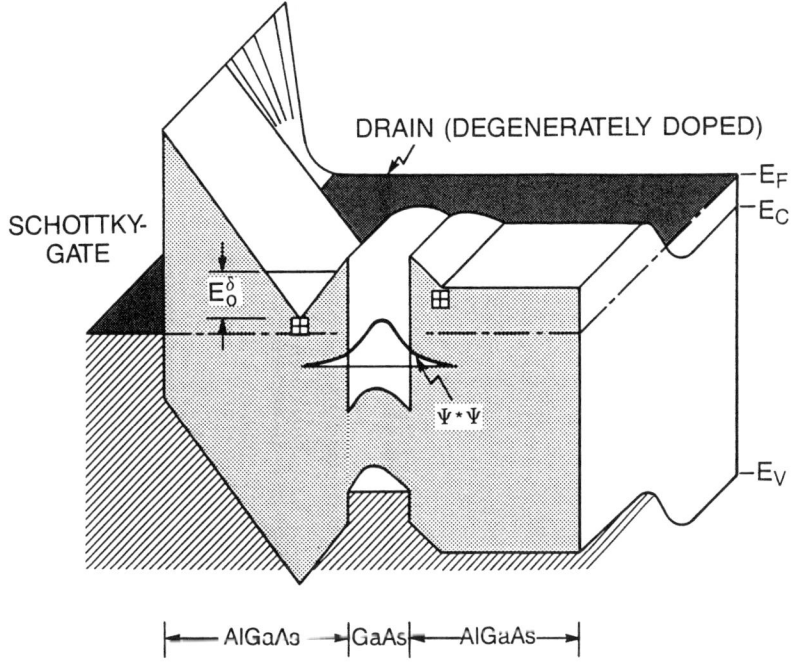

FIG. 64. Band diagram of a selectively δ-doped quantum well transistor below the Schottky gate and the degenerately doped drain region. Size quantization occurs in the $Al_xGa_{1-x}As$ with a quantization energy of E_0^δ.

superlattice potential can result in novel electronic properties not observed in conventional semiconductors.

Quantum-confined interband transitions were readily observed in the compositional $Al_xGa_{1-x}As/GaAs$ superlattices by Dingle *et al.* (1975). However, in doping superlattices, quantum-confined interband transitions were not observed. Doping superlattices grown using the δ-doping technique resulted in a significant improvement of their optical properties. Quantum-confined optical interband transitions were observed for the first time in absorption spectroscopy (Schubert *et al.* 1988e) as well as in photoluminescence spectroscopy (Schubert *et al.* 1989b).

Shortly after the initial proposal of doping superlattices published work was exclusively of theoretical nature (Ovsyannikov *et al.* 1971, Romanov 1972, Döhler 1972a, 1972b, Romanov and Orlov 1973). First experimental work started in the 1980s and included the observation of the tunability of the energy gap in doping superlattices (Döhler *et al.* 1981). Extended reviews

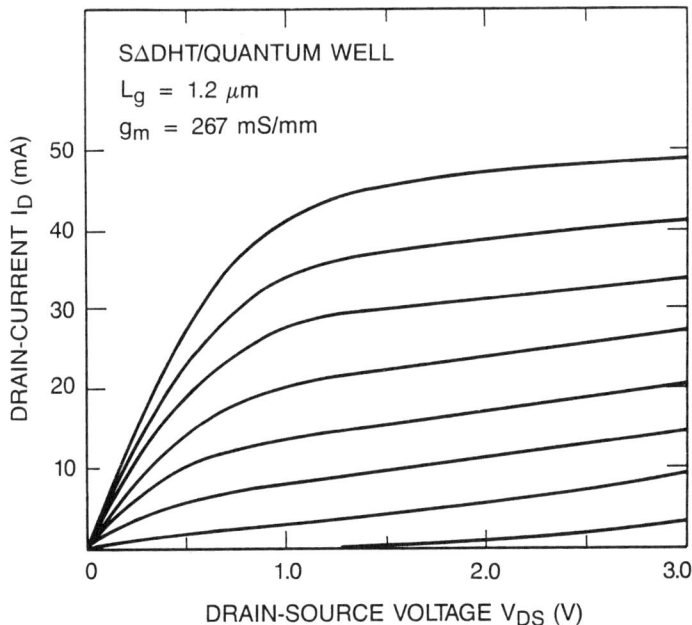

FIG. 65. Drain–current versus drain–source voltage of a selectively δ-doped quantum well transistor.

on homogeneously doped doping superlattices have become available (Döhler 1986, Ploog and Döhler 1983).

The improved δ-doped doping superlattice structure consists of a train of alternating n-type and p-type δ-doping sheets separated by intrinsic (undoped) layers. A periodic n-i-p-i sequence results, whose band diagram consists of linear sections. The δ-doped superlattice has several important advantages over the homogeneously doped structure including larger superlattice modulation, the feasibility of shorter periods, and the minimization of potential fluctuations (Schubert *et. al.* 1988d).

A comparison between the structure of the originally proposed doping superlattice and the δ-doped doping superlattice is shown in Fig. 66. The original proposal (Esaki and Tsu 1970) consists of alternating n-type and p-type regions. The band diagram is then a periodic series of parabolic sections. The δ-doped doping superlattice (Schubert *et al.* 1985d) confines the dopants in thin, highly doped sheets separated by undoped (i = intrinsic) material.

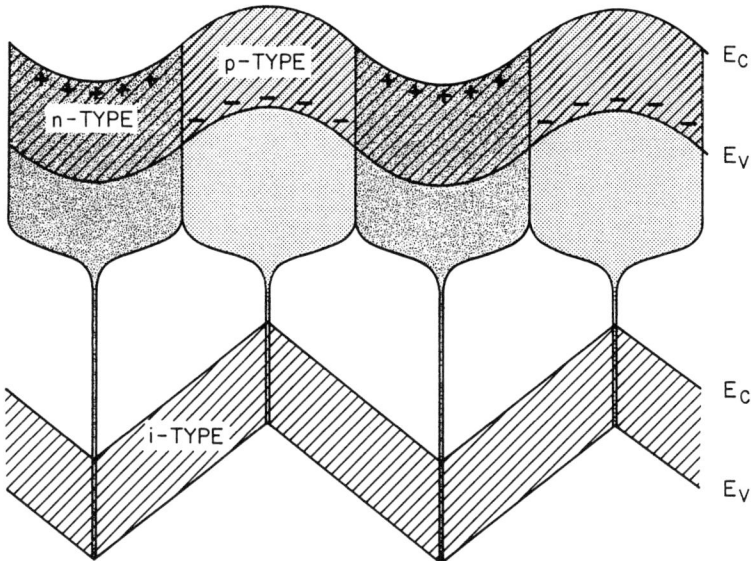

FIG. 66. Comparison of the (top) original proposal for doping superlattices and of the δ-doped doping superlattice. The δ-doped structure is advantageous due to a larger potential modulation, reduced potential fluctuations, and the feasibility of smaller superlattice periods.

14. QUANTUM-CONFINED ABSORPTION

The doping profile of a δ-doped doping superlattice is shown in Fig. 67. The superlattice has a period of z_p and the density of dopants per sheet is N_D^{2D} and N_A^{2D} for donors and acceptors, respectively. Under balanced conditions (i.e., $N_D^{2D} = N_A^{2D}$) electrons recombine with holes, and the superlattice is depleted from free carriers. The corresponding band diagram is shown in the lower part of Figs. 67, obtained by twofold integration of Poisson's equation. Since the band diagram consists of linear sections, it is also referred to as a *sawtooth superlattice*.

Sawtooth superlattices, and in general doping superlattices, have a number of unique properties. Three of the most prominent characteristics of doping superlattices are as follows. First, the bandgap of the superlattice is smaller than the energy gap of the host material (e.g., GaAs). The superlattice band-gap energy is defined as

$$E_g^{SL} = E_g - eV_z + E_0^e + E_0^{hh} \qquad (117)$$

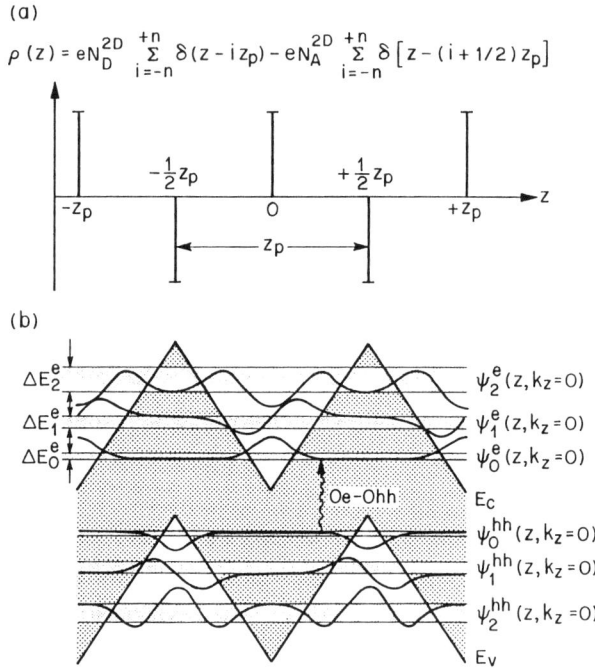

FIG. 67. (a) Doping profile of a δ-doped doping superlattice consisting of a train of alternating n-type and p-type δ-functions. (b) Schematic band diagram and wave functions of the δ-doped doping superlattice. The wave functions $\psi_n(z)$ are shown for $k_z = 0$. Minibands have a width of ΔE_n.

where E_g is the gap energy of the host lattice, V_z is the superlattice potential modulation given by $V_z = (1/4)eN^{2D}z_p/\varepsilon$ and E_0^e and E_0^{hh} are the lowest subband energy of electrons and heavy holes, respectively. It is thus possible to extend the energy gap of any semiconductor to lower energies by using the doping superlattice concept. Second, the oscillator strength of the superlattice transition is *weaker* than that of the host lattice. The maximum electron concentration is shifted with respect to the maximum hole concentration by half a superlattice period, as illustrated in Fig. 67. The transition matrix element which is proportional to the overlap of electron and hole wave functions is thus reduced in magnitude. Third, the doping superlattices have a *tunable* energy gap, i.e., their gap energy depends on excitation intensity. The tunability of doping superlattices originates in an accumulation of free carriers in the dopants planes, where they screen the ionized dopant charge. A reduced potential modulation thus results upon photo-excitation.

Optical transitions in sawtooth superlattices are not governed by conventional selection rules; i.e., optical dipole matrix elements are finite and nonzero for all transitions. The matrix element involves initial and final states that have an exponentially decaying part and a spatially oscillating part, as shown in Fig. 67, yielding a finite, nonzero transition probability for all transitions. This property of the sawtooth superlattice is in contrast to compositional superlattices where conventional selection rules do apply (e.g., the selection rule $\Delta n = 0$ for (AlGa)As/GaAs superlattices with no electric field present).

The GaAs epitaxial layers used for this study were grown by gas-source molecular-beam epitaxy on undoped semi-insulating GaAs substrates. The growth temperature was kept below $T = 550°C$ to avoid diffusion of n-type (Si) and p-type (Be) impurities. The design parameters of the superlattice include a period of $z_p = 150$ Å and a two-dimensional doping density of $N_D^{2D} = N_A^{2D} = 1.25 \times 10^{13}$ cm^{-2}. The samples have 10 periods of 20 dopant sheets separated by $\frac{1}{2}z_p = 75$ Å. The samples have a closely balanced impurity concentration, i.e., $N_D^{2D} \cong N_A^{2D}$. Such a balance is essential, because its absence would blue shift the absorption edge according to the Burstein–Moss shift. Absorption measurements were performed on polished, 0.25 cm^2 samples. A dual-beam Perkin–Elmer Model 330 spectrophotometer and a variable-temperature cold-finger cryostat were used.

Results of absorption measurements on GaAs sawtooth superlattices measured at $T = 6$ K are shown in Fig. 68. The gap energy of the undoped GaAs substrate corresponds to a wavelength of $\lambda = 820$ nm and is shown by a double arrow. The substrate absorbs light at energies slightly below the fundamental gap; this absorption of bulk material is known as the Urbach tail (Urbach 1953). We determined the corresponding Urbach-tail energy to be $E_U = 6$ meV for our undoped GaAs samples. A typical absorption spectrum of an undoped GaAs sample is shown as a dashed curve in Fig. 68.

The absorption spectrum shown in Fig. 68 shows strong absorption below the fundamenal gap of GaAs in a range of 400 meV below the band gap of the GaAs host lattice. The most striking aspect of the absorption spectrum are four distinct features: an absorption maximum (peak) at $\lambda = 1090$ nm and three shoulders at wavelengths of $\lambda = 1000$, 920, and 865 nm. The structure is attributed to transitions between quantum-confined states in the valence and conduction bands. Such quantum-confined interband transitions have not been observed since the invention of doping superlattices. Furthermore, the absorption is not monotonically increasing with energy, but has a clear peak at $\lambda = 1090$ nm. Unlike the absorption spectrum shown in Fig. 68, the joint density of states is increasing monotonically with energy. The occurrence of such an absorption peak therefore shows the presence of excitonic or electron–hole correlation effects.

The formation of excitons necessitates a novel, extended understanding of

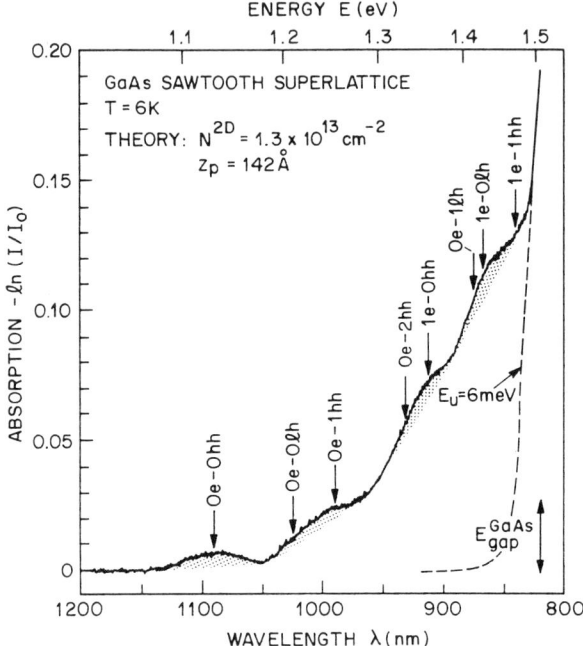

FIG. 68. Optical interband absorption spectrum of a GaAs sawtooth superlattice at $T = 6K$. Theoretical transition energies are indicated by arrows. The lowest electron to lowest heavy-hole transition is referred to as 0e–0hh. The parameters used for the calculation are a period of 142 Å and a dopant concentration of $1.3 \times 10^{13}\,\text{cm}^{-2}$. The energy gap of the substrate is marked by a double arrow. The absorption tail of the substrate is characterized by an Urbach-tail energy of $E_U = 6\,\text{meV}$.

physical properties of doping superlattices. According to previous beliefs, electron–hole separation naturally objects to the formation of excitons. However, the absorption peak shown in Fig. 68 elucidates exciton formation in sawtooth superlattices with appropriate design parameters. Furthermore, excitonic absorption increases the absorption coefficient by several orders of magnitude over nonexcitonic absorption, which in turn has been observed in conventional doping superlattices.

Proetto (1989) calculated properties of the ground-state exciton in a sawtooth superlattice structure using a variational wave function. He calculated the binding energy, lateral, and vertical extension as a function of the superlattice period and the dopant density. This calculation revealed that the binding energy is of a magnitude (e.g., $E_b = 6\,\text{meV}$ for a period of 100 Å) similar to compositional GaAs quantum wells, despite the spatial separation of electrons and holes.

The built-in electric field in the sawtooth superlattice is given by $E = eN^{2D}/2\varepsilon$, which equals $E \gtrsim 5 \times 10^5$ V/cm. At such high fields excitonic absorption is not observable in homogeneous semiconductors or square-shaped quantum wells due to a field-induced ionization of excitons. In contrast to homogeneous semiconductors or square-shaped quantum wells, the sawtooth structure, even though an extremely high field is present, opposes a field-induced separation of carriers over more than half a superlattice period. Thus, electron-hole correlation effects are observed even at field strengths exceeding 10^5 V/cm. Excitonic enhancement of the absorption can, however, occur in doping superlattices, only if the spatial electron–hole separation (which is approximately $z_p/2$ for the $n = 0$ states, see Fig. 67) is smaller than the electron–hole interaction length (excitonic diameter). This condition is indeed satisfied for our samples with $z_p/2 = 75$ Å.

We now compare the experimental absorption data to theoretical transition energies inferred from Eq. (117). The arrows shown in Fig. 68 are calculated energies of quantum-confined transitions. The lowest electron ($n = 0$) to lowest heavy hole ($n = 0$) transition is referred to as the 0e–0hh transition. Very good agreement between calculated quantum-confined transition energies and experimental ones is observed over a wide range of energies. For the calculation a period of $z_p = 142$ Å and a doping concentration of $N_D^{2D} = N_A^{2D} = 1.3 \times 10^{13}$ cm^{-2} has been used.

In Fig. 69 the absorption is plotted on a logarithmic scale versus energy. A linear relationship (straight line) between the envelope of $\ln(\alpha l)$ and E is obtained. Thus, an exponentially increasing transition probability with energy is inferred from Fig. 69. The envelope of the absorption is exponentially increasing due to an increasing overlap of wave functions and, accordingly, an exponentially increasing matrix element.

Additional absorption spectra for temperatures $6 \leqslant T \leqslant 300$K are shown in Fig. 70. The absorption edge of the substrate shifts from 820 nm at $T = 6$K to 875 nm at $T = 300$K. The quantum-confined interband transitions of the sawtooth superlattice exhibit the identical qualitative shifts to longer wavelength as shown in Fig. 70. Such a shift of peak energies of the superlattice absorption is expected to track the shift of the band edge of the host material according to Eq. (117).

15. PHOTOLUMINESCENCE SPECTROSCOPY

The optical properties of doping superlattices strongly depend on the period of the superlattice. Doping superlattices with short periods and long periods are displayed schematically in the inset of Figs. 71(a) and (b), respectively. In short-period (or Type A) sawtooth doping superlattices, the

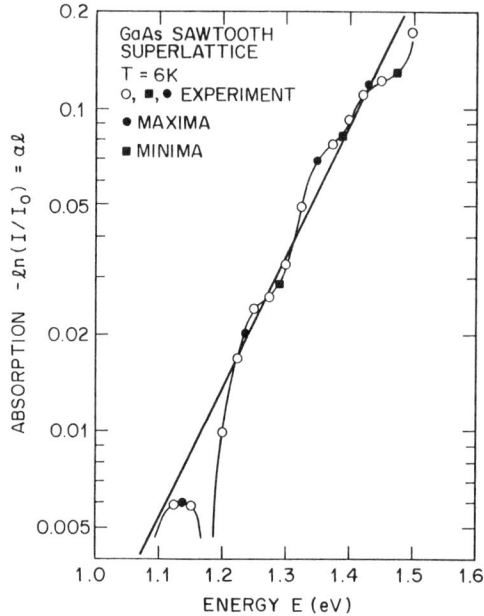

FIG. 69. Logarithmic absorption of a GaAs sawtooth superlattice at $T = 6K$ versus photon energy. The quasilinear functional dependence (solid line) indicates a transition probability that increases exponentially with energy. Solid circles and solid squares represent maxima and minima of the absorption spectrum.

barriers between adjacent quantum wells are thin (e.g., 100 Å), and coupling between adjacent wells becomes significant. Furthermore, the overlap between electron and hole wave functions is largest for short-period doping superlattices. Long-period (or Type B) doping superlattices with thick barriers between adjacent quantum wells (e.g., 600 Å) result in negligible tunneling between adjacent quantum wells. They also result in negligible overlap between electron- and hole-wave functions; i.e., long recombination lifetimes of carriers.

Drastic differences in the optical properties of long-period and short-period doping superlattices were indeed found in low-temperature photoluminescence experiments (Schubert et al. 1987d). The photoluminescence spectra of a short-period (Type A) and a long-period doping superlattice are shown in Figs. 71(a) and (b), respectively. The periods of the superlattices are 150 Å and 600 Å. The doping density is $1 \times 10^{13} \text{cm}^{-2}$. The photoluminescence spectrum shown in Fig. 71(a) reveals an emission peak energy at 1.37 eV. Note that the emission energy is smaller than the energy gap of GaAs, which is $E_g = 1.512 \text{eV}$ at low temperatures. Three different excitation

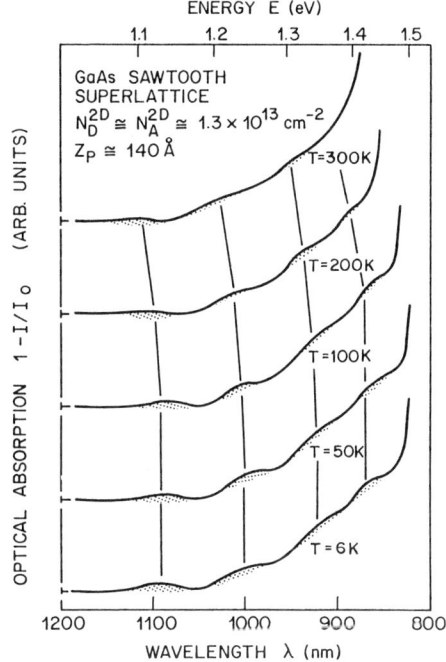

FIG. 70. Optical interband absorption of a GaAs sawtooth superlattice for sample temperatures $6 \leq T \leq 300$K. A shift of the fundamental gap as well as the quantum-confined transitions is found with increasing temperature.

intensities $I_1 < I_2 < I_3$ are used for excitation. The peak energy of the spectra slightly shifts to higher energies with increasing excitation intensity. However, the shift is small, if compared to the width of the spectra. Thus, negligible tunability of the peak energy as a function of excitation intensity is found in short-period doping superlattices.

Strong tunability is found in long-period doping superlattices as shown in Fig. 71(b). The wavelength is continuously tunable from 1.1 eV to 1.4 eV. The tuning range is larger than the full-width half-maximum of the luminescence line and confirms earlier observations of tunability in doping superlattices (for a review, see Döhler 1986).

Quantum-confined interband transitions in photoluminescence spectra are shown in Fig. 72 (Schubert et al. 1989c). The samples have a carefully balanced donor and acceptor density of 1.25×10^{13} cm^{-2} and a period of 150 Å. Three clearly resolved photoluminescence peaks are observed at $\lambda \cong 0.98$, 1.02, and 1.09 μm. Furthermore, a shoulder is observed at the high-energy side of the spectrum at $\lambda \cong 0.95$ μm. We attribute the clearly

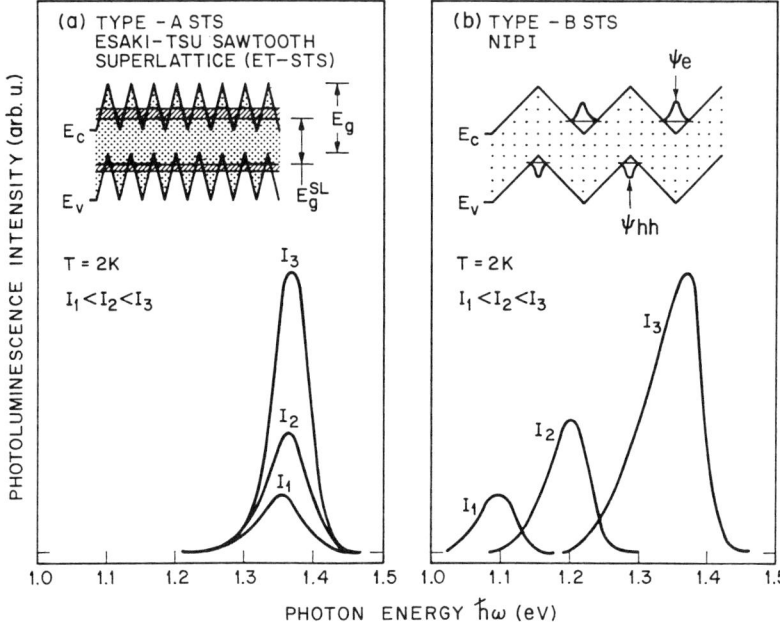

FIG. 71. Low-temperature photoluminescence spectra on (a) short-period (Type A) sawtooth superlattices and (b) long-period (Type B) sawtooth superlattices at three different excitation intensities. The emission wavelength is approximately constant for the Type A superlattice, although it is clearly tunable for the Type B superlattice.

resolved luminescence peaks to transitions between quantum-confined conduction- and valence-band states. The assignment of luminescence peaks is confirmed by calculation of transition energies and comparison with experimentally observed peak energies. A very good fit between experimental and calculated peak energies is obtained by using $N^{2D} = 1.3 \times 10^{13} \text{cm}^{-2}$ and $z_p = 142 \text{Å}$. Five transitions can be identified; namely the 0e → 0hh, 0e → 01h, 0e → 1hh, 0e → 2hh, and 1e → 0hh transitions.

Most striking, however, the photoluminescence spectrum of Fig. 71 displays not only the *lowest* transition but also transitions via excited states, e.g., the 0e → 1hh transition. Furthermore, excited-state transitions (e.g., 0e → 01h or 0e → 1hh are more intense than the ground-state (0e → 0hh) transition. The light-hole transition is stronger than the heavy-hole transition, even though the density of states of the light-hole subband is much smaller (approximately a factor of $m_{hh}^*/m_{lh}^* \cong 7$). Such photoluminescence spectra were hitherto not observed in any quantum well structure. We will now show that the specific characteristics of the photoluminescence spec-

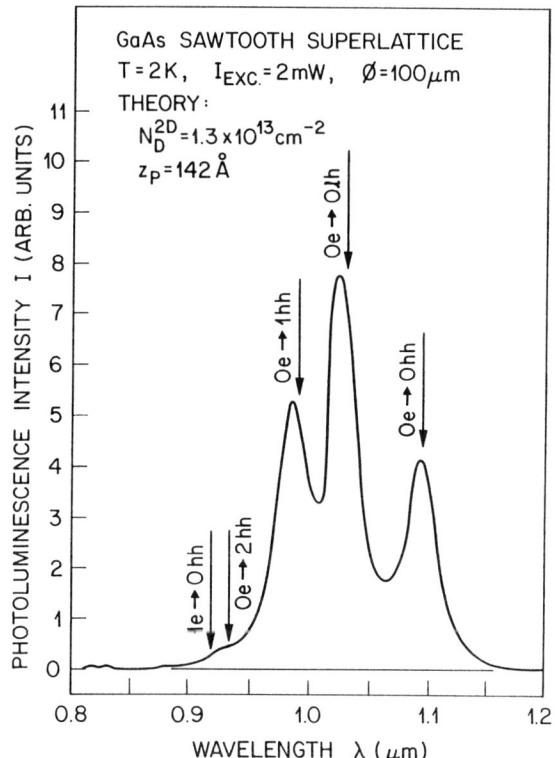

FIG. 72. Low-temperature photoluminescence spectrum on a sawtooth doping superlattice. The arrows indicate theoretical transition energies calculated for a superlattice with $z_p = 142$ Å and $N_D^{2D} = 1.3 \times 10^{13}$ cm^{-2}.

trum can be consistently explained in terms of the unique energy dependence of the oscillator strength of the sawtooth structure.

For completeness, we note that multisubband transitions were observed in compositional quantum well structures at very high excitation intensities (Miller *et al.* 1981). However, the intensity of those transitions was found to decrease exponentially with energy. Multisubband photoluminescences were not observed in conventional homogeneously doped *n-i-p-i* structures.

The photoluminescence line shape in semiconductors and semiconductor quantum well structures can be determined by the product of the joined density of states and the thermal distribution of carriers. The latter is usually modeled in terms of a Boltzmann distribution and a carrier temperature. The carrier temperature depends on the photoluminescence excitation intensity and is typically in the range $10 \leq T_c \leq 50$K at a lattice temperature of $T_1 = 2K$. The transition-matrix element depends weakly on energy in homogeneous semiconductors or compositional semiconductor quantum wells. In contrast, the oscillator strength increases exponentially with energy in the sawtooth superlattice. Thus, the oscillator strength has an opposite dependence on energy, as compared to the thermal distribution. Consequently, transitions via excited states can be observed in the sawtooth structure, even though they may be sparsely populated.

The energy dependence of the oscillator strength can be visualized with Fig. 73, which shows an overlap of wave functions which increases with energy. The matrix element of a quantum-confined transition in a sawtooth superlattice involves mostly the exponentially decaying part of the wave functions. Therefore, the limits of the integration are chosen in such a way that integration is limited to the region beyond the classical turning points z_{it} and z_{ft} of the initial (electron) and final (hole) state, as shown in Fig. 73. Evaluation of the overlap integral yields (Schubert et al. 1989c)

$$\int_{z_{ft}}^{z_p/2 - z_{it}} \psi_i(z)\psi_f(z)dz \cong \frac{\psi_i(z_{it})\psi_f(z_{ft})}{\alpha_f - \alpha_i} e^{-\alpha_i(z_p/2 - z_{it} - z_{ft})} \qquad (118)$$

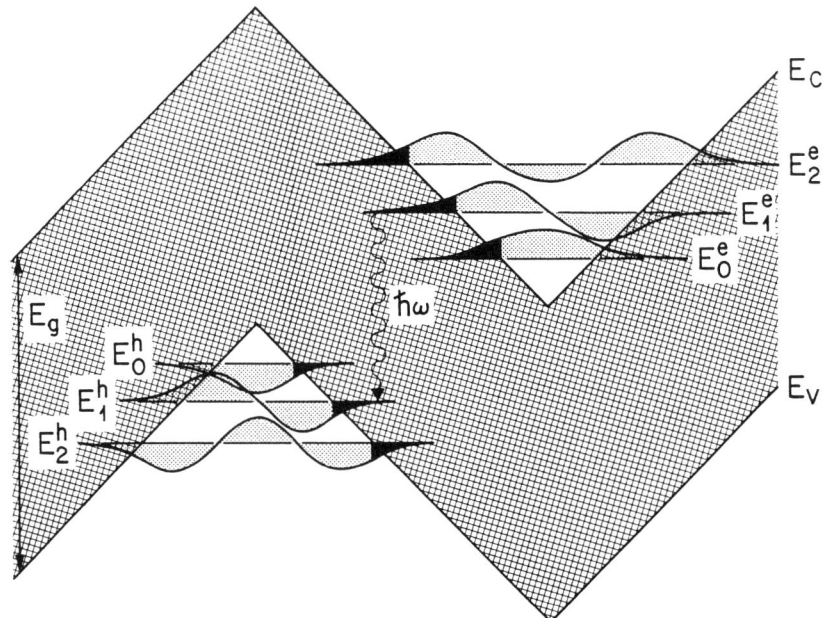

FIG. 73. Schematic band diagram of a sawtooth superlattice. The overlap of electron and hole wave functions increases exponentially with subband index.

where α_i and α_f are the decay constants of the initial and final wave functions, respectively. For the calculation of this result we use the fact that the electron mass is lighter than either heavy- or light-hole mass, i.e., $\alpha_i < \alpha_f$. Since the turning points depend on the eigenstate energies according to $z_{it} = E_i/e\mathbf{E}$ and $z_{ft} = E_f/e\mathbf{E}$, it is evident that the oscillator strength depends exponentially on the eigenstate energy due to an increasing overlap of wave functions.

The interplay between thermal distribution of carriers, density of states, and oscillator strength of the transitions is illustrated in Fig. 74. Thermal distribution of carriers and the oscillator strength have an opposite (exponential) dependence on energy. The transition probability is then a function with multiple peaks.

Next, a quantitative analysis is provided to understand the shift of the quantum-confined photoluminescence transitions as a function of excitation intensity. In this analysis we assume an exponential decay of the radiative recombination. In continuous-wave photoluminescence experiments, the generation rate and recombination rate coincide, i.e.,

$$\frac{dn}{dt} = \frac{n}{\tau} \qquad (119)$$

We will now show that the radiative lifetime τ can be determined from our experiments, since both the generation rate dn/dt is known at a given excitation intensity and the free-carrier concentration n can be determined from screening caused by the free carrier. To determine n we consider the shift of quantum-confined transitions in Fig. 75, which is further illustrated by the dashed line. For an increase of the excitation intensity from 2 mW (25 W/cm^2) by a factor of 10 to 20 mW (250 W/cm^2) the 0e → 01h transition, which is the strongest transition, increases in peak energy from 1.208 to 1.240 eV. We attribute this change of the peak energy to screening of dopant charge by the photogenerated carrier. The result of screening is easily illustrated by considering the photoluminescence energy given by

$$E = E_g^{GaAs} - \frac{1}{2}e\mathbf{E}z_p + E_n^e + E_n^h \qquad (120)$$

where E_g^{GaAs} is the gap energy of GaAs, $\frac{1}{2}e\mathbf{E}z_p$ is the modulation of the superlattice potential, and E_n^e and E_n^h are the electron and hole eigenstate energies. The electric field $\mathbf{E}$ is screened under photoexcitation and changes by an amount of

$$\Delta\mathbf{E} = \frac{e}{2\varepsilon}\Delta n_{2D} \qquad (121)$$

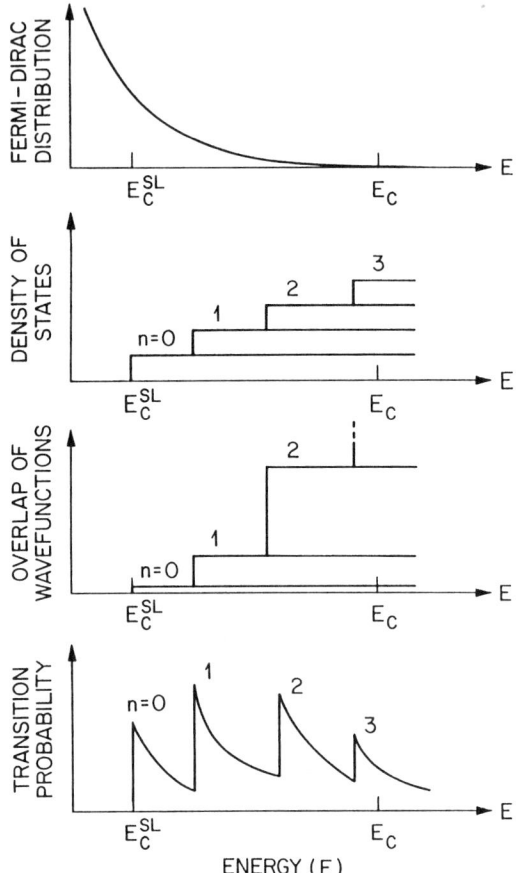

FIG. 74. Schematic illustration of thermal carrier distribution, density of states, and the overlap of wave functions as a function of energy. The lowest diagram shows the transition probability for optical transitions, which is the product of the top three diagrams. Note that the thermal carrier distribution and the overlap of wave functions have an *opposite* exponential dependence on energy.

where n_{2D} is the excitation-induced density of carriers per quantum well. Thus, the recombination energy increases due to the reduction of potential modulation, as inferred from Eq. (120). Upon screening, the potential modulation is reduced, and the individual V-shaped quantum wells become shallower. The eigenstate energies of the confined states reduce accordingly.

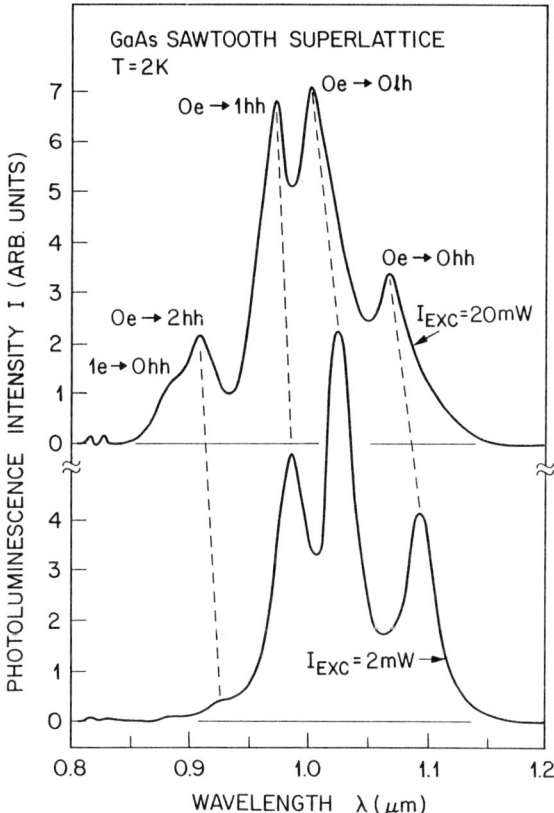

FIG. 75. Low-temperature photoluminescence spectrum of a sawtooth superlattice at two different excitation intensities of 2 mW (25 W/cm²) and 20 mW (250 W/cm²). The dashed lines indicate the shift of quantum-confined transitions with excitation intensity. At high excitation intensity a new transition arises, namely the 1e → 0hh and 0e → 2hh transitions.

The change in eigenstate energies is smaller than the change in superlattice potential modulation given by

$$\Delta E/e = \Delta \mathbf{E} \frac{1}{2} z_p \qquad (122)$$

and we therefore neglect the changes in eigenstate energies. Insertion of Eq. (122) into Eq. (121) allows us to determine Δn_{2D} as a function of the shift of the luminescence energy ΔE.

Finally, the generation rate can be determined from the exciting laser power P according to

$$\frac{dn}{dt} = \frac{P}{\hbar\omega} \frac{\alpha}{r^2\pi} \qquad (123)$$

where $\hbar\omega$ is the energy of exciting photons, $\alpha (\cong 4 \times 10^4 \text{cm}^{-1})$ is the absorption coefficient, and $r (= 50 \mu\text{m})$ is the radius of the laser beam on the sample surface. The lifetime of carriers obtained from Eq. (119) using the data provided earlier is $\tau \cong 10$ ns at $P_L = 20$ mW. Note that the lifetime of $\tau \cong 10$ ns is an upper limit, because changes in the eigenstate energies are not taken into account. This lifetime is slightly longer than lifetimes in homogeneous GaAs. We attribute the increase in lifetime due to a smaller overlap of wave functions, as illustrated in Fig. 73. The photoinduced carrier density inferred from this lifetime is $4 \times 10^{17} \text{cm}^{-3}$ at $P_L = 20$ mW.

Figure 75 further illustrates that high-energy transitions (e.g., 0e → 1hh and 0e → 01h) gain intensity relative to the lowest transition. In addition, a new peak and shoulder arise at the high-energy side of the spectrum. We attribute the peak and shoulder to the 0e → 2hh and 1e → 0hh transition in agreement with absorption measurements on the same sample. The occurrence of the new transitions cannot be understood solely on the basis of screening. The new, high-energy transitions can be explained by band filling or a higher effective carrier temperature at increased excitation intensity.

16. Transmission Spectroscopy

Transmission spectroscopy on δ-doped doping superlattices, using semiconductor diodes to detect the transmitted light, are advantageous due to much reduced noise of the signal as compared to (spectrophotometer) absorption measurements. Transmission experiments were carried out with a cooled Ge-photodiode, which detects the light transmitted through the doping superlattice. Using the "lock-in" technique for the detection, the transmission signal carries very little noise.

For the transmission measurements a sample holder with a 1.0 mm diameter hole was used, as illustrated in Fig. 76. The Ge photodiode detector was thermally decoupled from the sample holder. The sample transmission was measured with the phase-sensitive "lock-in" technique. The chopping frequency was 313 Hz. For monochromatic illumination of the sample we used a 250-W halogen lamp with a double monochromator

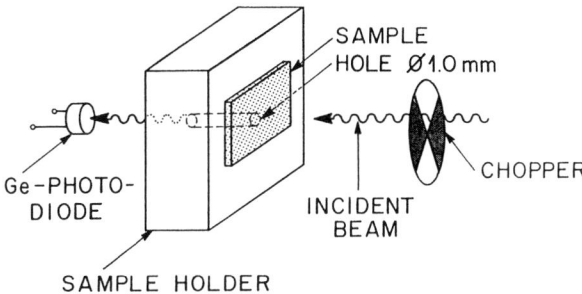

FIG. 76. Schematic illustration of the transmission experiment made from a sample illuminated by chopped monochromatic light. The transmitted light is detected with a Ge-pin detector.

(HRD 600 Jobin Yvon) incorporating 1200-lines/mm and 600-lines/mm gratings. The holographic 1200-lines/mm grating was used for the measurements due to the smooth optical response. In addition optical filters with cutoff wavelengths of 665 and 780 nm were used.

The observed spectra are normalized to a reference measurement performed on a semi-insulating substrate with a thickness of 300 μm. In this way one can obtain transmission spectra without detailed knowledge of the response of the measuring setup. From these corrected curves we obtain the first derivative numerically. The transmission spectra and its derivative of a δ-doped doping superlattice are shown in Fig. 77. The transmission spectrum shows four shoulders, which are identified as quantum-confined interband transitions. The structure is stronger in the derivative of the transmission spectrum. The arrows indicate calculated transition energies using the exact Airy-function solutions of the V-shaped potential well. The agreement between calculated and measured data is best if a doping density of 1.3×10^{13} cm^{-2} and a period of 142 Å is used.

The derivative of the transmission spectrum of a superlattice with longer period and smaller doping density is shown in Fig. 78. The number of peaks has significantly increased in the same wavelength interval as compared to the sample shown in Fig. 77. A decrease in doping density leads to a decrease in the subband spacing. More peaks with narrower spacing are therefore plausible for superlattices with smaller doping density.

The energies of quantum-confined interband transitions are calculated using the exact Airy-function solution. For a period of $z_p = 178$ Å and a doping density of 9×10^{12} cm^{-2}, one obtains the theoretical transition energies shown by the arrows in Fig. 78. Good agreement is found between measured and calculated transition energies.

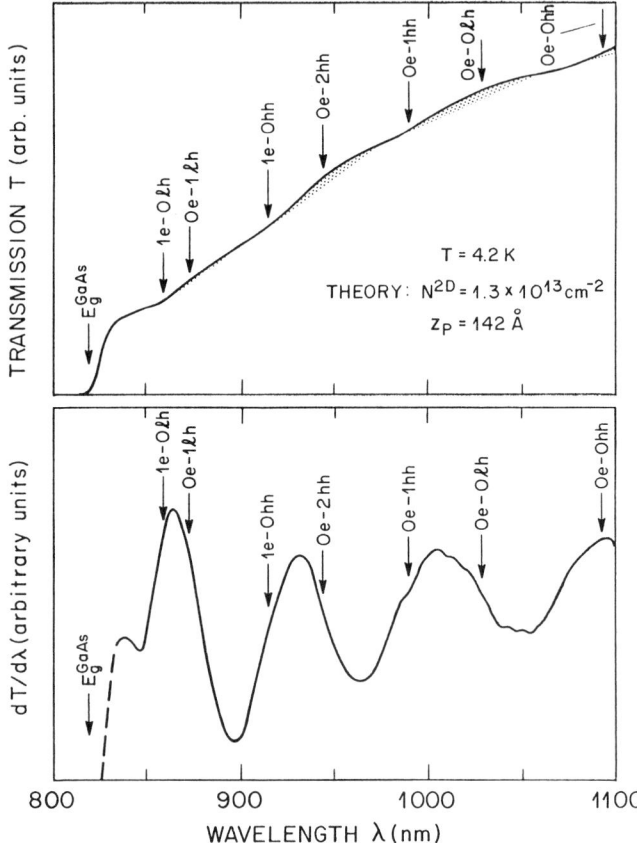

FIG. 77. Transmission and derivative of transmission versus wavelength of a sawtooth doping superlattice at low temperatures. The arrows represent calculated transition wavelengths using the superlattice parameters $N^{2D} = 1.3 \times 10^{13}\,\text{cm}^{-2}$ and $z_p = 142\,\text{Å}$.

The absorption coefficient for one-photon absorption can be obtained in terms of Fermi's golden rule (Elliot 1957)

$$\alpha(\omega) = \frac{4\pi e^2}{\omega c n_r m^2} \sum_{n \neq n'} |\langle F | p\varepsilon | I \rangle|^2 \delta(\omega - \omega_{nm'}) \quad (124)$$

where $\langle F |$ and $| I \rangle$ are the final and initial states, p is the dipole operator, c the speed of light, ε the polarization of the photon, and n_r the index of

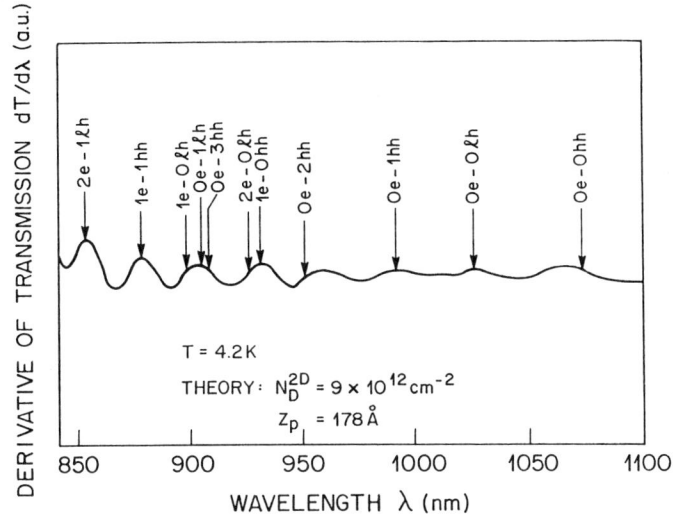

FIG. 78. Low-temperature derivative of transmission versus wavelength of a sawtooth doping superlattice. The superlattice has a smaller doping density and a longer period than the spectrum shown in Fig. 77. The arrows represent calculated transition wavelengths using the superlattice parameters $N^{2D} = 9 \times 10^{12}\,\text{cm}^{-2}$ and $z_p = 178$ Å.

refraction. Taking into account the multiple internal reflections within the sample, we write (Pankove 1975)

$$T(\omega) = \frac{(1-R)^2 e^{-\alpha(\omega)t}}{1 - R^2 e^{-2\alpha(\omega)t}} \qquad (125)$$

where R is the reflectivity given by $R = (n_r - 1)^2/(n_r + 1)^2$ in the long-wavelength limit; t is the thickness of the sample. The estimated order of magnitude of the absorption coefficient for our sample is $< 1000\,\text{cm}^{-1}$, and so the denominator can be treated as unity. We thus have $T(\omega) = (1 - R)^2 e^{-\alpha(\omega)t}$ and can write its derivative with respect to wavelength as

$$\frac{dT(\omega)}{d\lambda} = \frac{tc}{\lambda^2} \frac{d\alpha(\omega)}{d\omega} (1 - R)^2 e^{-\alpha(\omega)t} \qquad (126)$$

The absorption coefficient for subband transition is a broadened step function, therefore its derivative has a peak when the photon energy is equal to any transition energy and has a dip when the photon energy lies in the middle of two transition energies.

17. Perpendicular Transport

Transport in semiconductor structures in which the dopant plane is perpendicular to the current density vector were reported by Malik et al. (1980a, 1980b). They used a n-i-p-i-n doping profile that results in a triangular barrier structure and obtained a rectifying majority carrier diode. Subsequently, Kazarinov and Luryi (1981) calculated the charge injection over such a triangular barrier. They obtained an exact expression for the current–voltage characteristic of the triangular barrier diode, which was in excellent agreement with results published by Malik et al. (1980a). Further experimental results and a comparison with theoretical results were reported by Gossard et al. (1982).

Another interesting application of highly doped regions in semiconductors are interface dipoles to tune the barrier heights of band discontinuities of semiconductor heterostructures (Capasso et al. 1985). The dipoles use doped layers of thicknesses < 100 Å and obtained modified electro-optic transport characteristics. Recalling that the δ-doping technique results in minimized potential fluctuations, the realization of interface dipoles seems to be an interesting possibility.

The study of electron transport in δ-doped doping superlattices revealed drastic differences between the Type A (short-period) and the Type B (long-period) doping superlattices (Schubert et al. 1987a). A new type of negative differential conductivity (NDC) is found in long-period doping superlattices. In short-period doping superlattices, the NDC has not been observed. Instead, quite efficient electron transport is found in perpendicular direction to the superlattice planes in short-period superlattices.

The layer sequence of the GaAs samples used in this study is shown schematically in Fig. 79(a). The superlattice has 10 periods. The two-dimensional doping concentration is $N_A^{2D} = N_D^{2D} = 10^{13}\,\text{cm}^{-2}$. The superlattices have the periods $z_p = 150\,\text{Å}$ and $z_p = 600\,\text{Å}$. The superlattice is grown on an n^+-type substrate. Nonalloyed ohmic contacts were used for contacting the uppermost n-type GaAs layer. Alloyed ohmic contacts were not used to avoid harmful diffusion of impurities into the superlattice. After evaporation of circular Ti/Au (500 Å/2000 Å) contacts (diameter = 250 μm), the GaAs not covered by the contact metal is etched 2.0 μm deep to form mesas as shown in Fig. 79(b). The current–voltage measurements at 77 and 300K are performed with a Hewlett-Packard 4145A parameter analyzer.

Schematic band diagrams of doping superlattices with the same doping concentration but various periods are shown in Fig. 79(c). This figure reveals the sawtooth-shaped potential modulation of magnitude V_z, and the bottom of the lowest subband, E_0. Vertical transport depends sensitively on both barrier thickness and barrier height. The latter one is effectively given by the

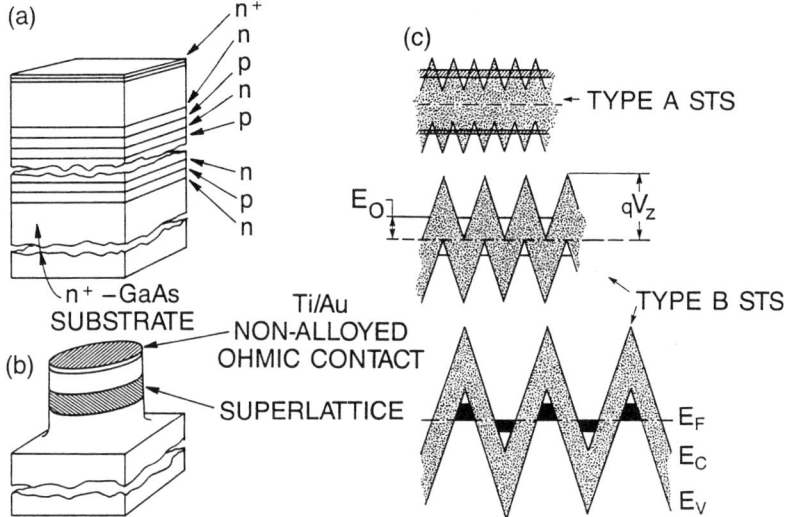

FIG. 79. Sketch of (a) a doping superlattice consisting of alternatingly n- and p-type δ-doped layers, (b) a mesa structure used for perpendicular transport measurements, (c) band diagrams of short-period (Type A) and long-period (Type B) sawtooth superlattices.

energy modulation of the superlattice minus the lowest subband energy; i.e., $eV_z - E_0$. The energy diagram and vertical transport depend strongly on the period of the superlattice. As shown in Fig. 79(c) short-period or Type A superlattices have thin barriers and small energy modulation. However, the energy modulation (>300 meV) is significantly larger than the thermal energy (<30 meV). As the period increases, carriers are strongly localized within the wells resulting in an array of purely two-dimensional systems. Such Type B sawtooth superlattices can be either depleted of free carriers (middle of Fig. 79(c)) or, for even longer periods, nondepleted (bottom of Fig. 79(c)).

The current–voltage (I–V) characteristic at $T = 77$ and 300K of a short-period ($z_p = 150$ Å) sawtooth superlattice is shown in Fig. 80. The I–V curve at 300K is linear and the total resistance has an areal resistance of $2.3 \times 10^{-2} \Omega \, cm^2$. This resistance is attributed entirely to the superlattice region. Neither the top ohmic contacts nor thee Si-doped substrate contributes significantly to the measured I–V characteristic. A linear I–V characteristic is also found for the short-period doping superlattices at $T = 77$K. The total resistance increases at low temperature to 19 Ω yielding an areal resistance of $3.8 \times 10^{-2} \Omega \, cm^2$.

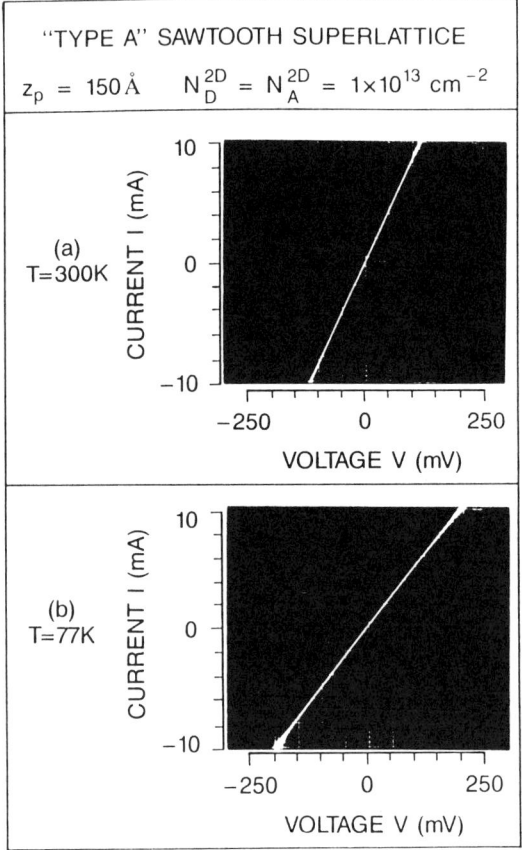

FIG. 80. Current–voltage characteristics of a Type A ($z_p = 150\,\text{Å}$, $N^{2D} = 1 \times 10^{13}\,\text{cm}^{-2}$) sawtooth superlattice at (a) $T = 300\text{K}$ and (b) 77K. The current–density vector is perpendicular to the doping planes.

Resonant tunneling (elastic tunneling) occurs if the energy drop per period (also called *Stark ladder energy*) is smaller than the miniband width of the superlattice. Thermally assisted (phonon-assisted) sequential tunneling occurs if the Stark ladder energy exceeds the miniband width of the superlattice. The decrease of the perpendicular transport efficiency at low temperature shows, however, that tunneling of carriers through the thin barriers cannot solely account for the vertical transport. In addition, thermionic emission of carriers over the barriers contributes to vertical transport in Type A superlattices.

The current–voltage characteristics of a long-period, Type B sawtooth superlattice ($z_p = 600$ Å) at the temperatures $T = 300$ and 77K are shown in Fig. 81. For each temperature the I–V curves are displayed on two scales. Figure 81(a) reveals a strongly blocking, symmetric characteristic at 300K. Voltages below 1.0 V yield currents in the nA range, indicating the drastically reduced tunneling probability through the thick barriers. An exponentially increasing current is measured at absolute values of voltage exceeding 1.5 V. The same qualitative shape is found at liquid-nitrogen temperature, as shown in Fig. 81(c). Comparison of Figs. 81(a) and 81(c) reveals an

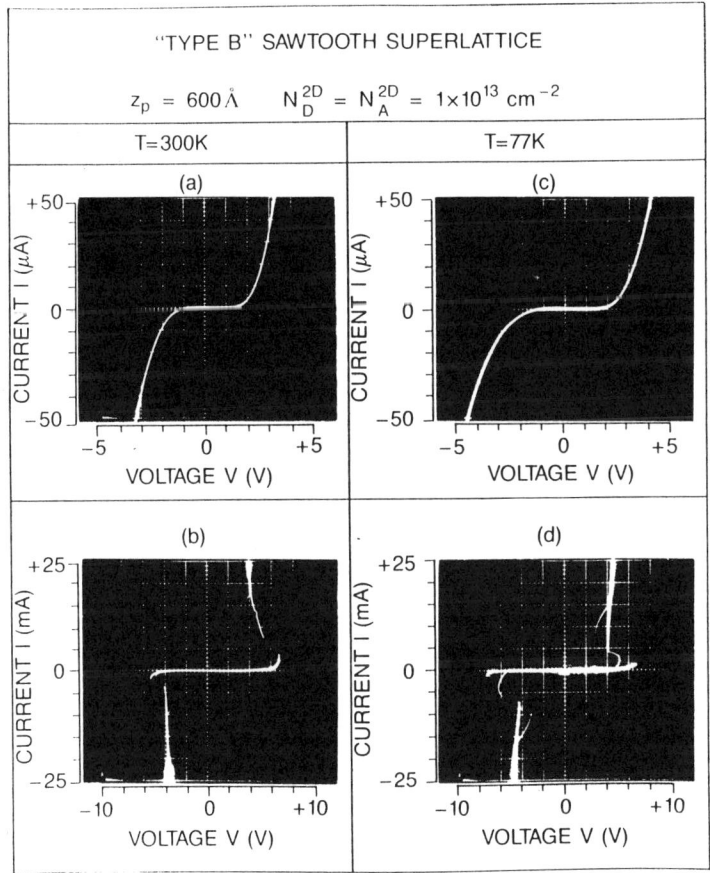

FIG. 81. Current–voltage characteristics of a Type B ($z_p = 600$ Å, $N^{2D} = 1 \times 10^{13}$ cm^{-2}) sawtooth superlattice at ((a) and (b)) $T = 300$K and ((c) and (d)) 77K. The I–V curves are shown on different scales to reveal negative differential conductivity (NDC). The current–density vector is perpendicular to the doping planes.

increased differential resistance at low temperature, showing again the relevance of thermally assisted tunneling and thermionic emission for perpendicular carrier transport.

At even larger voltages, a novel type of NDC is observed in the doping superlattice. This double-S-shaped NDC is approximately symmetric with respect to the applied voltage. At room temperature the I–V characteristic shows two NDC regions. The first NDC occurs at small currrent ($-5\,\text{mA} < I < +5\,\text{mA}$) at voltages of 5–7 V. The second NDC region occurs at higher currents above 5 mA. This second NDC regime is observed only at room temperature, but not at $T = 77\text{K}$ as the comparison of Figs. 81(b) and 81(d) reveals. The basic symmetric double-S-shaped NDC is, however, also observed at $T = 77\text{K}$.

The band diagram of a long-period sawtooth superlattice is shown in Fig. 82. The high barriers of the superlattice are difficult to overcome by thermionic emission. Furthermore, the barriers are thick ($z_p = 600\,\text{Å}$), yielding an extremely small tunneling probability. Suppressed vertical transport is therefore expected at small deviations from equilibrium, which is indeed measured experimentally. Application of an electric field in the superlattice region causes the superlattice to be tilted uniformly as shown in Fig. 82(b). Consequently, the barrier height and the barrier width are reduced, which makes possible a small current density. A key characteristic of the superlattice is the high built-in electric field. If the superlattice is not fully depleted, and electrons as well as holes are present in the ground state, the electric field is given by

$$\mathbf{E} = \pm E_g/(ez_p/2) + \mathbf{E}_{\text{ext}} \qquad (127)$$

where $\mathbf{E}_{\text{ext}}$ is the field due to the applied voltage. The electric field $\mathbf{E}$ can have a magnitude beyond 5×10^5 V/cm, yielding an avalanche ionization rate of $\alpha > 3 \times 10^4\,\text{cm}^{-1}$. Therefore, at some voltage V_A, avalanche breakdown in one or several periods of the superlattice will occur, as shown in Fig. 82(b). The occurrence of an avalanche effect leads to a potential redistribution, which manifests itself as negative differential conductivity: Holes, created by avalanche breakdown, will accumulate in the potential maxima of the valence band, due to their heavier effective mass, resulting in the screening of acceptors. The acceptor screening in turn lowers the potential barrier for electron injection into the superlattice region, leading to an increase of the total current density. The redistributed potential is shown in Fig. 82(c) with $V_{\text{on}} < V_A$, i.e., negative differential conductivity.

This interpretation is supported by the observation of infrared light emitted from the samples in the NDC domain. Radiative recombination of

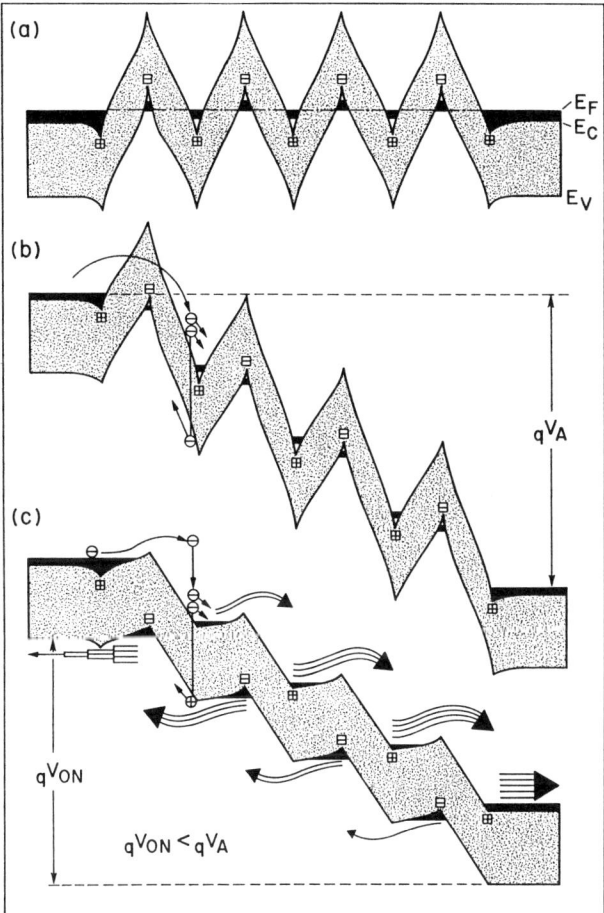

FIG. 82. Schematic sketch of band diagrams of a Type B sawtooth superlattice under (a) zero bias, (b) bias causing onset of avalanche effect, and (c) in the negative differential conductivity domain.

holes requires generation of holes by the avalanche effect, since no holes are injected in this (n-type GaAs) superlattice–(n-type GaAs) structure.

NDC was also observed in a structure similar to the Shockley diode (Wood et al. 1982). Subsequently, a three-terminal switching device was realized using δ-doped barriers. (Baillargeon et al. 1989). The authors found a controllable S-shaped NDC in a n^+-i-p^+-i-p^+-i-n^+ structure, which they propose as a high-frequency oscillator. Analysis of the oscillation yielded frequency components up to 21 GHz.

VIII. Optoelectronic Devices

Some properties of doping superlattices make possible semiconductor devices not feasible in other semiconductor systems. Specifically, the reduced superlattice energy gap and the tunability of the gap open intriguing possibilities for device application. The reduced energy gap allows one to extend the wavelength range of light-emitting devices. Thus, GaAs doping superlattice devices can operate at wavelengths $\lambda > 0.9\,\mu$m. The tunability of the energy gap allows one to continuously tune the emission energy of light-emitting diodes or lasers.

In this section, δ-doped doping-superlattice light-emitting diodes and lasers are reviewed. Furthermore, the use of doping superlattices as light modulators is reviewed.

18. Light-Emitting Diodes

Two properties of light-emitting diodes (LEDs) made from GaAs doping superlattices are unique; namely, the emission of radiation below the bandgap of GaAs and the tunability of the emission energy. The magnitude of the tunability depends on the period of the doping superlattice.

A sketch of an edge-emitting sawtooth doping superlattice diode is shown in Fig. 83 along with the δ-function-like doping profile and the sawtooth shaped band diagram (Schubert et al. 1985b). The edge-emitting diode has $Al_xGa_{1-x}As$ confinement layers. The p–n junction is located at the superlattice region, which results in carrier recombination in the superlattice region.

Electroluminescent spectra of the LEDs operating at 300K are presented in Fig. 84 for two representative diodes. The peak wavelengths of the two samples are at $\lambda = 925$ nm and 965 nm, well below the bandgap of GaAs. These wavelengths were inaccessible to conventional GaAs LEDs. At the wavelength corresponding to the band-gap energy of the GaAs host material, no luminescence signal is detected. This result demonstrates the superlattice character of the active region of the LED.

The full width at half-maximum (FWHM) value of the electroluminescence spectrum is 60 meV for sample 4670. This line width, approximately 2.2 kT, is larger than both the expected line widths of 1.8 kT for the square-root density of states in homogeneous semiconductors and 0.7 kT for the step-function density of states in two-dimensional semiconductors. The broader experimental line width is most likely due to doping density fluctuations. The emission intensity of the superlattice depends in linear

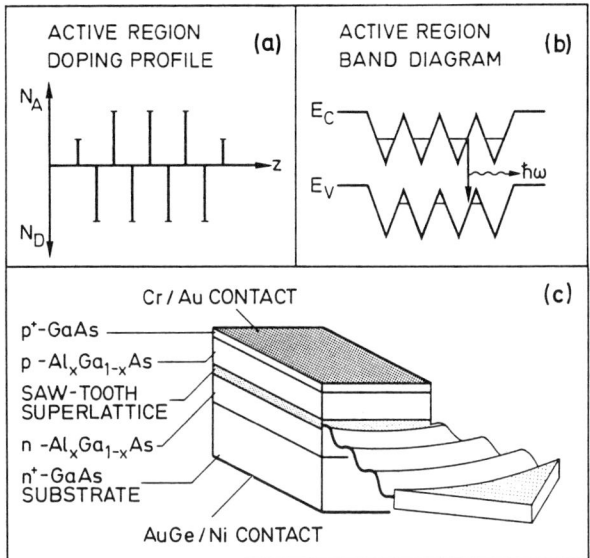

FIG. 83. (a) Doping profile, (b) band diagram, and (c) basic structure of the edge-emitting sawtooth superlattice light-emitting diode. The active doping superlattice region is sandwiched between $Al_xGa_{1-x}As$ confinement layers.

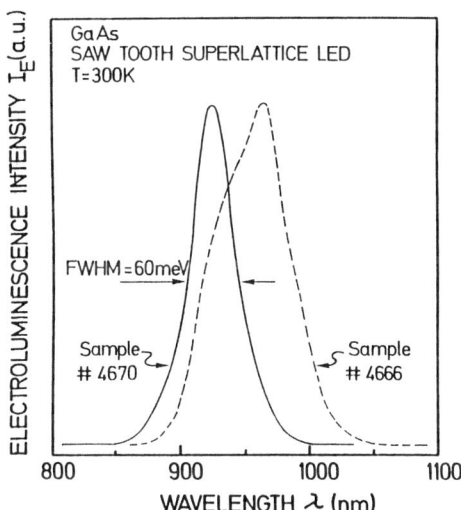

FIG. 84. Room-temperature spectrum of two doping superlattice light-emitting diodes. The electroluminescent peaks occur at wavelengths >900 nm, i.e., longer than the bulk wavelength of GaAs ($\cong 870$ nm).

manner on the injection current density as expected for the spontaneous nature of the emission.

The shift of luminescence energy with excitation intensity is rather small. Figure 85 shows the electroluminescence spectra at three different excitation intensities. The peak wavelength shifts from 966 nm at low excitation intensity to 959 nm at high excitation intensity. The total shift in energy is rather small compared to the total line width of the liminescence line ($\Delta\lambda > 50$ nm). The relatively small shift is due to the small period of the superlattice ($z_p = 200$ Å, $N^{2D} = 5 \times 10^{12}$ cm^{-2}). Even smaller shifts of the luminescence energy were observed for superlattice periods of $z_p = 150$ Å (Schubert et al. 1986d). Nevertheless, Fig. 85 shows that tunable spontaneous emission can be achieved in doping superlattices.

Subsequently, larger tuning ranges were observed in doping superlattice light-emitting diodes with longer periods (Hasnain et al. 1986). The authors used selective contacts for the n-type and p-type regions. The period of the superlattice reported by the authors is 600 Å and 900 Å.

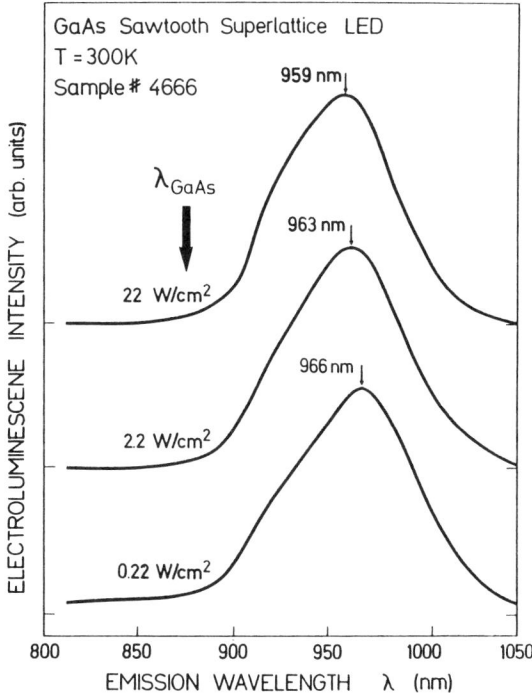

FIG. 85. Room-temperature electroluminescence spectra of a GaAs sawtooth superlattice light-emitting diodes at three injection current densities. The current densities equal a power density of 0.22, 2.2, and 22 W/cm².

19. DOPING SUPERLATTICE LASERS

Lasers are one of the most interesting device applications for doping superlattices. Tunable lasers are highly desirable for optical wavelength multiplex communication systems. On the other hand, optical transitions in doping superlattices are inherently of weaker oscillator strength as compared to bulk material. Thus, the realization requires high-quality epitaxial material. First attempts to realize a doping superlattice laser revealed optical amplification, i.e., gain (Jung et al. 1983). However, laser emission was not achieved.

The first doping superlattice current injection laser (Schubert et al. 1985a) was realized in GaAs using the δ-doping technique. The spontaneous and stimulated emission spectrum is shown in Fig. 86 for different injection currents. Below threshold (Fig. 86(a)) a wide, spontaneous spectrum is observed. As the current density is increased above threshold (Fig. 86(b–d)) a narrow line arises and completely dominates the spectrum. Longitudinal Fabry–Perot modes are found under high-resolution detection, as shown in the inset of Fig. 86(d). The peak energy of the laser emission occurs at $\lambda = 905$ nm well below the band gap of GaAs. Although laser emission was achieved, the emission energy was found to be constant, i.e., independent of the injection current density. The lack of tunability in the stimulated emission regime is due to the very short carrier lifetimes in the stimulated emission regime. Short lifetimes prevent the accumulation of carriers in the V-shaped wells of the superlattice and thus do not result in a reduction of the band-edge modulation via screening of dopant charges.

Laser emission in doping superlattices was also reported by Vojak et al. (1986). Their photo-pumped lasers emitted below the band gap of GaAs with no evidence of tunability.

The first tunable doping superlattice laser was realized by inhomogeneous excitation of the Fabry–Perot cavity (Schubert et al. 1989g). A schematic sketch of the layer sequence, the active region doping profile, and the corresponding band diagram are shown in Fig. 87. After epitaxial growth, the layers were cleaved into bars of nominal 250 μm width and 1 cm length. A frequency-doubled Nd-doped Q-switched YAG laser ($\lambda = 532$ nm) was used for optical excitation. Light emission from the sample was detected with a Si detector using gated detection. The laser samples were cooled in a variable-temperature He cryostat.

The spontaneous and stimulated emission spectra are shown together with the light output versus excitation intensity curve in Figs. 88 and 89 for 5K and 150K, respectively. In the spontaneous emission regime at low excitation intensity, the peak wavelength moves to a shorter wavelength with increasing excitation intensity. At higher excitation intensities laser

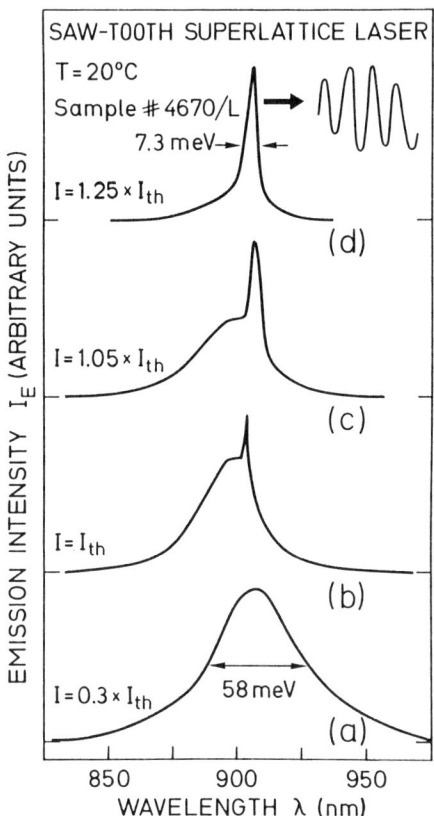

FIG. 86. Optical spectrum of the current–injection sawtooth doping superlattice laser below, at, and above threshold current. At an injection current of $1.25 \times I_{th}$ the narrow laser emission dominates the spectrum. Fabry–Perot modes, as shown in the inset, are observed.

emission occurs, which is accompanied by the characteristic kink in the light output curve of Figs. 88 and 89 and a narrowing of the emission spectrum below values of the thermal energy kT. However, the peak wavelength does not change in the lasing regime as clearly revealed in Figs. 88(b) and 89(b). Such a constant emission energy is not unexpected, since upon reaching the laser threshold the Fermi level remains constant and additional carriers undergo stimulated recombination with a correspondingly very short lifetime. Thus, the emission energy remains constant with excitation energy in the stimulated emission regime.

The dependence of the emission wavelength on the excitation intensity is shown in greater detail in Fig. 90 for three different temperatures. In the

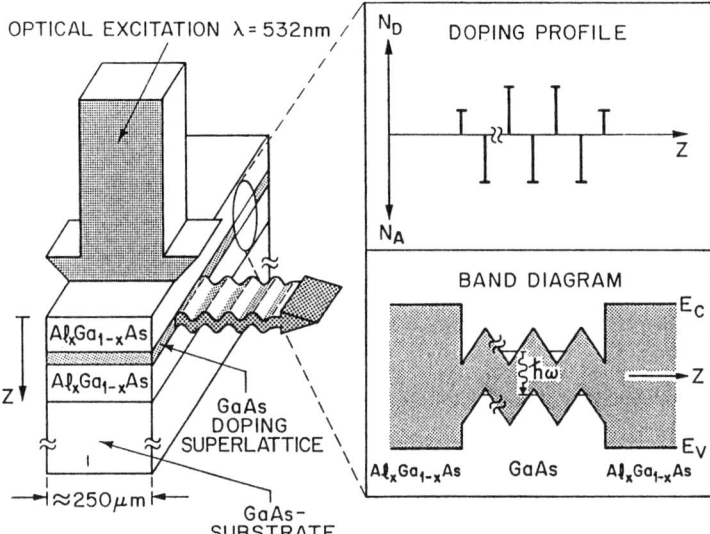

FIG. 87. Geometry used for optical excitation of the optically excited tunable sawtooth superlattice laser. The active region doping profile and its band diagram are shown as well.

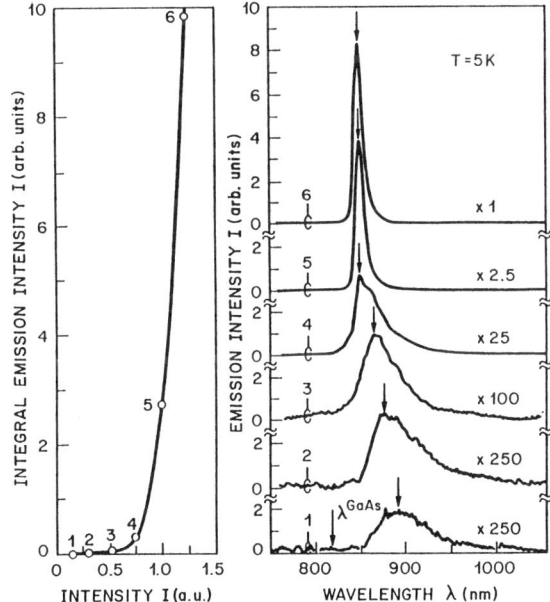

FIG. 88. (Left) Low-temperature emission intensity versus excitation intensity of the tunable sawtooth superlattice laser. (Right) Optical spectra below, at, and above threshold intensity. The laser wavelength remains constant when the lasing threshold is reached.

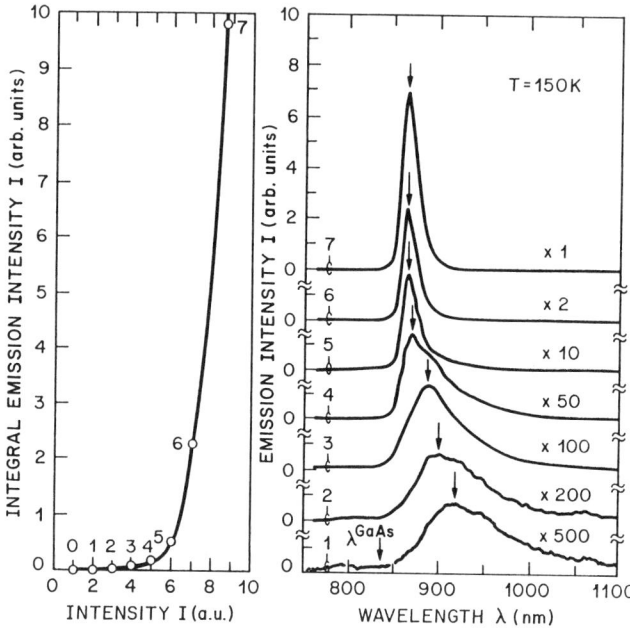

FIG. 89. (Left) Emission intensity versus excitation intensity of the tunable sawtooth superlattice laser at $T = 150$K. (Right) Optical spectra below, at, and above threshold intensity. The laser wavelength remains constant when the laser threshold is reached.

spontaneous regime the emission shifts to shorter wavelengths. However, upon reaching the laser threshold intensity, the emission remains constant. The stimulated emission energy is typically 30 nm below the band gap of GaAs.

The emission energy can be tuned continuously by inhomogeneous excitation of the Fabry–Perot cavity. Such inhomogeneous excitation is achieved by displacing the exciting beam from its centered position, as shown in the top part of Fig. 91. The inhomogeneous excitation results in a laser emission energy higher than the symmetric excitation. Figure 91 reveals that the tuning range of the laser is approximately 35 Å. This tuning range does not represent a fundamental limit. The current tuning range is limited by the intensity distribution of the exciting source. We expect a wider tuning range for a more inhomogeneous excitation, which could be also achieved in a two- or three-section current-injection laser.

Simultaneously, as the peak of the stimulated emission shifts to a shorter wavelength, the excitation intensity required to reach the threshold increases, as illustrated in Fig. 91(b). However, it is important to visualize that,

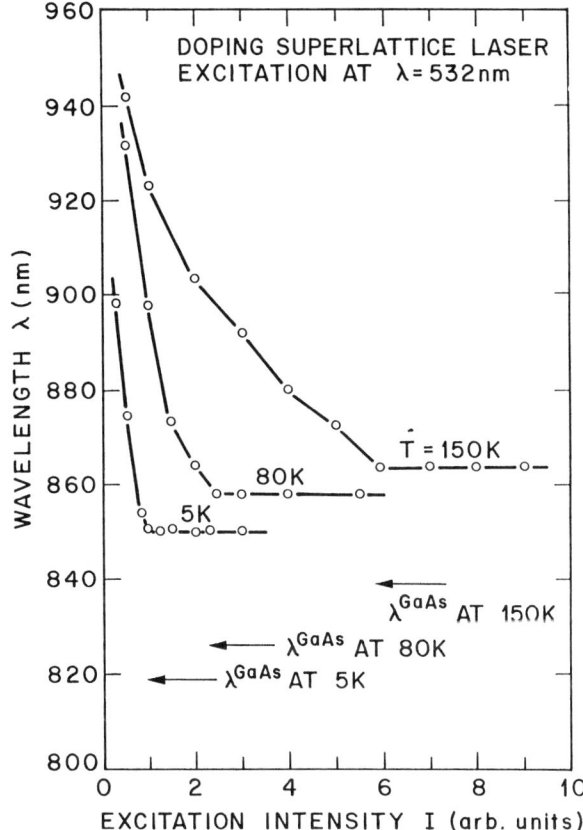

FIG. 90. Emission wavelength versus excitation intensity of the tunable sawtooth superlattice laser for the three temperatures: 5K, 80K, and 150K. Upon reaching the threshold intensity, the wavelength remains constant. Note that the stimulated emission occurs at wavelengths below the bulk band gap of GaAs, which is marked by the three arrows.

upon displacement, the sample is excited only by a small part of the exciting beam, as shown in the inset to Fig. 91. Thus, the increase of threshold intensity is overestimated, and the true increase in threshold intensity is not as pronounced as that suggested by Fig. 91(b).

A high-resolution spectrum of the doping superlattice laser is shown in Fig. 92 for the transverse electric (TE) and transverse magnetic (TM) modes. In analogy to conventional $Al_xGa_{1-x}As/GaAs$ double-heterostructure lasers with the cleavage planes being a $\{110\}$ plane, the TE mode dominates the laser emission, while the TM mode is comparatively weak.

Under homogeneous excitation conditions the stimulated emission energy

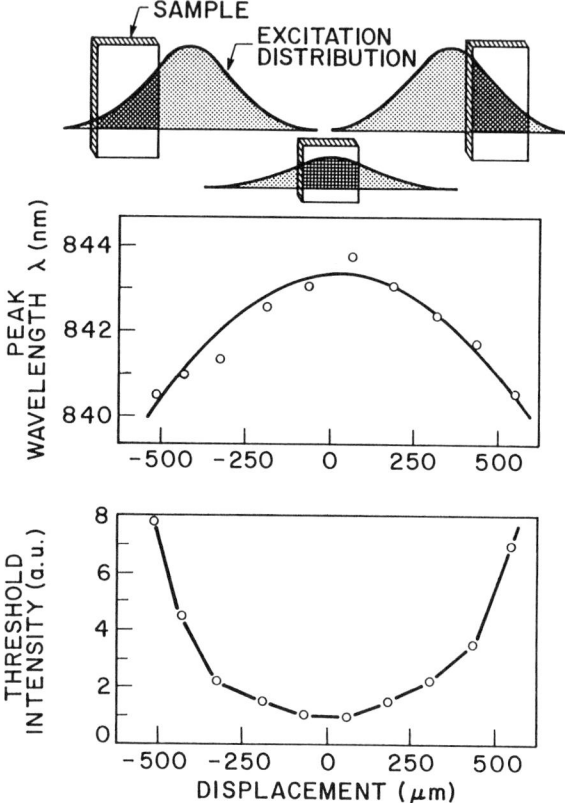

FIG. 91. (a) Peak wavelength of the tunable laser versus displacement of the optical excitation beam. Tunability of the laser emission is achieved over $\cong 35\,\text{Å}$. (b) Threshold intensity of the laser versus displacement of the exciting beam. The top part of the figure shows the displacement of the exciting beam with espect to the laser bar.

is below the band gap of GaAs. However, the emission energy is much higher than the spontaneous emission energy at low excitation intensities ($E \cong 1.35\,\text{eV}$). Thus, the superlattice modulation is reduced by photoexcited electrons and holes, which screen the ionized dopant charges of donors and acceptors, respectively. Even though the modulation is reduced, a residual band modulation is still maintained, as suggested by the low emission energy. Thus, stimulated emission is achieved before the bands are completely flat; that is, for incomplete screening of dopant charges.

The physical mechanism leading to the tunability of the semiconductor laser can be understood on the basis of increased loss induced by in-

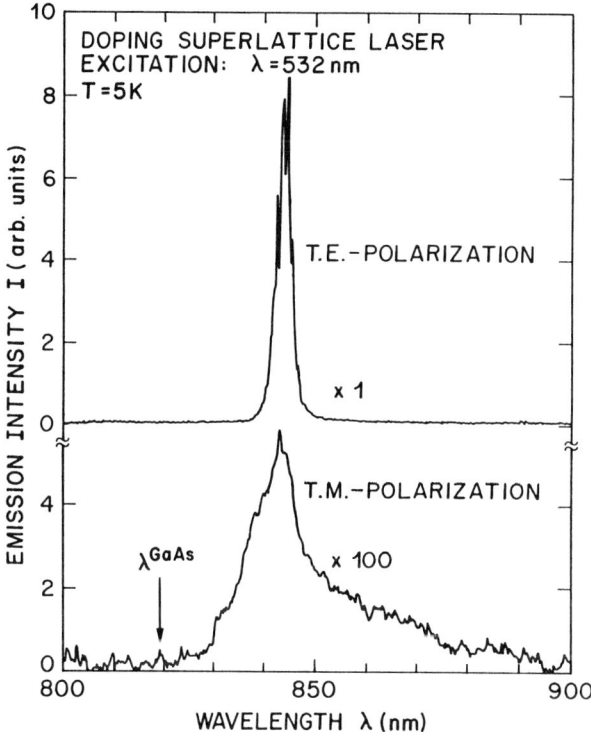

FIG. 92. Laser emission spectrum of the doping superlattice laser with transverse electric (TE) and transverse magnetic (TM) polarization. The TE polarization spectrum reveals a number of longitudinal modes.

homogeneous excitation. As a result of the inhomogeneous excitation, i.e., reduced excitation intensity in one part of the Fabry–Perot cavity, the optical loss is enhanced in this section. To obtain stimulated emission the other section must be subjected to higher excitation. As a result, the band modulation decreases and the superlattice energy gap increases in this section due to an enhanced density of carriers. Once the intentionally induced loss is overcome, stimulated emission occurs. However, the corresponding energy is increased as compared to the homogeneously excited cavity. Thus, tunability of the stimulated emission wavelength is achieved by a different excitation in the two sections of the laser. The principal limit of the tuning range is reached when flatband condition is achieved in one part of the laser. The corresponding tuning range is approximately 250 Å at low temperatures for the samples studied here.

The temperature dependence of the threshold intensity is shown in Fig.

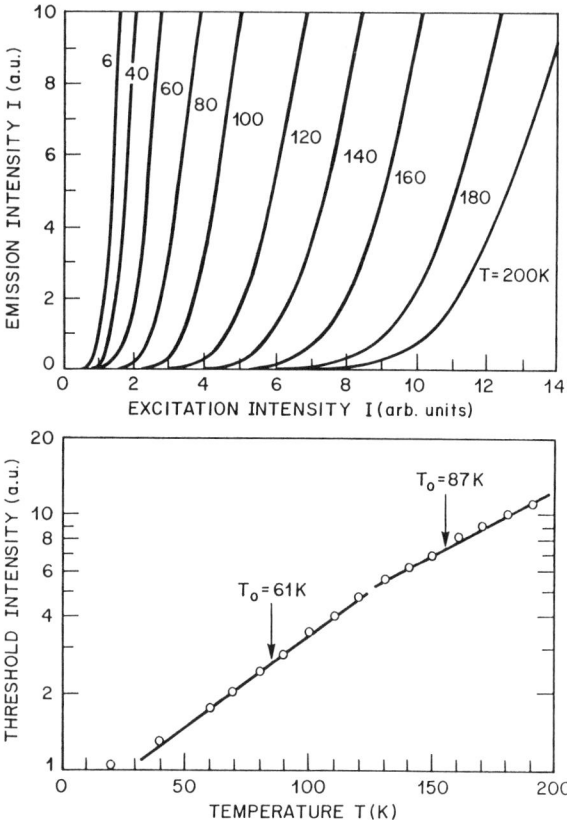

FIG. 93. Light output versus excitation intensity of the tunable sawtooth superlattice laser at different temperatures. The T_0 is determined to be 61K and 87K at low and intermediate temperatures, respectively.

93. The threshold intensity increases at higher temperature and is characterized by the following relation:

$$I_{th} \propto \exp(T/T_0) \qquad (127)$$

where T is the absolute temperature. The parameter T_0 describes the increase of threshold current with temperature. Evaluation of T_0 from the slope of I_{th} versus T yields a T_0 of 61K and 87K for low and intermediate temperatures, respectively. The relatively low T_0 is probably due to the low growth temperatures employed for crystal growth. The low growth temperatures are chosen to reduce impurity diffusion and segregation during growth.

20. Modulators

Modulation and switching of near-infrared radiation by means of electrical control will be an important functional device characteristic in future photonic switching systems. It is desirable that such photonic switching devices (i) have a broad wavelength range in which the light intensity can be modulated, (ii) have a large contrast ratio in the transparent and opaque states, (iii) can operate at voltages compatible to electronic integrated circuits, and (iv) have high-speed capability.

Modulators with such characteristics can be fabricated from doping superlattices with long periods. Contacting the n-type and p-type regions of the superlattice, allows one to change the modulation, i.e., the internal electric field of the superlattice, by means of an external bias. A schematic sketch of the band diagram is shown in Fig. 94 for different voltages. Forward bias of the p-i-n structure results in a decrease of the band

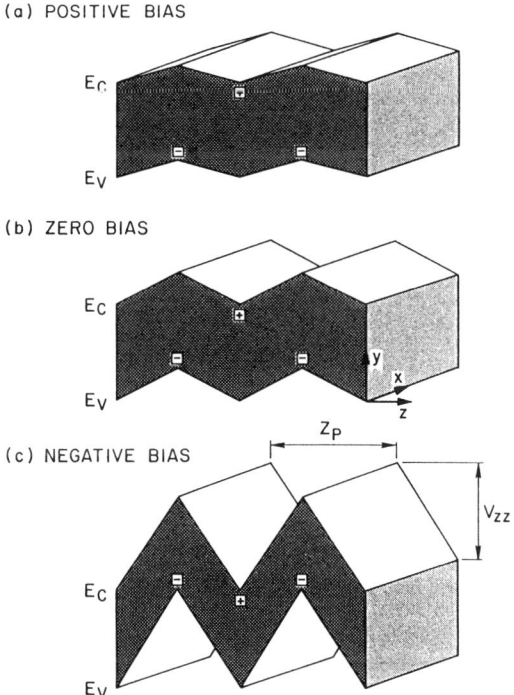

FIG. 94. Schematic illustration of the band diagram of a long-period sawtooth doping superlattice, in which the individual n-type and p-type layers are contacted, for positive, 0, and negative bias. The superlattice modulation and the electric field depend on the applied voltage.

modulation and of the internal electric field. Reverse bias results in an increase of the band modulation, i.e., the internal electric field.

Tunneling-assisted absorption (Franz–Keldysh absorption), which occurs at energies below the band gap of the semiconductor, depends exponentially on the electric field. Thus, doping superlattices can be used to modulate the intensity of transmitted light by tunneling-assisted absorption.

Experimental results of intensity modulation experiments are shown in Fig. 95 (Schubert and Cunningham 1988a) for different voltages applied to the modulator. Incident light is absorbed by the substrate at energies higher than the fundamental gap; that is, $\lambda < 870$ nm. The long-wave decay of the transmission signal ($\lambda \gtrsim 1000$ nm) is due to the decreasing sensitivity of the Si photodetector. The most striking feature of the modulator is the wide wavelength range ($\Delta\lambda > 100$ nm) in which a significant modulation is achieved. Such a wide wavelength range is desirable for photonic switching

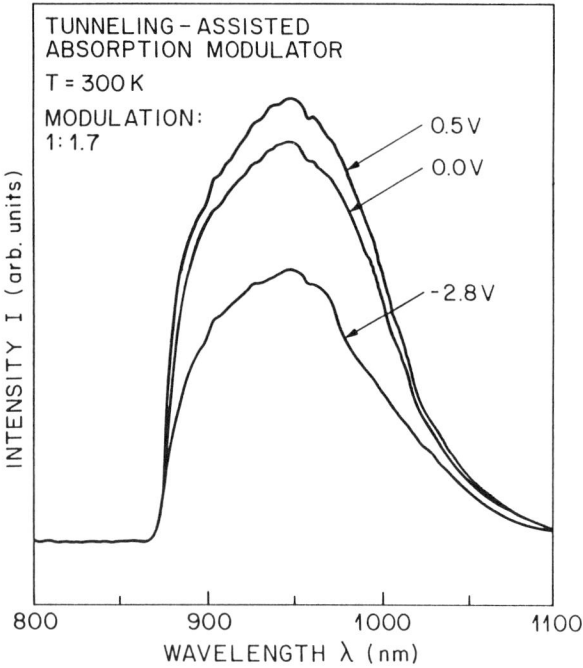

FIG. 95. Optical transmission through the doping superlattice modulator versus wavelength at different biases. A very large range of wavelengths (900 nm $\leqslant \lambda \leqslant$ 1000 nm) is accessible to modulation with the modulator. For $\lambda < 870$ nm the transmitted intensity approaches 0, since the GaAs substrate absorbs all light. For $\lambda > 1050$ nm, the Si-detector response becomes insensitive.

systems, because no accurate matching of light source wavelength and modulator wavelength is required. Furthermore, the wavelength range suitable for modulation (900 nm $\leqslant \lambda \leqslant$ 1000 nm) can be shifted towards short wavelengths (e.g., 800 nm $\leqslant \lambda \leqslant$ 900 nm) by using $Al_xGa_{1-x}As$ rather than GaAs as a host material for the sawtooth structure. The maximum contrast ratio between the opaque and transparent states is 1:1.7 or 70%, and occurs at $\lambda \simeq 950$ nm. Modulation experiments with the conventionally doped *n-i-p-i* structure yielded a transmission change of 22% using selective side contacts to the superlattice region (Chang-Hasnain *et al.* 1987).

The voltage difference between the transparent and opaque states is 3.3 V. Such small voltage swings are desirable for integration of the sawtooth modulator and additional electronic circuits.

The sawtooth modulator has potential for high-speed modulation. The total capacitance of the sawtooth structure with the design parameters used in this study and an area of 10 μm × 10 μm equals $C < 2$ pF. Together with the voltage swing of $V < 5$ V, the sawtooth modulator is potentially suitable for modulation in the GHz range.

IX. Concluding Remarks

Research and development in the field of semiconductor technology has always been focused on the fundamental physical limits such as limits of miniaturization, speed, and integration of electronic or optical circuits. The realization of δ-function-like doping distributions represents such a fundamental limit. It was shown in this chapter what δ-doped semiconductors are and how they are realized. Furthermore, a number of device applications were reviewed.

The field of δ-doping is truly interdisciplinary. It involves material science, solid-state physics, quantum mechanics, and device physics. Contributions from each of these fields are essential to the realization and understanding of δ-doped semiconductors. The interdisciplinary nature of the subject makes the field exciting and filled with a wealth of knowledge.

Employment of the δ-doping technique results in a significant improvement of a number of semiconductor structures and devices. For many purposes the δ-function-like doping distribution represents an optimized doping distribution. Future research and development of semiconductors will make use of the advances and knowledge gained in the course of research in the field of δ-doping. Certainly, the intention is the further improvement of semiconductor structures and devices.

However, a long way is still ahead of us. The employment of δ-function-like doping distributions requires improved crystal growth and processing technology. Many challenges will be met and problems must be solved, before the development and the production of semiconductor devices can take full advantage of the δ-doping technique. This is what continues to make semiconductor physics such an attractive field: continuing research effort, new results, surprises, and excitement in the research field; and a more efficient way to process and distribute information in the development and production of semiconductor integrated circuits.

Acknowledgments

The author has greatly benefitted from collaborations with colleagues at AT&T Bell Laboratories. The author wishes to thank T. Y. Chang, T. H. Chiu, A. Y. Cho, J. E. Cunningham, D. Deppe, J. L. de Miguel, L. C. Feldman, A. M. Glass, T. D. Harris, R. L. Headrick, L. C. Hopkins, R. F. Kopf, J. M. Kuo, A. F. J. Levi, H. S. Luftman, R. J. Malik, L. Pfeiffer, S. Schmitt-Rink, H. L. Störmer, B. Tell, W. T. Tsang, K. W. West, and J. P. van der Ziel for support, many valuable discussions, and significant contributions to this work.

References

Abe, M., Mimura, T., Nishiuchi, K., Shibatomi, A., and Kobayashi, M. (1986). *IEEE J. Quant. Electron.* **QE-22**, 1870.
Abernathy, C. R., Pearton, S. J., Caruso, R., Ren, F., and Kovalchik, (1989). *Appl. Phys. Lett.* (Oct. 23).
Abramowitz, M., and Stegun, L. A., Eds. (1972). *Handbook of Mathematical Functions.* National Bureau of Standards, Washington DC.
Allyn, C. L., Gossard, A. C., and Wiegman, W. (1980). *Appl. Phys. Lett.* **36**, 373.
Ando, T. (1982a). *J. Phys. Soc. Jpn* **51**, 3893.
Ando, T. (1982b). *J. Phys. Soc. Jpn.* **51**, 3900.
Ando, T. (1985). *J. Phys. Soc. Jpn.* **54**, 2676.
Ando, T., Fowler, A. B., and Stern, F. (1982). *Rev. Mod. Phys.* **54**, 437.
Baba, T., Mizutani, T., and Ogawa, M. (1983). *Jpn. J. Appl. Phys.* **22**, L627.
Baba, T., Mitutani, T., Ogawa, M., and Ohata, K. (1984). *Jpn. J. Appl. Phys.* **23**, L654.
Baillargeon, J. N., Cheng, K. Y., Laskar, J., and Kolodzey, J. (1989). *Appl. Phys. Lett.* **55**, 663.
Bardeen, J. (1947). *Phys. Rev.* **71**, 717.
Bass, S. J. (1979). *J. Cryst. Growth* **47**, 613.
Bastard, G. (1981). *Phys. Rev. B* **24**, 5693.
Beall, R. B., Clegg, J. B., and Harris, J. J. (1988). *Semicond. Sci. Technol.* **3**, 612.
Beall, R. B., Harris, J. J., Clegg, J. B., Gowers, J. P., Joyce, B. A., Castagnè, J., and Welch, V. (1989). In *GaAs and Related Compounds 1988* (J. S. Harris, ed.). IOP Publishing Ltd, Bristol.

Beltram, F., Capasso, F., Luryi, S., Chu S. N. G., Cho, A. Y., and Sivco, D. L. (1988). *Appl. Phys. Lett.* **53**, 219.
Capasso, F., Luryi, S., Tsang, W. T., Bethea, C. G., and Levine, B. F. (1983). *Phys. Rev. Lett.* **51**, 2318.
Capasso, F., Mohammed, K., and Cho, A. Y. (1985). *J. Vac. Sci. Technol. B* **3**, 1245.
Casey, H. C., Jr. (1973). In *Atomic Diffusion in Semiconductors* (D. Shaw, ed.), p. 351. Plenum Press, New York.
Cahi, Y. G., Chow, R., and Wood, C. E. C. (1981). *Appl. Phys. Lett.* **39**, 800.
Chang-Hasnain, C. J., Hasnain, G., Johnson, N. M., Döhler, G. H., Miller, J. N., Whinnery, J. R., and Dienes, A. (1987). *Appl. Phys. Lett.* **50**, 915.
Chattopadhyay, D., and Queisser, H. J. (1981). *Rev. Mod. Phys.* **53**, 745.
Chen, Y. K., Radulescu, D. C., Tasker, P. J., Wang, G. W., and Eastman, L. F. (1987). *GaAs and Related Compounds* (W. T. Lindley, ed.), p. 581. Institute of Physics, Bristol.
Chiang, T.-C., Ludeke, R., Aono, M., Landgren, G., Himpsel, F. J., and Eastman, D. E. (1983). *Phys. Rev. B* **27**, 4770.
Chiu, T. H., Tsang, W. T., Schubert, E. F., and Agyekum, E. (1987). *Appl. Phys. Lett.* **51**, 1109.
Chiu, T. H., Cunningham, J. E., Tell, B., and Schubert, E. F. (1988). *J. Appl. Phys.* **64**, 1578.
Cho, A. Y. (1975). *J. Appl. Phys.* **46**, 1733.
Cho, A. Y. (1976). *J. Appl. Phys.* **47**, 2841.
Cunningham, J. E., Tsang, W. T., Timp, G., Schubert, E. F., Chang, A. M., and Owusu-Sekyere, K. (1988). *Phys. Rev. B* **37**, 4317.
Delagebeaudeuf, D., and Linh, N. T. (1982). *IEEE Trans. Electron Dev.* **ED-29**, 955.
de Miguel, J. L., Shibli, S. M., Tamargo, M. C., and Skromme, B. J. (1988). *Appl. Phys. Lett.* **53**, 2065.
Di Lorenzo, J. V. (1971). *J. Electrochem. Soc.* **118**, 1645.
Dingle, R., Gossard, A. C., and Wiegman, W. (1975). *Phys. Rev. Lett.* **34**, 1327.
Dingle, R., Feuer, M. D., and Tu, C. W. (1985). In *VLSI Electronics*, (N. G. Einspruch and W. R. Weisman, ed.), p. 216. Academic Press, New York.
Döhler, G. H. (1972a). *Phys. Stat. Sol.* **52**, 79.
Döhler, G. H. (1972b). *Phys. Stat. Sol.* **52**, 533.
Döhler, G. H. (1986). *IEEE J. Quant. Electron.* **QE-22**,1682.
Döhler, G. H., Künzel, H., Olego, D., Ploog, K., Ruden, P., Stolz, H. J., and Abstreiter, G. (1981). *Phys. Rev. Lett.* **47**, 864.
Eisele, I. (1989). *Superlattices and Microstructures* **6**, 123.
Elliot, R. J. (1957). *Phys. Rev.* **108**, 1384.
English, J. H., Gossard, A. C., Störmer, H. L., and Baldwin, K. W. (1987). *Appl. Phys. Lett.* **50**, 1826.
Esaki, L. and Tsu, R. (1970). *IBM J. Res. Develop.* **14**, 61.
Fang, F. F., and Howard, W. E. (1966). *Phys. Rev. Lett.* **16**, 797.
Gillman, G., Vinter, B., Barbier, E., and Tardella, A. (1988). *Appl. Phys. Lett.* **52**, 972.
Glass, A. M., Schubert, E. F., Wilson, B. A., Bonner, C. E., Cunningham, J. E., Olson, D. H., and Jan, W. (1989). *Appl. Phys. Lett.* **54**, 2247.
Gossard, A. C. (1985). In *Molecular Beam Epitaxy and Heterostructures* (L. L. Chang and K. Ploog, ed.), p. 499. Martinus Nijhoff, Boston.
Gossard, A. C., Kazarinov, R. F., Luryi, S., and Wiegmann, W. (1982). *Appl. Phys. Lett.* **40**, 832.
Greiner, M. E., and Gibbons, J. F. (1984). *Appl. Phys. Lett.* **44**, 750.
Greiner, M. E., and Gibbons, J. F. (1985). *J. Appl. Phys.* **57**, 5181.
Hamm, R. A., Panish, M. B., Nottenburg, R. N., Chen, Y. K., and Humphrey, (1989). *Appl. Phys. Lett.* **54**, 2586.

Harris, J. J., Ashenford, D. E., Foxon, C. T., Dobson, P. J., and Joyce, B. A. (1984). *Appl. Phys. A* **33**, 87.
Hasnain, G., Döhler, G. H., Whimery, J. R., Miller, J. N., and Dienes, A. (1986). *Appl. Phys. Lett.* **49**, 1357.
Headrick, R. L., Feldman, L. C., and Robinson, I. K. (1989a). *Appl. Phys. Lett.*
Headrick, R. L., Robinson, I. K., Vlieg, E., and Feldman, L. C. (1989b). *Phys. Rev. Lett.*
Heine, V. (1965). *Phys. Rev.* **138**, A1689.
Hilsum, C. (1974). *Electron. Lett.* **10**, 259.
Hobson, W. S., Pearton, S. J., Schubert, E. F., and Cabaniss, G. (1989). *Appl Phys. Lett.*, submitted.
Hower, P. L., and Bechtel, G. (1973). *IEEE Trans. Electron. Dev.* **ED-20**, 213.
Huijser, A., van Laar, J., and van Rooy, T. L. (1977). *Surf. Sci.* **62**, 472.
Ihm, J., Chadi, D. J., and Joannopoulos, J. D. (1983). *Phys. Rev. B* **27**, 5119.
Ilegems, M. (1977). *J. Appl. Phys.* **48**, 1278.
Ishibashi, A., Funato, K., and Mori, Y. (1988). *Electronics Letters* **24**, 1035.
Ishikawa, T., Ogasqwara, K., Nakamura, T., Kuroda, S., and Kondo, K. (1987). *J. Appl. Phys.* **61**, 1937.
Johnson, W. C., and Panousis, P. T. (1971). *IEEE Trans. Elec. Dev.* **ED-18**, 965.
Joyce, B. A. (1985). In *Molecular Beam Epitaxy and Heterostructures* (L. L. Chang and K. Ploog, eds.), p. 37. Martinus Nijhoff, Boston.
Jung, H., Döhler, G. H., Göbel, E. O., and Ploog, K. (1983). *Appl. Phys. Lett.* **43**, 40.
Kahng, D. (1964). *Bell. Syst. Tech. J.* **42**, 215.
Kazarinov, R. F., and Luryi, S. (1981). *Appl. Phys. Lett.* **38**, 810.
Kennedy, D. P., Murley, P. C., and Kleinfelder, W. (1968). *IBM J. Res. Dev.* **12**, 399.
Kobayashi, N., Makimoto, T., and Horikoshi, Y. (1986). *Jpn. J. Appl. Phys. Lett.* **25**, L746.
Kobayashi, N., Makimoto, T., and Horikoshi, Y. (1987). *Appl. Phys. Lett.* **50**, 1435.
Koenraad, P. M., Blom, F. A. P., Langerak, C. J. G. M., Leys, M. R., Perenboom, J. A. A. J., Singleton, J., Spermon, S. J. R. M., van der Vlenten, W. C. Voncken, A. P. J., and Wolter, J. H. (1989). Proc. 4th Int. Conf. Modulated Semicond. Struc., Ann Arbor, MI.
Kroemer, H., Chien, W.-Y., Harris, J. S., Jr., and Edwall, D. D. (1980). *Appl. Phys. Lett.* **36**, 295.
Kuo, T. Y., Cunningham, J. E., Schubert, E. F., Tsang, W. T., Chiu, T. H., Reh, F., and Forstad, C. G. (1988). *J. Appl. Phys.* **64**, 3324.
Lanzillotto, A.-M., Santos, M., and Shayegan, M. (1989). *Appl. Phys. Lett.* **55**, 1445.
Lee, H., Schaff, W. J., Wicks, G. W., Eastman, L. F., and Calawa, A. R. (1985). *Inst. Phys. Conf. Ser.* **74**, 321.
Levine, B. F., Tsang, W. T., Bethea, C. G., and Capasso, F. (1982). *Appl. Phys. Lett.* **41**, 470.
Luryi, S. (1988). *Appl. Phys. Lett.* **52**, 501.
Maguire, J., Murray, R., Newman, R. C., Beall, R. B., and Harris, J. J. (1987). *Appl. Phys. Lett.* **50**, 516.
Malik, R. J., AuCoin, T. R., Ross, R. L., Board, K., Wood, C. E. C., and Eastman, L. F. (1980a). *Electron. Lett.* **16**, 836.
Malik, R. J., Board, K., Eastman, L. F., Wood, C. E. C., AuCoin, T. R., and Ross, R. L. (1980b). Gallium Arsenide and Related Compounds, *Inst. Phys. Conf. Ser.* **56**, 697.
Malik, R. J., and Dixon, S. (1982). *IEEE Elec. Dev. Lett.* **EDL-3**, 205.
Malik, R. J., Nottenburg, R. N., Schubert, E. F., Walker, J. R., and Ryan, R. W. (1988). *Appl. Phys. Lett.* **53**, 2661.
McLevige, W. V., Vaidyanathan, K. V., Streetman, B. G., Ilegems, M., Comas, J., and Plew, L. (1978). *Appl. Phys. Lett.* **33**, 127.
Miller, R. C., Kleinman, D. A., Munteanu, O. and Tsang, W. T. (1981). *Appl. Phys. Lett.* **39**, 1.
Mott, N. F. (1938). *Proc. Cambridge, Philos. Soc.* **34**, 568.

Nanichi, Y., and Pearson, G. L. (1969). *Solid State Electron.* **12**, 341.
Nichols, K. H., Goldwasser, R. E., and Wolfe, C. M. (1980). *Appl Phys. Lett.* **36**, 601.
Omura, E., Wu, X. S., Vawter, G. A., Coldren, L., Hu, E., and Merz, J. L. (1986). *Electron. Lett.* **22**, 496.
Overbury, S. H., Bertrand, P. H., and Somorjai, G. A. (1975). *Chem. Res.* **75**, 547.
Ovsyannikov, M. I., Romanov, Y. A., Shabanov, V. N., and Loginova, R. G. (1971). *Sov. Phys. Semicond.* **4**, 1919.
Panish, M. B. (1973). *J. Appl. Phys.* **44**, 2659.
Pankove, J. I. (1975). *Optical Processes in Semiconductors.* Dover Books, New York.
Pei, S. S., and Shah, N. J. (1989). In C. T. Wang, ed.). Wiley, New York.
Petritz, R. (1958). *Phys. Rev.* **110**, 1254.
Pfeiffer, L., Störmer, H. L., West, K. L., and Baldwin, K. W. (1989). *Appl. Phys. Lett.*
Ploog, K., and Döhler, G. H. (1983). *Adv. Phys.* **32**, 285.
Proetto, C. R. (1989). *Physical Review B.*
Price, P. J. (1981). *J. Vac. Sci. Technol.* **19**, 599.
Price, P. J. (1982). *Surf. Sci.* **113**, 199.
Price, P. J. (1984). *Surf. Sci.* **134**, 145.
Romanov, Y. A. (1972). *Sov. Phys. Semicond.* **5**, 1256.
Romanov, Y. A., and Orlov, L. K. (1973). *Sov. Phys. Semicond.* **7**, 182.
Saito, K., Tokumitsu, E., Akatsuka, T., Miyauchi, M., Yamada, T., Konagai, M., and Takahashi, K. (1988). *J. Appl. Phys.* **64**, 3975.
Santos, M., Sajoto, T., Zrenner, A., and Shayegan, M. (1988). *Appl. Phys. Lett.* **53**, 2504.
Sasa, S., Muto, S., Kondo, K., Ishikawa, H., and Hiyamizu, S. (1985). *Jpn. J. Appl. Phys. Lett.* **24**, L602.
Schottky, W. (1938). *Naturwissenschaften* **26**, 843.
Schottky, W. (1940). *Zeits. Physik* **41**, 570.
Schrieffer, J. R. (1957). In *Semiconductor Surface Physics* (R. H. Kingston, ed.), pp. 55–69. University of Pennsylvania Press, Philadelphia.
Schubert, E. F., Göbel, E. O., Horikoshi, Y., Ploog, K., and Queisser, H. J. (1984). *Phys. Rev. B* **30**, 813.
Schubert, E. F., Fischer, A., Horikoshi, Y., and Ploog, K. (1985a). *Appl. Phys. Lett.* **47**, 219.
Schubert, E. F., Fischer, A., and Ploog, K. (1985b). *Electron. Lett.* **21**, 411.
Schubert, E. F., Fischer, A., and Ploog, K. (1985c). *Phys. Rev. B* **31**, 7937.
Schubert, E. F., Horikoshi, Y., and Ploog, K. (1985d). *Phys. Rev. B* **32**, 1085.
Schubert, E. F., and Ploog, K. (1985e). *Jpn. J. Appl. Phys. Lett.* **24**, L608.
Schubert, E. F., Cunningham, J. E., and Tsang, W. T. (1986a). *Appl. Phys. Lett.* **49**, 1729.
Schubert, E. F., Fischer, A., and Ploog, K. (1986c). *IEEE Transactions on Electron Devices*, **ED-33**, 625.
Schubert, E. F., Hauser, M., Ullrich, B., and Ploog, K. (1986d). In *Two-Dimensional Systems: Physics and New Devices* (G. Bauer, K. Kuchar, and H. Heinnich, eds.). Springer, New York.
Schubert, E. F., and Ploog, K. (1986e). *Jpn. J. Appl. Phys.* **25**, 966.
Schubert, E. F., Cunningham, J. E., and Tsang, W. T. (1987a). *Appl. Phys. Lett.* **51**, 817.
Schubert, E. F., Cunningham, J. E., and Tsang, W. T. (1987b). *Solid State Commun.* **63**, 591.
Schubert, E. F., Cunningham, J. E., and Tsang, W. T. (1987c). *Phys. Rev. B* **36**, 1348.
Schubert, E. F., Cunningham, J. E., Tsang, W. T., and Timp, G. L. (1987d). *Appl. Phys. Lett.* **51**, 1170.
Schubert, E. F., and Cunningham, J. E. (1988a). *Electron. Lett.* **24**, 980.
Schubert, E. F., Harris, T. D., and Cunningham, J. E. (1988b). *Appl. Phys. Lett.* **53**, 2208.
Schubert, E. F., Stark, J. B., Chiu, T. H., and Tell, B. (1988c). *Appl. Phys. Lett.* **53**, 293.

Schubert, E. F., Stark, J. B., Ullrich, B., and Cunningham, J. E. (1988d). *Appl. Phys. Lett.* **52**, 1508.
Schubert, E. F., Ullrich, B., Harris, T. D., and Cunningham, J. E. (1988e). *Phys. Rev. B* **38**, 8305.
Schubert, E. F., Cunningham, J. E., Chiu, T. H., Stark, J. B., Tell, B., and Tu, C. W. (1989a). *GaAs and Related Compounds 1988* (J. S. Harris, ed.). Institute of Physics, Bristol.
Schubert, E. F., Harris, T. D., Cunningham, J. E., and Jan, W. (1989b). *Phys. Rev. B* **39**, 11011.
Schubert, E. F., Harris, T. D., Cunningham, J. E., and Jan, W. (1989c). *Phys. Rev. B* **39**, 11011.
Schubert, E. F., Kuo, J. M., Kopft, R. F., Jordan, A. S., Luftman, H. S., and Hopkins, L. C. (1989d). *Phys. Rev.* **42**, 1364 (1990).
Schubert, E. F., Pfeiffer, Loren, West, K. W., and Izabelle, A. (1989e). *Appl. Phys. Lett.* **54** 1350.
Schubert, E. F., Tu, C. W., Kopf, R. F., Kuo, J. M., and Lunardi, L. M. (1989f). *Appl. Phys. Lett.* **54**, 2592.
Schubert, E. F., van der Ziel, J. P., Cunningham, J. E., and Harris, T. D. (1989g). *Appl. Phys. Lett.* **55**, 757.
Schubert, E. F., Kuo, J. M., Kopf, R. F., Luftman, H. S., and Garbinski, P.A. (1990a). *Appl. Phys. Lett.* **57**, 497 (1990a).
Schubert, E. F., Kuo, J. M., Kopf, R. F., Luftman, H. S., Hopkins, L. C., and Sauer, N. J. (1990b). *J. Appl. Phys.*, **67**, 1969.
Shockley, W. (1961). *Solid State Electronic* **2**, 35.
Shklovskii, B. I., and Efros, A. L. (1984). In *Electronic Properties of Doped Semiconductors*, p. 303. Springer, New York.
Solomon, P., and Morkoc, H. (1984). *IEEE Trans. on Electron Devices* **ED-31**, 1015.
Spicer, W. E., Gregory, P. E., Chye, P. W., Babalola, I. A., and Sukegawa, T. (1975). *Appl. Phys. Lett.* **27**, 617.
Spicer, W. E., Lindau, I., Skeath, P., Su, C. Y., and Chye, P. W. (1980). *Phys. Rev. Lett.* **44**, 420.
Spitzer, W. G., Kahan, A., and Bouthillette, J. (1969). *J. Appl. Phys.* **40**, 3398.
Stern, F. (1974). *J. Vac. Sci. Technol.* **11**, 962.
Stern, F. (1983). *Appl. Phys. Lett.* **43**, 974.
Stern, F., and Das Sarma, S. (1984). *Phys. Rev. B* **30**, 840.
Störmer, H. L. (1983). *Surf. Sci.* **132**, 519.
Störmer, H. L., Dingle, R., Gossard, A. C., Wiegmann, W., and Logan, R. A. (1978). *Inst. Phys. Conf. Ser.* **43**, 557.
Sze, S. M. (1980). *Physics of Semiconductor Devices*, 2nd ed. Wiley, New York.
Tatsuta, S., Inata, T., Okamura, S., Muto, S., Hiyamizu, S., and Umebu, I. (1985). *Mat. Res. Soc. Sympl Proc.* **37**, 23.
Tejwani, M. J., Kanber, H., Paine, B. M., and Whelan, J. M. (1988). *Appl. Phys. Lett.* **53**, 2411.
Theis, T. N., Mooney, P. M., and Wright, S. L. (1988). *Phys. Rev. Lett.* **60**, 361.
Tsang, W. T. (1987). *J. Cryst. Growth* **81**, 261.
Tsang, W. T. (1989). In *VLSI Electronics: Microstructure Science*, **21**, p. 255. Academic Press, New York.
Tsang, W. T., and Swaminathan, V. (1981). *Appl. Phys. Lett.* **39**, 486.
Tu, C. W., Jones, W. L., Kopf, R. F., Urbanek, L. D., and Pei, Shin-Shem (1986). *IEEE Electron Device Letters*, **EDL-7**, 552.
Tuck, B. (1988). *Atomic Diffusion in III–V Semiconductors*. Adam Hilger, Bristol.
Ullrich, B., Schubert, E. F., Stark, J. B., and Cunningham, J. E. (1988). *Appl. Phys. A* **47**, 123.
Ullrich, B., Zhang, C., Schubert, E. F., Cunningham, J. E., and van Klitzing, K. (1989). *Phys. Rev. B* **39**, 3776.
Urbach, F. (1953). *Phys. Rev.* **92**, 1324.
van Gorkum, A. A., Nakagawa, K., and Shiraki, Y. (1987). *Jpn. J. Appl. Phys. Lett.* **26**, L1933.
Vinter, B. (1984). *Appl. Phys. Lett.* **44**, 307.

Vojak, B. A., Zajac, G. W., Chambers, F. A., Meese, J. M., Chumbley, P. E., Kaliski, R. W., Holonyak, N., Jr., and Nam, D. W. (1986). *Appl. Phys. Lett.* **48**, 251.
Webb, C. (1989). *Appl. Phys. Lett.* **54**, 2091.
Wood, C. E. C., and Joyce, B. A. (1978). *J. Appl. Phys.* **49**, 4854.
Wood, C. E. C., Judaprawira, S., and Eastman, L. F. (1979). *IEDM Tech. Dig.*, 388.
Wood, C. E. C., Metze, G. M., Berry, J. D., and Eastman, L. F. (1980). *J. Appl. Phys.* **51**, 383.
Wood, C. E. C., Eastman, L. F., Board, K., Singer, K., and Malik, R. (1982). *Electron. Lett.* **18**, 676.
Zrenner, A., Koch, F., and Ploog, K. (1988a). In *GaAs and Related Compounds 1987* (H. Rupprecht, ed.), IOP Publishing Ltd, Bristol.
Zrenner, A., Koch, F., and Ploog, K. (1988b). *Surface Science* **196**, 671.

CHAPTER 2

Wide Graded Potential Wells

Arthur C. Gossard, Mani Sundaram, and Peter F. Hopkins

DEPARTMENT OF ELECTRICAL AND COMPUTER ENGINEERING,
AND MATERIALS DEPARTMENT, UNIVERSITY OF CALIFORNIA,
SANTA BARBARA

I. INTRODUCTION 153
II. EPITAXIAL GROWTH OF GRADED WELLS 158
   1. Graded Well Growth: General 158
   2. MBE Graded Well Growth Using Digital Alloys 161
   3. Analog versus Digital Alloy 166
III. ELECTRONIC STRUCTURE AND OPTICAL PROPERTIES OF UNDOPED WELLS . . . 167
IV. MODULATION-DOPED WIDE PARABOLIC WELLS 176
   4. Electron Density Profile: Theory 176
   5. Capacitance Measurements 181
V. ELECTRON TRANSPORT 184
   6. Electrical Conductivity 184
   7. Magnetotransport 189
VI. ELECTRON EXCITATIONS 195
   8. Far Infrared Absorption 195
   9. Interband Optical Transitions 204
VII. SUPERLATTICES IN PARABOLIC WELLS 207
   10. Transport 207
   11. Magnetoplasmon Dispersion 211
VIII. CONCLUSIONS 212
   ACKNOWLEDGMENTS 214
   REFERENCES 214

1. Introduction

With the advent of atomic layer-by-layer epitaxial crystal growth, the study and use of compositionally graded semiconductor structures has developed rapidly. The electrical and optical properties of semiconductors can be strongly altered by tailoring their composition profile. The optical and electronic properties of graded structures, structures whose composition profile is changed gradually, are especially sensitive to the profile and give access to a broad new range of phenomena (in the THz regime). Thus, they extend the range of capability of compositionally abrupt structures such as superlattices and heterojunctions. In this chapter we consider the growth and the physics associated with graded semiconductor structures.

A number of applications of graded structures already exist. These include devices in which compositional grading can produce an internal electric field, for example, graded base transistors (where a graded energy gap accelerates electrons across a base layer) and graded index lasers (where electrons and light are concentrated into an active region by the guiding effect of the graded layers). Graded layers can also serve the function of providing buffer layers between lattice mismatched materials. Spatially asymmetric graded heterostructure devices can be used for electrical current rectification or second harmonic generation. A topic of special significance that we will treat here is the parabolically graded quantum well structure in which composition varies quadratically with distance along the growth direction, which in turn can lead to electron gases with wider spatial extent than in other modulation-doped structures (Halperni, 1987).

The potential profile and energy levels for a square well and a parabolic well are compared in Fig. 1. In the limit of an infinitely high, square, one-dimensional potential profile, the energy levels E_n of an electron with effective mass m^*, in a square well of width L, are quantized with energies $E_n = n^2\hbar^2\pi^2/2m^*L^2$, where $n = 1, 2, 3, \ldots$. The energy levels of an infinitely

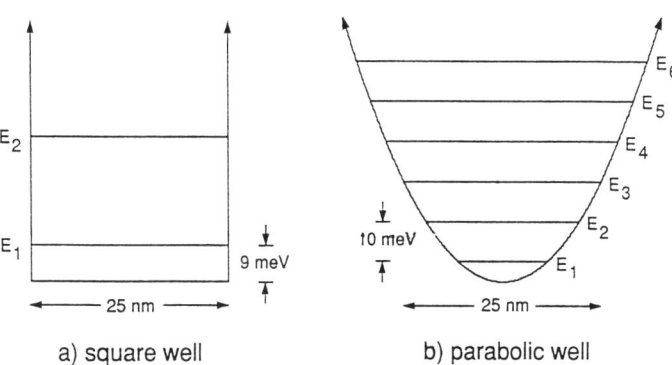

FIG. 1. Potential profiles and energy levels in (a) an infinite square potential well, and (b) a parabolic potential well for electrons with effective mass $m^* = 0.067\, m_0$.

2. WIDE GRADED POTENTIAL WELLS

wide, infinitely high parabolic potential with quadratic potential $V(z) = \frac{1}{2}Kz^2$ are uniformly spaced and given by

$$E_n = \left(n - \frac{1}{2}\right)\hbar\sqrt{\frac{1}{m^*}\frac{d^2V}{dz^2}} = \left(n - \frac{1}{2}\right)\hbar\sqrt{\frac{K}{m^*}} \quad (1)$$

The square well potential with finite height provides a fairly adequate description of electron and hole energy levels observed in many heterostructures, especially in lightly doped or undoped GaAs quantum wells with $Al_xGa_{1-x}As$ barriers (Bastard 1988). Electrons in these structures can be described with effective masses and so-called envelope wave functions. The potentials for the electrons are determined by the composition of the material. If, for example, the semiconductor energy gap at point z in a graded semiconductor is $E_g(z)$ and varies linearly with Al composition $x(z)$, the energy gap profile becomes

$$E_g(z) = E_g(z = 0) + ax(z) \quad (2)$$

If the energies of the top of the valence band and the bottom of the conduction band $E_i(z)$ also vary approximately linearly with composition, as is often the case in undoped semiconductors, the band edge profile becomes

$$E_i(z) = E_i(z = 0) + b_i x(z) \quad (3)$$

These tailored band edge profiles are used to realize potential wells of the corresponding shapes for electrons in the conduction band or holes in the valence band. Deviations of the energy levels in these wells from the ideal case of the infinitely deep square and parabolic potential wells come from band nonparabolicity that causes the effective mass m^* to be energy dependent, from interactions with other electrons or holes in the well, from penetration of carrier wave functions into the surrounding barrier layers, from nonlinearity in the relationship between band edge position and composition, and from atomic band structure effects.

In graded quantum wells the energy levels depend sensitively on the shape of the potential well, putting special demands on the precision with which the structures are grown. When grown by molecular beam epitaxy (MBE), two approaches have been used for the variation in molecular beam flux needed for formation of the profiles. One approach is to vary the flux by changing a molecular beam source oven temperature, and the other approach is to pulse the molecular beam flux by shuttering the beam generated from an oven at constant temperature. We refer to the continuously varying alloy formed by the former technique as an *analog alloy* and the short-period superlattice formed by the shuttering method as a *digital alloy*. After a discussion of the growth of wide graded wells and the digital alloy growth technique in Section II, the energy levels and properties of undoped parabolic wells are discussed in Section III. Among the experimental techniques that have been used to examine the undoped parabolic quantum wells are photoluminescence (Miller *et al.* 1984a), photoluminescence excitation (Miller *et al.* 1984a), and inelastic (Raman) light scattering (Menéndez *et al.* 1987). The energy levels have also been probed electrically by resonant tunneling measurements of electrons tunneling through undoped parabolic quantum wells from adjacent doped layers (Sen *et al.* 1987).

In Section IV, the doping and charge distribution in wide graded wells are discussed. In a modulation-doped parabolic well, the mobility of charge carriers in the well is enhanced by confining doping to the barrier layers outside the well. The doping of parabolic wells can produce structures with wider charge carrier distributions than could be obtained in modulation-doped square quantum wells and with higher carrier mobilities than could be obtained in uniformly doped semiconductors (Halperin 1987). In addition, the modulation-doped parabolic wells avoid the carrier freeze-out that occurs in uniformly lightly doped semiconductors at low temperatures. Each of these features is a prerequisite for study of the fundamental properties of interacting electron systems in three dimensions.

In wells wide enough (Fig. 2(c)) that electron quantization energies are small and in which there are enough electrons that several of these quantized energy levels (subbands) are filled, electrons will distribute themselves in a way that will screen the parabolic potential (Fig. 2(d)). In fact, since the quadratic potential of the bare well, $V = \frac{1}{2}Kz^2$, is of the same form as the electrostatic potential of a slab of uniform positive charge of density $\rho = K\varepsilon$, the classical electron distribution that will exactly screen the bare potential is a slab of uniform negative charge of density $\rho = -K\varepsilon$. When the wells are sufficiently wide that quantization effects are small, the electron gas will approach this classical limit of uniform density.

The electron distributions in the wide graded quantum wells can be probed experimentally by measurement of the capacitance and magnetocapacitance as a function of voltage between the quantum well electron gas and a nearby highly doped layer or metal gate, as also discussed in Section IV. These measurements have shown additionally the effects of unintentional background doping of the wells and interfaces and have been used to verify growth techniques (Sundaram 1991). As discussed in Section IV, they have also been used to probe the charge densities in wells containing intentional perturbations from perfect parabolas. Among the

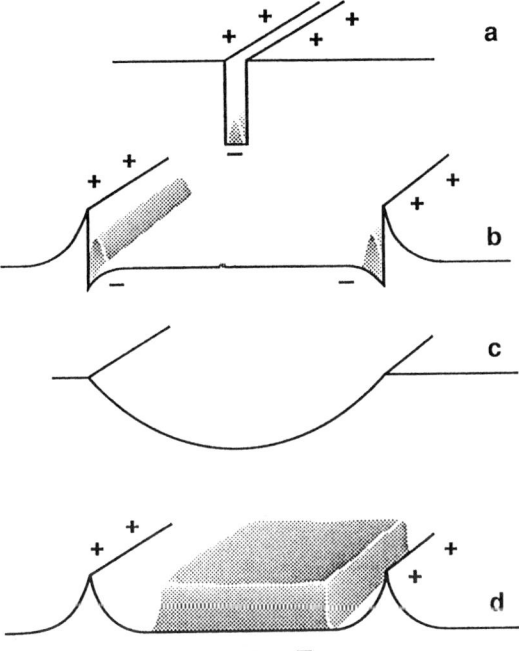

FIG. 2. Potential profile and charge distributions in (a) a narrow square well with doped barriers, (b) a wide square well with doped barriers, (c) an undoped parabolic well, and (d) a parabolic well with doped barriers.

purposely perturbed distributions that have been investigated are square superlattice potentials within the wide wells. These provide a powerful means of creating coupled high-mobility two-dimensional electron gases that have doping impurities neither in the two-dimensional electron gas wells nor in the coupling barriers.

Electron transport phenomena in parabolic wells are quite rich because of the wide variation in the thickness of the electron gas that is possible and because of the possibility of occupation of multiple electron subbands. Controlled variation of the width of the electron gas (such as with a front gate) also permits one to tune the number of occupied subbands and investigate the properties of the electron gas in the transitional regime from 2D (only 1 occupied subband) towards 3D in the same sample. In Section V, we discuss the conductivity and the magnetotransport in doped parabolic wells. The low-temperature mobilities in the modulation-doped wide parabolic wells are several orders of magnitude larger than the mobilities in uniformly doped structures. Several scattering mechanisms may be identified that are either unique to the three-dimensional electron gas or that are stronger than in two-dimensional electron gas structures. The magnetotransport shows quantum Hall effect phenomena that are strongly affected by the multiple subband occupation. The fractional quantum Hall effect is also strongly modified in parabolic quantum wells (Shayegan *et al.* 1990).

The graded potential shapes and the enhanced electron mobilities in the modulation-doped structures provide the basis for an extraordinary new generation of low-loss, solid state electron resonators. The bases for these electron resonators are intraband and plasma excitations of electrons in wide graded wells, which are discussed in Section VI. Remarkably, the electron plasma frequency of a perfectly parabolic well is independent of the number of electrons in the well. Also of interest are the cyclotron resonance and the coupling of the cyclotron and plasma resonances that occur in magnetic fields tilted with respect to the surface normal. Wells that deviate from a perfectly parabolic shape show new electron oscillation modes. We also discuss the nonlinear responses of electrons in wide wells excited with far infrared (FIR) radiation. These nonlinear responses can be strong, especially in the submillimeter wavelength range, where enhanced harmonic generation has been observed from wide asymmetric wells.

II. Epitaxial Growth of Graded Wells

1. Graded Well Growth: General

In principle, the formation of graded wells is straightforward, simply requiring temporal control of the arriving constituents that form the layers to be graded. In practice, however, useful structures require control, purity, and smoothness that are at the limits of growth technology. Current epitaxial growth techniques produce structures in which growth rates, layer thicknesses, and layer compositions can be closely controlled and layer smoothness is at the atomic level. These features are usefully exploited in the growth of quantum wells and superlattices and are also valuable for the growth of compositionally graded samples. An added factor required for growth of graded structures, however, is the ability to continuously vary flux with time.

In MBE, growth rates are determined by arrival rates of thermally emitted beams of atoms and molecules. The arrival rates of the thermally emitted beams are controlled in turn by the temperature and position of the beam sources. In chemical vapor deposition (CVD) techniques, growth rates are determined by flow rates of gases over surfaces and can be quite accurately controlled and reproduced by mass flow controllers and pressure regulators. In MBE, growth can be abruptly started and stopped by shutters that block the molecular beams, whereas in CVD, flow can be abruptly started and stopped in appropriately designed reactors by valves placed a short distance upstream from the surface on which growth occurs. In MBE, there are typically regimes of flux and substrate temperature, where the sticking coefficients of the rate-determining species at the epitaxial surface approach unity; i.e., conditions where nearly all of a particular rate-determining species stick to the surface. In either growth technique, subatomic monolayer coverages can be accurately controlled. In growth with either beam techniques or gas flow techniques, compositions can be changed both gradually or abruptly.

Alloys or mixed compounds of different compounds or elements may be formed in two different ways. The first and most widely used means is codeposition of the constituents of the alloy. The alloy or mixed compound may also be formed by alternate sequences of short depositions of the constituent materials. For example, GaAs and AlAs may be either codeposited to form $Al_xGa_{1-x}As$, or alternately deposited to form a short-period superlattice. If each period of the superlattice consists of m monolayers of AlAs followed by n monolayers of GaAs, it will have an average composition of $(AlAs)_m(GaAs)_n$ or $Al_{m/(m+n)}Ga_{n/(m+n)}As$. We refer to the codeposited alloy as an *analog alloy* and the short-period superlattice as a *digital alloy*. The two types of "alloys" may have significantly different microstructures and may have substantially different physical properties, as discussed later in this chapter. For sufficiently small m and n, though, the properties of the short-period A_mB_n superlattices should approach those of the analog alloy. Besides (Al, Ga)As, other systems of elements for which the above techniques for growth might be used include the (In, Ga, Al)As, (Al, Ga)Sb, and Ga(As, P) systems.

Growths of graded analog alloy structures involve a slow, controlled variation in beam flux or gas pressure that produces a gradual change in composition. The analog alloy approach in MBE requires changing the temperature of at least one molecular beam furnace to produce a highly accurate profile of beam flux intensity versus time (Harbison *et al.* 1987; Shayegan *et al.* 1988). Due to thermal time constants, thermal gradients, and heat capacities of the furnaces, there will be time delays between the set point of the control element of the furnaces and the achievement of the set temperature at the effusing surfaces of the cells; these delays can be partially compensated for by use of iteratively learned control loops for the furnaces (Harbison *et al.* 1987). Passing flux through a zero value presents a special problem because of the finite, although exponentially small, evaporant flux at any reduced temperature. For very small values of aluminum mole fraction $x_{Al} \ll 1$, to get a parabolic variation of x_{Al} in the presence of constant Ga and As fluxes, it is necessary to make flux proportional to the square of the time t; i.e., to make $f(t) \sim at^2$. When flux depends exponentially on temperature T, ($f(T) \sim be^{-c/T}$), this requires that $be^{-c/T} = at^2$ or $T(t) = -c/\ln(at^2/b)$, which has a logarithmic singularity at $t = 0$ and constitutes a problem for very small values of aluminum mole fraction $x_{Al} \ll 1$. Other factors that may affect the graded analog alloy are alloy disorder, alloy clustering, incorporation of unintentional background impurities, and surface roughness. CVD and MOCVD (metal-organic-CVD) growth will be limited by similar factors with the exception that flow rate is the controlled quantity rather than temperature, a feature that permits the flux to be truly ramped to 0 in a finite time (i.e., no singularity exists as in the thermal evaporated case in MBE).

In comparison, digital alloys formed by molecular beam shuttering or gas flow on/off switching, can be precisely controlled and graded purely by shutter or switch timing, assuming that the growth rate during "on" portions of the growth can be maintained at a constant rate and that the shuttering can be precisely timed. One of the first applications to growth of graded alloys by this technique was the growth of $Al_xGa_{1-x}As$ by MBE with arbitrary composition profiles by alternation of GaAs and $Al_xGa_{1-x}As$ simply by pulsing only an aluminum source (Kawabe et al. 1983). The growth rate and timing factor accuracies in the graded digital alloy formation are limited by the precision with which beam fluxes or gas flows are controlled and by the speed and precision of shutter and valve movement. A phenomenon that limits profile accuracy is the flux transient behavior that is observed after a molecular beam shutter is opened. The flux transients are caused principally by changing temperatures and temperature distributions of effusing surfaces after a shutter motion opens the furnace to radiative heat loss. The transients have time constants that are on the order of 10 to 100 seconds and magnitudes of approximately 5% to 10% for typical ovens and shutters. Noninstantaneous shutter motion is another element that specifically affects the composition profile of digital alloys. A shutter takes a finite time to move from closed to open and open to closed positions. As a shutter opens, the cell orifice is uncovered progressively from one side to another during the time of the motion. This will produce a gradient in integrated flux across a wafer surface for the case of a back-and-forth shutter motion. For example, for a 10 millisecond shutter motion time, one side of the cell orifice may be exposed for 20 msec longer than the opposite side of the orifice during one complete opening and closing cycle. Depending on details of the shutter penumbra, this may introduce an effective exposure difference per shutter cycle of order 10 msec between the two opposite sides of a typical 2-inch or 3-inch diameter substrate. For shutters operated with ac actuators, an error in the opening time of as much as half of a 60 Hz period (± 8 msec) can occur. For very short exposures, these factors are of increased relative importance and produce a composition uncertainty $\Delta x = \pm(\Delta t/t)x_{\max}$ for cycle period t and alloy composition $x_{\max}$.

As in the case of the analog alloy, alloy disorder, alloy clustering, surface roughness, and unintentional impurity incorporation are of concern. Since the growth of the alternating layers of superlattices generally produces a cleaner and smoother surface (Petroff *et al.* 1982) than does a random alloy, the digital alloy enjoys an advantage with respect to the analog alloy in surface smoothness and purity. It also has an advantage over the analog technique because almost any graded alloy profile can be approximated digitally. Alloy clustering may be more severe in the digital case, though, because growth by an island or terrace growth mode means that, particularly for submonolayer exposures, an alloy species may occur in monolayer islands or terraces rather than uniformly across the surface (Pfeiffer, private communication).

Calibration of the deposition cycles in digital or analog alloy growth by MBE can be obtained by ion-gauge measurement of the molecular beam fluxes. This measurement has given the most precise comparison of the two growth techniques (Sundaram *et al.* 1991c). With a high-speed nanoammeter, the shape of the individual pulses can be monitored, while by measurement of the time-averaged flux, the overall composition profiles are readily measured. Other kinds of beam flux monitors, for example quartz crystal monitors, quadrupole mass analyzers, and optical emission or absorption beam probes, are also applicable to the measurement of the flux profile. The correspondence of the beam flux measurements to the true composition profile of grown samples assumes that the beam flux signal is linear with flux and that the sticking coefficient of the deposited atoms on the actual sample is independent of flux and composition, in the flux and composition ranges of interest.

Electrical or optical measurements, as discussed later, can also be used to deduce the shape of graded alloy profiles, but they are less direct than the beam flux monitoring method. Profiling can also be accomplished by auger or secondary ion mass spectroscopy (SIMS) techniques (Karraï *et al.* 1990), but these techniques have poorer depth resolution.

2. MBE Graded Well Growth Using Digital Alloys

A typical graded quantum well structure design may involve $Al_xGa_{1-x}As$ alloys in which the Al content varies from $x = 0$ to $x = 0.30$, which covers a substantial part of the direct band gap region of the $Al_xGa_{1-x}As$ system. As illustrated in Fig. 3, a digital alloy can be formed by shuttering only the

a) analog alloy

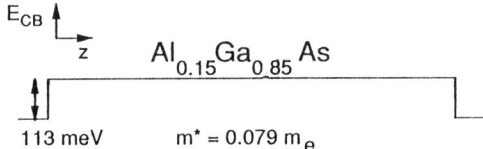

b) digital alloy

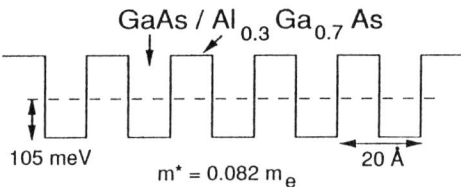

FIG. 3. Composition profile for (a) an $Al_{0.15}Ga_{0.85}As$ analog alloy, and (b) a 2 nm period $Al_{0.3}Ga_{0.7}As/GaAs$ digital alloy with average Al concentration $x = 0.15$.

Al source, which, when open along with the Ga and As sources, produces, say, an $Al_y Ga_{1-y} As$ alloy. Shuttering the Al source creates a superlattice of GaAs and $Al_y Ga_y As$ whose average composition is given by

$$x_{Al} = y \frac{d(Al_y Ga_{1-y} As)}{d(Al_y Ga_{1-y} As) + d(GaAs)} \quad (4)$$

where $d(Al_y Ga_{1-y} As)$ and $d(GaAs)$ are the (varying) layer thicknesses of $Al_y Ga_{1-y} As$ and GaAs in each period of the superlattice.

Grading the composition of the digital alloy is accomplished by varying the thickness of the barrier $Al_y Ga_{1-y} As$ layers in the superlattice. As an example, for a parabolic well of width L, having $x_{Al} = 0$ at the center and $x_{Al} = y$ at the edges, and containing one GaAs layer and one $Al_y Ga_{1-y} As$ layer in each period of a superlattice with M periods, the effective thickness of the $Al_y Ga_{1-y} As$ layer in the Nth period from the well center is $[(2N-1)/M]^2 L/M$, corresponding period being centered at a distance $(N - 0.5)L/M$ from the center of the well. The relative thicknesses of the $Al_y Ga_{1-y} As$ layers thus increase quadratically with distance from the well center while the GaAs layer thicknesses decrease. An example of such a digital alloy is seen in Fig. 4a. Calculated thicknesses of less than one monolayer ($\sim 2.83 \text{ Å}$), which can occur for the $Al_y Ga_{1-y} As$ layers near the center of the parabolic well, correspond to a partial monolayer of coverage.

If the intent is for the digital alloy to mimic an analog alloy, then it is

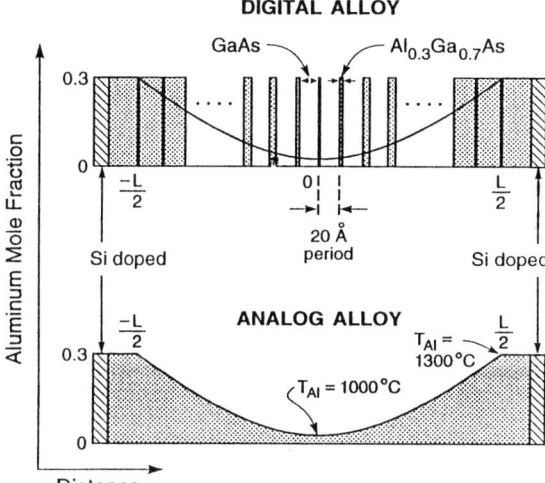

FIG. 4. Composition profiles of $Al_xGa_{1-x}As$ parabolic well produced by (a) the digital alloy technique, and (b) the analog alloy technique. The digital alloy is s superlattice made from GaAs and $Al_{0.3}Ga_{0.7}As$ with a 2 nm period in which the pulse duration of $Al_{0.3}Ga_{0.7}As$ is varied parabolically so that the average Al mole fraction varies quadratically with distance. The graded analog alloy is produced by varying the temperature of the Al molecular beam oven. Carriers are provided by Si doping of the barriers.

important that the periodic potential introduced by the digital alloy not affect the physical properties one wishes to study. In a digital alloy, the electrons (holes) will tend to be concentrated more in the well layers than in the barrier layers, creating a periodic modulation in the charge density profile and causing the energy of the lowest subband to be lower than the conduction band edge (valence band edge) energy in an analog alloy of equivalent average composition. The effective band gap for the digital alloy is therefore somewhat lower than the band gap of the equivalent analog alloy. Similarly, the motion of the carriers perpendicular to the digital alloy layers is impeded by the barriers, leading to a larger effective mass m^* for motion in that direction.

Rayleigh–Schrödinger perturbation theory can be used (Hopkins et al. 1993) to estimate the change in the effective mass, energy levels, and charge distribution for carriers in a constant potential with a sinusoidal perturbation $2U(G)\cos(Gz)$. The perturbed effective mass m^* in the growth direction and at the zone center $[E(k = 0)]$ is

$$m^* \sim m\left[1 + \frac{1}{2}\left(\frac{U(G)}{E_0(G/2)}\right)^2\right] \qquad (5)$$

where $E_0(G/2) = \hbar^2(G/2)^2/2m$ is the unperturbed energy for carriers with bulk effective mass m (assumed here to be the same in both the barrier and well bulk materials) at wave vector $G/2 = \pi/d$, where d is the period of the perturbation; note that the correction term is proportional to d^4. The Fourier transform power spectra for the conduction band edge potential of all digital alloys show a peak at the frequency $f = 1/d$ corresponding to the superlattice period and smaller amplitude peaks at the higher harmonics $f = n/d$, where $n = 2, 3, 4, \ldots$ Because of the strong dependence on d, we can approximate the digital alloy superlattice potential as a single wave vector perturbation; the correction in (5) due to the nth harmonic is smaller by at least a factor of n^{-4}.

For a $GaAs/Al_{0.3}Ga_{0.7}As$ digital alloy with $d = 2\,nm$, $U(G) = \Delta E_c/\pi = 83.7\,meV$ and $m = 0.0670\,m_0$, we estimate $m^* \sim 0.0671\,m_0$. The energy spectrum of the conduction band $E(k)$ is shifted down into the band gap by 2.5 meV. The first-order change in the ground state wave function creates 2 nm period ($= d$) ripples in the charge density $n(z)$ of size $(n_{peak} - n_{avg})/n_{avg} \sim 6\%$ due to localization of the carriers in the GaAs wells (in the region of density in the work discussed here). The approximation given by (5) breaks down sharply for superlattice periods d above $\sim 8\,nm$. Because of the heavier effective mass for holes, the effects of the superlattice potential are more severe for holes than for electrons.

For an electron gas in a parabolic well, the potential for a single electron is essentially a constant plus a quasi-periodic modulation of period d. Thus, to a first approximation, this perturbation approach should give correct values for the change in effective mass, charge density, and energy levels due to construction of the graded well with the digital alloy technique.

The energy levels for an empty digital parabolic well has been calculated in the effective mass envelope function approximation using a resonant tunneling method. The five lowest energy levels for such a calculation for an undoped 200 nm $Al_xGa_{1-x}As$ parabolic well in which the composition is graded quadratically from $x_{Al} = 0.00$ to $x_{Al} = 0.20$ and then stepped to $x_{Al} = 0.30$ are shown in Fig. 5. For comparison, the dashed lines show energy levels for an infinite parabolic well with the same curvature. For a 2 nm period, each of the first five energy levels lies within $\sim 5\%$ of the energy level values for the infinite smooth parabola, with the largest deviation being for the highest lying energy levels. Quantitative calculations of the change in electron density profile and energy levels for filled wide parabolic wells when a superlattice potential is inserted have been done by Brey *et al.* (1990b).

Changes in the effective mass in (5) (as well as the charge density profile, energy levels, etc.) due to the digital alloy can all be reduced by shortening the period d. However, as the period of the superlattice is made finer, the

2. WIDE GRADED POTENTIAL WELLS

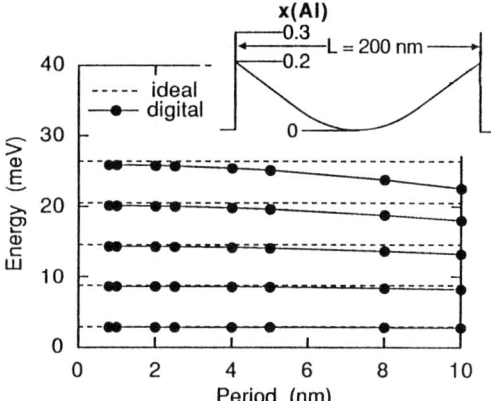

FIG. 5. Electron energy levels in an empty 200 nm parabolic well (inset) calculated for a digital alloy formation of the well (filled circles) as a function of the period of the digital alloy superlattice. Dashed lines are energy eigenvalues for ideal infinite parabolic well.

shuttering times of the Al source can become small, requiring precise calibration of the Al shutter opening and closing times. For a Varian Gen II MBE system with pneumatic shutter controls, errors in the shuttering times can be reduced to about 10 msec, putting a realistic lower bound on minimum shuttering times of ~ 100 msec. For typical graded structures shuttering times can easily be kept above this lower bound; however, for some graded structures this lower bound can be restrictive. For example, in $Al_xGa_{1-x}As$ parabolic wells, higher carrier mobilities are obtained if the amount of Al in the well is kept to a minimum; shutter times of typical high-mobility parabolic wells often approach 100 msec near the minima of the $Al_xGa_{1-x}As$ parabolas. Shutter times less than this can be avoided by lowering the growth rates, increasing the minimum average Al composition at the center of the parabola, using an additional Al source set at a lower growth rate, and increasing the period d.

Despite the strict design rules given in this section, one can come up with a satisfactory design for a digital alloy parabolic well; at typical GaAs and $Al_{0.3}Ga_{0.7}As$ growth rates of 1.0 μm/hr and 1.43 μm/hr respectively, a 200 nm wide parabolic well grown with a 2 nm period $GaAs/Al_{0.3}Ga_{0.7}As$ digital alloy, a minimum Al concentration $x = 0.006$ at the center of the well, and a maximum Al concentration of $x = 0.294$ at the edges is one example of such a design. According to (5), this design should be adequate for giving electrical and near-band-edge optical properties in close approximation to a smooth analog alloy parabolic well, without exceeding the performance limits of present shuttering systems.

3. ANALOG VERSUS DIGITAL ALLOY

In selecting a fabrication technique for graded potential wells, it is important to determine the accuracy of the potential profile that is required and the accuracy with which the desired profile can be formed. The results of a comparison between the analog and digital alloy growth techniques obtained by ion gauge measurement of Al beam fluxes for growth of 200 nm parabolic quantum wells are shown in Fig. 6. Figure 6(a) compares the average Al concentration versus distance in the growth direction for an analog-alloy and a 2 nm period digital-alloy parabolic well. The results are compared with the theoretical perfect parabola design profile. The Al flux

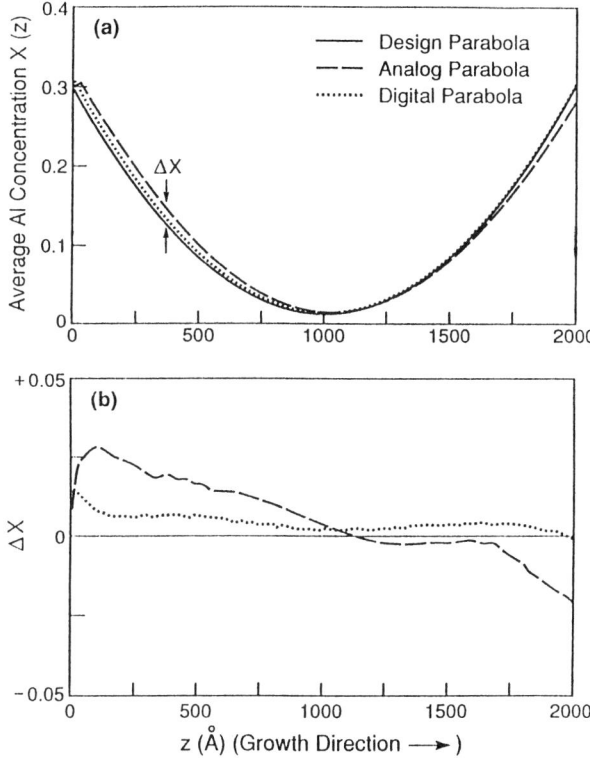

FIG. 6. (a) Comparison of ideal (design) 200 nm parabolic well composition profile with flux profiles measured by molecular beam flux monitor for analog-alloy parabola (dashed curve) and digital-alloy parabola (dotted curve). (b) Difference Δx in average aluminum concentration, x_{Al}, between analog alloy growth and perfect parabola (dashed curve) and between digital alloy growth and perfect parabola (dotted curve).

rate was set to give a AlAs growth rate of 0.32 μm/hr, and a GaAs growth rate of 0.75 μm/hr was assumed to convert Al flux to Al concentration versus distance. The greater phase lag relative to the design profile is seen in the analog-alloy version. The phase lag results from the time delays in flux changes as temperature settings are ramped. The deviations in Al composition, Δx, from the design profile are shown in Fig. 6(b). Improvement in the profile accuracy can be obtained by slowing the growth rate, and as mentioned previously, correcting for the thermal time constants. Thus, Harbison et al. (1987) made a series of growths in which the temperature versus time profiles were progressively modified to more accurately match the design profile. They ultimately obtained agreement with the design profile of a 50 nm well at the level of $\Delta x \leqslant 0.03$, as measured by RHEED oscillations. Karraï et al. (1990) have reported SIMS profiles of the Al composition upon sputter etching 150 nm analog-alloy parabolic wells and observed profiles that were parabolic to within the ~ 10 nm depth resolution of the SIMS–sputtering technique.

In comparing experimental profiles with design profiles for parabolic wells, it is important to compare the curvature of experimental and design profiles. The objective of modulation-doped parabolic wells is to produce a material that emulates a constant positive charge density, ρ, given by $\rho = \varepsilon d^2 V_{cb}/dz^2$ where $V_{cb}(z)$, the conduction band-edge energy for the undoped case, is proportional to the Al alloy content x_{Al}. Thus, a constant pseudo-charge background density requires constant $d^2 x_{Al}/dz^2$. Measurement of this quantity gives a sensitive test of the degree to which a constant pseudo-charge density is obtained. Thus, deviations of the experimental composition from the design composition are generally less important than deviations of the curvature from a constant value. Flat spots or nonquadratic terms in the well potential can produce large changes in the well charge density (Brey et al. 1990a). The curvatures versus depth for the data of Fig. 6, after averaging over ~ 25 nm, are shown in Fig. 7 and show curvatures that are constant to within $\pm 10\%$ over the central 150 nm for the digital well and $\pm 20\%$ for the analog well.

III. Electronic Structure and Optical Properties of Undoped Wells

In this section, we consider the quantized energy levels that occur in various well shapes and discuss the measurement of quantum phenomena based on these states. Exact solutions exist for a number of potential well shapes; for example, infinitely high square, sawtooth, parabolic, and hydrogenic shapes. For an infinitely high square well of width L, the energy levels

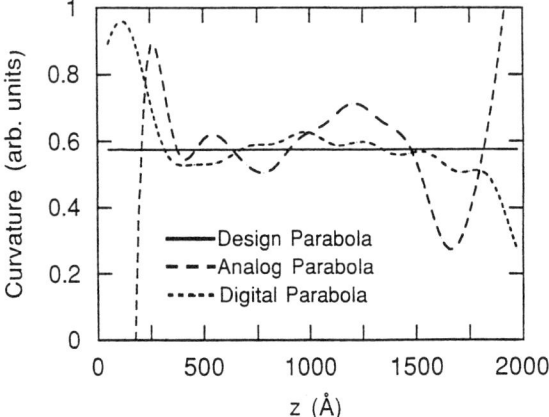

FIG. 7. Comparison of curvatures in effective composition profiles measured by a flux gauge monitor for analog alloy growth (long dashes) and digital alloy growth (short dashes) for 200 nm parabolic well. The averaging length is 250 Å.

are given by $E_n = n^2 h^2 / 8m^* L^2$. For an infinitely high parabolic well with $V(z) = Kz^2/2$, the energies are

$$E_n = \left(n - \frac{1}{2}\right) \hbar \sqrt{\frac{1}{m^*} \frac{d^2 V}{dz^2}} = \left(n - \frac{1}{2}\right) \hbar \sqrt{\frac{K}{m^*}}.$$

An infinitely high sawtooth well with $V(z) = eFz (z \geq 0)$ and $V(z) = \infty (z \leq 0)$, often used to approximate the confining potential of the 2D electron system, has levels $E_n = -A_n (e^2 F^2 \hbar^2 / 2m^*)^{1/3}$ for an electron of mass m^*, where A_n is the nth root of the Airy function (see, for example, Weisbuch and Vinter 1991). And an infinite potential of form $V = K|z^{2/3}|$ has energy levels E_n proportional to $(n - 1/2)^{1/2}$ (Sputz and Gossard 1988).

If the carriers confined in one-dimensional wells are free to move parallel to the plane with the remaining 2 degrees of freedom, then each of these energies includes, in addition, a kinetic energy term $\hbar^2 k^2 / 2m$ for motion parallel to the plane, where k and m are, respectively, the wave vector and effective mass in the plane.

In the case of finite barrier heights or in the case where the effective mass is dependent on energy, the levels cannot generally be determined from simple analytical expressions. However, they may be calculated by numerical methods. A particularly simple technique is the resonant tunneling calculation (Miller et al. 1984a), in which the transmission coefficients of electrons impinging perpendicular to the barrier layers is calculated as a function of the energy of the electrons. At incident electron energies close to the energy levels of the quantum wells with infinitely thick barriers, there are peaks in the electron transmission. For an arbitrary one-dimensional potential shape and arbitrary dependence of m^* on energy, the one-dimensional problem may be approximated by a series of thin slabs, and the Schrödinger equation is rapidly solved for transmission through each interface by use of appropriate boundary conditions (Miller et al. 1984a). Alternatively, self-consistent finite-element Schrödinger equation solving programs may be used for determination of the energy levels and charge density profiles (see, for example, Rimberg and Westervelt 1989).

More detailed calculations for parabolic potential wells have been made by Chu-liang and Qing (1988), who calculated energy levels, wave functions, and exciton binding energies; by Chuang and Ahn (1989), who investigated interband and intersubband optical transitions of parabolic potential wells in an applied field; by Ishikawa et al. (1990), who calculated the shift in the fundamental absorption edge in an electric field; by Juang (1991), who calculated electric-field-induced tunneling and energy level shifts; and by Herling and Rustgi (1991), who calculated the electric field dependence of energy levels, wave functions, electron–hole overlap integrals, and oscillator strengths.

For semiconductor quantum well materials in general, including graded wells, there are confining potentials for both conduction band and valence band electrons. For $Al_xGa_{1-x}As$ structures in the direct-gap region where $x_{Al} < \sim 0.45$, the sign of the conduction band and valence band potentials is such that conduction band electrons and valence band holes are both bound in the more Ga-rich regions of the structures. The offset in conduction band edge minima ΔE_{cb} at a heterojunction with energy gap discontinuity ΔE_g is approximately $\Delta E_{cb} = +Q_e \Delta E_g$, where Q_e is the band offset ratio for conduction electrons and is often approximated as independent of x. The offset in the top of the valence band ΔE_{vb} is likewise given by $\Delta E_{vb} = +Q_h \Delta E_g$. Then $\Delta E_{cb} + \Delta E_{vb} = (Q_e + Q_h)\Delta E_g$, requiring $Q_e + Q_h = 1$. This situation is illustrated in Fig. 8 for square and parabolic wells for the case in which $Q_e = 0.7$ and $Q_h = 0.3$, the approximate values used for $Al_xGa_{1-x}As$. Electron and heavy hole levels are schematically shown.

Interband optical transitions occur between states whose envelope functions (Bastard 1988; Weisbuch and Vinter 1991) overlap. For symmetric wells, this requires that the states have the same parity. In an infinite square well, only states with equal quantum numbers n overlap, and the optical selection rule is $\Delta n = 0$ for either absorption or emission of radiation. In square wells of finite height, on the other hand, states of different n penetrate the barriers differently and overlap can occur even between states with

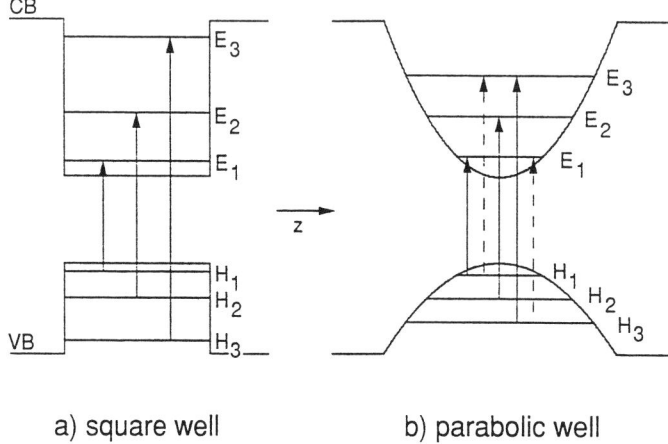

FIG. 8. Electron and hole energy levels and allowed interband transitions in infinite (a) square and (b) parabolic quantum wells, composed of semiconductor materials with direct straddling band gaps. In parabolic wells, additional transitions between electron and hole states with unequal quantum numbers are allowed and are illustrated by the dashed lines.

$\Delta n \neq 0$. These $\Delta n \neq 0$ transitions are particularly strongly allowed in graded wells, where there is more barrier penetration than in square wells.

Examples of the photoluminescence emission (PL) and photoluminescence excitation (PLE) spectra for a sample comprising a stack of 10 $Al_xGa_{1-x}As$ parabolic wells of width 51 ± 3.5 nm each are shown in Fig. 9. The wells were grown by the digital alloy MBE technique with a 2 nm period and a maximum Al content $x_{Al} = 0.30 \pm 0.06$ (Miller et al. 1984a). The PL emission spectra at $T = 5K$ consists of a single peak at 1.53 eV, whereas the PLE spectrum contains an extensive sequence of peaks with roughly equal peak spacing. Above the peaks are designated the absorption transitions that, when excited, lead to maxima in the emitted PL. E_{1h} refers to the excitation of an electron from the $n = 1$ heavy hole valence band state to the $n = 1$ electron state. E_{1l} refers to excitation from the $n = 1$ light hole state to the $n = 1$ electron state. E_{13h} refers to excitation from the $n = 3$ heavy hole state to the $n = 1$ electron state, etc. The uniform spacing between the observed transitions within each series (e.g., E_{1h}, E_{2h}, E_{3h},...; E_{13h}, E_{35h},... etc.) is in agreement with expectations based on the uniform spacing of energy levels of the electrons, heavy holes, and light holes, in their respective harmonic potentials. The numbers below the curves refer to the intensities calculated for the transitions in the envelope function approximation.

The observation of both the heavy hole and the light hole $\Delta n = 2$

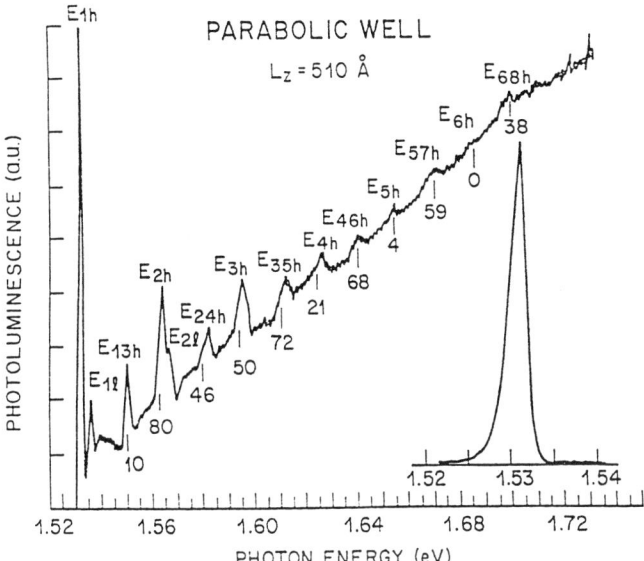

FIG. 9. Photoluminescence spectrum (inset) and photoluminescence excitation spectrum for 1.531 eV emission at $T = 5K$ for sample containing 10 $Al_xGa_{1-x}As$ quantum wells of width 51 nm. Exciton transition peaks are labeled. Theoretically calculated strengths are indicated relative to 100 for the E_{1h} peak (Miller et al. 1984a).

transitions is very helpful in establishing the spacing between energy levels within the conduction band; two transitions starting from a single hole level allow a direct measurement of the spacing between the electron energies for the final electron states. (In square wells, far fewer $\Delta n \neq 0$ transitions are observed, so determination of spacing between electron states from interband optical transitions is more difficult). Fitting initial observations on the parabolic wells to the conduction band to valence band offset ratio showed, in fact, that the previously assumed band offsets of $Q_e/Q_h = 0.85/0.15$ for the GaAs/Al_xGa_{1-x}As system were substantially too high (Miller et al. 1984b). The (In, Ga, Al)As system was also studied by PLE spectroscopy of parabolic quantum wells (Sandhu et al. 1987) and allowed measurement of the band offset ratio in this system in the composition range that lattice matches InP.

It is also possible to observe transitions between conduction band quantum states by Raman light-scattering experiments in which a photon is inelastically scattered by means of exciting an electron from an occupied conduction band state to a higher lying unoccupied state. Such measurements were made on a number of samples in which interband PL and PLE had been observed (Menéndez et al. 1987). The samples were not intentionally doped; optical pumping of the sample was done to generate enough electron–hole pairs to permit Raman scattering observations of the excitations of the conduction electrons. Examples of Raman spectra are shown in Fig. 10. Like the PLE measurements, the Raman measurements also showed a spacing between electron levels that was smaller than predicted from the previously accepted band offset ratios. A puzzling feature of the light-scattering results, however, is that the intersubband spacings determined from light scattering are larger than the intersubband spacings obtained from PL and PLE measurements. The source of this discrepancy is still not clear.

In asymmetric wells, the wave functions are no longer odd or even, and the parity-imposed rule that Δn be even for interband transitions no longer holds. A particularly simple asymmetric well is the half parabola in which one side of the well is abrupt and the other side is parabolically graded, as illustrated in Fig. 11. The energy levels for an infinite half parabolic well are just the $n = 2, 4, 6, 8, \ldots$ levels of the full parabolic well of the same curvature, corresponding to the wave functions of the full parabola having nodes at the center of the well. Such structures have been grown and studied by photoluminescence and PLE. As expected, transitions with $\Delta n = 0, \pm 1$, and ± 2 were observed (Miller et al. 1985).

The energy states in potential wells can also be observed in resonant tunneling measurements in which the wells are separated from the surrounding doped contact layers by undoped barriers. The situation for a parabolic $Al_xGa_{1-x}As$ well enclosed between n-type layers of GaAs is illustrated in Fig. 12. Under applied voltage between the doped GaAs layers, the conduction electrons on the negatively biased side of the junction become equal in energy to the quantum state energy levels in the potential well when the applied voltage is approximaely twice the energy of the corresponding bound states, measured with respect to the center of the potential well. More accurate calculations of the energy spacings include band bending from accumulation and depletion at the barrier edges in the contacting layers, increase in charge density in the well due to tunneling current, shifts in the Fermi level due to band filling in the contact layers, and shifts in the position and energy of the bottom of the parabolic well caused by the field. The effect of an applied electric field on the shape of a parabolic well is to maintain the curvature of the well but to shift the center of the well in energy and position. A well with potential $V(z) = \frac{1}{2}Kz^2$, upon application of an external field E, becomes $V(z) = \frac{1}{2}Kz^2 - Ez = \frac{1}{2}K(z - E/K)^2 - E^2/2K$, which is a parabola with its center shifted in the direction of the field by E/K and its bottom lowered by $E^2/2K$.

The experimental low-temperature resonant-tunneling current curves and

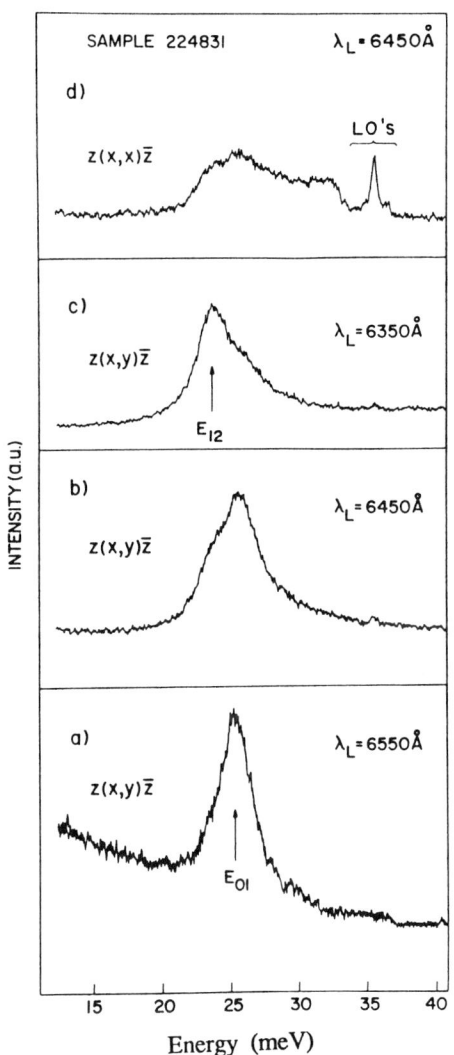

FIG. 10. Inelastic resonant (Raman) light scattering spectra from electronic excitations in optically pumped 51 nm undoped parabolic AlGaAs well: (a), (b), and (c) show spectra for perpendicular incident and scattered light polarizations, (d) shows a spectrum for parallel incident and scattered light polarizations in which peaks are shifted by electric fields due to charge density fluctuations. The E_{01} peak in spectrum (a) is assigned to the transition between lowest and first excited conduction electron states in the parabolic well. With increasing incident photon energy, spectra (b) and (c) show a shoulder and eventually a new peak that is assigned to the E_{12} transition between the first and second excited states of the electrons in the well. (Menéndez et al. 1987). (Reprinted from Menéndez et al. Solid State Communications, Copyright (1987), 61 602, with kind permission from Pergamon Press Ltd, Headington Hill Hall, Oxford OX3 OBW, UK).

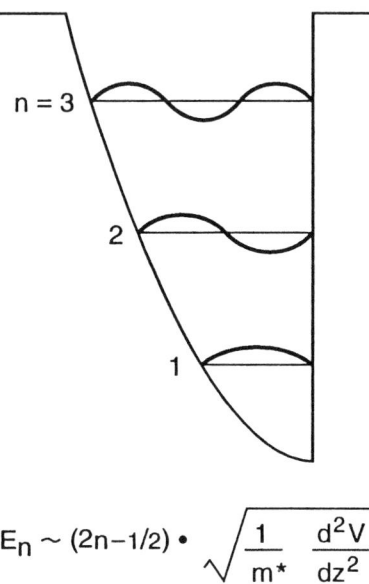

$$E_n \sim (2n-1/2) \cdot \sqrt{\frac{1}{m^*} \frac{d^2V}{dz^2}}$$

FIG. 11. Potential profile and electron energy levels for half parabolic potential well. Energy levels for infinite well are

$$E_n = \left(2n - \frac{1}{2}\right)\hbar \sqrt{\frac{1}{m^*} \frac{d^2V}{dz^2}}.$$

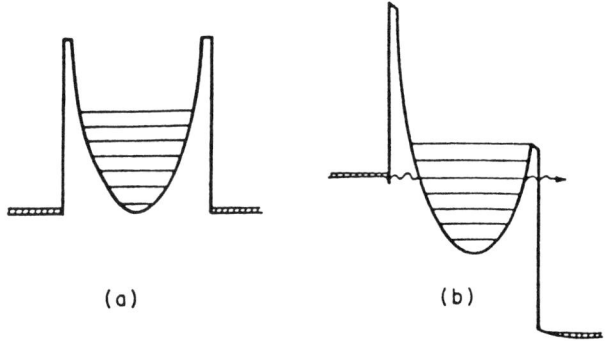

FIG. 12. Conduction band potential and energy levels for semiconductor tunneling structure containing a parabolic well between doped n-type layers. Zero-bias configuration is shown in (a). Under applied bias, resonant tunneling occurs when the energy of the electrons in the external layer coincides with the energy levels of the parabolic well, as shown in (b).

differential conductance as a function of applied voltage for a 43.2 nm undoped parabolic well sandwiched between two 3.5 nm AlAs undoped barriers are shown in Fig. 13 (Sen et al. 1987). The well was graded from $x_{Al} = 0.3$ at the edges to $x_{Al} = 0$ at the center, using a GaAs/Al$_{0.325}$Ga$_{0.675}$As digital alloy superlattice. The measurements were made on 50 μm diameter mesas and illustrate a uniform spacing between conductance peaks, as expected from the uniformly spaced parabolic well energy levels. The deduced spacing between energy levels, after correcting for depletion and accumulation effects, is ~31 meV, which is in agreement with the spacing that would occur for a conduction band offset $Q_c = 0.60$. The increased spacing at higher biases is due to the additional confinement resulting from the rectangular AlAs barriers sandwiching the parabolic well. The differences in shape between the curves for positive and negative

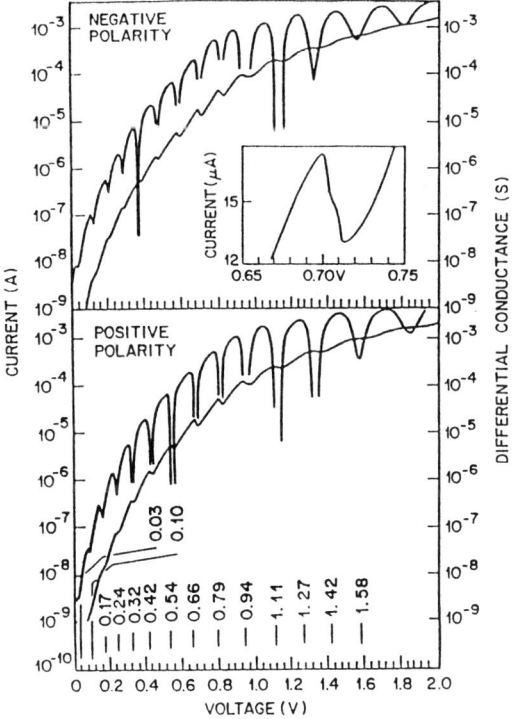

FIG. 13. Resonant tunneling characteristics of a 43 nm parabolic Al$_x$Ga$_{1-x}$As quantum well bounded by 3.5 nm AlAs undoped barriers. Temperature is 7.1 K. Both current and differential conductance are plotted on log scales. The inset shows the eighth resonance in current on a linear scale (Sen et al. 1987).

polarity are not understood. During the resonant tunneling process, electrons occupy states within the wells and have the possibility both of elastic resonant tunneling and of energy relaxation within the well before tunneling out. Capacitance measurements of the storage of electrons in the wells during resonant tunneling have been made as a function of bias for parabolic wells (Schubert et al. 1990). From the bias dependence of the density of electrons in the well, it is concluded that in wide parabolic wells, electrons tunneling into excited states of the well relax to the lowest subband before tunneling out. Clear maxima also occur in the capacitance near the same voltage bias as the resonant tunneling current peaks.

IV. Modulation-Doped Wide Parabolic Wells

4. Electron Density Profile: Theory

When dopant atoms are incorporated into barrier layers near the edges of potential wells, electrons transfer from the dopant atoms (if donors) into the wells, leaving positively charged donor atoms in the barrier and negatively charged electrons in the wells. By Poisson's equation ($d^2V/dz^2 = \rho/\varepsilon$), these charges contribute terms to the potentials within the positively and negatively charged regions, producing positive and negative curvatures. A solution for the charge distributions requires solving both Poisson's equation for the charged regions and Schrödinger's equation for the electrons in the wells, as well as knowledge of the shape of the initial potential well and barrier in the absence of space charges.

The simplest approach for calculating the electron density profile for a wide potential well is a classical treatment that completely ignores quantum mechanical effects of the electron wave functions and energies. In this classical limit, the electrons falling into the well exactly screen out variations in the band edge potential V_{cb} caused by the composition profile as shown in Fig. 14. The screening charge density in this case will be $\varepsilon d^2V_{cb}/dz^2$, which is proportional to $\varepsilon d^2x_{Al}/dz^2$ for an $Al_xGa_{1-x}As$ well in which V_{cb} is proportional to x_{Al}. The bare-well band curvature produced by the composition profile plays the role of a fixed background charge density. For an underfilled parabolic well with bare potential $V_{cb}(z) = Kz^2/2$, the screened potential is constant over a width $w_e = (dQ/dA)/dQ/dV)$ where $dQ/dV = \rho = \varepsilon d^2V/dz^2 = K\varepsilon$, is the charge per unit volume, and dQ/dA is the total charge per unit area within the well. Therefore, the electron gas occupies a width of $(dQ/dA)/K\varepsilon$. Outside the occupied width, the potential increases as $K(z - w_e/2)^2/2$. As the number of donors increases, the number of charge carriers in the well increases until the potential in the well rises to the point where no further transfer of electrons from barrier to well occurs.

2. WIDE GRADED POTENTIAL WELLS

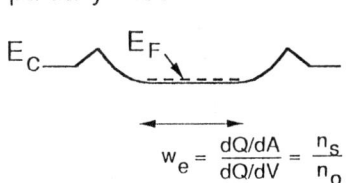

FIG. 14. (a) Potential of bare parabolic well, $V(z) = \tfrac{1}{2}Kz^2$. (b) Potential of screened parabolic well containing charge density $qn_0 = dQ/dV = \varepsilon d^2V/dz^2 = K\varepsilon$ charges per unit volume in a width

$$w_e = \frac{dQ/dA}{dQ/dV} = n_s/n_0.$$

Here, n_s is the total charge per unit area in the well.

A more realistic semiclassical calculation of the charge distribution in modulation-doped wells can be made by taking the Fermi statistics of the occupied electron states into account, but ignores quantization of energy levels by the well potential. In this case, the local charge density in the well is taken to be proportional to the Fermi integral of the local density of states. The Fermi energy with respect to the band edge is related to the local three-dimensional electron density N_Q at $T = 0K$ by $E_F = (h^2/2m^*)(3\pi^2 N_Q)^{2/3}$. At a finite temperature, the relationship between E_F and N_Q, as given by the Joyce–Dixon approximation (Joyce and Dixon 1977), has been used to evaluate the charge density distribution for different fractional amounts of filling of the well; i.e., 1/8, 1/4, 1/2, 1, 2, 4, and 8 times the filling needed to fill a 200 nm parabolic GaAs/Al$_x$Ga$_{1-x}$As well at its design density of $N_Q = 10^{16}$ cm^{-3} (Sundaram et al. 1988). The well constitutes a parabola with $x_{Al} = 0$ at the center, graded to $x_{Al} = 0.1$ at the edges, enclosed in square barriers with $x_{Al} = 0.3$. It is readily seen (Fig. 15) that the well fills from the center with a nearly constant charge density near the well center. As the well is increasingly filled with electrons, the excess charge

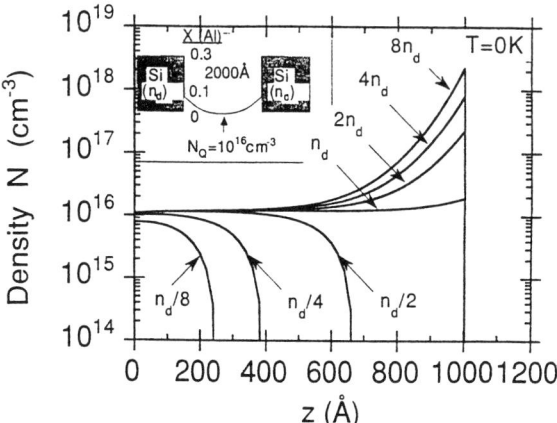

FIG. 15. Charge density profile for the right half of a symmetric 200 nm parabolic potential well with curvature $N_Q = 10^{16}\,\text{cm}^{-3}$. The inset shows the structure of the well. Here $n_d = 10^{16}\,\text{cm}^{-3} \times 200\,\text{nm} = 2 \times 10^{11}\,\text{cm}^{-2}$ is the number of charge carriers needed to classically fill the well exactly. Overfilling is seen to cause electron accumulation at the well edges.

accumulates at the edges of the well. As charge is added to the well, the potential changes from a parabola for the unfilled case to a flat-bottomed well for the fully filled case, with the potential increasing approximately quadratically from the edge of the flat-bottomed region containing the nearly uniform charge density. Insofar as the uniformity of the 3D electron density is concerned, therefore, the finite width of the well is unimportant in most cases.

For the full quantum mechanical calculation, both the Poisson equation and the Schrödinger equation must be solved. As electrons are added to a parabolic well, they initially occupy the first wave function of the parabola and start to fill a subband of electrons with increasing kinetic energy in the free direction of motion parallel to the layers. Concurrently, the potential at the bottom of the well is screened and flattens. On further filling, the second wave function starts to be occupied and most of the additional charge goes into the subband based on the second wave function, while the flat part of the potential at the center of the well broadens further. The total energy for each electron is the sum of the kinetic energy for the free motion parallel to the layers and the self-consistent energy of the state for the quantized motion perpendicular to the layers.

Several groups (Rimberg and Westervelt 1989; Stopa and Das Sarma 1989; Wixforth et al. 1990c; Sajoto et al. 1989b) have calculated the energies and wave functions of modulation-doped wells. Figure 16 shows the results for a 400 nm well with $N_Q = 5.5 \times 10^{15}\,\text{cm}^{-3}$, $\varepsilon = 12.87$, and $m^* = 0.0753\,m_0$,

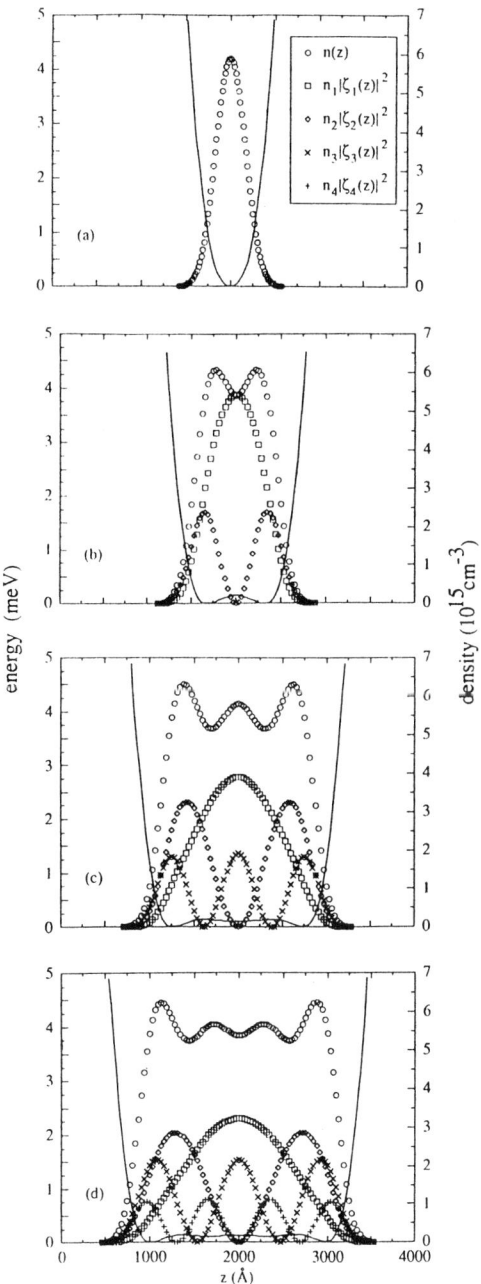

FIG. 16. Individual subband electron densities and total electron density for 400 nm parabolic well with $N_Q = 5.5 \times 10^{15}\,\text{cm}^{-3}$ calculated for well filling (a) $n_s = 0.24 \times 10^{11}\,\text{cm}^{-2}$, (b) $n_s = 0.5 \times 10^{11}\,\text{cm}^{-2}$, (c) $n_s = 1.0 \times 10^{11}\,\text{cm}^{-2}$, and (d) $n_s = 1.3 \times 10^{11}\,\text{cm}^{-2}$. The solid curves show the total potential for each case (Rimberg and Westervelt 1989).

with infinite square barriers at the edge of the parabolic region (Rimberg and Westervelt 1989). An envelope function approximation is used in the calculation. The calculation includes an exchange-correlation potential treated using the local density functional approximation. Figure 16 shows the computed electron density distribution (open circles) as a function of position within the well for four different values of total sheet charge density, as well as the charge density due to each occupied subband. The four choices of filling correspond to fractional filling ($=n_s/N_Q$) 0.11, 0.26, 0.465, and 0.59, respectively. One subband is occupied in the first case, two in the second, three in the third, and four in the fourth case. The total self-consistent potential is also shown, as the solid curve. As electrons are added to the well, the development of a nearly uniform charge distribution and flat potential profile are clearly seen for the case where two or more subbands are occupied. The full quantum mechanical calculations show that the classical calculations give a somewhat accurate simple picture of the filling of a parabolic well.

The total charge and the energy levels of the bottoms of the subbands as a function of well filling for a 400 nm well with $N_Q = 6.9 \times 10^{15}$ cm^{-3} are shown in Fig. 17 (Wixforth et al. 1990c). In this case, the structure contains a front gate electrode by which the total sheet charge is varied. As electrons are added, the higher subbands drop below the Fermi level and are successively populated. Up to a certain filling, only the lowest subband is occupied. Because of the energy independence of the density of states in two dimensions, the distance from the bottom of the subband to the Fermi level is linearly proportional to the total sheet charge n_s, causing the bottom of the lowest subband to drop linearly below the Fermi energy with more positive gate voltage. When the second subband drops below the Fermi level, most of the additional charge goes into it, so that the bottom of the first subband nearly stops dropping with respect to the Fermi level. As each additional subband is occupied, most of the additional charge goes into the new subband. The physical reason for this is that the new charge is added mainly at the edges of the charge distribution as it broadens to create a wider gas of uniform three-dimensional density. The energy levels of the bottoms of the subbands, which were originally the uniformly spaced simple harmonic oscillator levels of the unfilled-well case, drop closer together as the well is filled and evolve to approximate the quadratically spaced energy levels $E_n = h^2 n^2/8m^* L^2$ of a wide, infinite square well of width L, where $L \sim n_s/N_Q$. This is in contrast to the case of a triangular well (as in a modulation-doped single heterointerface) or a narrow square well, where relatively little difference in energy level spacing occurs as electrons are added to the well. The filling of the wide parabolic wells can be controlled either through doping in the original growth of the wells, through control

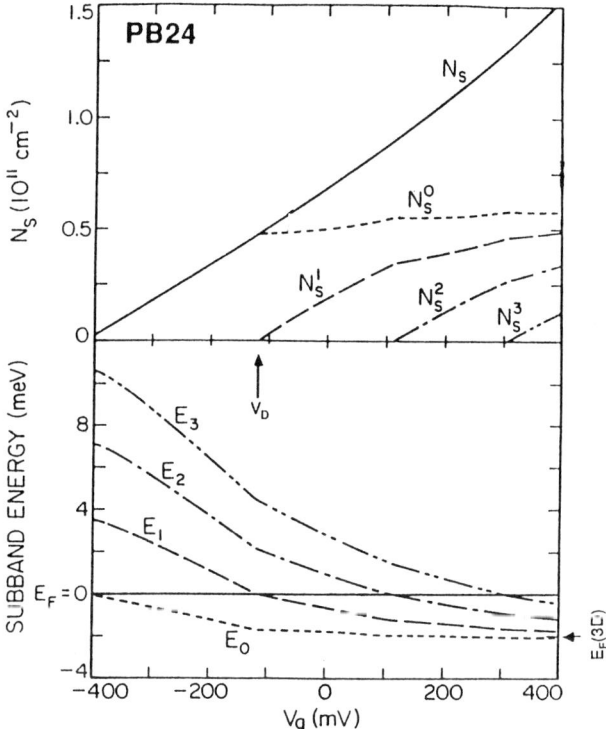

FIG. 17. (a) Calculated number of electrons per unit area, N_s^i, in individual subbands, and total density N_s for all subbands in a 464 nm parabolic well as a function of gate voltage, V, applied to a gate electrode on epilayer surface. (b) Subband energies as a function of gate voltage. In the empty well, the energies are those of a parabolic potential, whereas in the full well, the energies approach those of a square well. V_D marks the boundary between one-subband and multisubband occupancy. $E_F(3D)$ is the position of the conduction band edge below the Fermi energy for an ideal three-dimensional electron gas with the same density N_Q as the curvature of the parabolic well.

by a gate electrode, or by photogeneration of carriers from deep states in the barriers surrounding the well, as will be discussed further.

5. Capacitance Measurements

The distribution of free charge within a semiconductor material can be probed by a measurement of dQ/dV, the small signal capacitance between the semiconductor and an adjacent gate as a function of an external voltage applied between the two. The applied voltage gradually depletes charge from the edge of the charge-containing regions. The technique is routinely applied to homojunctions (constant band gap), but is also applicable to heterostructures to measure energy band offsets (Kroemer 1985) and was first applied to wide graded wells by Sundaram et al. (1991b). This technique has been applied to parabolic wells in two ways. In the first, a potential is applied between a gate electrode at the surface of the structure containing the well to be studied and a conducting substrate below the well. In the second, a potential is applied between the surface electrode and an ohmic contact to the buried electron layer in the well.

In each of these measurements it is not possible to deconvolve the experimental CV (Capacitance Voltage) data directly to get the explicit charge distribution, so the measured apparent charge density profiles, as read by a CV profiler, are compared with the calculated apparent charge density profiles for the measured structures. The result of such a comparison is shown in Fig. 18 for a modulation-doped parabolic well (Sundaram and Gossard 1993). The grown structure consisted of a 200 nm well with the Al content varying quadratically from $x = 0$ at the center to $x = 0.2$ at the edges, with barriers of $x = 0.3$. Using the conduction band offset $\Delta E_c = 0.60 \Delta E_g$, the potential of the undoped well varies from 0 meV at its center to 150 meV at its edges increasing to 225 meV at the barriers outside the parabolic region. The solution of Poisson's equation at 300K yields a true electron density profile (dotted curve). The CV profile that would be measured by the CV profiler is shown in the dot-dash curve. This plot of the ideal predicted profile is then compared with the experimentally measured CV profile, also shown in Fig. 18 for measurements on a sample with nominally the same design. The sample contained a superlattice spacer layer between the dopants and the well and was grown with a reduced substrate temperature for the barrier layer growths to prevent diffusion of silicon donor atoms toward the well. Samples grown without spacer layers and without the reduced temperature during the barrier layer growths showed CV profiles that resembled those calculated for asymmetric charge distributions. In addition to being a direct probe of the electron distributions in the wide parabolic wells and the surrounding barrier regions at $T = 300K$, these measurements also yielded the spatial width and the potential depth of the actually grown wells.

The second technique of CV profiling, i.e., measuring the capacitance as a function of a voltage bias applied between a front gate and the buried electron layer in the semiconductor, is another direct method of determining the carrier density profile in the parabolic well structure at low temperatures (Wixforth et al. 1990c). Data at $T = 4.2K$ on a 464 nm wide well demonstrated the existence of a wide slab of electron gas centered in the parabolic

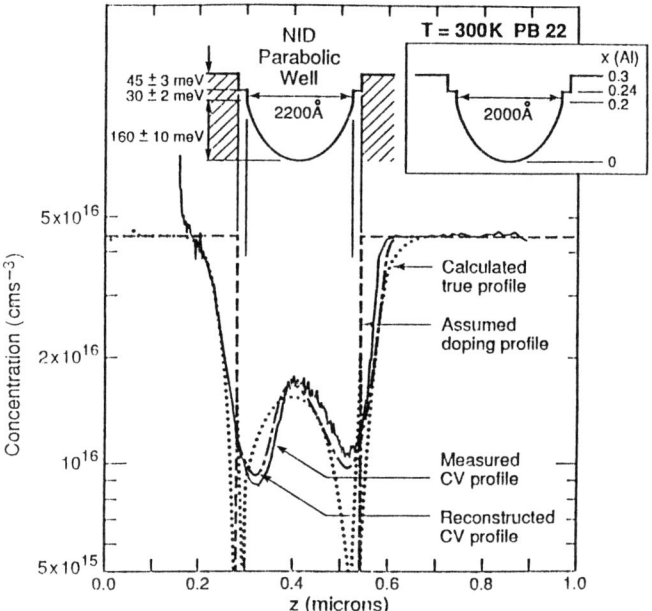

FIG. 18. Measured (solid) and reconstructed (dot-dashed) apparent carrier concentration profiles for a 220 nm parabolic well with $N_Q = 2.5 \times 10^{16}$ cm^{-3}. Measurements are from a CV capacitance profiler, and the reconstruction employs the parameters shown at the top of the figure for the nominal well design shown in inset. The reconstructed profile is the calculated CV profile that would be measured by the CV profiler and is shown as the dot-dashed curve. The apparent concentration is determined from $N_{app} = (2/q\varepsilon)[d(1/C^2)/dV]^{-1}$, and the plotted depth z is determined from $z = \varepsilon/C$, where C is the capacitance per unit area. The assumed doping profile (dashed) and the calculated true (dotted) electron profile are also shown.

well. The data also showed three broad bumps that have since been attributed to nonuniformities in the electron density profile in the well caused by unintentional nonparabolicities in the nominally parabolic potential; this demonstrated that this CV technique is one of the most sensitive methods for determining intentional and unintentional deviations from parabolicity; in fact, Sundaram et al. (1992) have used this technique to observe intentional periodic density modulations induced by superlattices in wide parabolic wells (see Section IV).

As successive subbands are occupied by carriers induced by external gate potentials, the capacitance between the gate and the electron gas is expected to show characteristic features (Wixforth et al. 1990c). These features in the CV curves due to subband filling can be distinguished from features due to nonparabolicity in the well by the difference in dependence of the capacitance features on an in-plane magnetic field (Kane et al. 1990).

Rimberg et al. (1993) observed large features, attributable to subband filling, in dC/dV versus V at 0 magnetic field. The features shift to lower gate biases with increasing field and disappear above fields on the order of 1 Tesla, at which field all the higher subbands are magnetically depopulated.

Measuring the capacitance between a front gate and the buried electron layer has been extensively used to gauge Landau level effects on the density of states in 2D electron structures in a transverse magnetic field (Smith et al. 1985; Mosser et al. 1986). Similar measurements have been made on parabolic wells (Sundaram et al. 1991a), and in transverse magnetic fields the data show minima in the CV curves corresponding to filling of integer numbers of Landau levels. The presence and absence of features at specific filling factors are explained by crossing of Landau levels from the different occupied subbands, in agreement with the explanation of Gwinn et al. (1989) of the quantum Hall effect transport measurements in wide wells.

V. Electron Transport

6. Electrical Conductivity

As in heterojunctions and narrow wells, the electrical conductivity of wide graded wells can be enhanced by modulation doping. The important advantage that modulation doping has in comparison with uniform doping is the reduction of scattering by ionized impurities in the wells. In uniformly doped wells, this scattering is unavoidable because the impurities occupy the same layers as the charge carriers. In the modulation-doped structures, on the other hand, the scattering is reduced because the dopant impurity ions are placed in the barriers, where they are spatially separated from the conduction electrons. Furthermore, the modulation-doped structures avoid the carrier freeze-out that occurs in uniformly doped semiconductors at low temperatures.

To date, modulation-doped wide parabolic wells have been grown with curvature design densities N_Q ranging from 2×10^{15} to 9×10^{16} cm^{-3} in the GaAs/AlGaAs system. The mobilities measured by Hall effect in a uniformly doped (at 6×10^{15} cm^{-3}) GaAs layer (Stillman, private communication) and in a wide parabolic well, also with a density N_Q of 6×10^{15} cm^{-3} in the occupied region of the well (Sundaram et al. 1991b), are shown in Fig. 19 as a function of temperature. Mobilities in excess of 200,000 cm^2/Vs are observed at low temperature in the modulation-doped parabolic well. By comparison, as seen in the figure, the mobility for uniformly doped GaAs at the same density is much lower, dropping at low temperatures because of ionized impurity scattering. The number of carriers in the modulation-doped sample is nearly constant for temperatures below ~ 70K and corre-

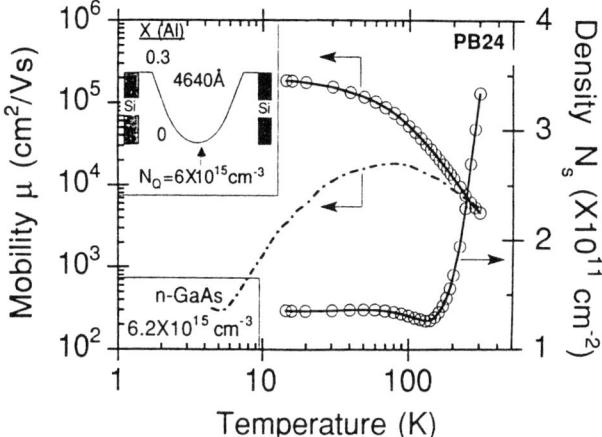

FIG. 19. Electron mobility versus temperature in uniformly silicon-doped GaAs with carrier density $n = 6 \times 10^{15}\,\text{cm}^{-3}$ (Stillman, private communication), and in a modulation silicon-doped 464 nm $\text{Al}_x\text{Ga}_{1-x}\text{As}$ parabolic well with carrier density $N_Q = 6 \times 10^{15}\,\text{cm}^{-3}$, filled to a total low-temperature charge per unit area of $N_s \sim 1.3 \times 10^{11}\,\text{cm}^{-2}$.

sponds to ~50% filling of the well. Hopkins et al. (1990a) showed that for a series of samples ranging in design density N_Q from $4 \times 10^{15}\,\text{cm}^{-3}$ through $3 \times 10^{16}\,\text{cm}^{-3}$ the carrier densities were independent of temperature below ~70K, and showed no signs of carrier freeze-out. Low temperature mobilities were nearly identical for this series of samples.

The filling of the parabolic well at low temperatures can be varied by the persistent photoconductivity (PPC) effect. After cooling the samples to low temperature in the dark, the samples are exposed briefly to illumination from a red light emitting diode (LED); this results in an increase in the number of carriers in the well that persists for days at low temperatures (Gwinn et al. 1989; Sundaram et al. 1991b). In contrast, Karraï et al. (1990) found that during the LED illumination, the carrier density in their wells decreased. The reasons for the decrease are not yet understood. In both cases the change in electron number is a large effect, amounting to as much as a + 100% change in well filling in the work of Gwinn et al. and a − 90% change in Karraï et al.'s work.

Data for the mobility as a function of well filling are shown in Fig. 20 for a 200 nm well (Sundaram et al. 1991b). In this case, the persistent photocarriers were introduced by a pulsed red LED, with the filling increasing from 35% to 82% of the well. In contrast to the case of the two-dimensional electron gas in which mobility generally increases with carrier concentration, the mobility of the wide parabolic well dropped by a factor of 2 as the well filled. The carrier concentration in the well can also be varied by means

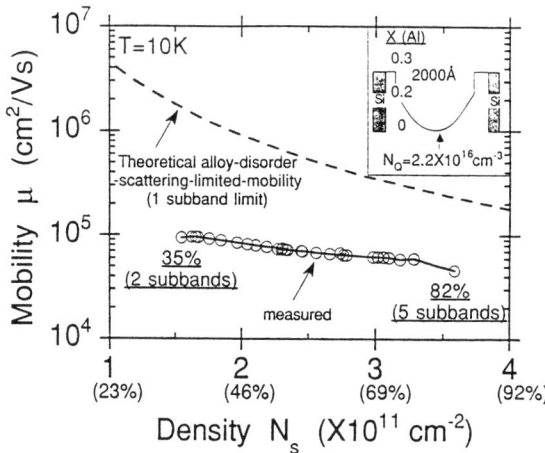

FIG. 20. Electron mobility versus total number of electrons per unit area measured after pulsed illumination of 200 nm $Al_xGa_{1-x}As$ parabolic well with a red light emitting diode.

of a surface gate; the resulting mobility dependence has been studied by Ensslin et al. (1993) for a 75 nm well with $N_Q = 8.8 \times 10^{16} \, cm^{-3}$. In this case nonmonotonic behavior for mobility versus carrier density was observed. For low fillings of the well, such that two or less subbands are filled, the mobility increased with carrier concentration. With the onset of occupation of the third subband, however, the mobility decreased. The mobilities of electron gases in parabolic wells are strongly enhanced with respect to mobilities in uniformly doped layers, but are substantially below those seen in narrow GaAs quantum wells and two-dimensional electron gases in modulation-doped heterojunctions.

This mobility behavior may be understood as follows. The electrons in wide graded wells are scattered by the same processes that are dominant for electrons in two-dimensional electron gases, particularly phonons at high temperatures and ionized impurities at low temperatures. But, in addition, several other processes are important in the wide parabolic wells: (1) alloy disorder scattering, (2) increased unintentional background impurity scattering, and (3) size effect scattering (Walukiewicz et al. 1991) at the edges of the electron gas.

Alloy disorder scattering is important in wide graded wells because the regions that contain electrons are composed of a mixed alloy. The alloy disorder scattering rate in a 2DEG is

$$\frac{1}{\tau_{alloy}} = \frac{m^* x(1-x)\Omega \langle V \rangle^2}{h^3} I_{alloy} \tag{6}$$

2. WIDE GRADED POTENTIAL WELLS

where x is the average mole fraction alloy composition, $\langle V \rangle$ is the alloy disorder scattering parameter, Ω is the unit cell volume, and

$$I_{\text{alloy}} = \int_{-\infty}^{+\infty} |\Psi_0|^4 \, dz,$$

where $\Psi_0(z)$ is the electron wave function in the alloy (Walukiewicz et al. 1984; Ando 1982). An estimate, based on an extrapolation from results for a 2D InGaAs heterostructure (Bastard 1983) of the alloy disorder scattering limited mobility for a full 200 nm parabolic well with 4.4×10^{11} electrons/cm^2 and a maximum Al content of 0.20, yields $\mu_{\text{alloy disorder}} \sim 1.5 \times 10^5$ cm^2/Vs, with higher values at lower filling of the well (Sundaram et al. 1991b). This predicted value is about 2.5 times higher than the observed low temperature mobilities for a filled well, showing that other scattering mechanisms are important in limiting the mobility.

Another mechanism that is likely to be important in wide wells is intersubband scattering (via electron–electron scattering, electron–background impurity scattering, etc.), which can occur by virtue of the fact that several subbands are occupied in the three-dimensional electron gas in the wide wells. Scattering by unintentional impurities in the wells is also a likely source of additional electron scattering. The concentration of these impurities is higher in Al$_x$Ga$_{1-x}$As than in GaAs because of the greater reactivity of Al than that of Ga with oxygen and carbon-containing molecules in the growth environment (Prior 1984). Thus, Al-containing wells will be subject to more of this scattering. A background impurity concentration of $\sim 10^{15}$ cm^{-3} in the well would give an impurity scattering limited low-temperature mobility of $\sim 3.0 \times 10^5$ cm^2/Vs in a 200 nm well with curvature $N_Q = 2.2 \times 10^{16}$ cm^{-3}. This mobility is equal to the observed mobility of the structure at $T = 4.2$K (Walukiewicz et al. 1991). However, the calculated mobility from this concentration of impurities falls off more slowly with temperature than does the experimentally measured mobility (the calculated mobility from 10^{15} ionized impurities/cm^3 would be $> 2.0 \times 10^5$ cm^2/Vs at 50K, whereas the experimentally observed mobility has dropped to less than 1.5×10^5 cm^2/Vs at this temperature). It is therefore probable that yet other, more temperature-dependent mechanisms of electron scattering are occurring in the wide parabolic wells.

The most probable candidate for this additional scattering is size effect scattering from the edges of the electron gas. At the observed mobilities in high-quality samples, the mean free path of the electrons (given by $\hbar k_F \mu_b/e$) considerably exceeds the width of the gas. At bulk mobility $\mu_b = 2.5 \times 10^5$ cm^2/Vs in an electron gas with 10^{16} electrons/cm^3, the mean free path is 1000 nm, or several times the width of a 200 nm well. Therefore, scattering from the edges of the gas is probably important. The special feature of the wide graded wells that makes this scattering important is the softness in the edges of the confining potential, which increase quadratically rather than abruptly. Small fluctuations in potential that would be expected to arise from random fluctuations of the remote ionized impurity concentration and fluctuations in alloy composition thus cause fluctuations in the position of the edge of the electron gas, leading to nonspecular reflection. The strongest scattering will come from fluctuations with wavelengths near the screening length of the electron gas (since longer range fluctuations are electrostatically screened by the gas and shorter range fluctuations are averaged out because of the finite wavelength of the electrons). Assuming this and following an approach by Cottey (1967), Walukiewicz et al. (1991) calculated the mobility and its temperature dependence, taking into account the size effect scattering, together with the background ionized impurity scattering, phonon scattering, and alloy disorder scattering. Results of the calculations are shown in Fig. 21. Good fits to the experimentally observed mobilities could be obtained by assuming surface roughness scattering

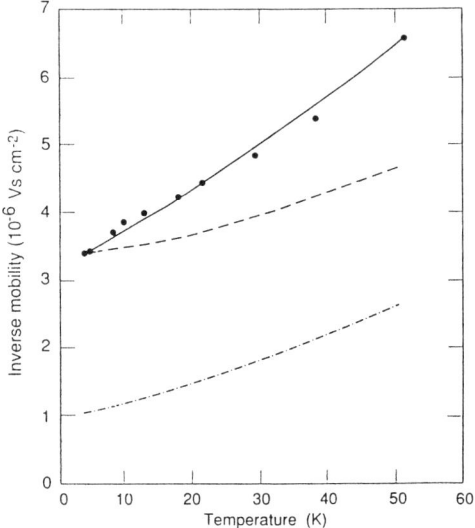

FIG. 21. Theoretically calculated values of inverse of electron mobility values expected for a parabolic well with ionized impurity scattering and size effect electron scattering. Solid curve is for ionized impurity density $n_{ii} = 2.3 \times 10^{14}\,\text{cm}^{-3}$ with size effect scattering included. Dashed curve is for $n_{ii} = 1.0 \times 10^{15}\,\text{cm}^{-3}$ and no size effect. Dash-dot curve is for $n_{ii} = 2.3 \times 10^{14}\,\text{cm}^{-3}$ and no size effect scattering. Points are experimental results from parabolic well with design density $N_Q = 2.5 \times 10^{16}\,\text{cm}^{-3}$ occupying a width of 90 nm (Walukiewicz et al. 1991).

together with a background ionized impurity concentration of 2.3×10^{14} cm^{-3} for the 200 nm well discussed previously. The nonspecular scattering reduces the low temperature mobility by more than a factor 2. The experimentally observed temperature dependence of the mobility is produced by acoustic phonon scattering, enhanced by surface roughness.

Perturbations in quantum levels due to interface roughness (Sakaki et al. 1987) are not a dominant source of scattering because these perturbations are small in wide wells. Consider first the case of a single electron in an empty parabolic well. The quantization energy E in the growth direction z varies as $E \sim (d^2V/dz^2)^{1/2} \sim V^{1/2}/W$, and $\Delta E \sim \Delta W/W^2$, where ΔW is the variation in well width produced by interface roughness. According to Sakaki et al. interface roughness scattering caused by well width fluctuations ΔW will vary as ΔE^2, which in this case is proportional to W^{-4}. Mobility limited by roughness scattering in this case for $W = 200$ nm parabolic well and ΔW = a few monolayers is far in excess of experimental values. Now consider the case of an electron in a filled parabolic well, where the z quantization energy levels can be approximated by a square well spectrum (Section IV) $E \sim 1/W^2$, with $W \sim w_e$, the width of the electron gas. In this case $\Delta E^2 \sim 1/w_e^6$, and for reasonable values of ΔW we again obtain a mobility estimate far in excess of experimental values.

The background impurities that limit the low temperature mobility are also the principal limiting factor for how wide and shallow a graded well can be made, and thus how dilute a three-dimensional electron gas can be formed. The unintentional background charges (e.g., negatively charged acceptors) produce a charge density that changes the potential within the well. For negatively charged acceptors, this produces a curvature of the opposite sign from the design curvature and leads to a conduction electron distribution that is different from the design value by an amount that is comparable to the unintentional background charge. Thus, a lower limit on the achievable electron densities is set by this background ionized impurity concentration, typically on the order of 10^{14} cm^{-3}.

7. MAGNETOTRANSPORT

The study of electrical conductivity in a magnetic field applied transverse to the plane of the well gives a great deal of information about the states of electrons in quantum structures. The free electron states for motion of electrons and holes parallel to the layers at zero magnetic field are quantized into Landau orbits at finite fields. The lowest energy electrons occupy the $i = 0$ Landau levels of energy $\hbar\omega_c/2$, where ω_c is the cyclotron frequency, and have the smallest cyclotron orbits (spin splitting is neglected in this analysis). The successively more energetic electrons occupy the higher Landau levels with energy $(i + 1/2)\hbar\omega_c$. As the magnetic field is increased, the Landau level energies increase and each Landau level can hold more electrons, so that progressively fewer levels are occupied. In magnetotransport, as the occupation of the Landau levels changes, the electrical resistance shows oscillations, called *Shubnikov–de Haas oscillations*.

In wide graded parabolic quantum wells, where quantum energy levels are closely spaced, it is common for several subbands to be occupied simultaneously at zero magnetic field. With increasing magnetic field, each subband is the basis of a set of Landau levels. With the emptying of successive Landau levels, several periods of oscillation occur simultaneously as the magnetic field is increased (Sajoto *et al.* 1989a). This is illustrated in Fig. 22, where several periods of oscillation beat against each other for a 464 nm modulation-doped parabolic quantum well in the field range below 0.6 T. The periods may be extracted from the data by Fourier analysis of the oscillations, where three distinct periods are resolved (Gwinn *et al.* 1990; Hopkins *et al.* 1990a). Each frequency f_i is directly related to the number of electrons in a subband, $f_i = N_s^i h/2e$. Assuming a 2D density of states $m^*/\pi\hbar^2 (= 2.8 \times 10^{10} \text{ meV}^{-1} \text{cm}^{-2})$ for $m^*(\text{GaAs}) = 0.067 m_0$) for each occupied subband, the frequency f_i gives the energy spacing $E_F - E_i = (\hbar e/m^*) f_i$

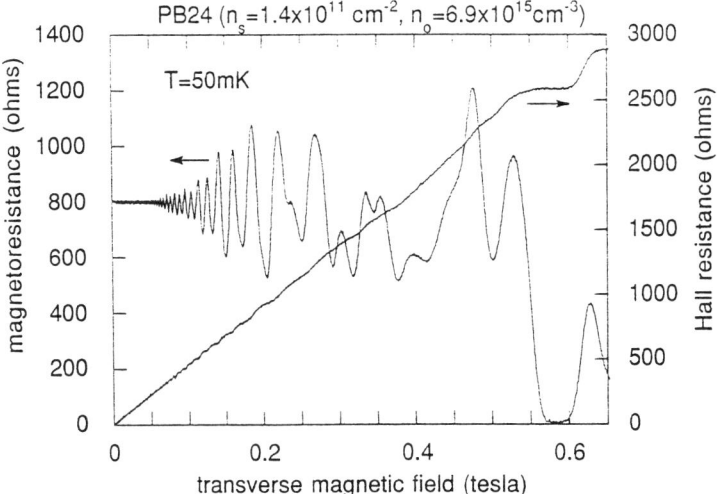

FIG. 22. Magnetoresistance and Hall resistance of a 464 nm $Al_xGa_{1-x}As$ parabolic well versus magnetic field at $T = 50$ mK, illustrating multiple periods in resistance ocillations and development of quantum Hall plateaus. The electron mobility in the sample is 240,000 cm^2/Vs. Here, $n_0 = N_Q$ is the well curvature, and n_s the electron sheet density.

between the Fermi energy E_F and the subband energy E_i; the highest frequency therefore gives the Fermi energy relative to the lowest subband $(E_F - E_1)$. These energy levels agree well with self-consistent calculations; differences between the experimental and calculated values have been used to determine deviations of the well potential from parabolicity and hence nonuniformities in the electron density profile (Hopkins et al. 1990b).

At higher magnetic fields, the quantum Hall effect is observed in this sample with broad resistance minima in the magnetoresistance and plateaus in the Hall resistance. However, not all quantum numbers v (the number of spin-split Landau levels filled) are observed in the quantum Hall effect for the parabolic wells. Because more than one subband is occupied, Landau levels from two subbands can simultaneously be filled. At some fields, Landau levels from different subbands will be nearly degenerate. If the Fermi level passes from above this degeneracy point to below it as the magnetic field is increased, the number of occupied Landau levels decreases by 2, and total occupancy including spin degeneracy decreases by 4. This was demonstrated (Ensslin et al. 1991) for a 75 nm wide parabolic well with

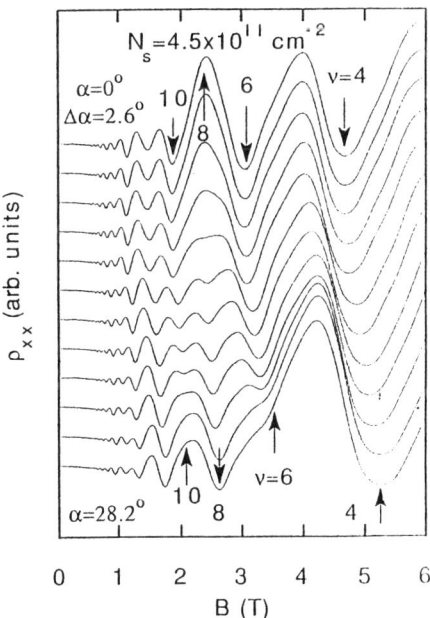

FIG. 23. Magnetoresistance ρ_{xx} versus magnetic field B for 75 nm $Al_xGa_{1-x}As$ parabolic well for angles of magnetic field from surface normal, α, from 0 degrees to 28.2 degrees. Disappearance of minima occurs when highest filled Landau levels cross or have a close anticrossing.

$N_Q = 8.8 \times 10^{16}$ cm^{-3}, in which $v = 4, 6$, and 10 resistance minima were seen but $v = 8$ was totally absent. Gwinn et al. (1989) first studied this effect and showed that the $v = 3$ resistance minima and Hall plateau could be suppressed and recovered as the number of electrons in the well was increased by LED illumination.

When the magnetic field is applied at an angle (i.e., tilted) with respect to the surface normal, Landau levels from different subbands no longer cross each other, but are repelled and anticross (Shayegan et al. 1989; Smrcka 1990; Ensslin et al. 1992). As the angle is increased from 0, some missing resistance minima and quantum Hall plateaus are regained and others lost. With the magnetic field at an angle of 28.2° from the surface normal, minima at $v = 4$ and 8 are seen, whereas $v = 6$ is weak and $v = 10$ is missing (Fig. 23) (Ensslin et al. 1992). Analysis of such measurements has confirmed the magnetic field dependence of the subband energy spectrum. This is a useful result because, as shown in Section VI, the subband spacing in a doped well cannot be measured directly spectroscopically. The intersubband excitations are coupled to the plasma excitations in such a way that only hybrid intersubband–plasma transitions are observed, whose frequencies are always exactly equal to the frequency of the intersubband transitions for an empty well, irrespective of the filling of the well (Brey et al. 1989).

At higher magnetic fields, the number of electrons that can be contained in each Landau level increases until only one Landau level is occupied (the quantum limit). In this limit, the diameters of the electron cyclotron orbits become less than the spacing between electrons and in pure samples at low temperatures we expect that Coulomb interactions between the electrons lead to correlated or ordered states of the electrons. The fractional quantum Hall effect is one manifestation of such states and results from correlated electron liquid state formation at fractional Landau level filling. The wide parabolic well gives a way of investigating the dependence of the ordering on the width of the electron gas layer. As the thickness of the layer in which the electrons are confined increases, the theoretical expectation (He et al. 1990) is that the interaction between the electrons will decrease and the ordering energy will also decrease, as evidenced by a more rapid disappearance of the fractional quantum Hall states with increasing temperature. Exactly this effect has been observed in wide parabolic wells (Shayegan et al. 1990) in the study of the fractional quantum Hall effect in a 300 nm Al$_x$Ga$_{1-x}$As parabolic well. The temperature dependence of the magnetoresistance minima at $v = 1/3$ and $v = 2/3$ filling were measured to obtain the energy gap Δ that separates the fractional quantum Hall correlated ground state from its excitations (Fig. 24). The energy gaps for both $v = 1/3$ and $v = 2/3$ states decrease rapidly for widths of the electron gas larger than about three magnetic lengths $l_0 = \sqrt{\hbar/eB}$ (cyclotron orbit

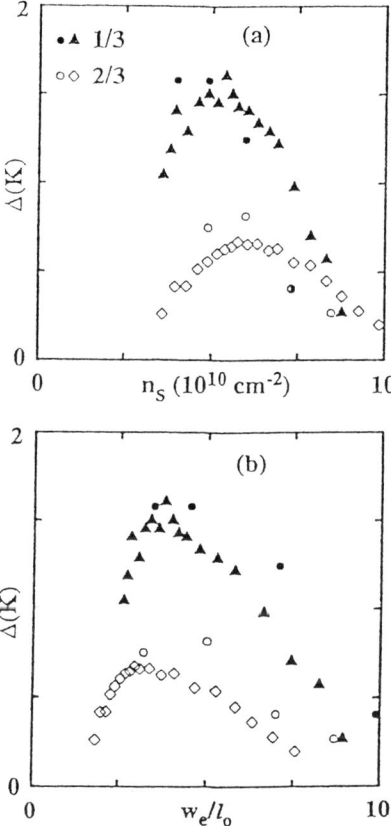

FIG. 24. The dependence of the measured energy gap Δ measured from the fractional quantum Hall effect, on electron density n_s and on the dimensionless parameter w_e/l_0 for $v = 1/3$ (closed symbols) and 2/3 (open symbols). Here, w_e is the width of the electron gas, and l_0 the magnetic length. The circles are from one sample and the rest of the data are from another. The accuracy is estimated to be $\pm 10\%$ for $\Delta \geqslant 0.5K$ and $\pm 30\%$ for $\Delta \leqslant 0.5K$ (Shayegan et al. 1990).

radii). Increasing widths of the gas in the measurements were obtained by varying the number of carriers by illumination with a red light emitting diode. Competing effects that could also reduce the energy gap for the correlated state when the width of the electron gas broadens are the decreased mobility and increased disorder that occur as the electrons spread into regions of higher aluminum content and the decreased separation between energy levels belonging to different subbands that becomes less than Δ as the electron gas spreads.

For a strong magnetic field applied perpendicular to the electron slab with one Landau level occupied ($v = 1$), Brey (1991) calculated that above a critical number of electrons, a soft mode develops in which the energy of a charge density excitation becomes 0. This implies that the static charge density response of the system diverges, with the accompanying development of an ordered state, probably a charge density wave state. Brey speculates, following MacDonald and Bryant's (1987) treatment of a three-dimensional electron gas, that the electrons form a crystalline structure in the x-y plane and a modulated distribution in the z direction.

At high magnetic fields and for orientation of the magnetic field parallel to the layers, the wide, quasi-three-dimensional electron gas is predicted to undergo low temperature ordering into a spin density wave state or an electron crystal state (Celli and Mermin 1965; Halperin 1987; Brey and Halperin 1989). The observation of these states was the initial motivation for growth and study of modulation-doped parabolic wells. In the highest magnetic fields, the ground state of an infinite-three-dimensional electron gas is asserted to be a Wigner crystal (Heinonen and Al-Jishi 1986). In the intermediate regime of field, a spin density wave state (Celli and Mermin 1965) is one of the theoretically possible states and has been predicted to have electrical resistance vanishing in the direction perpendicular to the magnetic field and diverging in the direction parallel to it (Halperin 1987). Brey and Halperin (1989) have argued that for an impurity-free finite parabolic well of thickness several magnetic lengths, the ground state in the limit of high magnetic field should also be a Wigner crystal since the electrons are easily localized relative to motion in the direction perpendicular to a high magnetic field and since they will have sufficiently small kinetic energy in the direction parallel to the field to allow localization in that direction also. The effect of impurities on the Wigner crystal in the wide parabolic well would be to broaden the transition into the Wigner crystal state and to destroy perfect periodicity.

For the intermediate field case, Brey and Halperin have calculated the spin density response function $\chi(z, Q_x)$ for a wide slab (>100 nm) of electrons in a wide (400 nm) parabolic well with low 3D design density $N_Q = 5.5 \times 10^{15} \, \text{cm}^{-3}$. The function $\chi(z, Q_x)$ gives the spin density at depth z in the parabolic well, induced by an infinitesimal stimulus of wavevector Q_x. They find that the electron density profile with an in-plane magnetic field of ~ 1 T remains a thick slab of nearly uniform density, as in zero field, and changes little if exchange correlation effects are included using a local density functional approximation. Although the magnetic field has little effect on the charge density, the response function $\chi(z, Q_x)$ without including exchange correlation effects develops a maximum at the wave vector $Q_x = k_F\uparrow + k_F\downarrow$, where $k_F\uparrow(k_F\downarrow)$ is the Fermi wave vector for spin-up

(-down) electrons. When exchange correlation effects are included, the maximum at $Q_x = k_F\uparrow + k_F\downarrow$ can become a divergence, indicating that a spin density wave instability with that wave vector is possible. For electron filling of the 400 nm well greater than 40%, the value of the exchange interaction needed to produce the divergence in the response becomes independent of well width. This suggests that, if the electron layer in this well exceeds 160 nm in width, it could sustain the spin density wave at sufficiently low temperatures.

To date, however, the predicted anisotropy (Brey and Halperin 1989) between the electrical resistances parallel and perpendicular to an in-plane magnetic field has not been seen for either a wide parabolic well or a uniformly doped semiconductor, so there is still no clear experimental evidence of spin density wave ordering. Resistance anisotropies for the two applied current directions, $I \parallel B$ and $I \perp B$, have been seen in wide (464 nm and 568 nm) wells (Hopkins 1990), and have also been seen in narrower (75 nm) wells (Ensslin et al. 1992). Low-field ($B \sim 1$ T) anisotropic features occur at fields at which subband depopulation shoud occur (Shayegan et al. 1988). High-field (up to 12 T) anisotropy can be explained by simple transport theory with δ-function scatterers in the quantum limit (Hopkins 1990). Nor has a strong isotropic increase in low temperature resistance in high magnetic fields been seen that would suggest an electron crystal. Temperature ranges searched have been down to 15 mK, and field ranges up to 30 T. Prediction of the critical temperature T_c for the spin density wave ordering is difficult, as T_c is exponentially dependent on interaction parameters that are not well known (Brey and Halperin 1989). Effects that could destroy an ordered state are scattering by impurities and phonons. In interpreting data, a competing effect that must be distinguished from ordering due to electron–electron interactions is the localization of electrons in spatial inhomogeneities at low temperature and strong in-plane magnetic field. Additional experimenal studies such as investigation of the dependence of transport on applied electric field will be required in differentiating localization in impurity potentials from formation of collective ordered states.

VI. Electron Excitations

8. Far Infrared Absorption

The parabolic potential well is of special interest because it allows a tractable theoretical analysis and comparison with experiment for many properties of a confined electron gas. The energy level spacings and plasma frequencies of electrons in parabolic wells are particularly simple. The spacing $\hbar\omega_0$ between electron energy levels in an empty or bare parabola is the same as the plasma resonane energy $\hbar\omega_p$ for an electron gas of density N_Q screening the parabolic potential. The frequency in each case is $\omega_0 = \omega_p = (4\pi N_Q e^2/\varepsilon m^*)^{1/2}$ for electrons of density N_Q and mass m^* in a medium of dielectric constant ε.

As electrons are added to the well, the bare parabolic potential is screened, the bottom of the well becomes flat, and the spacing between the single-particle electron energy levels decreases and becomes square-well-like. Although the energy levels are strongly dependent on the occupation of the well, it has been proven theoretically (Brey et al. 1989) that electrons in such a parabolic well will absorb light from a homogeneous radiation field only at the bare harmonic oscillator frequency ω_0 of the empty parabolic well, irrespective of the number of electrons in the well.

The wavelength corresponding to the bare harmonic oscillator frequency ω_0 is about 10^3 times longer than the width of the well for the wells discussed here. Thus, the theoretical result arises from the fact that a homogeneous radiation field acts only on the center of mass of the total electron system. Although the individual electrons see an approximately square potential well, the electron system absorbs radiation at the plasma or harmonic oscillator frequency. The theoretical result can be considered an extension of a theorem by Kohn that the cyclotron frequency of an electron gas in a magnetic field is independent of the interactions between electrons (Kohn 1961); the mathematics of the cyclotron motion and the harmonic oscillator motion are essentially identical. A practical consequence of this result is that structures or devices with specified far infrared frequencies can be built simply by tailoring the curvature of modulation-doped parabolic wells.

Coupling radiation to the plasma–intersubband excitations discussed previously requires a component of the electric field perpendicular to the potential well, i.e., perpendicular to the surface of the epitaxially grown layers. There are three ways in which this can be accomplished: (1) radiation can be propagated parallel to the layer with E field polarized perpendicular to it (Kneschaurek et al. 1976); (2) the radiation pattern of a wave normally incident on the surface can be perturbed by a grating on the surface to have components of E field perpendicular to the surface (Koch 1976; Tsui et al. 1978; Kotthaus 1978; Heitmann and Mackens 1986); and (3) the response of electrons to a wave incident normally on the surface can be altered by a tilted magnetic field that induces cyclotron motion at an angle to the surface that have components perpendicular to the surface (Schlesinger et al. 1983). All three techniques have been used on parabolic wells.

The first technique requires the construction of a guiding structure to focus the incoming radiation on the edge of the parabolic well since the wavelength (1000 to 100 μm) of the 10 cm^{-1} to 100 cm^{-1} radiation of interest for the wells considered here is much longer than the well width (500 to 100 nm). It is difficult in this coupling geometry to avoid leakage of the radiation outside the sample. This technique has been used for studies of nonlinear propagation and harmonic generation in intense far infrared beams (Bewley 1992).

The second technique (Fig. 25) can fairly readily be accomplished by the evaporation and lithographic definition of metallic gratings on the surface of the sample. The perpendicular components of field penetrate the surface region down to a depth of approximately the grating spacing, with periodic lateral modulation due to the grating. Since the grating spacing, a ($\sim 4\,\mu$m), is much larger than the depth of the electron layer from the surface ($<0.4\,\mu$m), the decay of E_z through the depth of the electron gas should be acceptably small. The grating allows excitations of the electron gas at finite in-plane wave vectors $k = 2n\pi/a$ ($n = 1, 2, \ldots$). By lithographically changing the grating period, a range of wave vectors can be accessed, allowing the

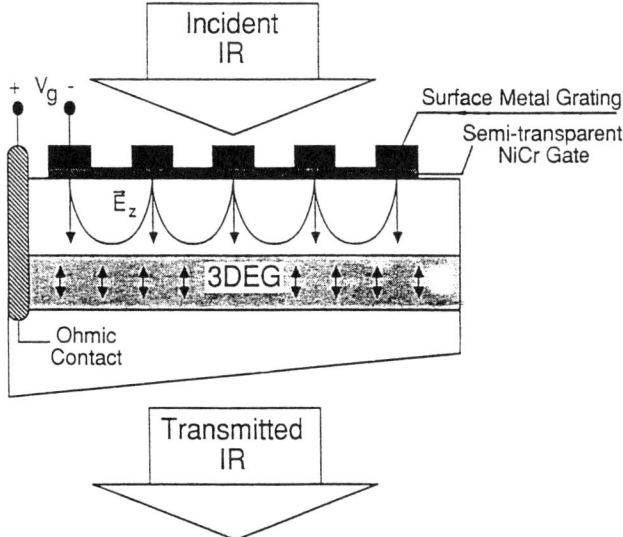

FIG. 25. Grating coupler on wide well structure for generating components of electromagnetic electric field perpendicular to surface. These components of electromagnetic field couple to plasma excitations with polarization perpendicular to the layers. The electric field lines in the near field of the grating are shown, with only the z component present at the grating–semiconductor interface.

frequency of absorption versus wave vector dispersion curve to be mapped out.

In the third (tilted magnetic field) technique for coupling to perpendicular motion of the electrons, the amplitude of the electron motion perpendicular to the surface is frequency dependent and strongest near the cyclotron frequency. The cyclotron resonance modes and the plasma modes are coupled in the tilted magnetic field geometry, shifting the frequencies of the cyclotron and plasma resonances and resulting in coupled modes and an anticrossing of the modes at the field where the cyclotron resonance and plasma resonance frequencies would be equal. This anticrossing results from an exchange of oscillator strength between the cyclotron and plasma resonances, the oscillator strengths of the two modes being the same when the two frequencies are equal.

Experimentally, the first observation of the far infrared response was reported by Karraï et al. (1989a) in a 200 nm parabolic well ($N_Q \sim 2.1 \times 10^{16}$ cm^{-3}) containing 2.5×10^{11} electrons/cm^2. Karraï et al. observed the two branches in the absorption spectrum in a magnetic field, resulting from the coupling between the cyclotron resonance and the plasma oscillations perpendicular to the layers. For both magnetic fields and incident light applied normal to the layers (Faraday geometry), only the cyclotron resonance was seen, whereas with tilted fields, two absorption lines, corresponding to the two coupled branches were seen. The thickness of the electron gas, n_s/N_Q, was about 10 times the magnetic length $l_0 = \sqrt{\hbar/eB}$ ($=250$ Å at $B = 1$ T) in their measurements. The measurements gave the first experimental confirmation that the resonance frequency, when extrapolated to zero field, corresponds to the plasma frequency rather than the energy separation between the subbands in the self-consistent electrostatic potential. The frequencies $\omega_\pm$ of the observed absorption peaks for magnetic field inclined at angle θ with respect to the surface normal, obeyed the expected theoretical relationship between the cyclotron frequency ω_c and a frequency Ω nearly equal to the plasma frequency ω_p,

$$1 = \frac{\Omega^2 \sin^2\theta}{\omega_\pm^2 - \omega_c^2} - \frac{\Omega^2 \cos^2\theta}{\omega^2} \tag{5}$$

In other experiments, Karraï et al. (1989b) studied the cyclotron resonance in four samples in the Voigt geometry in which the magnetic field is applied parallel to the layers. In this geometry, the carriers in their cyclotron motion are coupled to one another by the macroscopic depolarization field, leading to a collective or plasma-shifted cyclotron resonance. The number of electrons in the wells was changed by illumination with a red light emitting diode, which decreased the number of conduction electrons. The cyclotron resonance in this case is observed at a frequency $\omega = \sqrt{\omega_c^2 + \omega_p^2}$, where ω_p is the plasma frequency and ω_c the single-particle cyclotron frequency. The measurement thus allowed an independent measurement of ω_p, which was found to be independent of the filling of the well, in confirmation of the theoretical expectation (Brey et al. 1989) and in agreement with the result expected for constant well curvature. Line shapes of the Voigt geometry resonances, however, suggested a discrepancy in carrier concentrations, and relaxation times compared to values obtained from Faraday geometry measurements. They were interpreted in terms of inhomogeneous electron density near the edges of the occupied region of the electron gas, leading to an artificial broadening of the plasma shifted cyclotron resonance.

Similarly, coupled resonances in a tilted magnetic field and plasma shifted cyclotron resonances in magnetic fields parallel to the surface were observed in parabolic wells (Fig. 26) by Wixforth et al. (1990b). The plasma frequency

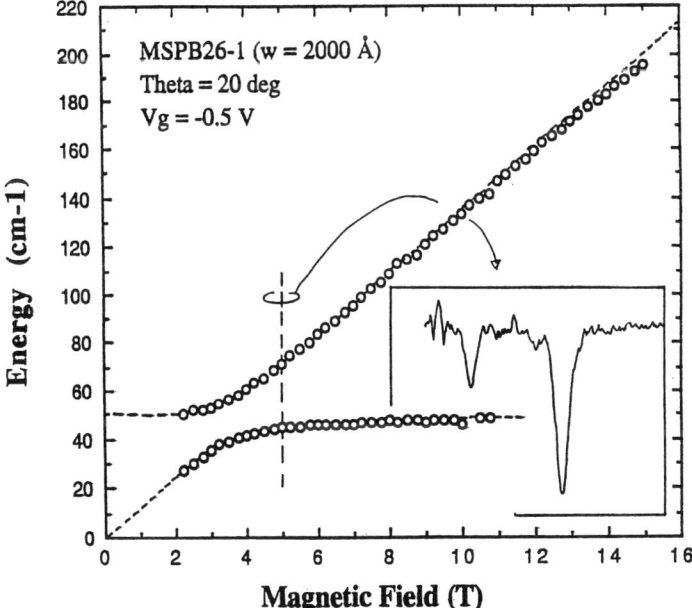

FIG. 26. Measured frequencies of coupled cyclotron and plasma oscillations versus magnetic field for 200 nm parabolic well ($N_Q = 2.5 \times 10^{16}$ cm^{-3}) in a magnetic field inclined 20 degrees from surface normal. Dashed lines are from a simple model of two coupled harmonic oscillators. Inset is the relative transmission spectrum $T(B = 5 \text{ Tesla})/T(B = 0 \text{ Tesla})$ versus energy.

determined from these measurements was found to be nearly independent of the total charge in the parabolic well when the charge in the well was varied by means of a semitransparent gate electrode (Fig. 27). For a narrow parabolic well with steep sidewalls the absorption spectra, for the different geometries and as a function of gate bias, were complex, the deviations from parabolicity causing the incident radiation to couple not only to the center of mass mode but also to internal modes of oscillation of the electron gas (Wixforth et al. 1990b).

Cyclotron resonance absorption in the Faraday geometry showed an increased effective mass with increased filling of parabolic wells (Karraï et al. 1990), which is attributable principally to the dependence of the effective

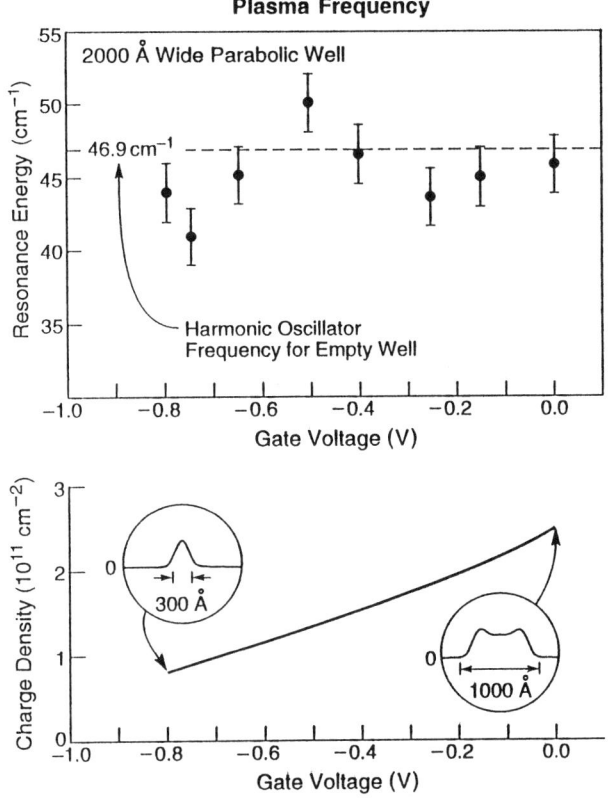

FIG. 27. (a) Plasma frequency and (b) total charge density per unit area versus gate voltage in 200 nm parabolic well with $N_Q = 2.5 \times 10^{16}$ cm^{-3} and with semitransparent gate. Insets in (b) are calculated electron distributions for the respective sheet densities.

2. WIDE GRADED POTENTIAL WELLS

mass on aluminum content in the $Al_xGa_{1-x}As$ wells. Wixforth et al. (1990a) observed an asymmetric cyclotron resonance line shape with a shoulder on the low energy side attributable to electrons in the second occupied subband, which, because of their wider spatial extent, experience a higher average aluminum content than electrons in the lowest subband and thus have a higher effective mass ($0.0724 m_0$ versus $0.0685 m_0$ in the studied sample).

Plasma resonances (also known as *dimensional resonances*) with nonzero in-plane wave vectors k have been observed directly with normally incident radiation without the use of a magnetic field by using the second technique, the grating coupler, to couple to the plasma resonances (Wixforth et al. 1991; Pinsukanjana et al. 1992). Wixforth et al. (1991) studied parabolic wells with widths 75 nm ($N_Q \sim 8.8 \times 10^{16}\,cm^{-3}$) and 200 nm ($N_Q \sim 2.5 \times 10^{16}\,cm^{-3}$) in which the number of electrons in the well was varied by a semitransparent front gate electrode. Coupling to the electron gas was accomplished by a silver grating of periodicity 6 μm, and far infrared transmission through the electron gas was studied by Fourier transform infrared techniques. The depletion of the 75 nm well by negative gate bias is illustrated in Fig. 28, which shows the results of self-consistent subband calculations for the potential and the charge distribution of the 75 nm well. The experimentally observed resonances for this well are shown in Fig. 29. The resonance positions for the fundamental absorption modes at $\sim 86\,cm^{-1}$ and $\sim 47\,cm^{-1}$, for the 75 nm and 200 nm wells respectively, are nearly independent of gate voltage and carrier densities, in accordance with the generalized Kohn theorem (Brey et al. 1989), which predicts complete independence for an ideal harmonic oscillator. An additional weak mode near $103\,cm^{-1}$ in the 75 nm sample is a satellite resonance due to deviations from parabolicity associated with the steep sidewalls of the experimentally studied well.

Pinsukanjana et al. (1992) showed that additional absorption modes (surface plasmon modes) occur in the wide electron layers using the grating coupler technique. Surface plasmon modes have been seen previously in thin metal films and in 2D electron layers in heterostructures and MOSFETs. However, coupling of surface modes on either side of the wide electron layers in the parabolic well changes the frequency of these modes as compared to the single isolated surface plasmon modes. By changing the width of the electron layer via a front gate, Pinsukanjana et al. could systematically change the coupling and hence the observed frequency of absorption for these modes. Absorption at the highest observed frequency plasmon mode, corresponding to the $k = 0$ fundamental Kohn mode, was also detected as in the preceding work (Wixforth et al. 1991) at a frequency nearly independent of well filling.

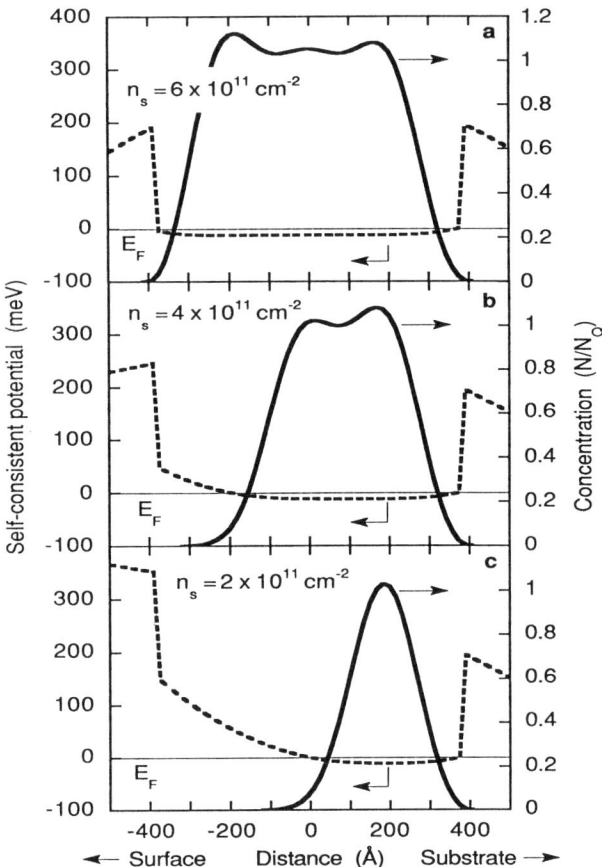

FIG. 28. Self-consistent potential and charge distribution in 75 nm parabolic well ($N_Q \sim 10^{17}\,\text{cm}^{-3}$) with abrupt barriers, at electron sheet densities N_s equal to (a) $6 \times 10^{11}\,\text{cm}^{-2}$ (completely filled well) at a gate voltage $V_G = 0.3$ V, (b) $4 \times 10^{11}\,\text{cm}^{-2}$ (partially filled well) at a gate voltage $V_G = -0.2$ V, and (c) $2 \times 10^{11}\,\text{cm}^{-2}$ (nearly depleted well) at a gate voltage $V_G = -0.65$ V.

For electrons in a perfect parabolic well with a magnetic field in an arbitrary direction, as discussed earlier, long wavelength light is absorbed only at the two frequencies (plasma and cyclotron or admixtures thereof) corresponding to the motion of the center of mass of the electron system. Because of the simplicity of this result, the optical absorption by the conduction electrons becomes a sensitive test of the ideality of parabolic wells. For wells that are not purely parabolic, the plasma and cyclotron frequencies are shifted and transitions between other states of the electron

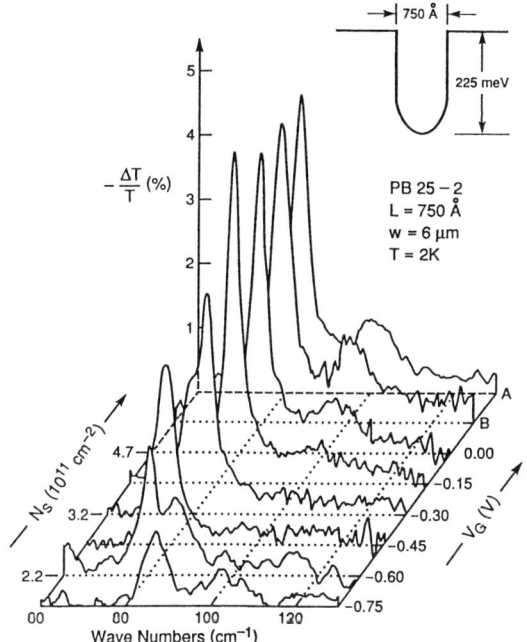

FIG. 29. Absorption versus frequency at various gate voltages at $T = 2K$ for 75 nm parabolic well with abrupt barriers and with a 6 μm period grating on a semi-transparent gate. Traces A and B were taken after illumination to increase the number of carriers. Satellite peaks are associated with deviations from the harmonic oscillator potential.

system are allowed. Brey et al. (1990a), and Stopa and Das Sarma (1992) calculated the shifts in the plasma frequency and the frequencies of new allowed plasma modes in zero magnetic field in imperfect parabolic quantum wells. The former considered the effects on absorption from three types of imperfections: the effects of overfilling a well with sharp sidewalls, the effects of nonquadratic terms in the potential, and the effect of a flat bottom in the potential. In general, for small perturbations, the position of the main peak is shifted slightly (due to a shift in the frequency of oscillation of the center of mass of the electrons) and new satellite peaks appear. The satellite peaks arise from transitions that are forbidden in perfectly parabolic potentials but that become allowed when the parabolic symmetry is lifted. With stronger perturbations, more new lines appear. The small satellite resonances observed by Wixforth et al. (1991) demonstrate the coupling of the far infrared radiation to the internal oscillations of the electron system. The high frequency line associated with the sharp sidewalls at the edges of the electron gas at high filling of the well disappears at intermediate filling where only one line is observed. At low filling of the well, a high frequency satellite associated with the lower sidewall of the parabolic well is seen. The experimental observations are reasonably well explained by the results of numerical studies of the effects of nonparabolic perturbations in wide parabolic wells (Brey et al. 1990a).

In the presence of a magnetic field, the situation in imperfect wells is more complicated. Experiments in magnetic fields on imperfect parabolic wells, where the electron density is perturbed by abrupt sidewalls, have shown extra peaks (Wixforth et al. 1990b). The magnetoplasma excitations were calculated theoretically for the first time by Dempsey and Halperin (1992a, 1992b) for perfect (Fig. 30(a)) and for imperfect (Fig. 30(b)) parabolic wells using a classical hydrodynamic model and later with a quantum mechanical treatment. The presence of modes with frequencies different from the center of mass modes occurs because the parabolic symmetry is broken by the presence of the well walls and by a gate electrode that controls the carrier concentration in the well. For small deviations from parabolicity, the energies of the magnetoplasmon excitations are shifted only slightly from those of the perfect wells, but the nonparabolicity allows excitations to be observed in optical absorption that are inaccessible in the perfect well case. The knowledge of the magnetoplasmon excitations in the perfect well helps in understanding the new peaks observed in the imperfect systems. Calculations of the frequencies of the magnetoplasmon modes in an imperfect well with an abrupt wall and with bulk carrier density $5.9 \times 10^{16}\,\text{cm}^{-3}$, plasma frequency $\omega_0 = 80\,\text{cm}^{-1}$, occupied width $\sim 70\,\text{nm}$, in a magnetic field tilted at $23°$ with respect to the growth axis are shown in Fig. 30(b). The center of mass mode frequencies of the corresponding perfect parabola are indicated by the lines in Fig. 30(a). Absorption in the imperfect well is represented by the dots in Fig. 30(b), with the area of each dot proportional to the oscillator strength of the corresponding transition. The agreement between the theory and the experimental points (square points in Fig. 30(b)) suggests that the experimentally observed modes are the calculated magnetoplasmon modes.

9. Interband Optical Transitions

As mentioned in the discussion of the optical properties of undoped wide wells, the photoluminescence, photoluminescence excitation, and the Raman scattering of undoped wells is understood in terms of the energy levels and transition probabilities for electrons and holes in the undoped well potentials. The situation for doped wells is more complex. Experimental measurements of the PLE data for wide (400 nm–600 nm) undoped and remotely doped parabolic wells have been done by Burnett et al. (1991). The wide

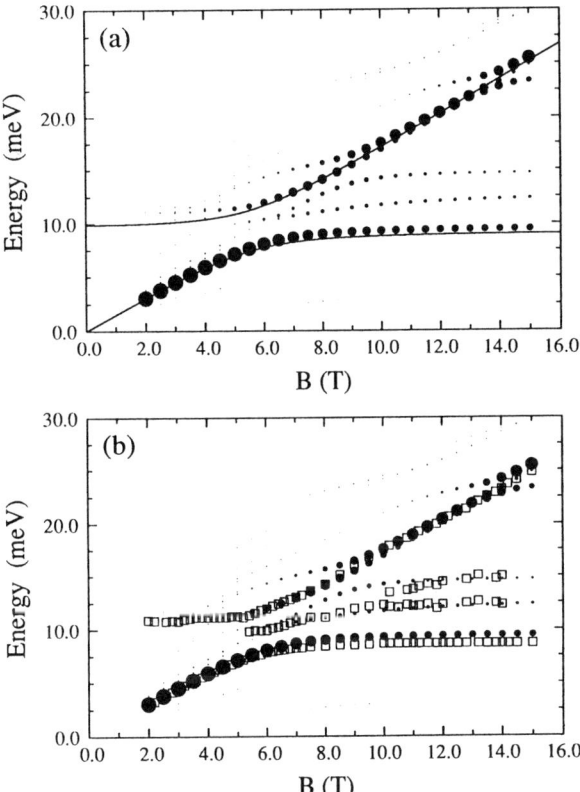

FIG. 30. (a) Calculated and (b) observed frequencies of the magnetoplasmon modes in a parabolic well with abrupt sidewalls, with bulk carrier density $5.9 \times 10^{16}\,\text{cm}^{-3}$, plasma frequency $\omega_0 = 80\,\text{cm}^{-1}$, occupied width between 60 and 75 nm, placed in a magnetic field inclined at an angle 23 degrees from the surface normal. The absorption peaks for the perfect parabola are the upper and lower heavy lines in (a) and are the center of mass modes. The dots in (a) are the absorption peaks for the imperfect parabola, with the dot area indicating the relative absorption strength. The squares in (b) show the positions of the peaks experimentally observed in the study of the transmission of infrared radiation through the sample at $T = 2\,\text{K}$ and are compared to the same dots shown in (a) (Dempsey and Halperin 1992b).

undoped wells show PLE peak positions and intensities that agree well with those calculated for exciton transitions between simple harmonic oscillator levels, in agreement with the data on narrower undoped parabolic wells (Miller *et al.* 1984a). The photoluminescence excitation spectra for the doped parabolic wells is quite different. For energies $\geq 10\,\text{meV}$ above the lowest energy peaks, the PLE spectra are composed of a series of peaks that are nearly equally spaced in energy with a spacing approximately equal to the plasmon energy (Burnett *et al.* 1991). The current explanation of the spectra is that they are dominated by successive allowed transitions from the equally spaced heavy hole states in the resultant parabolic shaped valence band. The lower energy spectra ($\leqslant 10$ meV above E_F) of the doped wells have not been systematically studied.

When the conduction band is flattened by screening, the valence band becomes proportionately more curved but still parabolic, as shown in Fig. 31, and leads to uniformly spaced arrays of heavy hole and light hole energy levels. The heavy hole levels have a spacing about half the observed spacing between the spectral peaks as shown in Fig. 32. Because of parity selection rules, transitions to a given conduction band level would be allowed only from alternate valence band levels, i.e., $\Delta n =$ even, thus giving a spectra with peaks spaced by twice the heavy hole energy spacing, in agreement with the experiment. The $\Delta n \neq 0$ transitions from valence band levels to a single electron level above the Fermi energy should be much stronger than in narrow square wells, due to the wide spatial extent of the conduction band energy levels. Each set of transitions from the heavy hole states to a given electron energy level will have the same spacing; the superposition of spectra resulting from all the transitions will preserve this periodicity.

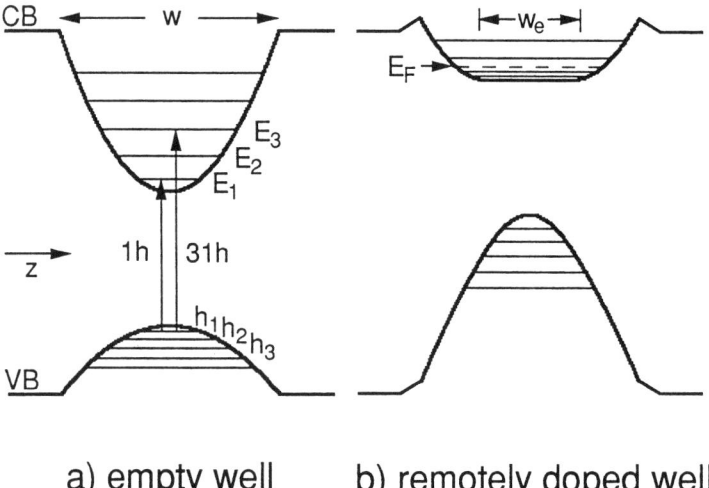

FIG. 31. Schematic comparison of electron and heavy hole energy levels in (a) an empty parabolic well and (b) a remotely doped parabolic well with electron layer width w_e and Fermi energy E_F. Arrows show two allowed interband optical transitions originating from the $n = 1$ heavy hole state that are allowed for the empty well.

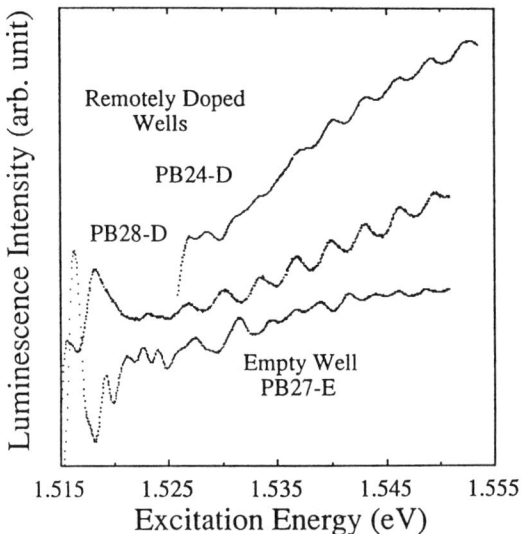

FIG. 32. Photoluminescence excitation spectra for two remotely doped 464 nm parabolic wells ($N_Q = 6.9 \times 10^{15}$ cm^{-3}) at $T = 2$K. Spectrum from an empty well of the same curvature is also shown (Burnett et al. 1991).

VII. Superlattices in Parabolic Wells

10. TRANSPORT

Stacks of two-dimensional electron layers separated by thin barrier layers have a special interest in that they present an intermediate case between a two-dimensional system and a three-dimensional system. If the barrier layers between the two-dimensional electron layers are wide and high, the two-dimensional layers will behave independently. In the limit that the barriers become low or thin, electrons tunnel increasingly freely between the layers, and the system approaches a three-dimensional electron gas with three degrees of motional freedom. Fundamental changes occur in the conductivity and magnetoconductivity of the electrons in the transition from two to three dimensions. For example, Shubnikov–de Haas oscillations in the resistance of a sheet of electrons pass from a situation in which only the perpendicular component of magnetic field determines the oscillations in the resistance to one where the full magnetic field determines the period. Similarly, the quantum Hall effect is expected to show profound changes and ultimately to disappear as the transition from two to three dimensions occurs (Störmer et al. 1986).

Parabolic wells with a superimposed superlattice potential have the advantage that the electron energy levels and density profile can be controllably altered from the case of the simple parabolic well. Interactions between the electron layers in the wells can be adjusted with the height and width of the barriers and the period of the superlattice. Coupling between layers will profoundly change the electron–electron interactions that produce fractional quantization of the Hall effect. In addition, this coupling can be altered with an in-plane magnetic field, particularly at field strengths where the magnetic length becomes comparable to or less than the superlattice period.

The fabrication of arrays of two-dimensional layers and barriers that show the two- to three-dimensional transition presents problems, however, that could not be solved before the availability of wide graded wells. The specific problem that limited previous arrays of coupled two-dimensional electron gases in semiconductors was the strong scattering of the electrons by impurities. Electron occupation of the wells required either dopant atoms in the wells themselves or in the barriers in very close proximity to the wells. In either case, the mobility was severely limited by Coulomb scattering from the ionized impurities (Störmer et al. 1986). Placing the dopant atoms at a greater distance from the coupled wells was not a possibility because of the band bending that widely separated dopant ions and conduction electrons would introduce. The band bending would deplete electrons in wells far from the dopants and accumulate electrons in wells closer to the dopants.

The wide parabolic well offers a means to circumvent these difficulties. The curvature of the parabolic potential can be used to cancel the undesired band bending and to produce arrays of layers with equal widths and equal electron populations. The principle of multiple two-dimensional electron gases within a wide parabolic well is illustrated in Fig. 33. A series of barrier layers is placed within a wide parabolic well that contains dopant layers set back in only the wide, outermost layers. The three-dimensional electron gas of the wide parabolic well is now interrupted by the series of thin internal barriers while the modulation doping advantages of high mobility and relatively large dopant setback are preserved.

Structures of this type have been studied theoretically and fabricated by molecular beam epitaxy using either the digital alloy technique (Sundaram et al. 1992) or the analog alloy technique (Santos et al. 1991). The computed electron density and potential profile for a 200 nm parabolic well containing a 10 period superlattice are shown in Fig. 34, along with the solutions for a 200 nm parabolic well not containing a superlattice. The near equivalence of the electron density in the interior eight wells is readily apparent, along with the decreased electron density in the outermost wells that is attributable to only partial filling of the well.

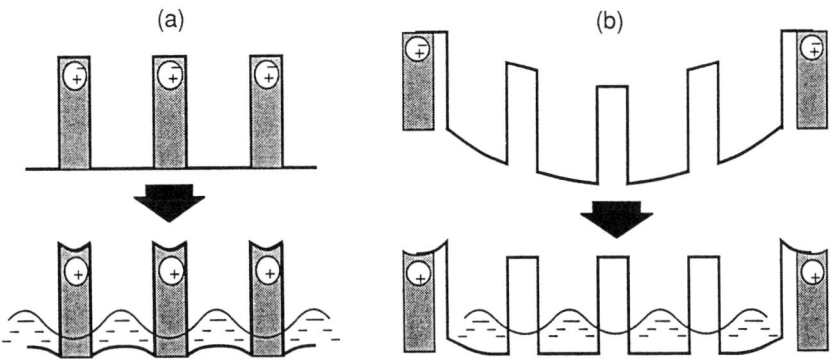

FIG. 33. Superlattice structures for producing coupled sheets of electron gases. The top of this figure illustrates band edges before transfer of electrons from dopants to wells, and the bottom shows band edges and carrier distributions after electrons transfer to the wells for (a) a modulation-doped superlattice, and (b) a superlattice in a parabolic well with only the exterior barriers doped.

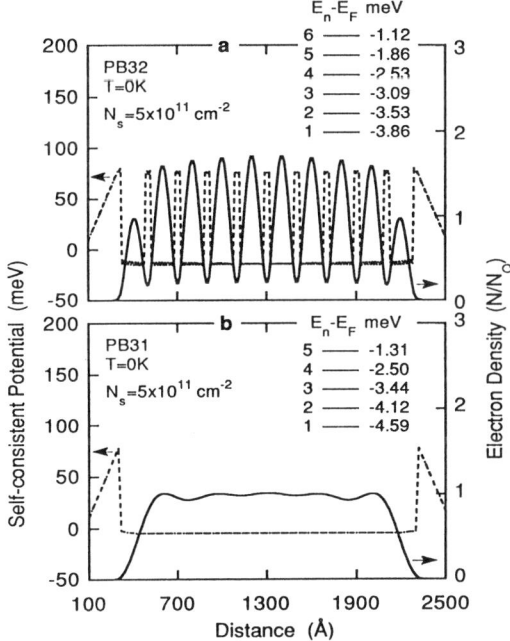

FIG. 34. Computed electron density and potential profile for a 200 nm parabolic well containing a 10 period superlattice with 16 nm wells and 4 nm barriers of height $\Delta x = 0.1$, and doped external barriers (top) and solutions for a 200 nm parabolic well with doped external barriers but not containing a superlattice (bottom). The parabola curvature $N_Q = 2.5 \times 10^{16}$ cm^{-3}.

Electrons in a parabolic well with a strong ($\Delta x_{Al} = 0.1$) 20 nm period superimposed superlattice potential showed low temperature ($T = 4.2$ K) mobilities of 1.1×10^5 cm^2/Vs and a total number of carriers $N_s = 2.0 \times 10^{11}$ cm^{-2} (Sundaram et al. 1992). This mobility is more than 10 times higher than was obtained in ungraded superlattices in which each barrier was doped (Störmer et al. 1986). The mobility of the electrons in the parabolic well plus superlattice structure is probably limited predominantly by alloy disorder scattering and unintentional impurities associated with the increased Al concentration in the superlattice barriers. With increasing carrier concentration, the mobility decreased, as in the simple wide parabolic well, again probably the result of higher disorder scattering and impurity scattering in wider electron distributions. Figure 35 shows measurements of the CV profile between a gate electrode and the electron gas, demonstrating the controllability of the number of electrons in the superlattice by the gate voltage, as well as prominent steps in the CV trace that results from oscillations in the apparent electron density profile from successive depopulation of the wells of the superlattice (Sundaram et al. 1992; Baskey et al. 1992). Low temperature (down to 50 mK) transport measurements in a transverse magnetic field were made on a different sample from the same wafer (Rimberg et al. 1992). Fourier analysis of the Shubnikov–de Haas oscillations in the magnetoresistance at low magnetic fields ($B < 0.3$ T) reveals four frequencies, indicating that four subbands are occupied at zero magnetic field. The experimental energy level splittings and the Fermi energy obtained from this analysis are dramatically reduced from

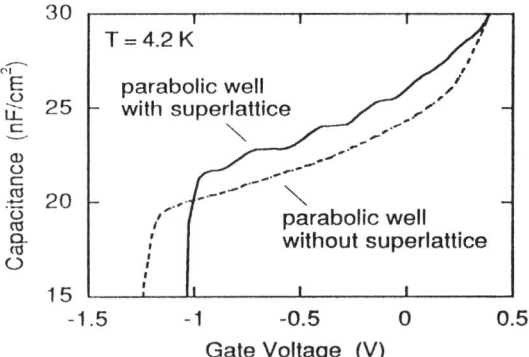

FIG. 35. Capacitance versus gate voltage for 200 nm parabolic wells ($N_Q = 2.5 \times 10^{16}$ cm^{-3}) *with* a 20 nm period superlattice (solid) and *without* (dashed). Steps in the solid CV trace correspond to depletion of electrons from successive periods of the superlattice and are evidence of electron density modulation.

the case of a parabolic well without the superimposed superlattice and are in good agreement with self-consistent calculations.

Jo et al. (1990) and Santos et al. (1991) have studied a 300 nm parabolic well with a superimposed 20 nm period sinusoidally modulated potential. The modulated potential was created by ramping the temperature of the Al furnace, with $\Delta x_{Al} = 0.05$ (peak to peak). The well contained 1.5×10^{11} electrons/cm^2 with a mobility of 1.1×10^5 cm^2/Vs. The integer quantum Hall effect was observed in this structure, in which calculations predicted that five subbands should be occupied.

Brey et al. (1990b) calculated the electron energy levels, effective masses, and expected optical absorption for superlattices in parabolic wells as a function of superlattice period. With thin barriers such that the superlattice bandwidth is much larger than the spacing between ideal parabolic well energy levels, the superlattice behaves like a parabolic well, but with an effective mass equal to the mass of the lowest miniband of the superlattice. As the barriers become thicker (or taller), the superlattice bandwidth becomes comparable to the energy level spacing of the ideal parabolic well, and the energy levels approach the two-dimensional quantum well levels of isolated single wells. For thin (short) barriers, the optical absorption spectrum consists of a single absorption line, independent of electron concentration and interactions between electrons. With thicker barriers, the calculated optical absorption peak shifts to lower frequencies and satellite lines appear.

11. Magnetoplasmon Dispersion

Measurement of the electron plasma response in parabolic wells containing superlattices offers an exceptional opportunity to measure plasmon dispersion in an electron gas, i.e., the dependence of the energy of electron plasma waves on the wave vector. The plasmon dispersion is sensitively dependent on the interactions between electrons in the gas. More specifically, the repulsive interactions between electrons causes the electron pair correlation functions to peak near the electron mean separation r. This lowers the energy of plasma excitations having wavelengths of this magnitude. Measurement of the collective excitation spectra may therefore be expected to show minima at wave vectors $q = 1/r$ (Prange and Girvin 1987).

By inserting a superlattice in the wide well, it is possible to couple not only to the Kohn mode but to additional collective electron excitations that have the same wavelength as the superlattice period. The superlattice wavelength corresponds to a larger wave vector q than can readily be probed with any other existing technique. In addition to measurements with superlattices in the wells, measurements at different filling widths also provided a selective access to plasmons with particular wavevectors. In each case, the deviation from a perfect infinite parabola, due either to the superlattice or to the edges of the well, results in the observability of new magnetoplasmon modes in addition to the $q = 0$ center of mass mode. Thus, the measurements in the wide parabolic wells were the first measurements of the dispersion of magnetoplasmons (Karraï et al. 1991). The measurements at different filling widths utilized δ-function potential spikes that allowed coupling of the incident radiation to standing magnetoplasma waves (also known as *dimensional resonances*) having wave vectors $q = n\pi/W$, where n is an integer and W is the width of the electron gas, and also $q = 2n\pi/a$, where a is the period of the perturbation due to the periodic potential spikes. The observed satellite resonance frequency ω is given by $\omega^2 = \omega_{ps}^2 + \omega_c^2$ for magnetic field parallel to the layers. The plasma component ω_{ps} of the observed hybrid plasma–cyclotron resonance is given by $\omega_{ps} = \sqrt{\omega^2 - \omega_c^2}$, where ω is the observed satellite frequency. A plot of this quantity ω_{ps} divided by ω_p, the Kohn mode center of mass frequency, is shown in Fig. 36 and shows a minimum for $ql_0 = 2$, where l_0 is the magnetic length $(\hbar/eB)^{1/2}$. The frequency of the plasma satellite mode ω_{ps} is thus lower than the $q = 0$ Kohn mode frequency ω_p and shows the minimum characteristically known as the *roton minimum*, in analogy with a similar minimum observed in the excitations of liquid ^{4}He. It occurs, as expected, near $q \sim 2/l_0$ and, in fact, near the value predicted from a single mode approximation calculation also shown in the figure. The minimum is not as deep as the single mode approximation model predicts, however. It may be concluded from these studies that potential modulations inserted in wide graded wells give a good opportunity to observe short wavelength electron excitations that are highly sensitive to electron–electron interactions and provide an important step along the way to understanding and observing the ordered electron states of the three-dimensional electron gas.

VIII. Conclusions

It is clear that wide graded potential wells offer a means to extend the study of high-quality electron gases to new geometries with broader spatial extents than had previously been possible. Growth is a substantially important issue in the formation of compositionally graded structures. Since the local effective positive background charge of a bare graded well is proportional to the local second derivative of the composition profile, a uniform constant charge density profile requires a precisely controlled

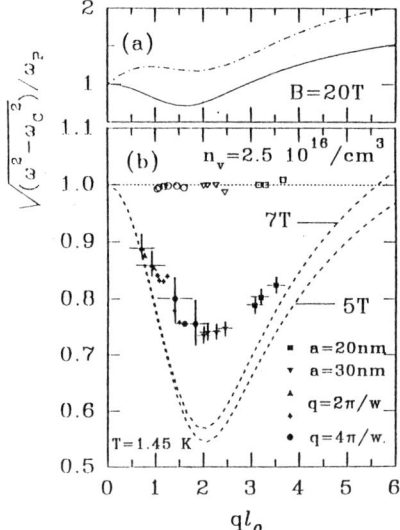

FIG. 36. Deduced dispersion of plasma satellite mode frequency ω_{ps} divided by ω_p, the Kohn mode center of mass frequency for superlattice in parabolic well. ω_{ps} is equal to $\sqrt{\omega^2 - \omega_c^2}$, where ω is the observed satellite frequency. The plot shows a minimum near $ql_0 \approx 2$. The frequency of the plasma satellite mode ω_{ps} is lower than the $q = 0$ Kohn mode frequency ω_p and shows a "roton minimum." The dashed lines are values calculated in a single mode approximation. The filled symbols are experimental points for the satellite modes and the open circles are experimental points for the Kohn mode frequencies. Here ql_0 is taken to be $2\pi l_0/a$ for the grating modes (symbols denoted by a) and $N\pi l_0/w_e$ for the dimensional resonances (symbols denoted by q). The curvature of the well N_Q is denoted by $n_v = 2.5 \times 10^{16}\,\text{cm}^{-3}$. The traces in (a) are calculated dispersions at higher magnetic field (Karraï et al. 1991).

composition profile. Small ripples or flat spots in the profile are expected to give strongly nonuniform charge densities. Molecular beam epitaxy provides a precisely controlled growth technique suitable for producing these closely controlled composition profiles. Although most of the work done to date has been to obtain a uniform electron density profile, arbitrary electron and hole density profiles are possible with suitably grown wells.

The intial motivation for creating the high-quality quasi-three-dimensional electron gas was to observe electron ordering in the form of spin density waves or electron crystallization; although these phenomena have not been observed (and it is still hoped that they will be), the behavior of the electron system has nonetheless proved to be quite rich. Our knowledge of the physics of the three-dimensional electron gas is by no means complete. The work on the approach to the three-dimensional electron gas in wide parabolic quantum wells is perhaps the most advanced and probably the most promising approach presently available to observe the range of phenomena that are expected for the ideal three-dimensional electron system.

Applications of the wide parabolic quantum wells are still in their infancy. Since the materials form resonant structures in the submillimeter and far infrared frequency ranges, it seems probable that they will find applications in these areas.

Acknowledgments

We have benefited a great deal from extensive discussions and collaborations with many colleagues. It is a pleasure to thank B. Halperin, R. Westervelt, W. Paul, A. Rimberg, J. Burnett, J. Dempsey, J. Baskey, S. Yang, and H, Cheong of Harvard University; J. Kotthaus, A. Wixforth, K. Ensslin, M. Kaloudis, and C. Pistitsch of the University of Munich; L. Brey of the University of Madrid; N. Johnson of Oxford University; H. Kroemer, J. Merz, P. Petroff, S. J. Allen, Jr., E. Gwinn, M. Sherwin, K. Campman, B. Bewley, C. Felix, J. Plombon, K. Craig, P. Pinsukanjana, E. Yuh, A. Markelz, N. Asmar, D. Donnelly, and V. Jaccarino of the University of California, Santa Barbara; M. Shayegan, M. Santos, T. Sajoto, and J. Jo of Princeton University; and K. Karraï, D. Drew, X. Ying, M. Stopa, and S. Das Sarma of the University of Maryland. Our thanks in particular to Achim Wixforth for a critical reading of the manuscript.

We would like to acknowledge and thank the United States Air Force Office of Scientific Research for supporting our research work in this field (contracts # AFOSR-88-0099 and AFOSR-91-0214). Our thanks also to the Office of Naval Research for its partial support through the Quantum Institute at UCSB (contract # N000 14-92-J-1452), and the National Science Foundation for its partial support through QUEST, a NSF Science and Technology Center for Quantized Electronic Structures, Grant # DMR-88-10430, at UCSB.

References

Ando, T. (1982). *J. Phys. Soc. Japan* **51**, 3900.
Baskey, J. H., Rimberg, A. J., Yang, Scott, Westervelt, R. M., Hopkins, P. F., and Gossard, A. C. (1992). *Appl. Phys. Lett.* **61**, 1573.
Bastard, G. (1983). *Appl. Phys. Lett.* **43**, 591.
Bastard, G. (1988). *Wave Mechanics Applied to Semiconductor Heterostructures.* Halsted Press, New York.

2. WIDE GRADED POTENTIAL WELLS 215

Bewley, W. W. (1992). Ph.D. dissertation, University of California, Santa Barbara.
Brey, L. (1991). *Phys. Rev. B* **44**, 3772.
Brey, L., and Halperin, B. I. (1989). *Phys. Rev. B* **40**, 11634.
Brey, L., Johnson, N. F., and Halperin, B. I. (1989). *Phys. Rev. B* **40**, 10647.
Brey, L., Dempsey, J., Johnson, N. F., and Halperin, B. I. (1990a). *Phys. Rev. B* **42**, 1240.
Brey, L., Johnson, N. F., and Dempsey, J. (1990b). *Phys. Rev. B* **42**, 2886.
Burnett, J. H., Cheong, H. M., Paul, W., Hopkins, P. F., Gwinn, E. G., Rimberg, A. J., Westervelt, R. M., Sundaram, M., and Gossard, A. C. (1991). *Phys. Rev. B* **43**, 12033.
Celli, V., and Mermin, N. David (1965). *Phys. Rev.* **140**, A839.
Chuang, S. L., and Ahn, D. (1989). *J. Appl. Phys.* **65**, 2822.
Chu-liang, Y., and Qing, Y. (1988). *Phys. Rev. B* **37**, 1364.
Cottey, A A (1967). *Thin Solids Films* **1**, 297.
Dempsey, J., and Halperin, B. I. (1992a). *Phys. Rev. B* **45**, 1719.
Dempsey, J., and Halperin, B. I. (1992b). *Phys. Rev. B* **45**, 3902.
Ensslin, K., Sundaram, M., Wixforth, A., English, J. H., and Gossard, A. C. (1991). *Phys. Rev. B* **43**, 9988.
Ensslin, K., Pistitsch, C., Wixforth, A., Kotthaus, J. P., Sundaram, M., Hopkins, P. F., and Gossard, A. C. (1992). *Phys. Rev. B.* **45**, 11407.
Ensslin, K., Wixforth, A., Sundaram, M., Hopkins, P. F., English, J.H., and Gossard, A. C. (1993). *Phys. Rev. B.* **47**, 1366.
Gwinn, E. G., Westervelt, R. M., Hopkins, P. F., Rimberg, A. J., Sundaram, M., and Gossard, A. C. (1989). *Phys. Rev. B* **39**, 6260.
Gwinn, E. G., Hopkins, P. F., Rimberg, A. J., Westervelt, R. M., Sundaram, M., and Gossard, A. C. (1990). *Phys. Rev. B* **41**, 10700.
Halperin, B. I. (1987). *Jpn. J. Appl. Phys.* **26** (Suppl. 3), 1913.
Harbison, J., Peterson, L. D., and Levkoff, J. (1987). *J. Cryst. Growth* **81**, 34.
He, S., Zhang, F. C., Xie, X. C., and Das Sarma, S. (1990). *Phys. Rev. B* **42**, 11376.
Heinonen, O., and Al-Jishi, R. A. (1986). *Phys. Rev. B* **33**, 5461.
Heitmann, D., and Mackens, U. (1986). *Phys. Rev. B* **33**, 8269.
Herling, G. H., and Rustgi, M. L. (1991). *J. Appl. Phys.* **69**, 2328.
Hopkins, P. F. (1990). Ph.D. dissertation, Harvard University.
Hopkins, P. F., Rimberg, A. J., Gwinn, E. G., Westervelt, R. M., Sundaram, M., and Gossard, A. C. (1990a). *Appl. Phys. Lett.* **57**, 2823.
Hopkins, P. F., Rimberg, A. J., Gwinn, E. G., Westervelt, R. M., Sundaram, M., and Gossard, A. C. (1990b). *Superlatt. and Microstruct.* **9**, 127.
Hopkins, P. F., Campman, K. L., and Gossard, A. C. (1993). *J. Cryst. Growth* **127**, 798.
Ishikawa, T., Nishimura, S., and Tada, K. (1990). *Jpn. J. Appl. Phys.* **29** (Part 1), 1466.
Jo, J., Santos, M., Shayegan, M., Suen, Y. W., Engel, L. W., and Lanzillotto, A. M. (1990). *Appl. Phys. Lett.* **57**, 2130.
Joyce, W. B., and Dixon, R. W. (1977). *Appl. Phys. Lett.* **31**, 354.
Juang, C. (1991). *Jpn. J. Appl. Phys.* **30** (Part 1), 827.
Kane, B. E., Boebinger, G. S., Pfeiffer, L. N., and West, K. W. (1990). *Bull. Am. Phys. Soc.* **35**, 682.
Karraï, K., Drew, H. D., Lee, M. W., and Shayegan, M. (1989a). *Phys. Rev. B* **39**, 1426.
Karraï, K., Ying, X., Drew, H. D., and Shayegan, M. (1989b). *Phys. Rev. B* **40**, 12020.
Karraï, K., Stopa, M., Ying, X., X., Drew, H. D., Das Sarma, S., and Shayegan, M. (1990). *Phys. Rev. B* **42**, 9732.
Karraï, K., Ying, X., Drew, H. D., Santos, M., Shayegan, M., Yang, S. R. E., and MacDonald, A. H. (1991). *Phys. Rev. Lett.* **67**, 3428.

Kawabe, M., Kondo, M., Matsuura, N., and Yamamoto, Kenya. (1983). *Jpn. J. Appl. Phys.* **22**, L64.
Kneschaurek, P., Kamgar, A., and Koch, J. F. (1976). *Phys. Rev.* B **14**, 1610.
Koch, J. F. (1976). *Surf. Sci.* **58**, 104.
Kohn, W. (1961). *Phys. Rev.* **123**, 1242.
Kotthaus, J. P. (1978). *Surf. Sci.* **73**, 472.
Kroemer, H. (1985). *Appl. Phys. Lett.* **46**, 504.
MacDonald, A. H., and Bryant, G. E. (1987). *Phys. Rev. Lett.* **58**, 515.
Menéndez, J., Pinczuk, A., Gossard, A. C., Lamont, M. G., and Cerdeira, F. (1987). *Sol. St. Comm.* **61**, 601.
Miller, R. C., Gossard, A. C., Kleinman, D. A., and Munteanu, O. (1984a). *Phys. Rev.* B **29**, 3740.
Miller, R. C., Kleiman, D. A., and Gossard, A. C. (1984b). *Phys. Rev.* B **29**, 7085.
Miller, R. C., Gossard, A. C., and Kleinman, D. A. (1985). *Phys. Rev.* B **32**, 5443.
Mosser, V., Weiss, D., Klitzing, K. V., Ploog, K., and Weimann, G. (1986). *Sol. St. Comm.* **58**, 5.
Petroff, P., Cho, A. Y., Reinhart, F. K., Gossard, A. C., and Wiegmann, W. (1982). *Phys. Rev. Lett.* **48**, 170.
Pinsukanjana, P. R., Yuh, E. L., Asmar, N. G., Gwinn, E. G., Sundaram, M., and Gossard, A. C. (1992). *Phys. Rev.* B **46**, 7284.
Prange, R. E., and Girvin, S. M., eds. (1987). *The Quantum Hall Effect.* (Springer-Verlag, New York.
Prior, K. A. (1984). *J. Cryst. Growth* **66**, 52.
Rimberg, A. J., and Westervelt, R. M. (1989). *Phys. Rev.* B **40**, 3970.
Rimberg, A. J., Baskey, J. H., Westervelt, R. M., Hopkins, P. F., Sundaram, M., and Gossard, A. C. (1992). *Superlatt. and Microstruct.* **11**, 317.
Rimberg, A. J., Yang, S., Dempsey, J., Baskey, J. H., Westervelt, R. M., Sundaram, M., and Gossard, A. C. (1993). *Appl. Phys. Lett.* **62**, 390.
Sajoto, T., Jo, J., Santos, M., and Shayegan, M. (1989a). *Appl. Phys. Lett.* **55**, 1430.
Sajoto, T., Jo, J., Wei, H. P., Santos, M., and Shayegan, M. (1989b). *J. Vac. Sci. Technol.* B **7**, 311.
Sakaki, H., Noda, T., Hirakawa, K., Tanaka, M., and Matsusue, T. (1987). *Appl. Phys. Lett.* **51**, 1934.
Sandhu, A., Nakata, Y., Sasa, S., Kodama, K., and Hiyamizu, S. (1987). *Jpn. J. Appl. Phys.* **26**, 1709.
Santos, M., Jo, J., Shayegan, M., and Lanzillotto, A. M. (1991). *J. Cryst. Growth* **111**, 366.
Schlesinger, Z., Hwang, J. C. M., and Allen, S. J. Jr (1983). *Phys. Rev. Lett.* **50**, 2098.
Schubert, E. F., Capasso, F., Hutchinson, A. L., Sen, S., and Gossard, A. C. (1990). *Appl. Phys. Lett.* **57**, 2820.
Sen, S., Capasso, F., Gossard, A. C., Spah, R. A., Hutchinson, A. L., and Chu, S. N. G. (1987). *Appl. Phys. Lett.* **51**, 1428.
Shayegan, M., Sajoto, T., Santos, M., and Silvestre, C. (1988). *Appl. Phys. Lett.* **53**, 791.
Shayegan, M., Sajoto, T., Jo, J., Santos, M., and Drew, H. D. (1989). *Phys. Rev.* B **40**, 3476.
Shayegan, M., Jo, J., Suen, Y. W., Santos, M., and Goldman, V. J. (1990). *Phys. Rev. Lett.* **65**, 2916.
Smith, T. P., III, Goldberg, B. B., and Stiles, P. (1985). *Phys. Rev.* B **32**, 2696.
Smrcka, L. (1990). *J. Phys.-Cond. Matt.* **2**, 8337.
Sputz, S. K., and Gossard, A. C. (1988). *Phys. Rev.* B **38**, 3553.
Stopa, M. P., and Das Sarma, S. (1989). *Phys. Rev.* B **40**, 10048.
Stopa, M. P., and Das Sarma, S. (1992). *Phys. Rev.* B **45**, 8526.
Störmer, H. L., Eisenstein, J. P., Gossard, A. C., Wiegmann, W., and Baldwin, K. (1986). *Phys. Rev. Lett.* **56**, 85.

Sundaram, M. (1991). Ph.D. dissertation. University of California, Santa Barbara.
Sundaram, M., Gossard, A. C., English, J. H., and Westervelt, R. M. (1988). *Superlatt. and Microstruct.* **4**, 683.
Sundaram, M., Ensslin, K., Wixforth, A., and Gossard, A. C. (1991a). *Superlatt. and Microstruct.* **10**, 157.
Sundaram, M., Gossard, A. C., and Holtz, P. O. (1991b). *J. Appl. Phys.* **69**, 2370.
Sundaram, M., Wixforth, A., Geels, R. S., Gossard, A. C., and English, J. H. (1991c). *J. Vac. Sci. Technol. B* **9**, 1524.
Sundaram, M., Wixforth, A., Hopkins, P. F., and Gossard, A. C. (1992). *J. Appl. Phys.* **72**, 1460.
Sundaram, M., and Gossard, A. C. (1993). *J. Appl. Phys.* **73**, 251.
Tsui, D. C., Allen, S. J., Jr., Logan, R. A., Kamgar, A., and Coppersmith, S. N. (1978). *Surf. Sci.* **73**, 419.
Walukiewicz, W., Ruda, H. E., Lagowski, J., and Gates, H. C. (1984). *Phys. Rev. B* **30**, 4571.
Walukiewicz, W., Hopkins, P. F., Sundaram, M., and Gossard, A. C. (1991). *Phys. Rev. B* **44**, 10909.
Weisbuch, C., and Vinter, B. (1991). *Quantum Semiconductor Structures*. Academic Press, Boston.
Wixforth, A., Sundaram, M., Donnelly, D., English, J. H., and Gossard, A. C. (1990a). *Surf. Sci.* **228**, 489.
Wixforth, A., Sundaram, M., English, J. H., and Gossard, A. C. (1990b). *Proc. 20th Int. Conf. Phys. Semicond.* (E. M. Anastassakis and J. D. Joannopoulos, eds.), p. 1705. World Scientific, Singapore.
Wixforth, A., Sundaram, M., Ensslin, K., English, J. H., and Gossard, A. C. (1990c). *Appl. Phys. Lett.* **56**, 454.
Wixforth, A., Sundaram, M., Ensslin, K., English, J. H., and Gossard, A. C. (1991). *Phys. Rev. B* **43**, 10000.

CHAPTER 3

Direct Growth of Nanometer-Size Quantum Wire Superlattices

Pierre M. Petroff

MATERIALS SCIENCE DEPARTMENT
UNIVERSITY OF CALIFORNIA, SANTA BARBARA

ABSTRACT	219
INTRODUCTION	220
I. GROWTH KINETICS AND INTERFACES IN CONVENTIONAL SUPERLATTICES AND QUANTUM WELLS	221
1. Structure and Kinetics of the Growth Front	222
2. The Surface Chemistry During MBE Deposition	230
3. Surface Kinetics Dependence on the Deposition Regime During Molecular Beam Epitaxy	233
II. THE GROWTH OF LATERAL SUPERLATTICES	235
4. Molecular Beam Epitaxy Method	235
5. Organo-Metallic Vapor Phase Method	240
III. THE SERPENTINE SUPERLATTICE GROWTH	241
IV. REMAINING ISSUES IN THE EPITAXY OF LATERAL SUPERLATTICE TYPE STRUCTURES: SEGREGATION AND SIZE UNIFORMITY	244
V. OPTICAL PROPERTIES OF LATERAL SUPERLATTICE BASED STRUCTURES	247
VI. OTHER LATERAL SUPERLATTICE BASED STRUCTURES	253
VII. ANOTHER DIRECT GROWTH METHOD OF QUANTUM WIRE STRUCTURES	254
VIII. CONCLUSIONS	255
ACKNOWLEDGEMENTS	255
REFERENCES	255

Abstract

In this chapter, we discuss the growth processes that are involved in the direct deposition of lateral superlattices (LSL) and quantum wire superlattices (QWS). In the LSL, a nanometer scale band gap modulation is introduced in a direction parallel to the original growth surface. The QWS consists of a thin LSL layer sandwiched between two wider band gap semiconductor layers. The LSL and QWS are grown on vicinal surfaces by using fractional submonolayer depositions of two III–V compound semiconductors with different band gaps. As an introduction to the LSL structural properties, we extensively discuss the growth kinetic issues relating to the interface structure and chemistry for growth on singular and vicinal surfaces. The critical parameters controlling the size fluctuations and compositions in the LSL and quantum wire structures have been identified. Quantum confinement effects measured by optical methods in serpentine quantum wire superlattices are presented.

Introduction

Modern crystal growth techniques have made it possible to grow ultra-high-quality layers of III–V and II–VI compounds semiconductors (Herman and Sitter 1989; Stringfellow 1989). The most important epitaxial deposition techniques for the growth of ultra-sharp heterostructure semiconductor interfaces are molecular beam epitaxy (MBE), organometallic vapor phase epitaxy (OMVPE), chemical beam epitaxy (CBE) and metalorganic MBE (MOMBE). They all have produced high-quality ultra-thin semiconductor layers with sharp interfaces and high-quality quantum wells or superlattice devices. The advent of quantum confinement effects in compound or elemental semiconductor quantum wells and superlattices has made possible a wide variety of novel quantum effects based devices. Technological developments have lead to ever cleaner growth conditions, which have resulted in record mobilities and ultra-long carrier diffusion length (50–100 μm) in two-dimensional electron gas (2DEG) using the modulation doping technique. Control over the interface quality and dopant positioning have led to dramatic discoveries such as the fractional quantum Hall effect, quantum point contacts, electron waveguides, and the magneto conductance properties of a 2DEG that suggest the existence of a Wigner crystal electron lattice.

It is now well established that the structure and chemical composition of interfaces in the complex heterostructures grown by MBE or OMVPE play a central role in determining the electronic properties of superlattices or quantum well structures. The majority of research efforts in the heterostructure field has been aimed at engineering of the interface and of the band gap in a direction parallel to the growth axis. With the recent emphasis on structures with reduced dimensionalities, the research on heterostructure interfaces has evolved in the direction of controlling the band gap chemistry and structure in the lateral direction (i.e., a direction parallel to the substrate surface). This new type of structure, i.e., the lateral superlattice (LSL) and its direct application to the direct growth of quantum wire superlattices (QWS) constitute the main topics of this chapter.

Two- (2D) and three-dimensional (3D) carrier confinement is achievable relatively easily when the confinement length scale is larger than 100 nm. This is the dimensional range of the so-called mesoscopic devices. All the technologies that permit this mesoscopic regime are based on the lithography methods used for the following carrier confinement schemes: (a) a depletion layer (Lorke et al. 1990; Wharam et al. 1988; Snider et al. 1991; Wiener et al. 1989; Kohl et al. 1989), (b) interdiffusion of a heterostructure (Cibert et al. 1986; Laruelle et al. 1990; Hirayama et al. 1985) or (c) a lateral band gap modulation with a strain field (Kash et al. 1991; Xu et al. 1992; Tan et al. 1991). Unfortunately in most III–V compounds, the interesting effects associated with the presence of a superlattice and quantization are not found until the structure dimensions are well below the 50 nm range. Presently few processing techniques will permit reaching these dimensions. Surprisingly these are primarily based on direct crystal growth (Petroff et al. 1984, 1989; Gaines et al. 1988; Fukui and Saito, 1988) or regrowth (Kapon et al. 1992).

Novel structures involving lateral band gap modulation have recently been manufactured either by direct growth (Petroff et al. 1984, 1989; Gaines et al. 1988). The combination of vertical and lateral band gap modulation offers infinitely richer possibilities for novel structures and devices. We therefore should expect an intensified effort in understanding the growth kinetics and chemistry at interfaces in lateral superlattices.

In this chapter we first discuss the surfaces of epitaxial films grown by MBE and OMVPE. The growth kinetics and structures of the semiconductor surfaces are important since they provide the template for the growth of the lateral superlattices. The structure and composition of interfaces in conventional superlattices (interfaces parallel to the substrate surface) and interfaces in lateral superlattices grown directly by MBE and OMVPE will be compared. A critical discussion of the interface characteristics for the tilted superlattices or the serpentine superlattices is presented. We also discuss the main application of the LSL to two-dimensionally confined structures and present some of the optical properties of quantum wire superlattices.

I. Growth Kinetics and Interfaces in Conventional Superlattices and Quantum Wells

The structural and chemical nature of the interface in conventional heterostructures and quantum well structures in a number of III–V compound semiconductors have been extensively investigated using in situ and ex situ techniques. This section reviews some of the major findings that are relevant to the direct growth of lateral superlattices and quantum wire structures.

1. STRUCTURE AND KINETICS OF THE GROWTH FRONT

a. Growth on Singular Surfaces

These surfaces have an orientation corresponding exactly to a low index crystal direction, e.g., (100), (110) and (111). Unless great care is taken in the preparation of such surfaces, a small misorientation of the surface will exist introducing a low step density. Most structural work on growing surfaces has been carried out on nominally singular surfaces using reflection high energy electron diffraction (RHEED) techniques on the growing surface (Joyce et al. 1990). More recently, in situ ellipsometry and reflectance difference spectroscopy (Aspnes et al. 1988) and surface sensitive X-ray diffraction (Fuoss et al. 1992) have recently been used for gathering information on the chemistry and structure of the growth front for the OMVPE growth.

Most polar compound semiconductors surfaces are reconstructed during epitaxial growth, and as will be seen later, the surface reconstruction plays an essential role in the nucleation kinetics by introducing a strong anisotropy in the diffusion of the group III elements and in the attachment kinetics of atoms at step edges.

The RHEED technique is based on the diffraction of a low energy (5–30 kV) electron beam by the surface during or after deposition. This technique has proved its usefulness by establishing correlations between the surface stoichiometry and surface reconstruction. For example, for the GaAs (001) surface, in a certain temperature and As/Ga pressure range, the As-rich surface is found to have a (2×4) reconstruction with a dimerization of the As surface atoms while the Ga-rich surface has a (4×2) reconstruction. The complete phase diagram of the reconstructed GaAs(100) surface as a function of deposition temperature and As/Ga flux ratio shows more than a dozen variety of reconstructed surfaces with unit cells having different sizes and degrees of symmetry (Daweritz et al. 1990).

On the other hand, the nonpolar GaAs(110) surface has a (1×1) reconstruction with a slight surface buckling (Chaddi 1987). Since one element of the compound is more volatile than the other, a stoichiometry and temperature dependence of the surface reconstruction is typical of most III–V binary compounds with $\{100\}$ or $\{111\}$ surfaces. Ternary compounds surfaces also exhibit surface reconstructions sensitive to stoichiometry and composition. Although the effects of the surface reconstruction should disappear as soon as the next epitaxial layer of the same material has been deposited, there is no evidence that the same will occur at the heterointerface (e.g., AlGaAs–GaAs). In this case, the reconstruction, by controlling the nucleation kinetics may indeed affect the interfacial structure and lead to exchange reactions and interface roughness on an atomic scale.

The main progress in following the growth kinetics during MBE deposition came from the discovery and use of the RHEED oscillations (Neave et al. 1985; Joyce et al. 1990). Briefly, these correspond to the time dependent oscillations of the specularly diffracted electron beam incident at a grazing angle on the growth front. Even though a precise understanding of the quantitative aspects of RHEED is still not available, it has been established by careful transmission electron microscopy (TEM) studies, computer simulations (Myers-Beaghton and Vvedensky 1991; Shitara et al. 1992a, 1992d), and luminescence studies that the time between two maxima in the RHEED intensity *approximately* corresponds to the growth of a complete monolayer for a layer by layer growth mode. Hence, growth rates can be measured with a fair degree of accuracy (3–5%) from the period of the RHEED oscillations. The sensitivity of the RHEED intensity to the presence of a small fraction of monolayer has recently been quantified (Sudjino et al. 1992).

At high enough temperature, the RHEED oscillations disappear since atom migration is rapid enough to ensure the existence of only large islands and minimize surface roughness. Above this critical temperature T_c, step flow growth is dominating. The step flow regime is characterized by the absence of stable island formation on the step terraces during growth.

Similar oscillations are observed during the surface sublimation of an epitaxial film and RHEED oscillations are then used to measure the sublimation rate as a function of temperature.

In OMVPE of GaAs on (100) GaAs surfaces, oscillations of the diffracted intensity have been observed using X-ray diffraction for specific diffracted beams that are sensitive to surface steps. Thus under specific deposition conditions, a layer-by-layer mode is dominant, but the same studies could not detect evidence of a reconstructed surface *during deposition* (Fuoss et al. 1992). However, the GaAs(100)–(2 × 4) was observed after the growth was stopped (Lamellas et al. 1992). The origin of these effects is not presently understood.

The preparation of a singular surface without a small amount of misorientation is difficult to achieve. This small vicinal angle will greatly influence the epitaxial growth process. The surface topography of the singular surface may evolve during growth towards a configuration that is no longer singular on a microscopic scale. The evolution of the surface is regulated by step–step interactions and other kinetic effects such as faceting. For this reason, vicinal surfaces are more easily understood since they constitute a periodic step array.

b. Growth on Vicinal Surfaces

We will illustrate the main characteristics of growth on vicinal surfaces, i.e., surfaces with a small misorientation (0.5°–4°) from a low index crystal directions, by using examples taken from a Monte Carlo computer simulation model developed by Shitara et al. 1992a, 1992d). Recent scanning tunneling microscopy (STM) investigations and RHEED studies of the GaAs vicinal (100) surface (Bressler-Hill et al. 1993) indicate that the main features of the model are experimentally reproduced.

For vicinal surfaces, the RHEED oscillations (see Fig. 1) are also observed under the layer-by-layer growth regime. In this regime, the film nucleation is occurring on the terraces between steps (Fig. 1(a)). As the substrate temperature T_s is raised above a critical temperature T_c, the RHEED oscillations disappear and epitaxy takes place by atom migration and capture at step edges (Fig. 1(b)). This happens for the so-called step flow growth regime. In this regime, islands are not stable on the terraces between steps and the main atomic sinks are the step edges. The value of T_c, will be function of the surface orientation, stoichiometry (As/Ga beam equivalent pressures) and deposition rate (Shitara et al. 1992a, 1992d). The interdependence of these various parameters on T_c, has been studied in detail for the GaAs (001) vicinal surfaces by T. Shitara et al. (Shitara et al. 1992b, Shitara 1992).

As seen in Fig. 2, for a (001) vicinal GaAs surface, the transition temperature T_c is function of the surface step orientation ([110] or [1$\bar{1}$0]), the As/Ga flux ratio and the misorientation angle.

The effects of surface reconstruction on the structure for the vicinal surface terraces are identical to those of the singular surfaces. However, it should be noted that, for the vicinal surface terraces, the step edge morphology may also be influenced by (a) the surface reconstruction and (b) the nature of the chemical bonds terminating the steps.

Pashley et al. (1988, 1991) produced beautiful scanning tunneling micrographs of the GaAs vicinal (001) surface that demonstrated these effects. Indeed, the step roughness appears to be controlled by the size of the reconstructed unit cell. In this case, the long dimension of the (2 × 4) unit cell oriented perpendicular to the step edges produces a kink size dimension of 16 Å. If a heterostructure is formed by depositing for example an AlGaAs layer, these superkinks should not disappear and remain frozen in the structure. The STM image in Fig. 3 shows for a 0.5° (001) vicinal GaAs surface, the anisotropy between the Ga terminated A steps parallel to [1$\bar{1}$0], which are straighter, and the rougher As terminated B steps parallel to [110]. The difference between the two step orientations may be due to reconstruction induced anisotropy in the diffusion of the Ga or to different

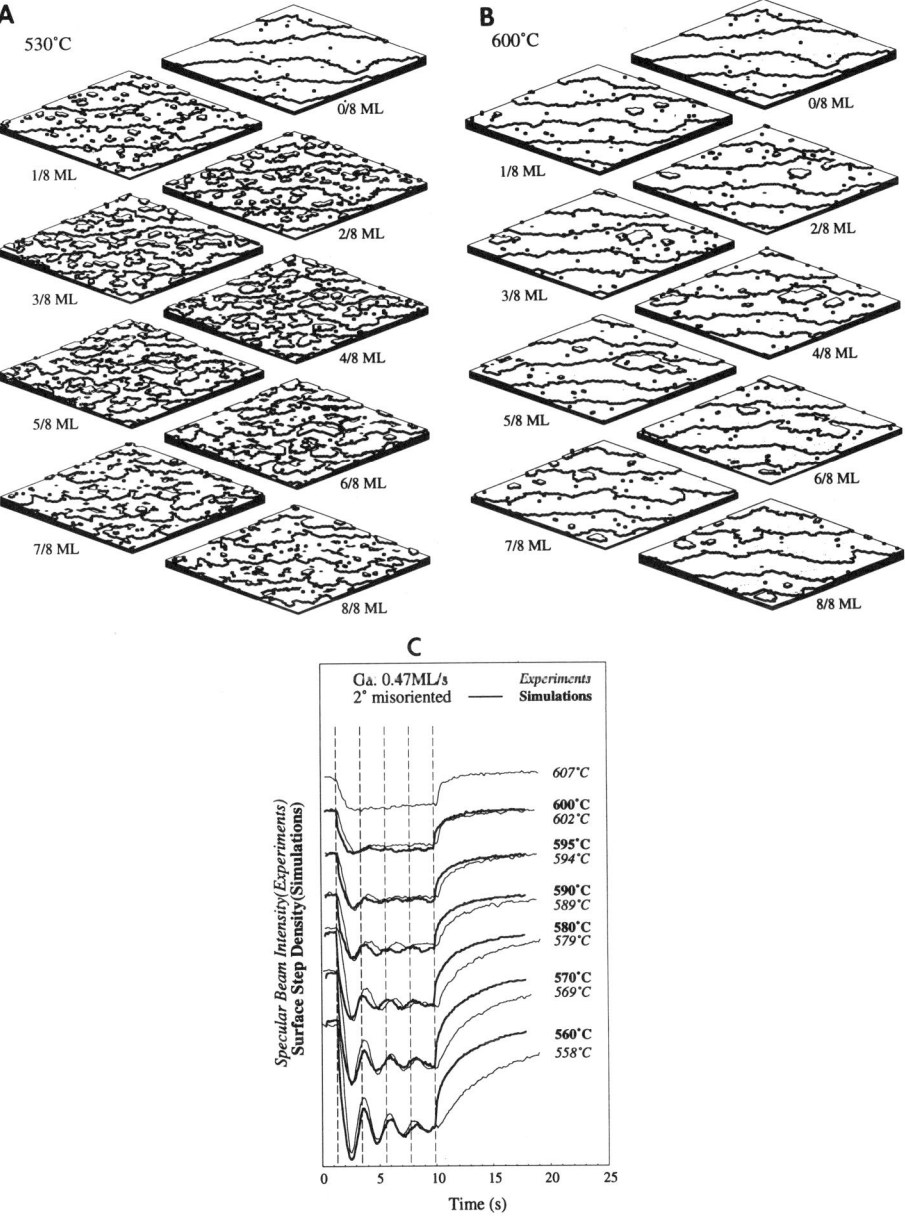

FIG. 1. (a) Computer simulation of the surface morphology of an 80 × 80 lattice sites section of a GaAs{100} surface lattice during the first layer deposition at 530°C. (b) Same as (a) but for a deposition temperature of 600°C. (c) Direct comparison between the measured and computed RHEED intensities as a function of temperature. The Ga deposition rate for the simulations and the RHEED measurements was 0.47 monolayer per second (after Shitara et al. 1992b).

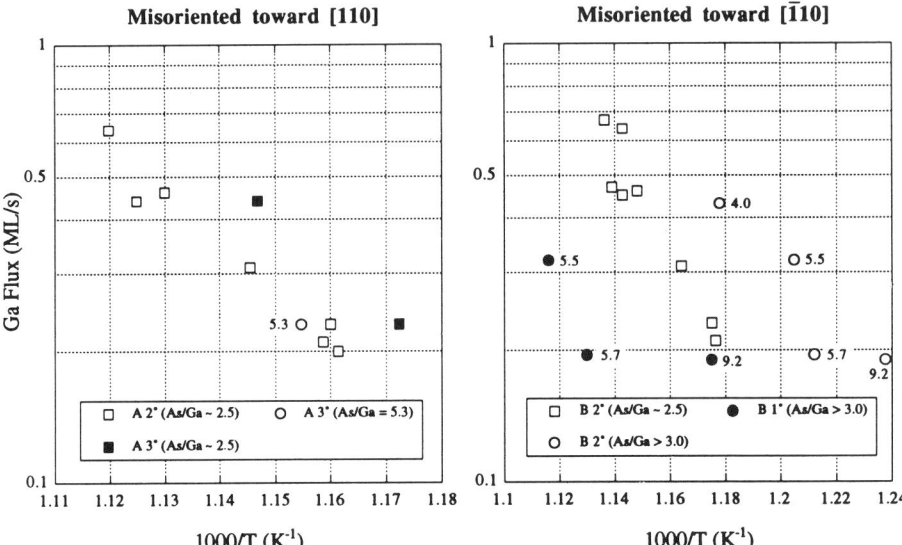

FIG. 2. Dependence of the transition temperature T_c on the As/Ga flux ratio, the misorientation angle and the misorientation direction. The A and B surfaces correspond respectively to vicinal surfaces with step edges parallel to the $[1\bar{1}0]$ and $[110]$ directions (after Shitara et al. 1992a, 1992d).

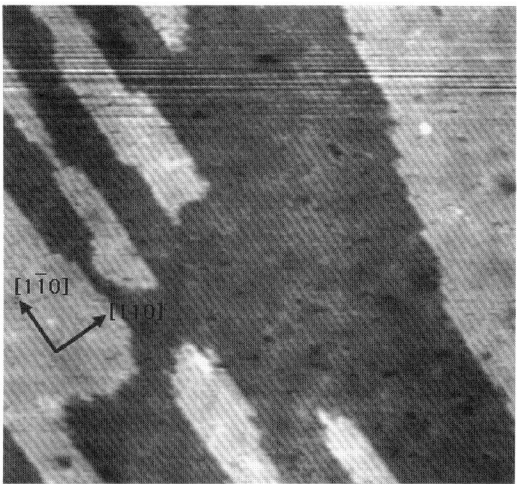

FIG. 3. Scanning tunneling micrograph of a 2×4 reconstructed GaAs $\{100\}$ surface with a smaller than $0.5°$ misorientation angle. Note the anisotropic step structure for the $[1\bar{1}0]$ (smooth steps) and $[110]$ (rough steps) directions (after Pond et al 1993).

attachment kinetics at step edges. As will be seen in Sections III and IV, these features are important in the discussion of an intrinsic interface roughness for lateral superlattices grown on vicinal surfaces.

The period between two maxima of the RHEED oscillations is *approximately* related to the completion of a monolayer (Sudjino *et al.* 1992). However for a vicinal surface, the RHEED oscillations measure only changes in the nuclei density on the terraces. The absolute completion of a monolayer exists when no islands or potholes are present on the terraces and in this case. Computer simulations of the growth process on (100) vicinal surfaces by Shitara *et al.* (1992a, 1992b, 1992d) and the correlation between the surface structure and the computed RHEED specular beam intensities for the surface during the simulated growth are shown in Fig. 1. The agreement between the simulated growth and the experimentally observed ones is remarkable considering the simplicity of the growth model used in the simulation and the use of kinematical diffraction theory. The period of the oscillations increases with the substrate temperature T_s, while the amplitude of the oscillations decreases. The latter result is expected since the density of nuclei on the terraces will decrease because of an increase in the Ga arrival rate at step edges with increasing T_s. The increase in the period is reflecting the change in the attachment rate of atoms at step edges of islands. A large island density at low temperature provides a larger step density and hence a faster monolayer completion, i.e., shorter RHEED oscillation period (Shitara *et al.* 1992).

The RHEED diffraction pattern in the case of a vicinal surface carries a great deal of information on the perfection and evolution of the vicinal surface. The specular beam will show, under "off-Bragg" diffraction conditions and for the incident electron beam perpendicular to the step edges, a double peak structure for the specular beam in the out of phase diffraction condition (Fig. 4) (Pukite *et al.* 1989; Chalmers *et al.* 1989, 1990a). The peak spacing gives a direct measure of the vicinal surface orientation and the full width at half-maximum of these peaks gives a measure of the step ordering. Indeed, surface faceting may be directly evaluated from these profiles.

Using the specular beam intensity profile method, a remarkable step ordering effect is taking place on the vicinal surface while growth is proceeding. The vicinal surface is initially composed of terraces with unequal width, thus giving rise to a broad lines in the specular beam intensity (Fig. 4(a)). However as the growth of a buffer layer is proceeding under step flow conditions, a double peak structure appears out of the broad intensity distribution of the specular beam (Figs 4(b) and 4(d)). This effect arises from the equalization, i.e., the ordering of the steps on the surface. A third possible configuration, illustrated in Fig. 4(c), is associated with surface faceting. Step bunching will produce local facets with a different step

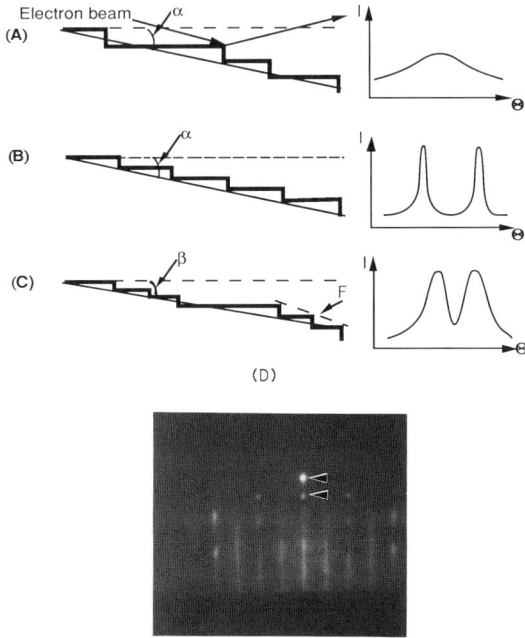

FIG. 4. Schematic of the vicinal surface configurations. (a) Disordered vicinal surface with an average misorientation α. (b) Ordered vicinal surface with misorientation angle α and (c) faceted vicinal surface with misorientation angle changed from α to β. A facet F is shown as a series of closely spaced steps. The intensity profile (I) of the RHEED specular beam is plotted as a function of the scattering angle θ for each case. (d) RHEED pattern of a 2° (100) ordered vicinal surface. The spots indicated by arrows correspond to the ordered array of steps with step edges parallel to the [1$\bar{1}$0] direction.

periodicity on the surface. As a consequence of faceting, the surface will acquire a new vicinal orientation.

This step ordering process has been simulated both by Monte Carlo method (Tokura et al. 1989) and analytically (Gossmann et al. 1990). The step period L on the vicinal surface is statistically controlled, and the step distribution is given by given by $L \pm \sqrt{L}$ (Williams and Krishnamuthy 1992). We note that this statistical modeling does not take into account the two-dimensional character of the surface and the effects of surface reconstruction.

The existence of an energy barrier and diffusion kinetics conditions that prevent atoms from jumping from one terrace to the other during growth appears to be essential to the step ordering process. Presumably the dimerization of bonds at the step edges provides this energy barrier. The step equalization under this assumption is visualized intuitively by considering that the longer terraces receive a larger atom flux than the shorter ones. Therefore, a short terrace adjacent to a long one will grow faster laterally since the growth kinetics under step flow conditions are dominated by surface diffusion. If atoms landing on the terrace have a probability greater than 0.5 to migrate towards the terrace step riser, step equalization will take place. If on the other hand the probability to move to the downward step is greater than 0.5 and they are not reflected by a potential barrier, step bunching will take place. This situation leads to the formation of microfacets (Krishnamurthy et al. 1992).

The characteristics of the RHEED specular beam spot have also been exploited to find the optimal temperature for obtaining a well-ordered $Al_xGa_{1-x}As$ (001) vicinal surface (Chalmers et al. 1989, 1990a). In Fig. 5, the phase field for the well-ordered vicinal surface is given for the $Al_{1-x}Ga_xAs$ surface. This phase diagram was constructed by following the full width at half-maximum of the specularly diffracted beam as a function of the substrate temperature, T_s and for migration enhanced epitaxy conditions (reduced group V elemental flux during the deposition of the group III elements). The points in this phase diagram are defined by the (AlAs) fraction of monolayer that is included on the 2° vicinal surface and the corresponding deposition temperature. The results indicate two phase fields: one corresponding to the formation of a well-ordered surface with no observed RHEED oscillations during deposition and the other corresponding to a poorly ordered surface with island formation on the terraces. In

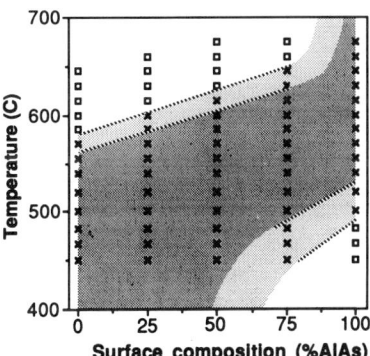

FIG. 5. Phase diagram indicating phase field of the ordered vicinal surface (opened squares) as a function of surface composition and growth temperature. The shaded areas indicate the phase field for a poorly ordered vicinal surface. Both the GaAs and AlAs fractional monolayers are deposited in the migration enhanced epitaxy mode. RHEED oscillations are observed only in the heavily shaded phase field (after Chalmers et al. 1989).

this shaded region of the phase diagram, RHEED oscillations are observed suggesting that layer-by-layer growth dominates. The existence of a low temperature phase field for the Al-rich AlGaAs surfaces is surprising and not presently understood. However we may speculate that the absence of RHEED oscillations in the unshaded part of the phase fields (Fig. 5) suggest the existence of a high density of small and unstable AlAs islands that will induce a step flow growth regime. The phase diagram (Fig. 5) indicates that growth at $T_s = 620°C$ would give rise to a step flow regime for both a GaAs surface and a $Al_{.5}Ga_{.5}As$ surface. However, at this temperature sublimation of the GaAs is also taking place thus making it difficult to control precisely the growth rate.

The surface step ordering process is strongly affected by the presence of surface roughening and microfaceting. This faceting can be detected by observing the shape and spacing of the specular beam peaks (Fig. 4(c)). In many instances faceting is related to surface contamination. Asom et al. (1991) have shown that for the (100) GaAs surface, the presence of oxygen and its gettering on the surface can cause interface roughening in the case of Al containing layer (e.g., AlAs). The insertion of a GaAs–AlGaAs gettering superlattice has been shown to improve the quality fo the interface by eliminating faceting. The desorption of a Ga_2O_3 is known to take place at a substrate temperature $T_s > 620°C$. The microfaceting of the surface is dependent on the Al content of the surface layer. By growing at a higher temperature, the microfaceting is eliminated, suggesting that surface kinetics and oxide desorption during growth are important in the process. Impurities at the surface such as C or O are also reported to promote this surface microfaceting.

2. The Surface Chemistry During MBE Deposition

The great majority of data on interface chemistry has been obtained for the MBE deposited films. Much remains to be done for the OMVPE growth, where in situ characterization methods are not readily available.

During MBE growth, the surface of the epitaxial film is most probably not under thermodynamic equilibrium. This makes it difficult to use conventional equilibrium phase diagrams to predict the chemistry of the surface or interface. It should be noted however that, for some elements, equilibrium thermodynamics have been successfully used (Ivanov et al. 1991) in predicting phase separation and dopant incorporation.

The surface composition is controlled by (1) the surface reconstruction or stoichiometry, (2) the species segregation and (3) exchange reactions at the surface.

In this context, it is worth examining the $(III)_a(III)_b(V)$ as well as $(III)(V)_{a'}(V)_{b'}$ alloys surfaces and interfaces that have been studied during MBE growth (the indices a, and b to group III elements and a', b' refer to different group V elements).

In the $(III)_a(III)_b(V)$ case, several surface and interfaces reactions may be occurring by exposing the $(III)_aV$ surface to a $(III)_bV$ flux. Surface and interface chemical analytical techniques have been used to show that surface segregation of the element that will form the binary (or nearly binary) compound with the *lowest binding energy* will occur (Moison et al. 1991). This compound that will continue forming on the surface during deposition until an equilibrium between the bulk and surface concentrations is established, is in fact responsible for the nonabrupt compositional changes at the heterojunction interface. This equilibrium is controlled by the growth kinetics. For example, Auger spectroscopy (Moison et al. 1991) indicate that at 600°C the AlAs layer deposited on top of a (100) GaAs surface shows the presence of Ga on the AlAs surface. Interface studies of the AlGaAs grown on top of a (100) GaAs epitaxial film using Raman spectroscopy (Jusserand et al. 1990) as well as high resolution transmission electron microscopy (HRTEM) indicate that, for this interface, the transition layer is not compositionally abrupt but rather Ga rich (Ourmazd et al. 1990). The reverse interface formed by growing GaAs on top of AlAs does not show the existence of a compositionally graded interface. A puzzling and unanswered question in this system is that the structurally abrupt interface from detailed photoluminescence measurements (Bimberg et al. 1992) is also the interface found to be not compositionally abrupt by Auger and Raman spectroscopy and HRTEM.

The direction and extent of this surface segregation process follow the In > Ga > Al order (Moison et al. 1991). This process may account for the compositional profile and intrinsic atomic roughness of these heterointerfaces. For example, an InGaAs chemically graded interface has been measured at the $(InAs)_{substrate}/GaAs$ heterointerface. The $(GaAs)_{substrate}/InAs$ reverse inteface on the other hand is found to be compositionally abrupt. Figure 6 schematically shows the surface reaction and the $(III)_a(III)_bV$ interface that will form during MBE deposition of a $(III)_bV$ film on a $(III)_aV$ substrate when the formation of a compound with a lower binding energy is favored (Moison et al. 1991).

The $(III)(V)_{a'}(V)_{b'}$ alloys surfaces that are formed by exposing a III–(V)a' surface to a $(V)_{b'}$ elemental flux are also relevant. These surfaces exhibit dramatic compositional changes as a function of temperature during deposition (Yano et al. 1991). When a $(III)(V)_{a'}$ surface is exposed to a flux of $(V)_{b'}$ atoms, a ternary alloy will tend to form if the growth kinetics are favorable (at high enough temperature). As shown in Fig. 6, the new compound that

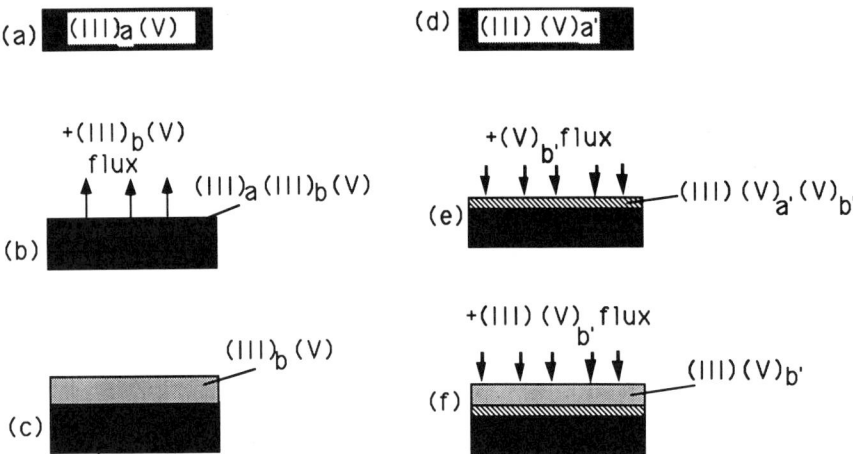

FIG. 6. Formation of an intermediate interface during growth of a heterostructure. Here, (a–c) describe the interface reaction sequence during deposition of a $(III)_b$–(V) film on a $(III)_a$–(V) substrate. The $(III)_a V$ compound is assumed to have a lower binding energy than that of the $(III)_b V$ compound; this will induce the formation of an interfacial layer with a $(III)_a (III)_b V$ composition. Figures, (d–f) describe the interface reaction during exposure of a $III(V)_{a'}$ binary compound to a group $(V)_{b'}$ flux. The $III(V)_{b'}$ compound is assumed to have a lower binding energy than that of the $III(V)_{a'}$ compound; this condition yields an interfacial layer with a $III(V)_{a'}(V)_{b'}$ composition.

will tend to form at the surface is the one with the *larger binding energy*. For example, if the GaSb surface is exposed to an As flux at 520°C, a GaAs-rich GaAsSb alloy will tend to form. The AlAs surface exposed to an Sb flux will be very stable up to 800°C. On the other hand, an InAs surface exposed to an Sb flux at $T > 500°C$ is transformed into an InAs-rich InAsSb alloy.

As shown schematically in Fig. 6, the two cases of interface *exchange reactions* discussed previously may lead to intermediate interface layers that could play a major role in modifying the band structure and changing the physical properties of the heterojunction structure. The driving force and the kinetics of these exchange reactions are not presently understood.

Attempts to control the interface chemistry through the choice of the terminating and starting layers at an interface may be completely nullified if these chemical effects are not well understood and controlled. Lower growth temperatures are a possible solution to these problems, however, in many instances, the growth kinetics will not favor the formation of a structurally flat interface or preserve the vicinal orientation of the surface.

3. SURFACE KINETICS DEPENDENCE ON THE DEPOSITION REGIME DURING MOLECULAR BEAM EPITAXY

Several deposition modes are possible during MBE and their influence on the growth kinetics has been the subject of numerous studies.

In the regular MBE deposition mode both the group III and group V elements are simultaneously deposited. In the alternate beam epitaxy (ABE) deposition mode, also called the *migration enhanced epitaxy* (MEE) mode, the group III and group V elements are deposited (Horikoshi et al. 1988; Yamaguchi et al. 1988; Yamaguchi and Horikoshi 1989). In the MEE deposition regime, the group III mobility has been found to be significantly enhanced when the group V elements are not incident on the surface. In practice, this condition is realized only approximately since there is always a small partial pressure of the group V elements in the MBE chamber. High-quality GaAs and AlGaAs can be deposited epitaxially in this mode at temperatures as slow as 300°C. Some evidence of the kinetics processes involved is provided by the micro-RHEED and scanning surface imaging experiments during growth (Inoue 1991). The deposited Ga on the GaAs $\{100\}$ surface initially forms small clusters or droplets that will subsequently dissolve and form a monolayer coverage of GaAs when the As beam is incident on the surface.

As demonstrated by photoluminescence measurements of GaAs–AlGaAs quantum well structures, the MEE growth kinetics are favoring a layer-by-layer growth mode at temperatures as low as 200°C (Horikoshi et al. 1988; Yamaguchi et al. 1988; Yamaguchi and Horikoshi 1989).

Growth interruption has also been used to improve the smoothness of interfaces in heterostructures. By interrupting the growth, both exchange interactions and coarsening of the islands are expected. The monolayer island coarsening has been observed along with improvements in the quality of the photoluminescence of GaAs quantum wells grown by this method (Petroff et al. 1987, Bimberg et al. 1990).

During deposition of a ternary III–V compound, the growth kinetics may also promote a phase separation. The occurrence of this phase separation is function of the surface orientation and deposition conditions (e.g., MBE or MEE). Such phase separations have been reported for the AlGaAs deposited by MBE or $\{110\}$ vicinal surfaces (Petroff et al. 1982), and AlInAs deposited by MBE on the $\{100\}$ vicinal surface (Hu et al. 1988). The kinetics and the processes involved in the phase separation are presently not well understood. However, the presence of steps on the vicinal surface and the occurrence of exchange reactions appear to be essential ingredients of the phase separation kinetics.

AlGaAs was also observed to phase separate when deposited in the MEE mode on the {100} vicinal surfaces (Lu et al. 1990; Lu and Metiu 1991; Petroff et al. 1990). To experimentally show the Al and Ga phase separation, a codeposition of the Al and Ga was carried out under MEE conditions. The Al + Ga coverage was adjusted using the RHEED oscillations to be close to one monolayer. The subsequent As coverage insured the formation of a GaAs + AlAs covered vicinal surface without islands since the growth was carried out in the regime of step flow growth. By repeating this deposition cycle, the self-organization of Al and Ga produced a GaAs–AlAs coherent lateral superlattice that could be imaged in transmission electron microscopy (TEM). TEM experiments showed the Al atoms preferential segregation at the step edges while the Ga atoms covered the remainder of the terrace (Lu et al. 1990; Petroff et al. 1990).

In this case, a plausible explanation of the self-organization kinetic process was obtained through a simulation of the growth using a kinetic Monte Carlo method. In this simulation, atoms are moving on the surface according to predetermined rates. These rates are function of the atomic paths (e.g., direction, nearest neighbor species, attachment or detachment at a step, attachment to a cluster of given size, etc.). The essential elements required in the modeling are (a) a stronger Al–Al interaction compared to the Al–Ga or Ga–Ga interactions and (b) the number and type of first and second nearest neighbor involved when an atom performs a jump on the surface. By considering that, on the {100} surface, there is a fast diffusing and a slow diffusing direction respectively perpendicular and parallel to the direction $[1\bar{1}0]$ of the As dimer on the reconstructed surface, the model is able to reproduce the essential experimental findings. These are (a) the preferential segregation of Al at the step edges and (b) the absence of self-organization if steps of the vicinal surface are parallel to the fast diffusing direction (steps parallel to [110]). In the latter case, the simulations show the Al atoms cluster into long islands perpendicular to the step edges. Fig. 7 shows the results of a kinetic Monte Carlo simulation of the self-organization of the Al at a $[1\bar{1}0]$ step edge for the deposition of a monolayer containing 1/3 Al and 2/3 Ga atomic fluxes.

Recently, the self-organization has been found in an even more striking case: the deposition of alternate $(GaAs)_1$ and $(AlAs)_1$ monolayers in the MBE and MEE regime respectively on a 0.5°(100) vicinal surface. In this case we are led to invoke vertical exchange reactions and a lateral segregation on the surface to explain these observations (Krishnamurthy et al. 1993).

The $(InP)_n(GaP)_m$ short period superlattice with m and n close to 1 deposited by MBE also was reported to self-organize into a pseudo lateral superlattice of InP-rich and GaP-rich layers with no definite periodicity

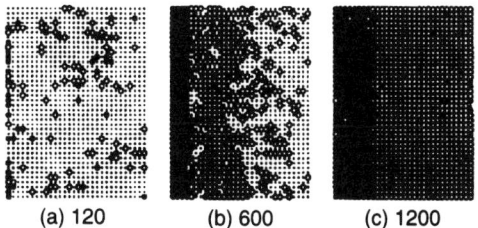

(a) 120   (b) 600   (c) 1200

FIG. 7. Computer simulation of the self-organization during the codeposition of Al (filled circles) and Ga (open circles) on a GaAs {100} surface. The step edge parallel to the [1$\bar{1}$0] is vertical for each frame. The sequence represents various coverage stages of the terraces viewed from the top. The Al and Ga are codeposited in a ratio of 1 to 2 (after Lu et al. 1990).

(Hsieh et al. 1990). A similar observation was also reported for the $(InAs)_n(GaAs)_m$ short period superlattice (Cheng et al. 1992).

Other ternary systems such as AlInAs or AlGaSb or AlInSb may also exhibit a similar self-organization feature when deposited on a vicinal surface. Understanding this self-organization is important since it may be the key to improving interfaces in lateral superlattice structures.

II. The Growth of Lateral Superlattices

Using the material developed in the preceding sections, we will now discuss the growth of lateral superlattices.

4. MOLECULAR BEAM EPITAXY METHOD

a. Growth on Vicinal {100} Oriented Substrates

The growth of lateral superlattices (Petroff et al. 1984; Gaines et al. 1988) relies on the deposition of fractional monolayers and an epitaxial growth controlled by an ordered array of step edges that form an additional template. The idealized LSL growth requires a layer growth mode and a complete segregation of atomic species at step edges. As shown schematically in Fig. 8 for a {100} surface, the LSL deposition is a sequential process in which the two III–V compounds are alternatively deposited. A well-ordered vicinal surface and a step flow growth mode are required. Note that, to achieve a true step flow mode during growth, no RHEED oscillations should be observed (Gaines et al. 1988).

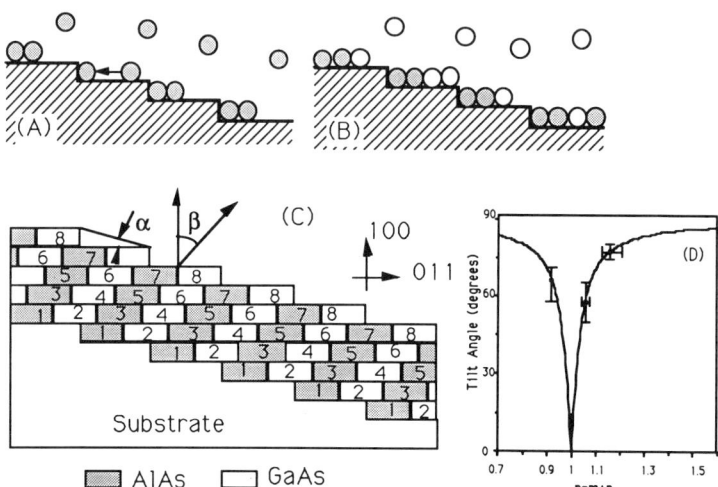

FIG. 8. (a and b) Step flow growth regime in the LSL growth. Only the two group III elements are shown. The group V elements are assumed to uniformly cover each group III fractional monolayer. (c) Schematic of the deposition for a tilted superlattice. The numbers indicate the sequence of deposition for the fractional monolayers. (d) Tilt angle as a function of coverage parameter for a 2° vicinal surface.

Measurement of the deposition rates is carried out using the RHEED oscillations at a substrate temperature that is as close as possible to that used during the LSL growth. Since during the LSL growth, no RHEED oscillations are present, the sticking coefficients and growth kinetics of the various species at $T_s > T_c$ are assumed to be identical to those existing at $T_s < T_c$. The errors involved in this procedure have been discussed in Section I.1.b. Transmission electron microscopy measurements of the LSL tilt angle indicate that these assumptions are valid to within 2–3%. The deposition times required for fractional monolayer depositions are extrapolated linearly from those measured to deposit a monolayer of each of the TSL semiconductors. Typical growth rates for GaAs and AlAs are between 0.3 and 1 monolayer per second for a substrate temperature of 610°C. The deposition temperature T_s is also chosen to ensure that a well-ordered step lattice is maintained at all time during growth. This can be verified by observing the presence of the double peak structure in the RHEED specular beam intensity profile (See Section I.1.b). Sequential deposition of two different III–V semiconductors is programmed as indicated for the example in Fig. 8. Cycle 1,3... correspond to the sequential deposition of a fraction m of $Al_xGa_{1-x}As$ monolayer with $0 < x \leqslant 1$ while those indicated by 2,4... correspond to a deposition of a fraction n of GaAs. The monolayer fractions

m and n are chosen to ensure that their sum $m + n$ is as close as possible to a monolayer coverage, i.e., $m + n = 1$. The LSL tilt angle, β, is controlled by the amount of material deposited in each sequence (Petroff et al. 1984; Gaines et al. 1988). The tilt parameter or coverage parameter $p = m + n$ is therefore related to the tilt angle by

$$\tan \beta = |p - 1|/\tan \alpha \qquad (2)$$

where the substrate misorientation angle from the singular orientation is α and the step height is d.

The period T of the superlattice is given by

$$T = pd/\{\tan^2\alpha + (1 - p)^2\}^{1/2} \qquad (3)$$

We readily see from the schematic in Figs. 8(c) and 8(d) that a small change (few percents) in coverage p will result in a tilt of the LSL interfaces. For $p = 1$, the interfaces are parallel to the [001] axis whereas $p > 1$ or $p < 1$ will produce a positive or negative tilt of the interfaces respectively (Gaines et al. 1988).

The period and tilt angle can be changed by adjusting the coverage parameter p. However, the great sensitivity of β and T to small variations of p is one of the main draw back of this type of structure and very often the LSL exhibits a large tilt angle with the [001] direction and hence the name *tilted superlattice* (TSL).

Indeed, two major sources of errors in controlling precisely the value of p are inherent to the present MBE technology. First, the flux variation over a 2" diameter wafer can be of the order of 3-4%, even for a rotated substrate. This will produce a tilt angle change of 30° for a 2° vicinal surface for two diametrically opposite points on the 2" diameter wafer. The second source of error is associated with the measurements of m and n carried out at a temperature T_s that is below the actual TSL growth temperature. As indicated earlier the errors in establishing the values of m and n from the RHEED oscillations can be as high as 2-3%. The need to limit these sources of error has restricted experiments on the TSL growth to small samples (5×5 mm).

In Fig. 9, we show the TEM cross section through a GaAs-AlAs TSL grown at 610°C on a 2° (100) GaAs vicinal surface. The imaging conditions in Fig. 9 reflect the composition in the structure: the GaAs-rich and AlAs-rich regions correspond respectively to the dark and light regions of this micrograph. The TSL periodicity in the direction parallel to the interface is that of the step array $\cong 80$ Å. The contrast from the GaAs- and AlAs-rich regions in the structure is not as large as what is expected for a

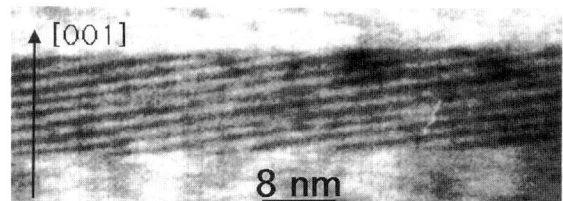

FIG. 9. Cross section transmission electron micrograph of a GaAs–AlAs tilted superlattice. The imaging conditions are such that the GaAs-rich regions appear as dark areas. The substrate misorientation is 2° and the TSL period is 80 Å.

pure GaAs and AlAs compositional fluctuations (Petroff 1989). This is an indication that the existing composition is not the intended one, and the origin of this effect is discussed in Section V.

A remarkable feature of the TSL structure that was revealed by TEM analysis is the difference in the TSL perfection as a function of the step edge orientation. It was found that only the A type $\{100\}$ vicinal surface with Ga terminated step edges parallel to $[1\bar{1}0]$ gave an ordered TSL. The B type $\{100\}$ vicinal surfaces showed a highly disordered and poorly defined TSL. This difference may be explained by the difference in the attachment kinetics at step edges and the intrinsic roughness of the step edge for this orientation (see Section V).

b. Growth on Vicinal $\{110\}$ Substrates

The most severe obstacle to the formation of a perfect TSL is the poor sharpness of the interfaces. Since both TEM analysis and optical properties indicate a poor segregation of Al at the step edges, it is worth attempting the TSL growth on a surface that promotes good segregation.

The initial motivation for exploring the $\{110\}$ vicinal surface for the TSL growth lies in the observation that the AlGaAs deposited in the MBE regime spontaneously phase separates into AlAs- and GaAs-rich regions (Petroff *et al.* 1982). The origin of this self-organization is not well understood.

A striking feature of the growth on the $\{110\}$ surface is that an ordered vicinal surface (as evidenced by two sharp diffraction maxima in the RHEED specular beam) is observed for both the GaAs and the AlAs vicinal surfaces (Krishnamurthy *et al.* 1991) in a large temperature range $480°C < T_s < 540°C$. This is in contrast with the $\{100\}$ GaAs vicinal surface where an ordered GaAs or AlAs vicinal surface did not exist at the same growth temperature (see phase digram in Fig. 5). Furthermore, the absence of RHEED oscillations during MBE growth of GaAs or AlAs on this surface over a wide temperature (450°C to 600°C) range further suggests that surface migration is taking place more easily on the {110} vicinal surface.

Transmission electron microscopy cross sections (Fig. 10) indicate that the GaAs {110} vicinal (for misorientation angles between 0.5° and 2°) surface tends to form a high density of quasi-periodic microfacets that are parallel to step edges. In the present study, vicinal surfaces with $\langle 110 \rangle$ or $\langle 100 \rangle$ step edges were investigated. In this structure, the AlAs layers (white layers) serve as a marker of the GaAs growth front. The microfacets period (0.5–1 μm) is dependent on both the surface orientation and the deposition conditions (As pressure and substrate temperature). These growth characteristics imply that a step flow growth regime is dominant and are also consistent with the absence of RHEED oscillations during deposition. The microfacets are presumably developing via step migration and bunching.

A Monte Carlo simulation of the growth process for this surface (Krishnamurty et al. 1992), shown in Fig. 11(a), is compared to experimentally measured facet densities. The model reproduces well the main surface features: the presence of a quasi-periodic array of facets with a periodicity dependent on the layer thickness. The main components of the modeling are the atomic diffusion length that is larger than half the terrace length and a probability of atomic motion downward the step array that is larger than 0.5. The modeled step configuration (Fig. 11(b)) shows the step bunching and the resulting reorientation of the original surface.

The segregation and TSL interface sharpness problem are not as severe

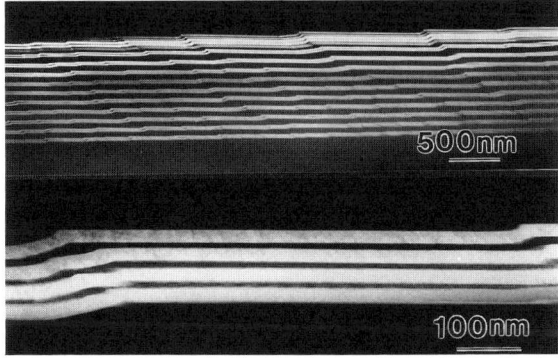

FIG. 10. Cross section transmission electron micrograph showing the evolution of microfacets formation on a 2° vicinal (110) surface. The growth rate was 0.5 ml/s and the growth temperature 500°C. The imaging conditions are such that the white and black regions correspond to the AlAs and GaAs layers, respectively.

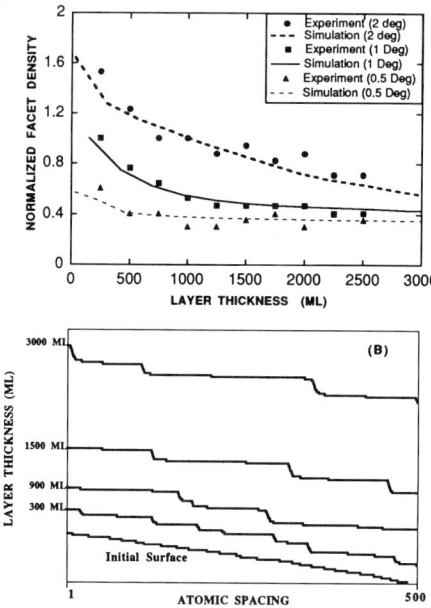

FIG. 11. (a) Variations of the normalized facet density with film thickness. The data are normalized to the maximum facet density for the 1° vicinal (110) surface. The data for several vicinal surfaces are shown along with the simulation results. (b) Formation and evolution of microfacets simulated from an initially uniform step distribution on a vicinal surface.

for this surface orientation as for the vicinal (100) surface. TEM measurements of the Al and Ga distribution were made using a chemically sensitive imaging mode (Petroff 1989) and a couple of random alloys with known compositions for calibration purpose. The results (Krishnamurthy et al. 1991) for the (110) vicinal surface indicate a 40–60% modulation in composition, i.e., the intended AlAs–GaAs TSL ended up with a composition: $Al_{.4}Ga_{.6}As–Al_{.6}Ga_{.4}As$. We will see later that this is an improvement over that measured optically for the same type of TSL deposited on a {100} vicinal surface.

5. Organo-Metallic Vapor Phase Method

The OMVPE growth technique has been used for the LSL growth using the principles described in the preceding section. The difficulty with this growth method lies in the absence of an in situ growth monitoring technique. The RHEED method is not practical, and only recently reflectance difference spectroscopy (RDS) (Aspnes et al. 1988) has become available for growth rate measurements. The second difficulty resides in the accurate control of the incident fluxes. The demands on the gas flow handling system are enormous since the fluxes should be turned on and off accurately and abruptly (on a millisecond scale).

In spite of these difficulties, Fukui and Saito (1988) have succeeded in growing AlAs–GaAs TSL structures by OMVPE. The growth rate calibrations are made in a series of runs in which the superlattice period for thick superlattice and submonolayer superlattices were measured using X-ray diffraction (Fukui and Saito 1988, 1990) and the position of the LSL superlattice spot (Petroff et al. 1984). The method is long and tedious and its use presumes a good reproducibility of the gas flux control from run to run over a long period of time.

An interesting difference between the OMVPE and MBE methods resides in the orientation of the vicinal orientation that gave the best LSL structure. In the OMVPE case the B type {100} vicinal surface with As terminated step edges parallel to [110] yielded the well-ordered LSL. This difference with the MBE growth has been attributed to the role of Hydrogen on the surface.

Fukui and Saito (1990) have argued that a small misorientation (3°) of steps from the [110] direction is in fact beneficial to the quality of the TSL structure. They propose a kink ordering process along the step edges that will eliminate terrace width fluctuations. However, these arguments do not take into account the possible existence of superkinks due to a surface reconstruction. TEM analysis of thick LSL structures indicate that under optimal growth conditions, the LSL uniformity is excellent over areas as large as 5 μm^2 (Fukui and Saito 1990).

The flux uniformity issues over large wafer areas are identical to those presented for the MBE growth and until the flux uniformity issue is solved, only small areas (1 cm^2) of uniformly controlled LSL will be available.

III. The Serpentine Superlattice Growth

To surmount the difficulties associated with the flux calibration accuracy and uniformity over a large wafer area, a novel approach derived from the LSL deposition scheme has been proposed. The so-called serpentine superlattice (SSL) (Miller et al. 1990, 1991, 1992; Petroff et al. 1991) requires that a linear variaton of the coverage p per cycle be maintained between two values p_{max} and p_{min} such that $p_{min} < 1 < p_{max}$. The values p_{max} and p_{min} will be chosen 10 or 20% smaller and larger than $p = 1$. With this method, the

3–5% flux variations across a wafer will not be a major problem since the p variations imposed in the SSL growth is larger than both these errors. Further, the linear ramping of p with time ensures a quantum well superlattices with interfaces in the shape of parabolas (Fig. 12) near the $p = 1$ value (Miller 1992). The interface equation is defined as

$$x = \frac{1}{2}Kz^2. \tag{4}$$

where the coordinate axis ox and oz are parallel the [001] and [110] axis respectively (Fig. 12). The curvature of the parabola is given by

$$K = \frac{2\dfrac{|p_{max} - 1|}{h}}{\tan \alpha} \tag{5}$$

The interface equation is

$$x = \frac{\Delta p z^2}{h \tan \alpha} \tag{6}$$

where h is the thickness of the SSL layer corresponding to a variation of p between p_{min} and p_{max} and $\Delta p = p_{max} - 1$.

This approach warrants that, over a large wafer area, there will always be

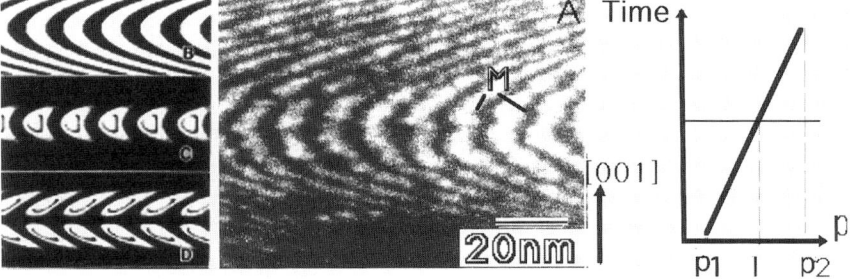

FIG. 12. (a) Cross section TEM micrograph of a single crescent SSL. Black represents areas that are GaAs rich and white regions correspond to the Al-rich digital alloy barrier layers. Arrows in M indicate a meandering of the interface. (b, c and d), the computer simulated SSL structure (b) and the density distributions of electrons confined to the ground state (c) and the first excited states (d) (after Yi et al.; Miller et al. 1991). The oz and ox axis are parallel to the [001] and [110] directions, respectively. The insert on the right shows the time dependence of the coverage parameter $p = m + n$.

a portion of the SSL parabolas with their apex within the film of thickness h. The 2–4% error inherently associated with the coverage determination or with the flux variation over the wafer area, shifts up or down the apex of the parabola by a few percent.

The radius of curvature of the SSL may be adjusted by ramping the coverage rate, i.e., choosing the appropriate Δp for a given h value. The choice of K controls the degree of vertical carrier confinement along the [001] axis.

Modeling of the density distribution function in such structures (Yi et al. 1991, and Yi Dagli 1992; Sundaram et al. 1991), clearly shows (Fig. 12) that the electron ground energy state is associated with electrons confined in the GaAs regions at the apex of the parabolas.

The SSL structures have been grown using barriers consisting of digital alloys. This type of barrier, as shown in Fig. 13, presents two benefits: (a) it eliminates the formation of the coherent tilted superlattice during the deposition of an $Al_{1-x}Ga_xAs$ alloy barrier that exists under MEE conditions (see Section I.3); and (b) it reduces the possible exchange reactions that would trap Al in the GaAs quantum well regions. An added advantage is that the superlattice reflections associated with the digital alloy presence can be used effectively to image the serpentine superlattice (Krishnamurthy et al.

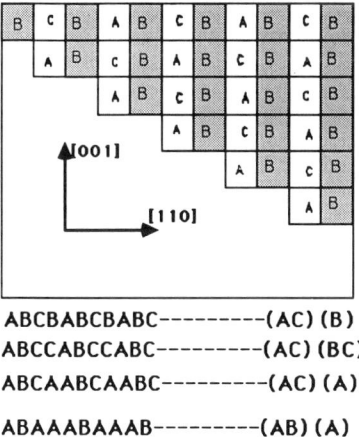

FIG. 13. Schematic of possible stacking sequences for depositing digital binary alloys (a, b and c). Half-monolayers are deposited at a time and no surface exchange reaction is taking place. The righthand side column gives the composition of the vertical barrier and wells that are created for each deposition sequence. The ABAAABAAA stacking sequence yields a digital alloy of the type that was used as a barrier layer in the growth of the serpentine superlattice. In this case, A and B correspond to GaAs and AlAs respectively yielding an average $Al_{.5}Ga_{.5}As$ digital alloy barrier (after Petroff et al. 1989).

1992). This is the case for the cross section TEM micrograph shown in Fig. 12(a) (Krishnamurthy et al. 1992). One notices that near the apex of the parabolas, the SSL interfaces show more irregular boundaries than near the start and finish of the SSL growth sequence. These interface fluctuations are tentatively attributed to meandering of the step edges that form the vicinal surface. These are less visible on the more tilted section of the SSL because of the smaller sensitivity of the tilt angle β to a change in the step density, i.e., a change in the angle α in equation (2). As discussed in the next section, the step meandering shows a memory effect that is frozen during the SSL growth. Figure 12(a) shows several interface fluctuations (M) that are present only along a few adjacent quantum wires.

In the SSL, as in the case of the LSL structure, the TEM contrast analysis indicates that some Al is present in the regions of the structure that should contain only GaAs. The possible reasons for this effect are discussed in the next section.

IV. Remaining Issues in the Epitaxy of Lateral Superlattice Type Structures: Segregation and Size Uniformity

Several issues remain to be solved with this type of structures before the measured properties can meet all the expected theoretical predictions.

The interface sharpness remains the main issue. Indeed, a contrast analysis of the SSL TEM micrographs and the optical measurements (see Section V) indicate that a large amount of the Al intended for the barrier is incorporated in the wire. The incorporation mechanism is not presently clear, and several causes may exist either separately or simulataneously.

As seen the schematic in Fig. 14(a), the trapping of Al may occur by cluster formation and incomplete Al segregation during deposition. Because of surface kinetics, these islands should have a maximum size near the center of the terrace, and a depleted zone should be present near the step riser. This is a likely process since a unique growth temperature is used to deposit both the GaAs and AlAs fractional monolayers. As seen in the phase diagram, Fig. 5, to obtain a good phase segregation two different substrate temperatures should be used.

The second process that may be responsible for the Al trapping is illustrated in Fig. 14(b). The segregation at step edges is complete but superkinks give rise to large-scale step fluctuations. As mentioned in Section I.1.b, the thermally induced superkinks arise from the existence of the surface reconstruction, which induces the formation of low-energy reconstructed (2×4) unit cells with dimensions of $16\,\text{Å} \times 8\,\text{Å}$. Since the optimal

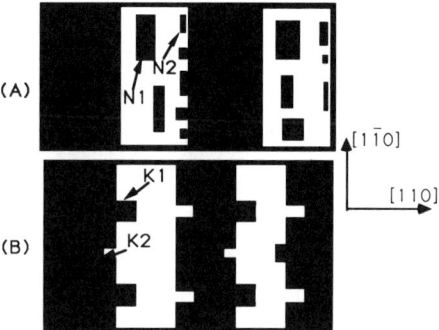

FIG. 14. Schematics of possible growth front configurations during epitaxy. The ordered vicinal surface is viewed from the top and the two different semiconductors epitaxially deposited are shown as shaded and white areas. (a) Epitaxy with island nucleation conditions dominant. Anisotropically elongated islands are shown as $N1$ and $N2$. (b) Surface kinetics are dominated by step flow mode. Superkinks ($K1$) induced by the surface reconstruction and thermal kinks ($K2$) at step edges are indicated.

step orientation for the LSL growth by MBE is orthogonal to the short dimension of the reconstructed cell, the step edge fluctuations induced by the reconstruction will be on average 16 Å. This step roughness will most probably remain embedded in the structure. In a 40 Å wide quantum wire, these superkinks will produce a chemically nonabrupt interface that will significantly reduce the lateral carrier confinement by a very large amount. The superkinks are indicated on the scanning tunneling micrograph of a 1.5° vicinal GaAs (100) surface (Fig. 15). This surface was prepared by MBE deposition. The vicinal surface topography may be completely different if the same surface had been prepared by MEE.

Recent TEM observations (Poudoulec *et al.* 1992) of vicinal surfaces grown by MBE at 595°C substantiate the importance of terrace width variations. The step width fluctuations are found to be worse for steps parallel to the [110] direction (As terminated steps). This observation is also in agreement with the reported poor quality of the LSL grown on the B type vicinal surface (Gaines *et al.* 1988). Note however that in most cases, the LSL are deposited under MEE conditions rather than the conventional MBE conditions, and there is a possibility that the step configuration is strongly affected by the surface stoichiometry during deposition.

The size fluctuation and size uniformity issues are of prime importance if true quantum wire properties are expected (Vahala 1988). Size fluctuations in the wires are due to growth accidents, step pinning or step bunching. Unfortunately, these are not easily detected by in situ techniques. A slow

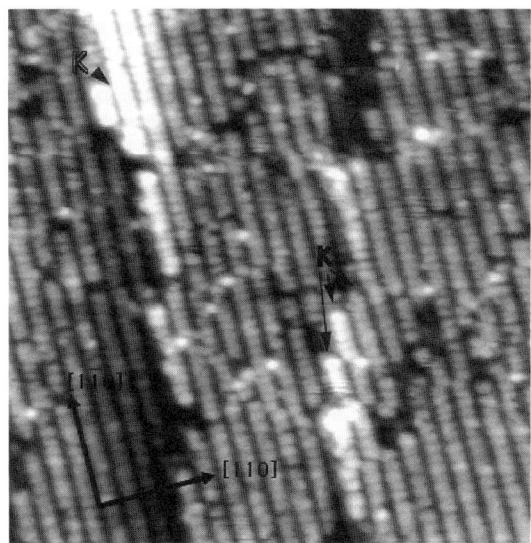

FIG. 15. Scanning tunneling micrograph of a GaAs [100] vicinal surface (1.5° misorientation) grown by MBE. The step edges are parallel to the [1$\bar{1}$0] direction and their average spacing is 120 Å. The superkink (K) size along the [011] direction is a multiple of 16 Å (after Bressler-Hill et al. 1993).

and tedious study using both RHEED and STM analysis will be required for controlling these growth accidents.

A major source of step size fluctuations will be related to the presence of superkinks. If the formation energy for the elemental superkink is W, and the deposition temperature is T, the equilibrium kink concentration per unit length of step edge is

$$n/n_0 = \exp(-W/kT) \tag{7}$$

where n_0 is the is the concentration of kink free sites per unit length of the step and k is Boltzmann's constant. If we assume an equal concentration of positive and negative superkinks, their spacing L is

$$L = 2a(1 + 2\exp(W/kT)) \tag{8}$$

where a is the interatomic distance along the step edge. Unfortunately, the superkink formation energy for a reconstructed step edge is not known. Equation (8) suggests that the lowest deposition temperature still compatible with step flow growth will be beneficial to the interface sharpness since these conditions will favor a lower superkink density.

To solve the step edge roughness problem, we may have to use other step edge orientations for the (100) vicinal surface or other surface orientations such as the {110} (see section II-2) or {111} surfaces. Finally, one should not exclude a possible intermixing through exchange interactions between impinging Al atoms with the surface layer atoms. This effect would prevent the complete segregation of Al species at step edges. The results discussed in Section I.3 suggest that this effect is important for some surface orientations and deposition conditions (Krishnamurthy et al. 1993).

The interface roughness and its amplitude relative to the exciton radius is potentially one of the most severe problem faced when processing utrasmall quantum wires using the LSL approach. Recent photoluminescence data on OMVPE deposited lateral superlattices (Chavez-Pirson et al, 1991) also indicate the existence of an interface roughness and intermixing problem for this growth method

V. Opical Properties of Lateral Superlattice Based Structures

a. Optical Properties of Serpentine Superlattices

Measurements of both the photoluminescence (PL) and resonantly excited photoluminescence (PLE) in the TSL and SSL quantum wire superlattice have recently been carried out (Miller et al. 1991, 1992; Weman et al. 1992a) at low temperature (1.4°K).

The linear polarization of the PL emission is detected at low temperature using a photoelastic modulation technique (Wassermeier et al. 1991). The polarized PL spectra of an SSL structure with light polarized parallel or perpendicular to the wire axis is shown in Fig. 16. The single crescent SSL in this structure had an intended GaAs–$Al_{.33}Ga_{.67}As$ quantum wire superlattice with a period of 80 Å. The linear polarization is defined as

$$P_z = (I_x - I_y)/(I_x + I_y) \qquad (9)$$

The subscript indices are referring to the directions of the polarization direction of the emitted intensity (I) or polarization (P) along the axis shown in the insert to Fig. 16. For an observation direction normal to the sample surface (oz in Fig. 16), the SSL emission shows a linear polarization $P_z = 23\%$ due to the lateral hole confinement (Miller et al. 1992). The measured PL polarization in the perpendicular directions (ox and oy axis as indicated in the insert Fig. 16 are respectively $P_x = 36\%$ and $P_y = 57\%$,

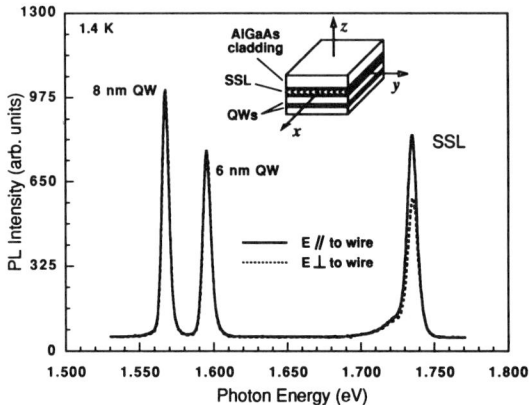

FIG. 16. Polarization dependent PL spectra for a single quantum wire SSL. The observation direction is along the z axis as defined in the inset, which also shows the SSL position and the quantum wells in the sample. The solid and dashed lines are for light polarized parallel and perpendicular to the quantum wires, respectively. The quantum wells emit unpolarized light in this direction whereas the SSL displays a polarization of 23% (after Miller et al. 1992).

where cyclically permuted indices in (9) define the other polarization ratios. This sample also contained two test quantum wells (thicknesses 60 Å and 80Å) that did not show any linear polarization for light emitted along the normal to the sample surface. The polarization for the 60 Å quantum well were respectively $P_z = 0\%$, $P_x = 67\%$ and $P_y = 76\%$. The in-plane linear polarization for the SSL photoluminescence measured along the three directions is a consequence of the heavy and light hole states mixing induced by the quantum wire confinement (Pryor 1991).

The PLE spectra in Fig. 17 were observed by detecting the luminescence in a direction normal to the sample surface. The detection energy was set on the low-energy side of the SSL emission peak. The exciting beam was polarized either in direction parallel or perpendicular to the wires. A Stokes shift of 4.6 meV from the detection energy is observed for the parallel polarization direction of the exciting light. The Stokes shift is 6.6 meV for the perpendicular direction. The 60 Å quantum well does not show such anisotropy effects. The heavy and light hole transitions are indicated in Fig. 17. The 2 meV energy difference between the 1*ehh* and 1*elh* line in the polarized PLE spectrum (Fig. 17) is also consistent with the 80 meV lateral barrier in the conduction band that was obtained from the PL measurements (Miller et al. 1992).

The intrinsic excitonic character of the emitted light has also been observed by following the quantum wire superlattice PL emitted intensity

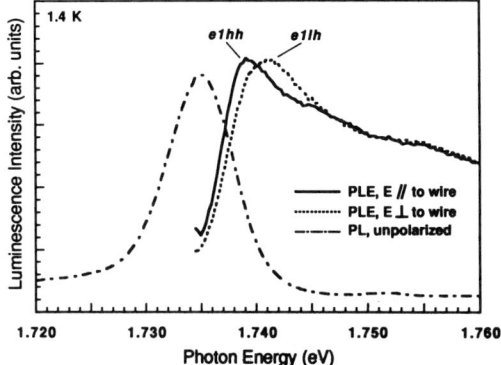

FIG. 17. Polarization dependent PLE spectra of light emitted normal to the sample in the z direction as shown in Fig. 16. The solid and dashed lines are for the existing light polarized parallel and perpendicular to the quantum wire axis, respectively. The dash-dotted line is the unpolarized PL (after Miller et al. 1992).

variations as a function of the laser pump power. The SSL emission is observed to vary linearly over three orders of magnitude whereas the quantum wells have an initial increase that is less than linear and becomes linear at higher pump power intensities (Weman et al. 1992a).

The main advantage of the SSL structure, namely, the good uniformity of the PL polarization, has also been demonstrated. A linear polarization variation in P_z of 2–3% has been measured across a 1.5 cm SSL structure.

The measured lifetimes for the quantum wire superlattice was 379 ps and those of the 80 Å and 60 Å quantum wells in the structures were 250 ps. The lifetimes measured here include both the radiative and nonradiative decay times. The lifetimes in quantum wire superlattices are found to be 1.5 to 2 times larger than those of quantum wells. There is presently no satisfactory explanation for this increase in lifetime in quantum wire structures.

It is worth mentioning also that the line width of the SSL structures are remarkably narrow considering that the well width variations are on the order of 16 Å at least. This effect is not presently understood.

The imperfect segregation of the Al has been estimated by comparing the linear polarization measured in the x, y and z directions with those computed in the framework of an envelope function approximation (Pryor 1991). The conduction band and four coupled valence band states and the optical polarization of the light emitted were calculated as a function of x_B, the Al concentration in the barrier. Theory (Pryor 1991) indicates that a value of $P_z = 100\%$ should be measured if no Al was present in the wire regions of the structure. A value of $P_z = 23\%$ yields a value of $x_B = 0.19$. The other measured polarizations, P_x and P_y, yield values of $x_B = 0.21$ and 0.22, respectively. A value of $x_B = 0.33$ was chosen for this SSL deposition. Thus the quantum wires contain about 10% Al. This value is in good agreement with that obtained from the PLE data that show a 2 meV shift in the PLE emission with the parallel and perpendicular polarization excitation (Miller et al. 1992; Weman et al. (1992a, 1992c). The calculation assuming a uniform redistribution of the Al in the wire and barrier yield a value of $x_B = 0.22$.

In PLE, higher electron subband states have been resolved with a polarization anisotropy due to the laterally induced heavy–light hole splitting (Weman et al. 1992b).

From the PL measurements, the lateral potential difference between the wells and barriers is estimated at 80 meV in the conduction band. The PLE measurements yield a value of 90 meV, in reasonable agreement with the PL measurements.

Two important points need to be made regarding the PL and PLE anisotropy:

1. The PL polarization is a direct proof of 1D confinement effects for the hole states in the SSL structures. However, the computed luminescence properties for this type of structures indicate that with the present lateral barriers and the Al concentration in the well, the 1D confinement applies only to the heavy holes. The conduction band electrons are forming a quasi-2D miniband.
2. The PL anistropy measurements made along the ox, oy and oz directions give consistent results for the amount of Al in the quantum wires. If instead of wires, we had a distribution of elongated AlAs islands, the polarization measurements would not give a consistent value for the Al content for the same measurements.

The low-temperature magneto-luminescence properties of these quantum wire arrays have also been studied (Weman et al. 1992c, 1993) in magnetic fields up to 13 Teslas. The diamagnetic shift from the SSL (Fig. 18) was found to be smaller than for a reference alloy quantum well with the same Al content, indicating an increase in the binding energy of the exciton due to the lateral confinement in the SSL structure. A diamagnetic shift anisotropy is observed when the magnetic field is applied in the three perpendicular directions of the wire. This is a direct observation of the one-dimensional properties of the excitonic wave function. The PL line width increase with the applied magnetic field is dependent on the direction

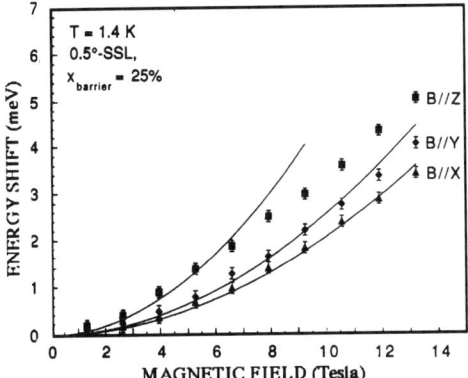

FIG. 18. Energy shift of the photoluminescence peak as a function of applied magnetic field in the x, y and z direction of the serpentine superlattice (see insert to Fig. 16). The experimental points are marked with error bars. The solid lines are quadratic fits expected for diamagnetic excitons at low magnetic fields (after Weman et al. 1992c, 1993).

of the applied field. These properties are consistent with the way the exciton volume is probing the potential variations in the SSL structure.

b. Optical Properties of Quantum Wire Superlattice and Tilted Superlattices

The optical properties of GaAs–AlAs tilted superlattices grown by MBE have also been measured using the photoelastic modulation technique (Weman et al. 1992a; Wassermeier et al. 1991). The TSLs show linear polarization effects similar to those reported for the SSL structure.

The MBE grown GaSb–AlSb TSL quantum wire superlattice structures (Chalmers et al. 1990b) with a structure similar to that shown in Fig. 19(a)

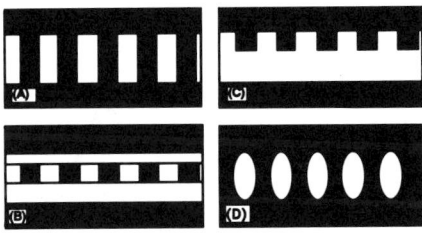

FIG. 19. Various types of structures that have or could be fabricated using the tilted superlattice epitaxial growth: (a) Quantum wire structure, (b) inserted grid structure, (c) corrugated interface structure and (d) the elliptical quantum wire structure. The shaded areas represent the wider band gap material in the structure.

have also been intensively investigated (Chalmers et al. 1992). Here also, an intermixing of the Sb in the quantum wire regions has been detected, both by transmission electron microscopy and photoluminescence polarization measurements. From the TEM, the barrier and wire thicknesses in this structures were both equal to 53 Å and the thickness of the quantum wire superlattice layer was 120 Å. The luminescence polarization and the TEM contrast measurements indicate that the intermixing is not as important in this material system as in the AlAs–GaAs system. From the measured luminescence polarization (Fig. 20) and the model computation (Chalmers et al. 1992), the lateral barrier confinement is 153 meV in the conduction band and the Sb content in the barrier and the wire are, respectively, $x_B = 0.42$ and $x_w = 0.255$. The ground state energy level was found to be at 90 meV above the conduction band minimum of the wire material whereas for the holes, the ground state was 27 meV below the valence band maximum. Thus, in this system both electrons and holes are confined to the wire although the confinement is not as good for the electrons (4/1) as for the holes (6/1).

In OMVPE grown TSL, the linear polarization and nonlinear absorption effects have also been observed at room temperature for thick $(AlAs)_{1/2}$–$(GaAs)_{1/2}$ and $(Al_{.5}Ga_{.5}As)_{1/2}$–$(GaAs)_{1/2}$ TSL structures (Kasu et al. 1991; Chavez-Pirson et al. 1991) but not for quantum wires. The polarized photoluminescence data for these structures indicated that the GaAs quantum wire region also contained a large concentration of Al. However, no precise estimate of the Al content in these superlattices was given. In addition, these thick TSL superlattices exhibit polarization-dependent optical nonlinearities at room temperature that are related to the presence of

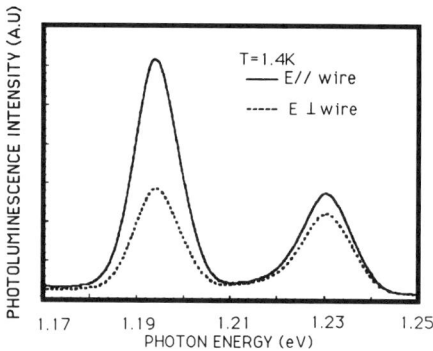

FIG. 20. Photoluminescence spectra for light polarized parallel (solid line) and perpendicular (dotted line) to the wire axis in the GaSb–AlSb quantum wire superlattice at 1.4°K (after Chalmers et al. 1992).

the lateral modulation induced by the TSL structure (Chavez-Pirson et al. 1991). Here again the data indicates a strong intermixing between the barrier and well regions.

VI. Other Lateral Superlattice Based Structures

Several structures based on the LSL growth principles are possible, and a few of them have been attempted. These are schematically shown in Fig. 19.

The inserted grid structure first proposed by Tanaka and Sakaki (1989) is based on a thin TSL layer located in a quantum well (Fig. 19(b)). The location and thickness of the TSL within the quantum well are optimized to introduce the maximum effects on the transport or optical properties of the quantum well (Sugawara et al. 1991). Anisotropy in the transport properties for directions parallel and orthogonal to the wire is expected and has been reported. However, such anisotropies are never as important as those expected from calculations and care should be taken to eliminate interface effects.

For example, a modulation-doped GaAs–AlGaAs heterostructure will often show anisotropic magnetoresistance and mobility. These effects could well be related to the interface structure that is itself anisotrophic. Interface studies, using cathodoluminescence (Petroff et al. 1987, Bimberg et al. 1992) or scanning tunneling microscopy (see Fig. 3), indicate the GaAs growth takes place through island formation on a (001) singular surface. These islands, which are preferentially elongated along the $[1\bar{1}0]$ direction, once buried at the AlGaAs–GaAs interface, could act as interface scattering centers.

The second type of TSL based structure is the corrugated interface structure shown in Fig. 19(c). A variant of it that can easily be realized with the proper deposition sequence is the staggered corrugated interface structure, where both interfaces contain corrugations that are either in phase or out of phase with each other. By choosing appropriately the III–V compound that is part of the corrugated interface, a number of novel structures are possible. In Fig. 19(d), we show a quantum wire structure that can be achieved using the LSL based deposition approach. Elliptical, triangular or circular cross sections can be achieved with appropriate programming of the deposition parameter p and of m and n.

In the present state of the crystal growth techniques, both the corrugated interface structure and the inserted grid structure suffer from the poor Al and Ga segregation at step edges. The lateral confinement in the AlAs–

GaAs system is reduced by the poor Al segregation at step edges (see Section V.a). In fact, the carrier lateral confinement reported for an MBE grown GaAs–AlAs corrugated interface quantum well structure (Tsuchiya et al. 1989) similar to that shown in Fig. 19(c) turned out to be in error (Weman et al. 1992a). The theoretical tight binding calculations of Citrin and Chang (1991) indicate that such a structure with perfect Al segregation should indeed confine carriers in two dimensions and show polarization effects similar to those shown by the quantum wire or serpentine superlattices. However, a small amount of intermixing rapidly reduces the lateral confinement by pushing the hole wave function outside the corrugation.

In conclusion, we see that a better control of the interface structure and composition is still desirable to fully exploit the potential of this crystal growth technique towards the processing of quantum wire structures.

VII. Another Direct Growth Method of Quantum Wire Structures

A novel approach that uses the high-temperature equalibrium surface morphology of an MBE deposited film has shown potentially important results. R. Nötzel et al. (1991) have recently observed the microfaceting of the {311} and {211} singular surfaces when the GaAs and the AlAs are deposited. On the {311} GaAs A surface, this faceting occurs on a very short scale (32 Å period) and gives rise to the formation of a GaAs–AlAs quantum wire superlattice with dimensions of $\approx 20 \times 20$ Å and a period of 32 Å. The formation of the quantum wire superlattice relies on microfaceting characteristics of the AlAs that are different from those of the GaAs. The growth kinetics involved in the faceting are not understood nor is the process that would make these facets periodic or extend over long distances. The evidence for the existence of quantum wire relies mostly on the analysis of the RHEED patterns from these surfaces and TEM cross sections. The strong red-shifted photoluminescence signal and PL polarization observed at room temperature are given as evidence for quantum wires. However, the small distances between quantum wires and the thin GaAs layer connecting them should make this structure behave more like a corrugated interface quantum well. The miniband overlap and their small separation (<3 meV) should make unlikely the observation of two-dimensional confinement effects at room temperature.

Nevertheless, the approach based on forming a template starting with the microfaceting of the surface could prove to be important if the quantum structures can be made periodic and the period and dimensions of the wires can be adjusted.

VIII. Conclusions

The direct growth of quantum wire superlattices remains a challenging topic for the crystal grower. Clearly, the interface control in the lateral superlattice structures is still the major problem, and its solution relies on a better understanding of the growth kinetics during epitaxy on vicinal surfaces. Progress will be achieved with better experimental techniques that permit analysis of the growth process in situ. The important issue of size uniformity remains difficult to solve, and perhaps one should aim initially at a very narrow size distribution to obtain interesting characteristics that use the two-dimensional confinement properties of the structures.

Quantum wire superlattices obtained by direct growth of lateral superlattices already have shown two-dimensional confinement characteristics in their optical and magneto-optical properties. Further work on their transport properties should utilize the relatively weak lateral periodic potential variations of the LSL. Even with noncompositonally abrupt interfaces, these LSL should exhibit novel and interesting transport characteristics and anisotropies.

Acknowledgments

The author is heavily indebted to several graduate students and postdoctoral visitors who contributed critical results to this research area. Special thanks to M. Miller, J. Gaines, S. Chalmers, H. Weman, M. Wassermeir, V. Bressler-Hill, K. Pond and M. Krishnamurthy for their invaluable and original research contributions and for making their data available to me. Fruitful and enlightening discussions with several of the faculty at UCSB, in particular, H. Metiu, H. Kroemer, A. C. Gossard and J. Merz, have moved this research forward. Finally the financial support of Quest, an NSF Science and Technology Center (Contract DMR#S92-000) and of an AFOSR research grant are gratefully acknowledged.

References

Asom, M. T., Geva, M., Leibenguth, R. E., and Chu, (1991). S.N.G. *Appl. Phys. Lett.* **59**, 976.
Aspnes, D., Harbison, J. P., Studna, A. A., Florez, L. T., and Kelly, M. K. (1988). *J. Vac. Sci. Technol. B* **6**, 1127.
Biegelsen, D., Briegens, R. D., Northrop, J. E., and Schwartz, L. E. (1990). *Phys. Rev. B* **41**, 5701; **42**, 3195.

Bimberg, D., Heinrichsdorff, F., Bauer, R. K., Gerthsen, D., Stenkamp, D., Mars, D. E., and Miller, J. N. (1992). *J. Vac. Sci. Technol.* B **10**, 1793.

Chaddim, D. J. (1987). *J. Vac. Sci. Technol.* A **5**, 834.

Chalmers, S. A., Gossard, A. C., Petroff, P. M., Gaines, J. M., and Kroemer, H. (1989). *J. Vac. Sci. Technol.* B **7**(6), 1357.

Chalmers, S. A., Gossard, A. C., Petroff, P. M., and Kroemer, H. (1990a). *J. Vac. Sci. Technol.* B **8**, 431.

Chalmers, S. C., Kroemer, H., and Gossard, A. C. (1990b). *Appl. Phys. Lett.* **57**, 1751.

Chalmers, S., Weman, H., Yi, J. C., Kroemer, H., Merz, J. L., and Dagli, N. (1992). *Appl. Phys. Lett.* **60**, 1676.

Chavez-Pirson, A., Yumoto, J., Ando, H., Fukui, T., and Kanbe, H. (1991). *Appl. Phys. Lett.* **59**, 2654.

Cheng, K. Y., Hsieh, K. C., and Ballargeon, J. N. (1992). *Appl. Phys. Lett.* **60**, 2892.

Cibert, J., Petroff, P. M., Dolan, G. J., Pearton, S. J., Gossard, A. C., and English, J. H. (1986). *Appl. Phys. Lett.* **49**, 1275.

Citrin, D. S., and Chang, Y. C. (1991). *J. Appl. Phys.* **70**, 867.

Fuoss, P., Kisker, D. W., Lamelas, F. J., Stephenson, G. B., Imperatori, P., and Brennan, S. (1992). *Phys. Rev. Lett.* **69**, 19, 2791.

Fukui, T., and Saito, H. (1988). *J. Vac. Sci. Technol.* B **6**, 1373

Fukui, T., and Saito, H. (1990). *Jpn. J. Appl. Phys.*, L 731.

Gaines, J. M., Petroff, P. M., Kroemer, H., Simes, R. J., Geels, R. S., and English, J. (1988). *J. Vac. Sci. Technol.* B **6**(4), 1378.

Gossmann, H. G., Siden, S. W., and Feldman, L. C. (1990). *J. Appl. Phys.* **67**, 745.

Herman, M. A., and Sitter, H. (1989). *Molecular Beam Epitaxy Fundamentals.* Springer-Verlag, Berlin and Heidelberg.

Hirayama, Y., Suzuki, Y., and Okamoto, H. (1985). *Jpn. P. Appl. Phys.* **24**, L 516.

Horikoshi, Y., Kawashima, M., and Yamaguchi, H. (1988). *J. J. Appl. Phys.* **27**, 169.

Hsieh, K. C., Baillargeon, J. N., and Cheng, K. Y. (1990). *Appl. Phys. Lett.* **57**, 2244.

Hu, Y. P., Petroff, P. M., Qian, X., and Brown, A. S. (1988). *Appl. Phys. Lett.* **53**, 2194.

Inoue, M. (1991). *J. Cryst. Growth* **111**, 75.

Ivanov, S. V., Kopev, P. S., and Ledentsov, N. N. (1991). *J. Cryst. Growth* **111**, 151.

Joyce, B. A., Neave, J. H., Zhang, J., Vvedensky, D. D., Clarke, S., Hugill, K. J., Shitara, T., and Myers-Beaghton, A. K. (1990). *Semicond. Sci. Technol.* **5**, 1147.

Jusserand, B., Mollot, F., Moson, J. M., and Le Roux, J. (1990). *Appl. Phys. Lett.* **57**, 560.

Kapon, E., Hwang, D. M., and Bhat, R. H. (1989). *Phys. Rev. Lett.* **63**, 430.

Kash, K., Van der Gaag, B. P. D., Mahoney, D., Gozdz, A. S., Florez, L. T., and Harbison, J. P. (1991). *Phys. Rev. Lett.* **67**, 1326.

Kasu, M., Ando, H., Saito, H., and Fukui, T. (1991). *Appl. Phys. Lett.* **59**, 301.

Kohl, M., Heitmann, D., Grabow, P., and Ploog, K. (1989). *Phys. Rev. Lett.* **63**, 2124.

Krishnamurthy, M., Wassermeier, M., Weman, H., Merz, J. L., and Petroff, P. M. (1991). *MRS Proceedings*.

Krishnamurthy, M., Miller, M., and Petroff, P. M. (1992). *Appl. Phys. Lett.*, **61**, 2990.

Krishnamurthy, M., Lorke, A., and Petroff, P. M. P. (1993). *J. Vac. Sci. Technol. B.*

Lamelas, F. J., Fuoss, P., Imperatori, P., Kisker, D. W., Stephenson, G. B., and Brennan, S. (1992). *Appl. Phys. Lett.* **60**, 2610.

Laruelle, F., Bagchi, A., Tsuchiya, M., Merz, J., and Petroff, P. M. (1990). *Appl. Phys. Lett.* **56**, 1561.

Lorke, A., Kotthaus, J. P., and Ploog, K. (1990). *Phys. Rev. Lett.* **64**, 2559.

Lu, Y. T., and Metiu, H. (1991). *Surf. Sci.* **245**, 150.

Lu, Y. T., Petroff, P. M., and Metiu, H. (1990). *App. Phys. Lett.* **57**, 2683.
Miller, M. S. (1992). Ph.D. dissertation, University of California, Santa Barbara.
Miller, M. S., Pryor, C. E., Somoska, L. A., Weman, H., Kroemer, H., and Petroff, P. M. (1990). *The Physics of Semiconductors* (E. M. Anastassakis and J. D. Joannopoulos, eds.), p. 1717. World Scientific Publ., Singapore.
Miller, M. S., Pryor, C. E., Weman, H., Kroemer, H., and Petroff, P. M. (1991). *J. Cryst. Growth* **111**, 323.
Miller, M. S., Weman, H., Pryor, C. E., Krishnamurthy, M., Petroff, P. M., Kroemer, H., and Merz, J. L. (1992). *Phys. Rev. Lett.* **68**, 13464.
Moison, J. M., Houzay, F., Barthe, F., Gerard, J. M., and Jusserand, B. (1991). *J. Cryst. Growth* **111**, 141.
Myers-Beaghton, A. K., and Vvdensky, D. (1991). *J. Cryst. Growth* **111**, 162.
Neave, J. H., Dobson, P. J., and Joyce, B. A. (1985). *Appl. Phys. Lett.* **47**, 100.
Nötzel, R., Ledentsov, N. N., Daweritz, L., Hohenstein, M., and Ploog, K. (1991). *Phys. Rev. Lett.* **67**, 3812.
Nötzel, R., Ledentsov, N. N., Daweritz, L., and Ploog, K. (1992). *Phys. Rev. B* **45**; Nötzel, R., Eissler, D., Hohenstein, M., Ploog, K., (1993), *J. Appl. Phys.* **74**, 431.
Ourmazd, A., Bauman, F. H., Bode, M., and Kim, Y. (1990). *Ultramicroscopy* **34**, 237.
Pashely, M. D., Haberern, K. W., and Woodall, J. M. (1988). *J. Vac. Sci. Technol.* **6**, 1468.
Pashley, M. D., Haberern, K. W., and Gaines, J. M. (1991). *Appl. Phys. Lett.* **58**, 406.
Petroff, P. M. (1989). *Ultra Microscopy* **31**, 67.
Petroff, P. M., and Cibert, J., (1987). *J. Vac. Sci. Technol. B* **5**, 1204.
Petroff, P. M., Logan, R. A., Cho, A. Y., Reinhart, F. K., Gossard, A. C., and Wiegmann, W. (1982). *Phys. Rev. Lett.* **48**, 170.
Petroff, P. M., Gossard, A. C., and Wiegmann, W. (1984). *Appl. Phys. Lett.* **45**(6), 620–622.
Petroff, P. M., Cibert, J., Gossard, A. C., Dolan, G. J., and Tu, C. W. (1987). *J. Vac. Sci. Technol. B* **5**, 1204.
Petroff, P. M., Gaines, J. M., Tsuchiya, M., Simes, R., Coldren, L., Kroemer, H., English, J., and Gossard, A. C. (1989). *J. Crystal Growth* **95**, 260.
Petroff, P. M., Tsuchiya, M., and Coldren, L. (1990). *Surf. Science* **228**, 24.
Petroff, P. M., Miller, M., Y., Lu, T., Chalmers, S., Metiu, H., Kroemer, H., and Gossard, A. C. (1991). *J. Crystal Growth* **111**, 360.
Pond, K., Maboudian, R., Bressler-Hill, V., Leonard, D., Wang, X. S., Self, K., Weinberg, H., Petroff, P.M., (1993) *J. Vac. Sci. Technol. B.* **11**, 4.
Poudoulec, A., Guenas, B., D'Anterroches, C., Auvray, P., Baudet, M., and Regeny, A. (1992). *Appl. Phys. Lett.* **60**, 2406.
Pryor, C. (1991). *Phys. Rev. B* **44**, 12912.
Pukite, P. R., Petrich, G. S., Batra, S., and P. I. (1989). *J. Cryst. Growth* **95**, 269.
Shitara, T. (1992). Ph.D. thesis, Imperial College, London.
Shitara, T., Vvdensky, D. D., Wilby, M. R., Zhang, J., Neave, J. H., and Joyce, B. A. (1992a). *Appl. Phys. Lett.* **60**, 1504.
Shitara, T., Vvedensky, D. D., Wilby, M. R., Zhang, J., Neave, J. H., and Joyce, B. A. (1992b). *Phys. Rev. B* **46**, 11, 6825.
Shitara, T., Vvedensky, D. D., Wilby, M. R., Zhang, J., Neave, J. H., and Joyce, B. A. (1992c). *Phys. Rev. B* **46**, 11, 6825.
Shitara, T., Xhang, J., Neave, J. H., and Joyce, B. A. (1992d). *J. Appl. Phys.* **71**, 4299.
Simhony, S., Kapon, E., Hwang, J., Colas, E., Stoffel, N. G., and Worland, P. (1991). *Appl. Phys. Lett.* **59**, 225.
Snider, G. L., Miller, M., Rooks, M., and Hu, E. (1991). *Appl. Phys. Lett.* **59**, 2727.

Stringfellow, G. B. (1989). *Orgamnometallic Vapor Phase Epitaxy*. Academic Press, New York.
Sudjino, J., Johnson, M. D., Snyder., C. W., Elowitz, M. B., and Orr, B. G. (1992). *Phys. Rev. Lett.* **69**, 2811.
Sugawara, H., Schulman, J. N., and Sakaki, H. (1991). *J. Appl. Phys.* **69**, 2722.
Sundaram, M., Chalmers, S., Hopkins, P., Gossard, A. C. (1991). *Science* **254**, 1326.
Tan, T. H., Lishan, D., Mirin, R., Jayaraman, V., Yasuda, T., Hu, E., and Bowers, J. (1991). *Appl. Phys. Lett.* **59**, 1875.
Tanaka, M., and Sakaki, H. (1989). *Appl. Phys. Lett.* **54**, 1326.
Tokura, Y., Saito, H., and Fukui, T. (1989). *J. Crystal Growth* **94**, 46.
Tsuchiya, M., Gaines, J. M., Yan, R. H., Simes, R. J., Holtz, P. O., Coldren, L. A., and Petroff, P. M. (1989). *Phys. Rev. Lett.* **62**, 466.
Vahala, K. (1988). *IEEE J. Quant. Electr.* **24**(3), 523.
Wassermeier, M., Weman, H., Miller, M. S., Petroff, P. M., and Merz, J. L. (1991). *J. Appl. Phys.* **71**, 2397.
Weman, H., Miller, M. S., Pryor, C. E., Petroff, P. M., Kroemer, H., and Merz, J. L. (Proceedings of the SPIE, The International Society for Optical Engineering. (1992a, vol. 1675, 120)
Weman, H., Miller, M. S., Pryor, C. E., Petroff, P. M., and Merz, J. L. (1992b). International conference on the physics of semiconductors, Beijing, China.
Weman, H., Jones, E. D., McIntyre, C. R., Miller, M. S., Petroff, P. M., and Merz, J. L. (1992c). Sixth International Conference on Superlattices, Microstructures and Microdevices, Xian, China. *J. of Superlattices and Microstructures*, (1993), V13, 1, 5.
Weman, H., Miller, M. S., and Merz, J. L. (1992c). *P.R.L. Comments* **68**, 3656.
Wharm, D. A., Thornton, T. J., Newbury, R., Pepper, M., Ahmed, H., Fros, J. E. F., Hasko, D. G., Ritchie, D. A., and Jones, G. A. (1988). *J. Phys. C: Solid State Phys.* **21**, L209.
Wiener, J. S., Danan, G., Pinzuck, A., Valladares, J., Pfeiffer, L. N., and West, K. (1989). *Phys. Rev. Lett.* **63**, 1641.
Williams, D., and Krishnamuthy, M. (1992). *Appl. Phys. Lett.*, (1993), V62, 12, 1350.
Xu, Z., Wassermeier, M., Li, Y. J., and Petroff, P. M. (1992). *Appl. Phys. Lett.* **60**, 586.
Yano, M., Yokose, H., Iwai, Y., and Inoue, M. (1991). *J. Cryst. Growth* **111**, 609.
Yamaguchi, Y., and Horikoshi, Y. (1989). *J. J. Appl. Phys.* **28**, 352.
Yamaguchi, Y., Kawashima, M., and Horikoshi, Y. (1988). *Appl. Surf. Sci.* **33**, 406.
Yi, J. C., and Dagli, N. (1992). *Appl. Phys. Lett.* **61**, 219.
Yi, J. C., Dagli, N., and Coldren, L. A. (1991). *Appl. Phys. Lett.* **59**, 3015.

CHAPTER 4

Lateral Patterning of Quantum Well Heterostructures by Growth on Nonplanar Substrates

*Eli Kapon**

BELLCORE, RED BANK, NEW JERSEY

I. INTRODUCTION	259
II. LATERAL PATTERNING MECHANISMS	264
1. Quantum Well Tapering	266
2. Alloy Composition and Doping Patterning	269
3. Low-Dimensional Quantum Structures	270
III. LATERAL PATTERNING BY MBE ON NONPLANAR SUBSTRATES	274
4. Effect of Growth and Pattern Parameters	275
5. One-Dimensional Lateral Band-gap Patterning	284
6. Two-Dimensional Lateral Band-Gap Patterning	286
IV. LATERAL PATTERNING BY OMCVD ON NONPLANAR SUBSTRATES	288
7. Effect of Growth and Pattern Parameters	289
8. Formation of Quantum Wires by Self-Ordering	296
9. Optical Properties of Quantum Wire Heterostructures	304
V. DEVICE APPLICATIONS	310
10. Patterned Quantum Well Lasers	311
11. Integrated Laser Structures	318
12. Quantum Wire Lasers	321
VI. CONCLUSIONS AND FUTURE DIRECTIONS	327
ACKNOWLEDGMENTS	330
REFERENCES	331

I. Introduction

The development of methods for patterning the basic properties of semiconductor crystals, particularly the band structure and related physical parameters, has been motivated by both the desire to explore novel structured materials and the potential for generating new device concepts. *One-dimensional* (1D) patterning of semiconductor heterostructures is readily accomplished by modern epitaxial growth technologies such as molecular beam epitaxy (MBE) and organometallic chemical vapor deposition (OMCVD). The excellent uniformity and control of thickness, composition and doping on a monatomic-layer level offered by these techniques allow

*Currently affiliated with the Institute of Micro- and Optoelectronics, Lausanne, Switzerland.

growth of multiple-layer structures exhibiting controlled 1D modulation (along the growth direction) of various material parameters, including the energy gap, built-in electric and magnetic fields and refractive index. These modulated structures have provided a wealth of information on the physics of 2D systems, 1D quantum confinement, superlattice (SL) formation and tunneling phenomena. They have also made possible a number of new device concepts, most notably the high-electron mobility transistor (HEMT) (Mimura et al. 1980), the quantum well (QWL) semiconductor laser (Holonyak et al. 1980) and the resonant tunneling diode (RTD) (Sollner et al. 1983).

Semiconductor structures patterned in *two or three dimensions* are a natural extension of the 1D modulated structures, but offer unique features and present new challenges for fabrication and crystal growth technologies. Controlled band-gap patterning in 2D or 3D would allow spatial confinement of charge carriers to extremely small volumes, which is desirable for developing electronic and optoelectronic circuits with high packing density and reduced power dissipation. For sufficiently small dimensions of the confining potential wells, 2D and 3D quantum confinement effects become significant. Of particular interest are *quantum wire* (QWR) and *quantum dot* (QD) (also called *quantum box*) heterostructures, in which the quasi-1D or 0D nature of the carriers leads to enhancement of various material properties such as optical absorption and gain (Arakawa and Yariv 1986) and carrier mobility (Sakaki 1980). Two- and three-dimensional SLs based on these quantum structures would provide insight into multidimensional tunneling and electron localization phenomena. Similar *photonic* localization effects in multidimensional periodic structures can be investigated in structures in which the refractive index is patterned in 2D or 3D (Yablonovitch 1987). A viable technology for patterning semiconductor heterostructures would also provide a basis for developing 3D integrated circuits with higher packing density and simpler interconnection configurations for future ultrahigh speed and large memory capacity computers.

The fabrication of 3D patterned heterostructures, however, presents considerable technological challenges. The planar nature of conventional epitaxial methods, which allows 1D patterning in the *transverse* (growth) direction, makes it more difficult to achieve *lateral* (i.e., in the substrate plane) patterning of material parameters. One of the most important features of 1D, transversely patterned heterostructures is the fact that all their (often numerous) layer interfaces are formed in situ, during epitaxial growth. This feature yields passivated (buried), defect-free interfaces, which has been essential for producing structures suitable for studying the intrinsic properties of layered materials and applying them in useful devices. On the other hand, most techniques for lateral patterning combine various processing steps with epitaxial growth to generate the desired material patterns, and this significantly complicates their fabrication and can severely degrade the quality of their lateral interfaces.

A representative group of demonstrated approaches for lateral patterning of semiconductor heterostructures is shown in Fig. 1. The lightly shaded areas in this figure represent low band-gap materials (e.g., GaAs) whereas the unshaded ones correspond to higher band-gap regions (e.g., AlGaAs).

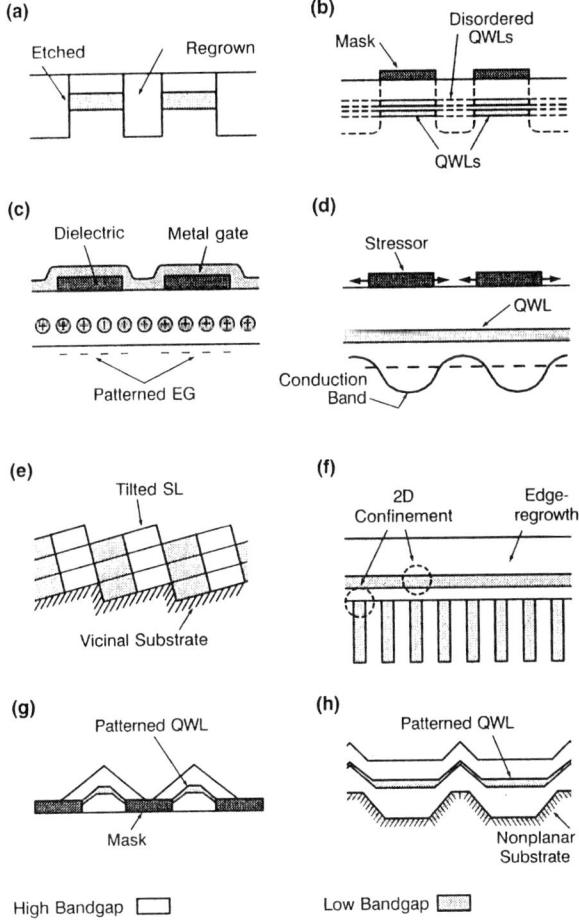

FIG. 1. Approaches for lateral patterning of semiconductor heterostructures: (a) etching and regrowth, (b) selective QWL disordering, (c) electrostatic gating, (d) strain-induced band-gap modulation, (e) growth on vicinal substrates, (f) regrowth on cleaved edge of a QWL heterostructure, (g) growth on masked substrates and (h) growth on nonplanar substrates.

The most direct patterning approach (Fig. 1(a)) consists of selective removal of parts of the grown heterostructure using lithography and etching techniques, followed by epitaxial regrowth for surface passivation (Kash et al. 1986; Marzin et al. 1992). This method provides flexibility in designing the patterned heterostructure, but suffers from the disadvantage of defect incorporation into the lateral interfaces during the etching and regrowth steps.

Selective disordering of QWL layers (Fig. 1(b)) induces by implanted or diffused impurities, introduced through a mask or using focused ion beams, has also been used to pattern the band-gap and refractive index of QWL and SL heterostructures (Laidig et al. 1981; Cibert et al. 1986; Zarem et al. 1989a; Kapon et al. 1988c). This technique avoids direct regrowth on etched surfaces, but can result in introduction of defects due to high-energy implantation or contamination due to incorporation of impurities.

A number of methods have been specifically aimed at lateral patterning of 2D electron gas structures. As an example, electrostatic gating has been employed to deplete electrons from the reverse-biased gate regions, leaving islands of electron gas underneath the patterned dielectric mask (see Fig. 1(c)) (Ford et al. 1988). Such methods are most useful for producing lateral potential wells for a single carrier type, and have provided the basis for numerous studies of electrical conduction in mesoscopic and low-dimensional systems (Beenaker and van Houten 1991). The gating technique, however, has limited spatial resolution, since the actual electric field distribution producing the patterning in the electron gas is usually wider than the dielectric mask.

Patterned "stressors" deposited adjacent to a QWL layer can produce lateral band-gap modulation via strain effects (Fig. 1(d)) (Kash et al. 1989, 1991). As in the case of electrostatic gating, this patterning method can produce smooth and defect-free lateral interfaces. One of the limitations of this approach stems from the reduction in the range of the strain field with decreasing stressor width, which sets a limit on the achievable width of the lateral potential wells.

In the approaches just described, the lateral dimensions of the patterned features are limited by the resolution of the lithography technique employed. This not only sets a lower limit on the achievable feature size, but also implies that imperfections in the lithographically defined features may directly affect the structural quality of the patterned semiconductor. This is a particularly important issue in the context of multidimensional quantum structures. In these structures, fluctuations in the size and shape of the potential wells can lead to inhomogeneous broadening effects in the energy spectrum, which can wash out the features of interest. Other approaches have attempted to overcome this limitation of by relying, to a varying degree, on the growth itself to generate the lateral patterning. Epitaxial growth of fractions of monolayers on vicinal substrates gives rise to preferential deposition of (e.g.) the group III atoms at the edges of the stepped surface (see Fig. 1(e)) (Petroff *et al.* 1984; Fukui and Saito 1988; Brandt *et al.* 1991). This leads to lateral modulation in the material band gap with typical periodicities in the sub-10 nm regime, well below the capabilities of lithography techniques available at present. Lateral SLs prepared in this way, however, suffer from nonuniformities due to incomplete phase segregation of the group III atoms, imperfections in the initial stepped surface and growth rate variations. Attempts to improve the uniformity of these lateral SLs using serpentine SL configurations have been reported recently (Miller *et al.* 1992; see also the chapter by P. Petroff in this volume).

Edge growth of QWL or modulation-doped heterostructures on cleaved, multiple QWL wafers relies on the lateral variation in potential energy induced by the (originally transverse) stack of heterostructure layers (Fig. 1(f)) (Pfeiffer *et al.* 1990). Arrays of QWRs can thus be formed in a QWL layer regrown on the cleaved edge via the interaction of the electronic waves function in the regrown layer with the potential of the edge QWLs. Alternatively, a 1D electron gas system can be formed in the edge QWLs by regrowing a modulation-doped heterostructure. This technique has the advantage of controlling the lateral dimensions of the patterned structures by features produced via epitaxial growth, i.e., with monatomic layer uniformity and accuracy. However, this is achieved at the expense of reduced flexibility for structure design and the greater difficulty of growth on (110) planes.

Lateral heterostructure patterning can also be achieved by epitaxial growth on substrates patterned with dielectric masks (Cho and Ballamy 1975; Kamon *et al.* 1985; Lebens *et al.* 1990; Fukui *et al.* 1991) (see Fig. 1(g)). In this approach, the substrate is coated with a dielectric film, typically SiO_2 or Si_3N_4, which is then patterned lightographically. Restriction of the epitaxial growth to the unmasked regions can then yield the desired lateral patterning of the grown heterostructure. Furthermore, faceting at characteristic crystallographic planes, common in such selective area growth, can be used to introduce lateral features within the growth islands. These latter patterning effects are closely related to those achieved by growth on nonplanar substrates, described next.

Growth of QWL heterostructures on nonplanar (NP) substrates is based on the lateral variations in QWL parameters, including thickness, alloy composition and doping, brought about by the growth on the patterned substrate (Tsang and Cho 1977; Miller 1985; Kapon *et al.* 1987; Meier *et al.* 1989b) (see Fig. 1(h)). All interfaces of the resulting lateral potential wells are formed in situ and can thus be defect and contamination free. Moreover, in some cases it is possible to obtain lateral potential wells of sub-10 nm dimensions whose size and shape depend on the growth conditions alone. In this sense, this approach combines the advantages of the techniques that utilize lithography-independent patterning with some flexibility gained from using lithography to define the initial pattern. Utilization of self-ordering effects to generate the smallest structures achievable with this method, however, relies on the generation of specific crystal facets and thus somewhat limits its flexibility.

This chapter reviews progress in lateral patterning of semiconductor QWL heterostructures by epitaxial growth on NP substrates. Lateral patterning of III–V compounds is specifically considered here, although many of the concepts described are applicable to other material systems as well. Section II summarizes the mechanisms for generating laterally patterned structures using controlled variation in QWL thickness, alloy composition and doping. Lateral patterning on a sub-μm scale for producing multidimensional quantum confinement effects is also discussed. The features of MBE growth on NP substrates and observation of lateral carrier confinement in these structures are described in Section III. Section IV summarizes the features of OMCVD growth on NP substrates, particularly formation of dense QWR arrays by self-ordering and their optical properties. Application of this lateral patterning technique in low-threshold current, patterned QWL lasers and laser arrays, laser–waveguide integration and QWR laser devices is described in Section V. Conclusions and future trends in this field are finally discussed in Section VI.

II. Lateral Patterning Mechanisms

Epitaxial growth on NP substrates can yield lateral patterning of the grown heterostructures via a number of different mechanisms. Prior to growth, the substrate is patterned by etching through a mask prepared, e.g., by photolithography or electron beam lithography. Transfer of the mask pattern onto the substrate and formation of the NP pattern is usually accomplished using wet chemical, reactive ion or ion beam etching methods. Surface profiles characterized by distinct crystalline facets can be achieved using anisotropic, preferential wet etchants (Tsang and Wang 1976), while profiles that are independent of the crystal structure are usually obtained with dry etching. Nonplanar surface profiles can also be achieved by growth on substrates patterned with dielectric masks. In any case, the initial stages of growth on the patterned substrate lead to the evolution of a specific set of facets that tend to minimize the surface energy of the crystal. The macroscpic (i.e., with a length scale much larger than the surface migration length) features of the resulting surface profile can be predicted using the Wulff construction, which makes use of a polar diagram of the surface energy to construct the evolving surface while accounting for the crystal anisotropy (Wulff 1901; Herring 1951). Equivalently, the faceted surface can be constructed using a polar diagram of the growth rate, by requiring minimization or maximization of the surface-integrated growth rate for convex or concave surface profiles, respectively (Jones et al. 1991). A nonplanar surface profile characterized by distinct crystal facets can be generated also by selective area growth on a planar substrate masked with a patterned dielectric film (Kamon et al. 1985; Lebens et al. 1990; Fukui et al. 1991).

Faceting at the surface can give rise to lateral patterning by itself via the dependence of various semiconductor parameters, particularly the band structure, on the crystal direction. Subsequent growth of layers on the faceted surface can further shape the heterostructure due to variations in the layer composition, thickness and doping across the different facets. Surface migration and preferential deposition of the growth species on specific crystallographic planes can also play an important role, particularly when the size of the pattern features is comparable to or smaller than the characteristic migration lengths. The morphology of the NP layered structure depends on the particular set of crystallographic planes that evolves and their relative growth rates h_i, as shown in Fig. 2. It is useful to describe the growth evolution in terms of the angles ϕ_{ij} subtended by the normal to plane i and the line drawn through the vertices between planes i and j, given by

$$\tan \phi_{ij} = |h_j - h_i \cos \theta_{ij}|/h_i \sin \theta_{ij} \tag{1}$$

where θ_{ij} are the angles between adjacent planes. For $h_j = h_i \cos \theta_{ij}$, $\phi_{ij} = 0$,

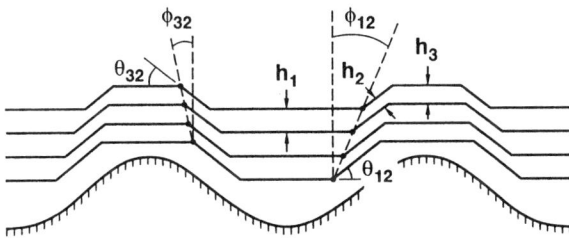

FIG. 2. Schematic cross section of a layered structure grown on a nonplanar substrate. Growth rates of the different facets are indicated by h_i. Tilt angles θ_{ij} and corner propagation angles ϕ_{ij} are also shown.

and the grown layers conform to the initial faceted surface. For $h_j > h_i \cos\theta_{ij}$, facet $j(i)$ consumes facet $i(j)$ for concave (convex) features, and the reverse occurs for $h_j < h_i \cos\theta_{ij}$; in both cases the surfaces eventually planarizes, and measurements of ϕ_{ij} can yield the relative growth rates of the intersecting planes (Smith et al. 1985). The exact nature of the lateral variations in the parameters of the patterned structures depends on the growth technique, the growth conditions and the details of the initial pattern, as illustrated in the subsequent sections. In the present section we consider the mechanisms of patterning via lateral variations in QWL thickness, alloy composition and doping, and their utilization in fabrication of low-dimensional quantum structures.

1. QUANTUM WELL TAPERING

The concept of lateral band-gap patterning by epitaxial growth of QWLs on NP substrates is illustrated in Fig. 3, which shows a schematic cross section of a (type I) heterostructure grown on a corrugated surface. Thin epitaxial layers grown on NP surfaces exhibit faceting and lateral thickness variations due to variations in the flux of the source atoms (in MBE and related methods), the growth rate on different crystallographic planes (primarily in OMCVD and related methods) and preferential migration between adjacent facets (in all growth methods) (Burnham and Scifres 1975; Tsang and Cho 1977; Nagata and Tanaka 1977; Nagata et al. 1977). For layers whose thicknesses are comparable to the optical wavelength, these thickness variations can be utilized for lateral patterning of the effective optical mode index. This provides a means for defining optical channel waveguides and other structures requiring 2D or 3D tailoring of the refractive index. For extremely thin layers, with thickness comparable to the de Broglie charge carrier wavelength, a QWL heterostructure is formed at each section of the patterned heterostructure. The strong dependence of the "transverse" confinement energy E_{conf} (i.e., the energy due to quantum confinement introduced by the QWL normal to its interfaces) then leads to lateral variation in the valence and conduction band edges (Kapon 1985; Kapon et al. 1987). For infinitely deep QWL potential wells, the lateral (y direction) variation in confinement energies is given by

$$E_{\text{conf}}^{c,v}(y) = \frac{\hbar^2 \pi^2 l^2}{2m_{c,v}^* t^2(y)} \tag{2a}$$

where $l = 1, 2, \ldots$ is the (transverse) index of the QWL energy subband, $m_{c,v}^*$ is the carrier effective mass for the conduction and valence bands and $t(y)$

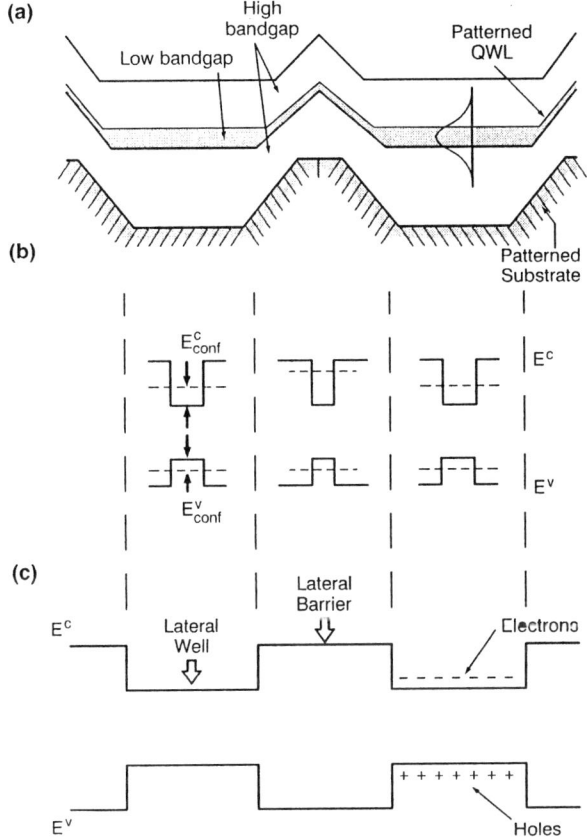

FIG. 3. Lateral band-gap patterning by growth of QWLs on nonplanar patterned substrates: (a) schematic cross section, (b) confinement energies for different QWL thicknesses and (c) resulting lateral variation in conduction and valence bands.

is the lateral thickness distribution. The effective band-gap variation is then given by

$$E_{\text{eff}}(y) = E_g + E_{\text{conf}}^c(y) + E_{\text{conf}}^v(y) = E_g + \frac{\hbar^2 \pi^2 l^2}{2t^2(y)\mu^*} \tag{2b}$$

where E_g is the band gap of the bulk semiconductor and $1/\mu^* = 1/m_c^* + 1/m_v^*$ is the reciprocal of the reduced effective carrier mass. The lateral thickness variations thus result in effective band-gap modulation, with the thicker QWL sections constituting lateral potential wells that can trap charge carriers. The thinner QWL sections function as effective lateral barriers, analogous to the higher band-gap layers that serve as the transverse barriers in conventional QWL heterostructures. Note that the shape and depth of these lateral potential wells is generally different for different carrier types due to the dependence of the confinement energy on the effective mass. Furthermore, the variation in band gap is larger for the higher energy QWL subbands (designated by their subband index l). In the structure of Fig. 3, the lateral QWL thickness distribution gives rise to carrier confinement in one lateral dimension, normal to the groove axis. Confinement in both lateral dimensions can be similarly achieved by utilizing 2D tapering effects in, e.g., QWL layers grown over etched craters.

The effect of well tapering is illustrated more quantitatively in Fig. 4, which shows the calculated E_{conf} versus thickness for electrons and holes in (100) oriented GaAs/Al$_x$Ga$_{1-x}$As QWLs at the Γ point. The confinement energies are evaluated in the effective mass approximation using a finite well model. (Note that the infinite well expression (2a) closely approximates the variation of E_{conf} for $x = 1$, in the thickness range shown in Fig. 4.) It can be seen that sufficiently large differences in QWL thickness can result in effective band-gap variations greater than 100 meV, which are sufficient for providing efficient carrier confinement at room temperature ($k_B T = 25.9$ meV at $T = 300$K). Additional contributions to the band-gap variation in such structures can result from the abrupt bending in the QWL

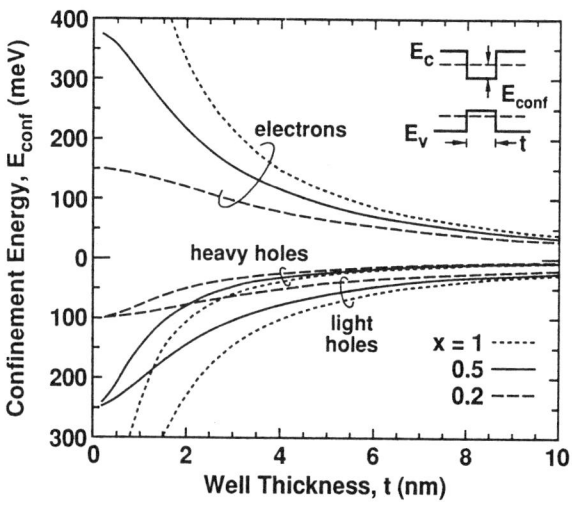

FIG. 4. Calculated variation of confinement energy versus well thickness (for the gamma valley) in GaAs/Al$_x$Ga$_{1-x}$As quantum wells for several Al mole fractions x in the barriers.

layers, variations in the composition of the barrier and well materials, as well as lateral strain effects.

2. ALLOY COMPOSITION AND DOPING PATTERNING

Alloys grown on NP substrates can exhibit lateral variation in composition across the different crystal planes or at their intersections (Meier et al. 1989b). These compositional variations arise from both the dependence of the incorporation rate of the group III and V atoms on the crystal plane and variations in the fluxes of the growth species due to geometrical or diffusion effects. In lattice matched systems such as GaAs/AlGaAs, lattice matching of the patterned heterostructure can be maintained even with large, lateral variations in alloy composition. In other material systems, such as InP/InGaAsP or GaAs/InGaAs, deviations from the perfect lattice matching condition can introduce additional patterning effects via lateral strain fields. However, large compositional variations inevitably lead to formation of defects, which may limit the usefulness of this patterning technique in these systems.

A special type of compositional variation over NP surfaces can arise from facet-dependent spontaneous ordering of the alloy into SL structures (Hoenk et al. 1989a, 1989b). Lateral variations in alloy composition can alter the effective band gap either directly (i.e., via the effect on the band gap of the bulk material), or through the effect on the confinement energy in QWLs incorporating alloys in the well or barrier layers. Whereas the lateral thickness distributions can be readily measured from cross-sectional transmission electron microscope (TEM) images of such patterned structures (see Sections III and IV), direct measurement of lateral composition distributions is more difficult to achieve. In some cases, information about such composition variations can be obtained by comparing the calculated effective band gap, based on well-thickness measurements, with measurements of the effective band gap obtained from, e.g., photoluminescence (PL) data (Walther et al. 1993).

Controlled lateral variation in doping profiles can be achieved via the dependence of dopant incorporation *rate and type* on the crystallographic orientation of the facets. For MBE growth, amphoteric dopants such as Si in GaAs/AlGaAs, have been shown to yield *p*- or *n*-type doping under the same growth conditions, depending on the crystal plane (Ballingall and Wood, 1982; Wang et al. 1985; Subbanna et al. 1986). These effects can be utilized in forming lateral *p-n* junctions in NP structures in which adjacent facets exhibit opposite doping types (Miller 1985; Miller and Asbeck 1987). Significant orientation dependence of doping in epitaxial layers grown by

OMCVD has also been reported (Bhat et al. 1991; Kondo et al. 1992) (see subsections 4 and 7 for more details).

3. Low-Dimensional Quantum Structures

Quantum confinement of charge carriers in more than one-dimension can be achieved by fashioning lateral potential wells of sufficiently small lateral extent. For the III–V compound semiconductor systems, these dimensions fall in the sub-100 nm range; potential well widths of 10 nm or less are desirable to maintain significant quantum size effects at room temperature. Quantum confinement in one or both lateral dimensions can lead to quasi-1D or 0D carriers trapped in QWR or QD heterostructures. The reduced dimensionality in these structures results in significant modification in the density of states (DOS) profile, which acquires increasingly sharp features at the bound state energies as the dimensionality is reduced. This, in turn, is expected to give rise to a variety of new physical effects, including modified optical absorption, refraction (Miller et al. 1988) and gain spectra (Arakawa and Yariv 1986), enhanced exciton and impurity binding energies (Banyai et al. 1987; Osorio et al. 1987) and reduced carrier scattering rates (Sakaki 1980). These effects should be useful in enhancing the performance of various electronic and optoelectronic devices such as transistors, lasers (Arakawa and Yariv 1986; Kapon 1992), optical modulators and switches (Matsubara et al. 1989) and photodetectors (Crawford et al. 1991).

Achievement of multidimensional quantum confinement, however, presents stringent requirements on the structural quality of the patterned structures (Kapon et al. 1992b). Since the dimensions of the potential wells are much smaller than typical carrier diffusion lengths, defects incorporated into the interfaces of the wells can serve as recombination centers, which may interfere with intrinsic recombination processes taking place at the core of the quantum structure. Imperfect interfaces are particularly detrimental for optical studies and optoelectronic device applications involving *radiative recombination* (e.g., lasers, amplifiers), since nonradiative recombination reduces carrier lifetimes and quantum efficiency. In addition to defect-free interfaces, the quantum structures should be extremely uniform, with size fluctuations much smaller than their dimensions, in order to minimize inhomogeneous broadening effects that can smear out the ideal DOS profile. Many applications also require dense, uniform arrays of QWRs or QDs in order to compensate for the small interaction volume provided by a single wire or dot.

Growth of QWL layers on NP substrates offers attractive advantages addressing these requirements for realization of useful quantum structures.

The modulation of both conduction and valence bands *with the same polarity* yields lateral confinement of both electrons and holes in overlapping regions (in type I heterostructures). Formation of the lateral interfaces during growth minimizes interface defects and allows preparation of quantum structures exhibiting intrinsic electronic and optical properties. As demonstrated in Sections III and IV, self-ordering effects during growth on the NP substrates can yield lateral potential wells whose size and shape are insensitive to lithography imperfections, thereby minimizing inhomogeneous broadening effects. At the same time, the use of lithography provides sufficient flexibility in designing quantum structures that suit specific applications. In particular, high densities of wires or dots can be fabricated using high-resolution electron beam lithography or vertically stacked array configurations.

Several examples of QWR structures achievable by growth on NP substrate are described by the schematic cross sections in Fig. 5. Tapered QWL layers grown in grooves yield 2D carrier confinement, and thus quasi-1D structures can be made in this way provided that the tapered structure is sufficiently narrow (Kapon *et al.* 1987). One way to accomplish this is by MBE growth in very narrow (<100 nm wide) grooves, as demonstrated in Section III. The lateral width of the QWR is then directly determined by the width of the etched channel (Fig. 5(a)). This approach

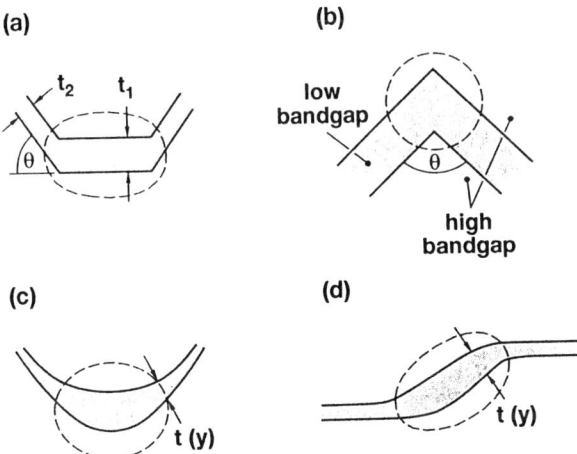

FIG. 5. Schematic cross sections of QWR structures formed by epitaxial growth on NP substrates. (a) tapered QWL in a narrow channel, (b) corner-QWL layer at the intersection of two crystallographic planes, (c) crescent-shaped QWR grown at the bottom of a V-groove and (d) asymmetric QWR formed at the corner between crystallographic planes. Dashed circles show position of a 2D quantum-confined wave function.

requires virtually perfect lithography to avoid fluctuations in size and shape of the resulting wires. A bound state confined in 2D can also be formed at the corner of two intersecting planes on which a QWL layer is grown (Fig. 5(b)) (Kojima et al. 1989). The size and shape of such QWRs is thus independent of the details in the substrate profile. The depth of the effective 2D potential well is determined by the angle of intersection, the layer thickness and the difference in well and barrier band gaps.

Particularly narrow and deep 2D potential wells can be accomplished using tapered QWLs grown at the corners between intersecting facets. The facets, which evolve during growth of the lower barrier layers, form a surface profile independent of the initial etched pattern. The final shape and size of the wire are largely determined by surface diffusion effects that shape the tapered QWL (Kapon et al. 1989b). Crescent-shaped QWR structures can be grown by OMCVD at the corner between $\{n11\}A$ planes, as described in detail in Section IV (Fig. 5(c)). Similarly, asymmetric, tapered QWRs can be formed at the corner of (100) and $\{n11\}A$ planes by OMCVD on vicinal, patterned substrates (Fig. 5(d)). In both structures, the bending as well as the tapering of the QWL layer may contribute to the lateral quantum confinement. In the actual structures we will discuss, however, the tapering of the QWL, described by the thickness distribution $t(y)$ along the lateral dimension y, is the more significant confinement mechanism. This "adiabatic" variation in QWL thickness can be utilized to model the formation of 1D subbands in these QWRs, as outlined later.

The tapered QWL heterostructure is represented by the 2D potential distribution $V(x, y)$, with V_w indicating the potential at the low band-gap well region and V_B the corresponding value at the barrier material (see Fig. 6(a)). The quasi-1D subbands in this structure can be found (in the effective mass approximation) by solving the 2D Schroedinger equation

$$\left[-\frac{\hbar^2}{2m^*}\left(\frac{\partial^2}{\partial x^2} + \frac{\partial^2}{\partial y^2}\right) + V(x, y)\right]\psi(x, y) = E\psi(x, y) \quad (3)$$

where ψ is the envelope wave function and E is the subband energy. However, for the crescent shaped (and other tapered) QWL structures, the potential variations in the lateral (y) direction are usually much slower than in the transverse (x) one, and this 2D equation can be approximated by the two coupled, 1D Schroedinger equations (Kapon et al. 1989b)

$$\left[-\frac{\hbar^2}{2m^*}\frac{\partial^2}{\partial x^2} + V_y(x)\right]\chi_y(x) = E_{\text{conf}}(y)\chi_y(y) \quad (4a)$$

$$\left[-\frac{\hbar^2}{2m^*}\frac{d^2}{dy^2} + E_{\text{conf}}(y)\right]\phi(y) = E\phi(y) \quad (4b)$$

where $\psi(x, y) \approx \chi_y(x)\psi(y)$, χ_y being a slowly varying function of y, and $V_y(x)$ is the 1D potential distribution in the transverse x-direction at each lateral position y. In this adiabatic approximation, the 1D transverse problem is first solved for each carrier type as a function of the QWL thickness distribution $t(y)$. This yields the confinement energy distributions $E_{conf}^{c,v}(y)$, which serve as the 1D lateral potential wells providing lateral quantum confinement (see Fig. 6(b)). Solving for the eigenstates of these lateral wells finally yields the lateral wave functions and subband energies. For the symmetric, crescent shaped QWRs the lateral potential wells can be approximated by parabolic profiles,

$$E_{conf}(y) = E_{conf}(0) + (1/2)py^2 \tag{5a}$$

which assumes harmonic-oscillator-like quasi-1D subbands of the form

$$E_l = E_{conf}(0) + \hbar\sqrt{p/m^*}(l - 1/2), \quad l = 1, 2, \ldots \tag{5b}$$

where the subband energies are measured with respect to the bulk semiconductor band edge.

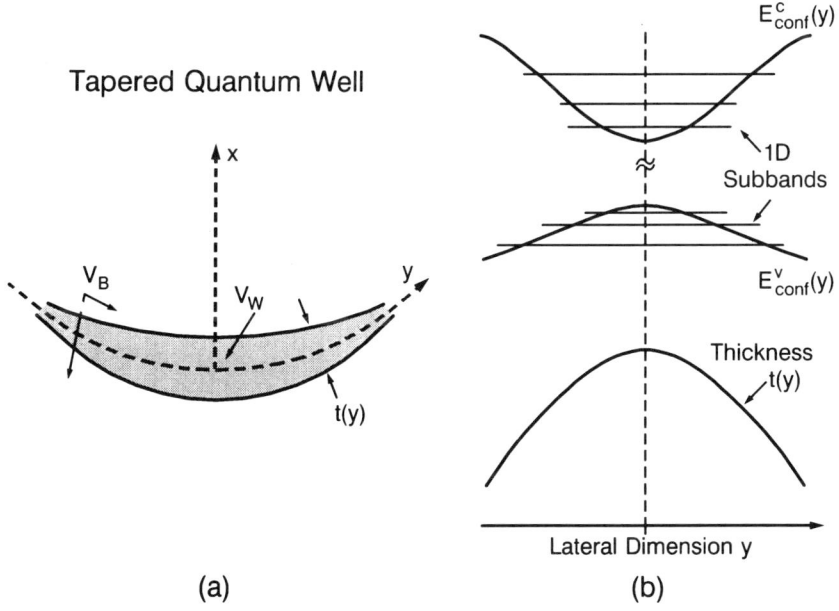

FIG. 6. Model of 2D quantum confinement in crescent-shaped QWRs. (a) schematic cross section, (b) lateral distributions of thickness and confinement energies for electrons and holes. Horizontal lines indicate energies of the resulting quasi-1D subbands.

III. Lateral Patterning by MBE on Nonplanar Substrates

The MBE growth process consists of three principal stages: (i) arrival of the growth species to the surface of the substrate via molecular beams, (ii) attachment of the growth species onto the substrate in the form of adatoms that can migrate on the surface (physisorption) and (iii) incorporation of the adatoms into the growing crystal (chemisorption). In the case of growth on NP substrates, the profiled surface can significantly modify each of these mechanisms. The different tilt angles across the patterned substrate lead to variation of the beam fluxes, which directly alters the deposition rates. For pattern features comparable to or smaller than the surface migration length of the adatoms, surface diffusion between neighboring facets can also change the concentration of adatoms available for deposition at a given crystal facet. The ultimate deposition rate is also determined by the effective sticking coefficient, which can depend on the crystallographic orientation.

Schematic cross sections of typical patterned heterostructures grown by MBE on corrugated (100) substrates are depicted in Fig. 7. Growth on

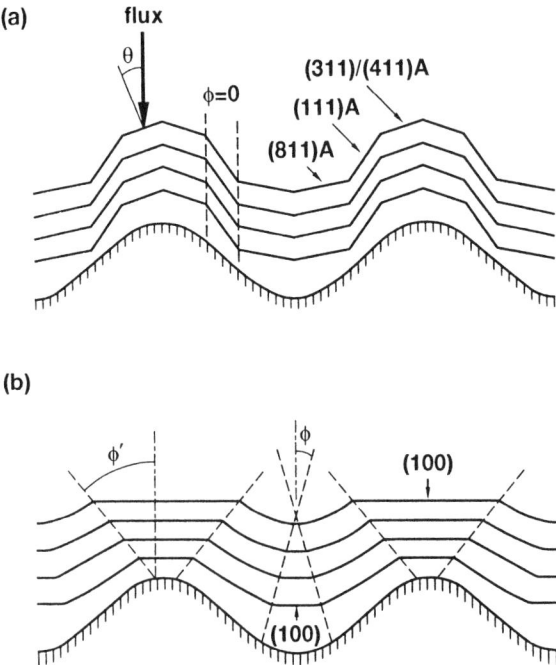

FIG. 7. Schematic cross sections of patterned heterostructures grown by MBE on corrugated (100) substrates. (a) [01$\bar{1}$]-oriented corrugations, (b) [011] corrugations.

[01$\bar{1}$]-oriented grooves leads to evolution of three types of crystal planes: $\{811\}A$ at the bottom of the grooves, $\{111\}A$ on the slopes and $\{311\}A$ or $\{411\}A$ at the peaks of the corrugations (Tsang and Cho 1977; Nagata and Tanaka 1977; Nagata et al. 1977; Smith et al; 1985, Kapon et al. 1987). For larger features, (100) planes may also develop at the centers of the grooves and mesas (e.g., Guha et al. 1990). Subsequent evolution of the surface profiles is determined by the growth rates at these facets. For pattern features much larger than the surface migration length, the facet growth rates are determined by the flux rates and the sticking coefficients. For identical sticking coefficients, the growth rates are proportional to the beam fluxes, which vary with $\cos\theta$, where θ is the tilt angle of the corresponding facet (see Fig. 2). In this case, it follows from (1) that the growth evolves conformally; i.e., with corners between neighboring planes propagating at an angle $\phi = 0$ with respect to the normal to the (planar) surface (see Fig. 7(a)). However, variations in sticking coefficients and surface migration effects can result in growth rates that deviate from this $\cos\theta$ rule and hence lead to finite values of ϕ.

Growth on [011] grooves, on the other hand, is characterized by formation of [100] facets both at the bottom of the channels as well as on top of the ridges, with smooth slopes forming on the edges of the grooves (Fig. 7(b)) (Smith et al. 1985; Kapon et al. 1988b, 1988c). The growth rate on the slopes is larger than that on top of the (100) planes, and this leads to a rapid planarization of the initially NP surface.

4. Effect of Growth and Pattern Parameters

These MBE growth features are illustrated in Figs. 8 and 9, which show dark-field transmission electron microscopy (TEM) cross sectional images of multiple-QWL GaAs/AlGaAs layers grown on [01$\bar{1}$]- and [011]-oriented corrugations, respectively (Kapon et al. 1988b). The grooves were prepared in this case by wet chemical etching in $H_2SO_4:H_2O_2:(30\%):H_2O$ ((1:8:40) by volume) through a photoresist mask made by conventional photolithography. The preferential nature of the chemical etchant resulted in V-shaped [01$\bar{1}$] grooves 2–3 μm wide and $\sim 3\,\mu$m deep, whereas the [011] ones were square-shaped and only $\sim 1\,\mu$m deep. However, rounded groove profiles developed during free etching in $H_2SO_4:H_2O_2:(30\%):H_2O$ (4:1:1)) employed prior to growth for surface cleaning. The heterostructure, grown at $T_s \sim 630°C$ under an As-rich atmosphere, consisted of several sets of $5X(10\,nm\ GaAs/10\,nm\ Al_{0.3}Ga_{0.7}As)$ layers, with the two NP samples with different groove orientation grown side by side (the layer thicknesses

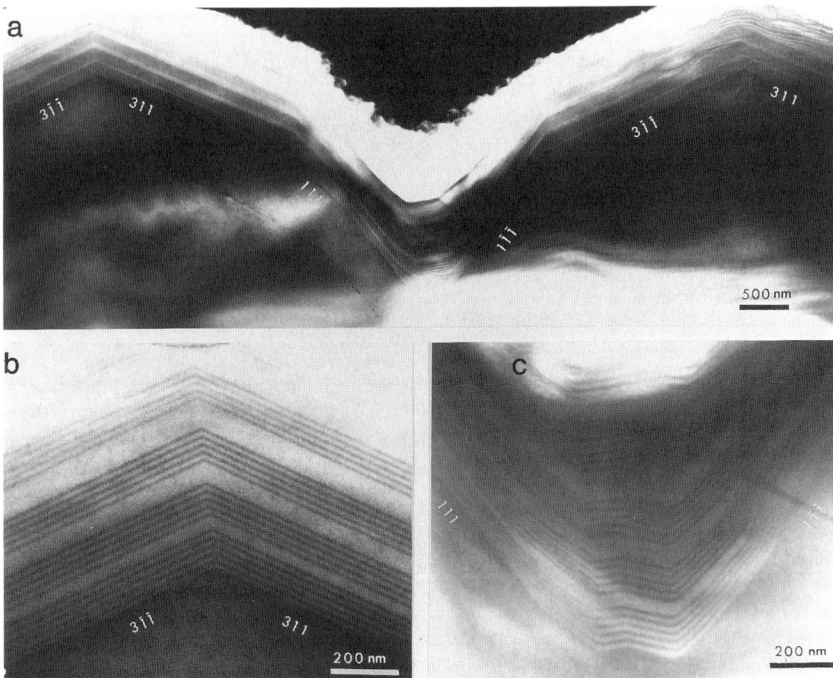

FIG. 8. Dark-field TEM cross sectional images of GaAs/Al$_{0.3}$Ga$_{0.7}$As multiple-QWL heterostructure grown by MBE on a corrugated, (100) GaAs substrate, with grooves oriented along the [01$\bar{1}$] directions. Dark stripes represent GaAs layers. Parts (b) and (c) shows magnified parts of (a) (Kapon et al. 1988b).

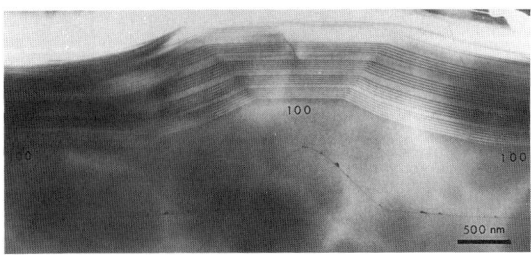

FIG. 9. Dark-field TEM cross sectional images of GaAs/Al$_{0.3}$Ga$_{0.7}$As multiple-QWL heterostructure grown by MBE on a corrugated, (100) GaAs substrate, with grooves oriented along the [011] directions. Dark stripes represent GaAs layers (Kapon et al. 1988b).

quoted correspond to growth rates on planar (100) substrates, as is the case for all structures described in this chapter).

For the [01$\bar{1}$] corrugations, {811}A, {111}A and {311}A facets are formed, with multiple-QWL periodicity of 18, 10 and 16 nm, in fair agreement with the $\cos\theta$ rule ($\theta = 10.0°$, 54.7° and 25.2°, respectively). This indicates that the surface diffusion length in this case is much smaller than ~1 μm. Smith et al. (1985) have studied in detail the growth profile on [01$\bar{1}$] grooves of similar dimensions as a function of substrate temperature and As pressure. Generally, the enhancement in growth on the {811} (and in some cases the (100)) facets at the bottom of the grooves) increases with increasing temperature and decreasing As pressure. This shows that preferential surface diffusion into these planes can also contribute to their higher growth rates.

Growth on the [011] corrugations is dominated by strong diffusion from the top (100) mesa (multiple-QWL periodicity of 14 nm) to the side walls (periodicity of 20 nm), which lie in a plane near {41 $\bar{1}$}. This may reflect growth in islands that exhibit higher growth rate along the [01 $\bar{1}$] direction compared to the [011] one on a (100) substrate (Neave et al. 1985). Smith et al. (1985) have observed growth on {41$\bar{1}$} facets on the side walls, with enhancement in growth rate relative to the (100) facets that increases with increasing substrate temperature between 580°C and 660°C.

Surface diffusion lengths for several group III adatom species that have been measured for MBE growth on nonplanar substrates under various growth conditions are summarized in Table I. Generally, the surface diffusion length increases with increasing substrate temperature (for conventional temperature values) as well as with decreasing group V/III flux ratio. Surface migration lengths as long as a few microns have been observed for Ga adatoms under lean As fluxes. The surface mobility of Al is usually much smaller, whereas diffusion lengths as long as 25–30 μm have been observed for In adatoms. Under similar growth conditions, surface diffusion lengths along the [011] direction are considerably smaller than along the [01$\bar{1}$] one, which explains the different impact of diffusion effects in the structures of Figs. 8 and 9.

The role of these surface diffusion effects is further illustrated in Fig. 10, which shows TEM cross sections of a GaAs/AlAs heterostructure grown over a 3.5 μm pitch, periodically corrugated GaAs substrate (grooves are aligned along the [01$\bar{1}$] direction) (Clausen et al. 1990). The grown structure consists of 0.1 μm GaAs buffer layer, five periods of 1 nm GaAs/1 nm AlAs superlattice, 10 nm AlAs barrier, 8 nm GaAs well, 10 nm AlAs barrier and 50 nm GaAs cap, all nominally undoped. The structure was grown at 680°C under As stabilized conditions.

The characteristic (100) plane at the bottom of the grooves and the

TABLE I

MEASURED DIFFUSION LENGTHS OF GROUP III ADATOMS DURING MBE GROWTH ON NONPLANAR SUBSTRATES

Adatom	Facets	Growth Conditions	Diffusion Length	Reference
Ga	(111)A to (100)	560°C, As-stabilized	$\sim 1\,\mu m$	Hata et al. 1990
Ga	(311)A to (100) ridge	700°C, V/III = 0.9	$> 5\,\mu m$	Meier et al. 1989b
Ga	(311)A to (100)	710°C, V/III ~ 1	a few μm	Meier et al. 1989a
Ga	(311)A to (100)	690°C, V/III ~ 2	$1-2\,\mu m$	Nilsson et al. 1989
Ga	(311)A to (100) ridge	610°C, As-rich	$> 0.9\,\mu m$	Guha et al. 1990
Ga, Al	(311)A to (100) ridge	610°C, As-rich	$< 0.3\,\mu m$	Guha et al. 1990
Ga, Al	(111)A to (811)A	680°C, V/III ~ 3	20–30 nm	Kapon et al. 1987
Ga	(100) to (111)B	560°C, As-stabilized	$\sim 8\,\mu m$	Hata et al. 1990
Ga	(100); along [01$\bar{1}$]	550°C, "As-rich"	< 20 nm	Nagata and Tanaka 1977
Ga	(100); along [01$\bar{1}$]	550°C, "Ga-rich"	190 nm	Nagata and Tanaka 1977
Ga, Al	(100) ridge to (41$\bar{1}$)	580–700 V/III = 2	$\sim 1\,\mu m$	Smith et al. 1985
Ga, Al	(100) ridge to (41$\bar{1}$)	630°C, "As-rich"	$\sim 1\,\mu m$	Kapon et al. 1988b
In	(311)A to (100) ridge	520°C, V/III ~ 2	$25-30\,\mu m$	Arent et al. 1989

$\{411\}A$ planes on their slopes develop during the growth of the GaAs buffer layer. Due to surface migration of Ga adatoms from $\{411\}A$ to (100), the GaAs QWL layer tapers off at a linear rate along the $\{411\}A$ slopes, from 9 nm at the peak to 2.5 nm near the (100) plane, and exhibits a fairly uniform thickness of 13 nm at the bottom of the grooves. The GaAs QWL thickness at the peak of the corrugations is larger than the thickness on unpatterned, planar substrates grown side by side with the corrugated one, which indicates migration of Ga adatoms from the $\{411\}A$ side walls to the apex as well. Note that the AlAs layers show much less pronounced thickness variations, being 10 nm thick at the apex, 8 nm near the (100) plane, and 11 nm at the bottom of the groove. This reflects a shorter diffusion length and smaller difference in sticking coefficient between the exposed crystal planes for the Al atoms. Strong migration effects from $\{311\}/\{411\}A$ planes to (100) planes *at the top of the grating ridges* (before complete formation of the $\{311\}/\{411\}A$ corner) have been observed in similar structures by Guha et al. (1990).

Shen et al. (1992) studied the MBE growth of AlAs on [01$\bar{1}$]-oriented V-grooves. For growth at $T_s = 580$C and V/III ~ 4 they observed no surface migration of Al from the $\{111\}A$ planes to the (100) planes. Furthermore, a very sharp corner, defined by the two $\{111\}A$ planes, developed at the bottom of the groove. Subsequent growth of GaAs QWL layers, for which

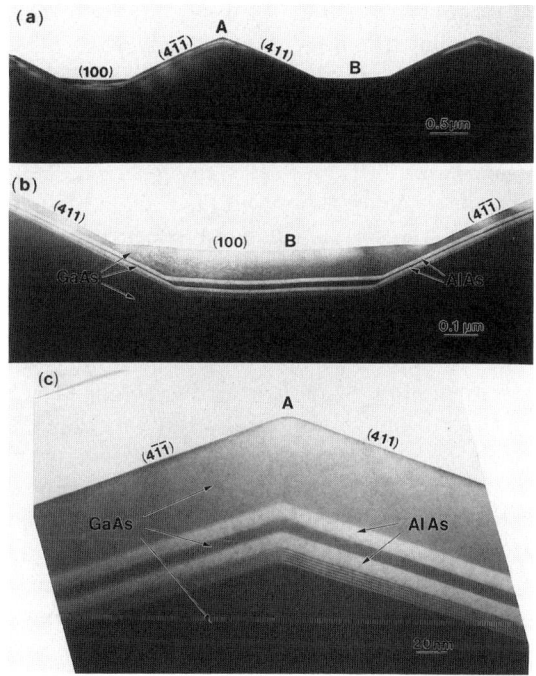

FIG. 10. TEM cross sectional images of GaAs/AlAs QWL heterostructures grown on periodic corrugations by MBE. Grooves are along the [011] direction. Parts (b) and (c) shows magnified views of the bottom and apex parts of the corrugations (Clausen et al. 1990).

the lateral surface migration was much more significant, resulted in formation of crescent-shaped QWRs at these corners, similar to the OMCVD-grown structures described in subsection 8.

The difference in surface diffusion lengths of the various group III adatoms can lead to considerable lateral variation in composition of alloys grown on NP substrates. Meier et al. (1989a) and Nilsson et al. (1989) have employed cathodoluminescence (CL) spectroscopy to measure variations in Al mole fraction in $Al_xGa_{1-x}As$ grown on ridges and grooves with $\{311\}A$ oriented side walls. For nominal $x = 0.33$ (far away from the side walls), reduction to $x = 0.29$ on 4 μm wide ridges and to $x = 0.31$ in 5 μm wide grooves, and concurrent increase to $x = 0.43$ and $x = 0.37$ on the side walls, respectively, were observed ($T_s = 710°C$, V/III ~ 1). These lateral variations are probably due to both the different diffusion lengths of Ga and Al and the (related) lower sticking coefficient of Ga on the side walls. Lateral variations in In mole fraction in strained $In_xGa_{1-x}As$/GaAs QWLs grown on [01$\bar{1}$]-oriented ridges (T_s = 530°C, V/III ~ 2.0) were measured by Arent et al. (1989) using a similar technique. Maximum increase of ~6% in x was observed near the side walls, possibly limited by strain effects.

Spontaneous formation of GaAs/AlAs SLs has been identified by Hoenk et al. (1989b) as one of the mechanisms through which patterning of AlGaAs alloys grown by MBE on NP substrates can proceed. MBE growth of $Al_{0.25}Ga_{0.75}As$ (nominal composition) at T_s = 600°C in [01$\bar{1}$]-oriented channels on (100) substrates resulted in an effective bandgap lower by 130 meV on the {111}A side walls, as determined by CL imaging. Transmission electron microscopy of the samples revealed quasi-periodic modulation in Al mole fraction along the [111] direction on these facets, with periods of 5–7 nm. The measured effective bandgap indicates nearly pure GaAs composition at the low-Al mole fraction regions. The SL is formed only on the {111}A facets and an abrupt SL-alloy junction was observed at the intersection with the other growth planes. Evidence for this SL phase was supported by the disappearance of the SL emission line after Zn diffusion across the sample's surface, which was attributed to impurity-induced disordering of the SL. Note that the lateral band-gap modulation arising from this SL formation is opposite in sign to that induced by difference in migration of group III atoms (e.g., Al and Ga) which produces a *higher* band gap at the {111}A planes.

Turco et al. (1990b) investigated MBE growth of InGaAs and InAlAs alloys on NP InP substrates patterned with [01$\bar{1}$]-oriented grooves. Compositional variations across the grooves were evaluated using energy-dispersive x-ray analysis (EDX). Relatively small (5%) variations in In mole fraction were measured in the InGaAs layers, whereas the InAlAs layers exhibited as much as 30% variations in In content between the side walls and the bottom of the grooves. This may be due to the larger difference in diffusion lengths of In and Al versus In and Ga. In addition, the InAlAs layers developed a convex bulge bounded by {311} planes at the bottom of the grooves. InGaAs/InAlAs QWL structures grown in such grooves exhibited red-shifted PL spectra (compared to structures grown side by side on flat substrates) as a result of variations in QWL thickness and composition at the recombination regions.

The patterned QWL structure of Fig. 10 illustrates the formation of narrow (~0.5 μm) lateral potential wells in [01$\bar{1}$] grooves. Still narrower potential wells can be obtained by growth on sub-μm pitch periodic corrugations. MBE growth of patterned QWLs on such gratings (0.3 μm pitch) has been studied by Turco et al. (1990a). The undoped (100) GaAs substrates were patterned in this case by holographic photolithography and wet chemical etching. The resulting [01$\bar{1}$]-oriented grooves were V-shaped, exposing {111}A sidewalls, and ~0.2 μm deep. Prior to growth, the corrugated substrates were cleaned by degreasing, dipping in HCl and rinsing in water, so that the groove profiles were preserved. Growth of various GaAs/AlAs and GaAs/AlGaAs QWL heterostructures on these patterned substrates was studied, all grown at $T_s = 600°C$, with V/III ratios of ~ 3.

Figure 11 shows TEM cross sections of a GaAs/Al$_{0.5}$Ga$_{0.5}$As heterostructure grown on these gratings. The nominal GaAs QWL thickness was 5–6 nm, and the AlGaAs barriers were ~ 24 nm thick, as measured from TEM cross sections of planar structures grown side by side. Strong migration of Ga adatoms from the $\{311\}A$ planes, which evolve during growth of the first layers, results in rapid planarization of the grooves. The GaAs QWLs at the bottom of the groove evolve into wires of increasing widths lying in the (100) plane, laterally bounded by much thinner $\{311\}A$ GaAs QWLs. The thickness of the latter is much smaller than that expected

(a)

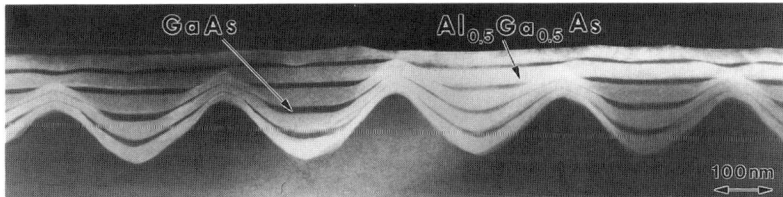

(b)

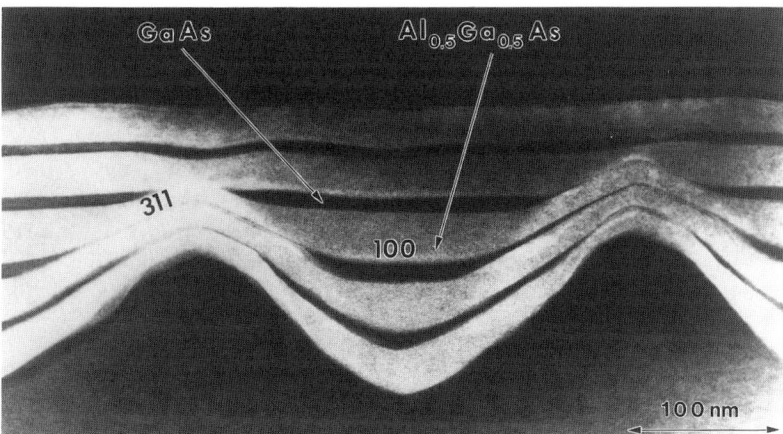

FIG. 11. Transmission electron microscope cross sections of GaAs/Al$_{0.5}$Ga$_{0.5}$As QWL heterostructures grown on sub-μm pitch gratings etched onto a GaAs substrate. Grooves are oriented along the [01$\bar{1}$] direction. Part (b) shows a magnified view of a grating groove (Turco et al. 1990a).

from just flux reduction on the tilted facets and is fairly uniform across the entire $\{311\}A$ facets. This indicates that the migration length of adatoms in this case is greater than the size of the pattern features ($\sim 100\,\text{nm}$). Planarization of the structures takes place also during growth of the AlGaAs barrier layers, albeit to a lower extent. This is due to the much smaller migration length of Al adatoms, compared to that of Ga. Separate growth experiments of GaAs/AlAs heterostructures on similar grating structures showed no migration of Al from the $\{311\}A$ planes to the (100) ones (Turco et al. 1990a). Significant lateral variation in the composition of the ternary alloy is expected from this difference in migration lengths; from the observed growth rates, the Al mole fraction was estimated to be only $\sim 35\%$ at the bottom of the grooves.

The sub-μm grating structure illustrates the formation of narrow, low potential energy wires during MBE growth on the NP substrate. Note that before the (100) GaAs QWL becomes fully developed, tapered QWL structures are formed at the bottom of the grooves. Thus, three types of wires are obtained. The tapered QWL layer constitutes a wire whose effective width is much smaller than the width of the groove, as a result of the transverse quantum size effects discussed in Section II. The width of the GaAs wire formed on the (100) plane at the center of the grooves is comparable to the groove width and becomes wider as the barrier layer thickness underneath it is increased. If such wires are grown just before complete planarization of the structure, the lateral wire barriers, defined by the $\{311\}A$ wells become very narrow, which provides a possibility of forming lateral wire SLs. The third type of wires is formed at the corner of the $\{311\}A$ planes, and its size and shape are independent of the initial surface profile (Kojima et al. 1989, 1990). However, the depth of the 2D potential well corresponding to this wire is considerably smaller than those of the other two types.

It is worth mentioning that the uniformity of the array of wires formed at the bottom of the grooves directly reflects the uniformity of the initial surface grating. Thus, application of this technique in forming dense arrays of QWRs requires high-quality lithography during preparation of the patterned substrates. However, an interesting effect apparent in the TEM cross sections of Fig. 11 is the high wire-to-wire uniformity in the *thickness* of the (100) wires, which is in contrast to the highly nonuniform growth of the GaAs QWLs deposited just above the point where complete planarization is achieved. This indicates the existence of a smoothing effect, which allows formation of uniform QWL layers grown in a groove even with relatively thin buffer layers ($<100\,\text{nm}$ in the structures discussed here). Growth of multiple QWL structures on planar substrates, on the other hand, requires much thicker buffer layers to achieve comparable smoothing

(Weisbuch 1987; Turco et al. 1990a). This smoothing effect may be due to the preferential migration of the group III atoms to the bottom of the grooves, which regulates the nucleation along the channels.

MBE of patterned GaAs/AlGaAs QWLs grown on sub-μm gratings has also been reported by Kojima et al. (1990), Ismail et al. (1991) and Marti et al. (1991). In the last work, growth on gratings with grooves along the [011] and [01$\bar{1}$] directions was compared. The wires at the bottom of the grooves were found to be considerably thicker for the [011] case, probably due to the enhanced diffusion along the [01$\bar{1}$] direction described previously (see Fig. 9). Mirin et al. (1992) reported the fabrication of strained InGaAs/GaAs wires by migration enhanced MBE on sub-μm pitch gratings. Thicker wires are obtained as a result of the enhanced InGaAs migration into the grooves.

Lateral patterning of doping (both type and level) has been obtained by MBE growth of Si-doped GaAs/AlGaAs heterostructures on NP substrates. Silicon behaves amphoterically for the III–V compound system, yielding p-type doping for $\{111\}A$, $\{211\}A$ and $\{311\}A$ planes and n-type for $\{111\}B$, $\{211\}B$, $\{311\}B$ and either of the $\{511\}$ and higher index planes (Ballingall and Wood 1982; Wang et al. 1985; Subbanna et al. 1986). The $\{111\}A$ surface is Ga terminated (or at least expose a higher fraction of the Ga than the (100) planes), and hence the Si impurities occupy As sites, resulting in acceptor behavior. The $\{211\}A$ and $\{311\}A$ surfaces have sufficiently high density of $\{111\}$-type single dangling bonds and therefore exhibit similar behavior. Hole concentrations of up to $6 \cdot 10^{19}$ cm^{-3} have been obtained on $\{111\}A$ planes (Okano et al. 1989). N-type doping with Si on $\{111\}A$ (Miller and Asbeck 1987) and $\{311\}A$ (Meier et al. 1988a) planes has been obtained as well using higher V/III ratios and lower substrate temperatures. These doping effects have been utilized to achieve lateral p-n junctions by growing Si-doped AlGaAs and GaAs layers on mesas and grooves oriented in the [01$\bar{1}$] directions and exposing $\{111\}A$ or $\{311\}A$ planes at their side walls (Miller 1985; Miller and Asbeck 1987; Meier et al. 1988a). Light emitting diodes employing lateral carrier injection across such p-n junctions have been demonstrated (Miller and Asbeck 1987; Meier et al. 1988). Lateral GaAs p-n junctions grown on NP $\{111\}A$ substrates patterned with equilateral traingles have also been reported (Fujii et al. 1992).

The lateral doping patterning is illustrated in Fig. 12, which displays cross sections of a GaAs/AlGaAs laser heterostructure grown by MBE on (100)-GaAs substrate patterned with a [01$\bar{1}$]-oriented groove (Kapon et al. 1988a). The lower, Si-doped AlGaAs cladding layer exhibits lateral tone variations in the SEM photograph, which represent lateral variations of the doping type. The doping is n-type on the (100) and $\{411\}A$ planes, and changes to p-type on the $\{111\}A$ slopes. A transition region between the $\{411\}$ and $\{111\}$ planes is also evident.

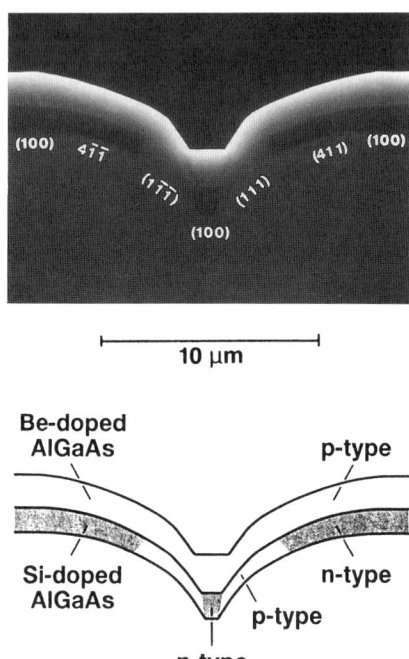

Fig. 12. SEM (top) and schematic (bottom) cross sections of a GaAs/AlGaAs heterostructure grown by MBE in [01$\bar{1}$]-oriented groove, showing lateral doping variations in the Si doped AlGaAs epitaxial layer (Kapon et al. 1988a).

5. One-Dimensional Lateral Band-Gap Patterning

Lateral variations in thickness and/or composition of QWL heterostructures grown on NP substrates lead to corresponding lateral variations of their effective band gaps. These band-gap variations can be evaluated by measuring the lateral distribution in the recombination energy of excited carriers using various luminescence techniques. This is illustrated in Fig. 13(a), which shows low-temperature CL spectra of the patterned QWL structure of Fig. 10 (Clausen et al. 1990). The lateral QWL thickness modulation in this case results in formation of *two* lateral potential wells per period, one at the bottom and the other at the peak of the corrugations. Excited carriers can transport and thermalize into these potential wells, recombining and emitting photons of energy corresponding to the local effective band gap of the patterned QWL region. The CL spectrum exhibits two peaks, at 1.538 and 1.570 eV, both red shifted compared to the single

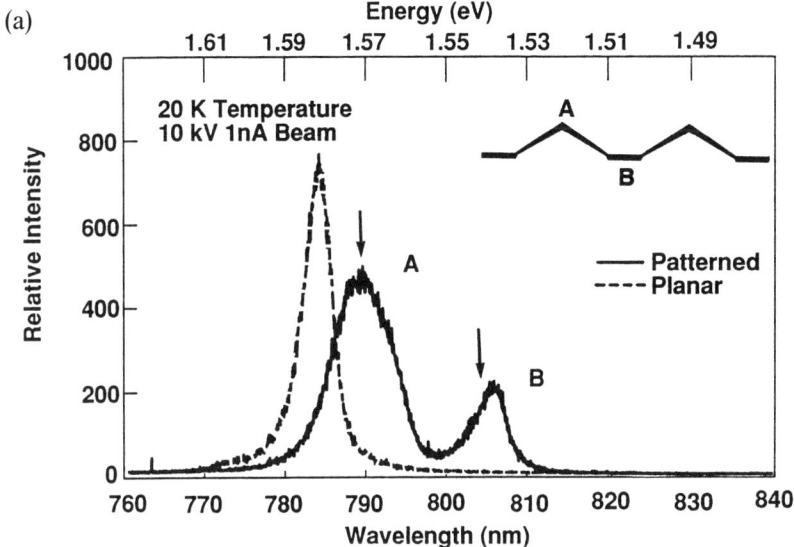

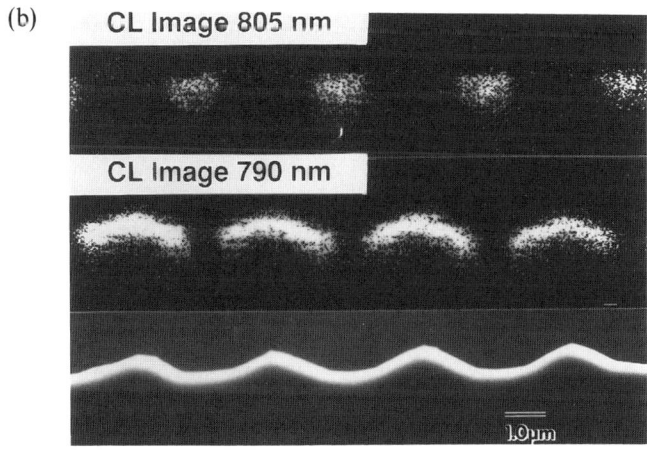

FIG. 13. (a) Cathodoluminescence spectrum of the patterned structure of Fig. 10 (solid line), compared with the spectrum of a planar, control structure grown on a flat substrate (dashed line). Arrows indicate the calculated emission energies for regions *A* and *B*. (b) Spatially resolved cross sectional images of the patterned structure taken at the two emission lines in (a). The bottom part is a secondary electron image of the structure (Clausen et al. 1990).

emission line (at 1.582 eV) obtained for a control, planar sample grown side by side with the nonplanar one. This is simply because the QWL at both the peak and the bottom of the corrugations is thicker than the uniform QWL grown on a planar substrate due to the lateral diffusion effects. The calculated effective band gaps corresponding to the two lateral potential wells, obtained using the QWL thickness measured from the TEM data and accounting for exciton binding energies, are in good agreement with the measured energies (see arrows in Fig. 13(a)).

The 1D periodic modulation in effective bandgap in this structure is directly visualized using spatially and spectrally resolved CL images, as shown in Fig. 13(b). The two CL images were recorded at the two CL lines of Fig. 13(a) and are compared with a conventional secondary electron image of the cross section. These images clearly display the periodic modulation in band gap and the consequent lateral carrier confinement at the two types of lateral potential wells created by the QWL thickness modulation.

Lateral carrier confinement in wire-like GaAs/AlGaAs potential wells grown by MBE on sub-μm gratings has also been observed in low-temperature photoluminescence spectra by Turco et al. (1990a), Kojima et al. (1990) and Marti et al. (1991). Polarization anisotropy in the PL spectra has been interpreted by Kojima et al. (1990) as evidence for 2D quantum confinement in the bent {311}A QWL grown at the peak of the corrugations. Polarization anistropy of the PL spectra from InGaAs/GaAs QWR arrays grown on sub-μm pitch periodic corrugations was reported also by Mirin et al. (1992).

6. Two-Dimensional Lateral Band-Gap Patterning

Two-dimensional lateral band-gap patterning in heterostructure parameters can be achieved by growth on NP substrates patterned with 2D features. In conjunction with the more conventional control of the band gap available along the growth direction, this can yield 3D patterned structures. In particular, 2D lateral band-gap patterning is useful for applications in integrated optoelectronics, where areas of different band gaps on a single substrate can be used for monolithically integrating different devices (see subsection 11). Two-dimensional lateral patterning is also required for forming the 3D potential wells for QDs and other heterostructures relying on 3D quantum confinement.

An example of InGaAs/GaAs strained layer dot structures formed by MBE growth on (100) NP substrate patterned with craters is shown in Fig. 14 (Krahl et al. 1992). The craters, of diameters ranging between 0.2 and

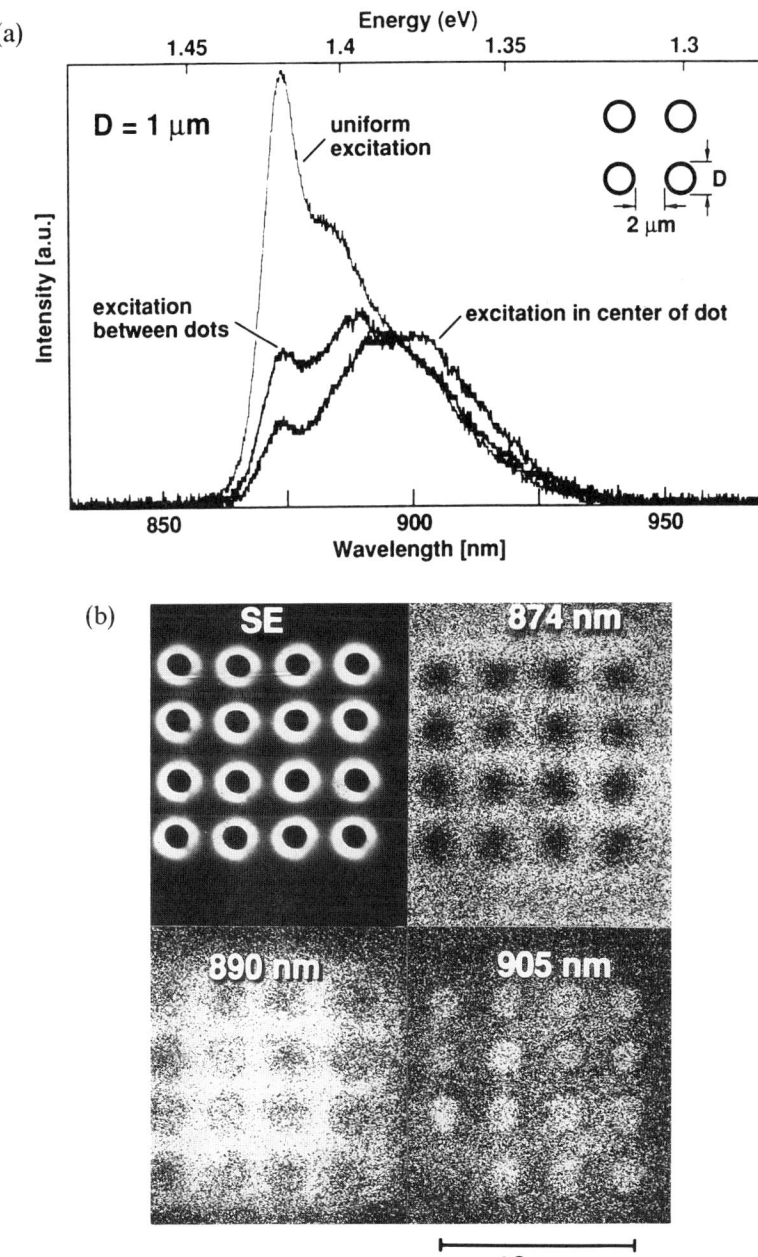

FIG. 14. (a) Low-temperature CL spectra of 1 μm-diameter InGaAs/GaAs dot structures measured under different excitation modes. (b) Secondary electron (SE) spectrally and spatially resolved CL images of 1 μm-diameter dots recorded at different emission wavelengths (Krahl et al. 1992).

5 μm, were dry-etched onto the (100) GaAs substrates using reactive ion etching through resist masks prepared by direct electron beam lithography. The MBE grown structure consisted of 100 nm GaAs, 5 nm $In_{0.2}Ga_{0.8}As$ and 20 nm GaAs, all nominally undoped. The growth temperature was 530°C, and the As_2 to Ga flux ratio was ~3 to 1.

Growth of the InGaAs layer on the NP substrate results in preferential migration of the In and Ga adatoms from the slopes of the etched dots to the (100) planes inside and outside the craters. Since the surface diffusion length of In is much larger than that of Ga (see Table I), 2D lateral band-gap patterning can occur due to thickness as well as In mole fraction variations. The CL spectra of an array of 1 μm-diameter dots, measured under different excitation conditions, indeed show features at wavelengths longer than 874 nm (1.419 eV) corresponding to emission from the flat, unpatterned parts of the sample (Fig. 14(a)). These longer wavelength features dominate the spectrum when the pump electron beam is focused between the dots and at their center, indicating effective energy gaps of 1.393 eV (890 nm) and 1.373 eV (905 nm) at the corresponding regions.

The 2D bandgap patterning effect is clearly displayed in the CL images of Fig. 14(b), which reveal ring-shaped potential wells surrounding the craters and dot-shaped potential wells at their centers. The ring-shaped potential wells correspond to band-gap reduction of 26 meV, whereas the dot-shaped ones correspond to 49 meV reduction. These values can be attributed to an increase in the In mole fraction of ~4 and 9%, or increase in QWL thickness of ~1.5 and 4 nm, respectively.

IV. Lateral Patterning by OMCVD on Nonplanar Substrates

In OMCVD growth of III–V semiconductors, the source materials are introduced in the form of organometallic compounds (e.g., trimethylaluminum (TMA), trimethylgallium (TMG)) for group III elements, and arsine (AsH_3) or phosphine (PH_3) for group V elements. The use of a gas-phase ambient has an important impact on the available transport mechanisms at the surface of the substrate, which are particularly relevant to epitaxial growth on NP substrates. The reactant molecules can diffuse in the low-velocity *boundary* (*stagnant*) *layer*, formed next to the substrate's surface, with typical *gas-phase* diffusion lengths as large as 10–20 μm at atmospheric pressures and ~100 μm at lower pressures. Subsequent adsorption and reaction of these molecules at the surface of the crystal results in epitaxial deposition. Surface diffusion of nonvolatile reaction products prior to deposition provides a second channel for surface migration, which is

4. LATERAL PATTERNING OF QUANTUM WELL HETEROSTRUCTURES   289

characterized by a much shorter range, typically 100 nm or less. The differences in migration processes and surface chemistry account for the vastly different growth behavior of MBE and OMCVD on NP substrates.

7. EFFECT OF GROWTH AND PATTERN PARAMETERS

Figure 15 depicts schematically the morphology of OMCVD growth on NP (100) substrates corrugated with µm-size grooves along the [01$\bar{1}$] and [011] directions (Hersee *et al.* 1986; Demeester *et al.* 1988; Bhat *et al.* 1988; Tate *et al.* 1988). Growth on the [01$\bar{1}$]-oriented corrugations is characterized by formation of two types of facets, (100) on the ridges and facets close to {111}A on the slopes of the grooves. Several examples of actual structures grown in this configuration are discussed later. Achievement of near-exact {111}A planes requires low growth rates to allow sufficient mass transport (in the gas phase) for facet construction. High growth rates can lead to

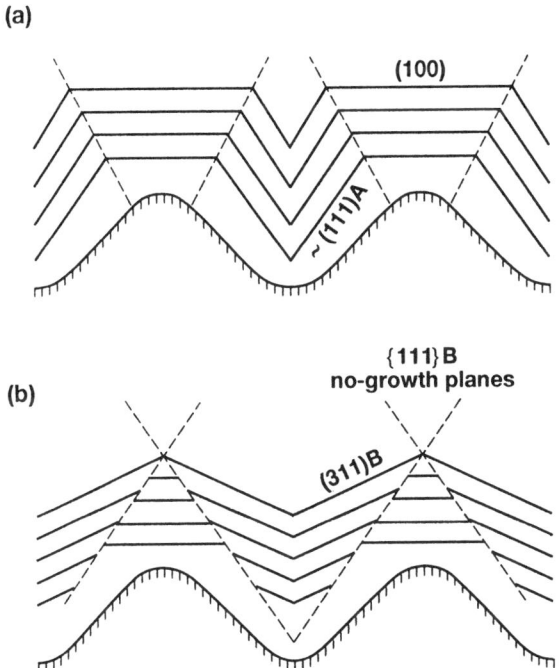

FIG. 15. Schematic cross sections of patterned heterostructures grown by OMCVD on corrugated (100) substrates, with corrugations along the (a) [01$\bar{1}$] and (b) [011] directions.

evolution of vicinal {111}A surfaces and, in extreme cases, unstable surface profiles, e.g., dips at the bottom of the grooves (Ratsch and Zangwill 1991).

Growth on the [011]-oriented corrugations exhibits *no-growth* planes lying in the {111}B planes, giving rise to discontinuity in the layers grown on the slopes of the grooves. The very slow growth rate on the {111}B facets is due to fast migration of Ga or Al hydrocarbon molecules on these planes (Zinke-Allmang *et al.* 1988). These no-growth facets yield sharp tips bounded by the {111}B planes when growth in the grooves is suppressed, e.g., in the case of growth on substrates patterned by dielectric masks (Kamon *et al.* 1985), or in the case of growth on sufficiently high mesas. Growth in the grooves proceeds at or near the {311}B planes, leading to gradual planarization of the structure. An example of a GaAs/AlGaAs multilayer structure grown on a [011]-oriented mesa is shown in Fig. 16. In this case growth proceeded past the point at which the {311}B side walls met with the {111}B corner, resulting in contiguous layer growth over the mesa and in (100) planes of increasing width.

The temperature dependence of the features of OMCVD of GaAs/AlGaAs over NP substrates has been investigated by Hersee *et al.* (1986) and Dzurko *et al.* (1989a). For [01$\bar{1}$] ridges, the ratio of the growth rate on the (100) ridge tops to that on the side walls was found to increase with substrate temperature, leading to narrowing of the ridge widths at temperatures above $\sim 750°C$. This provides a means for producing extremely narrow potential wells on top of the ridge and has been employed to demonstrate narrow, low-threshold QWL lasers (see subsection 10).

OMCVD growth on NP *vicinal* (100) substrates oriented a few degrees off the (100) plane has been investigated by Colas *et al.* (1989, 1990a, 1990b,

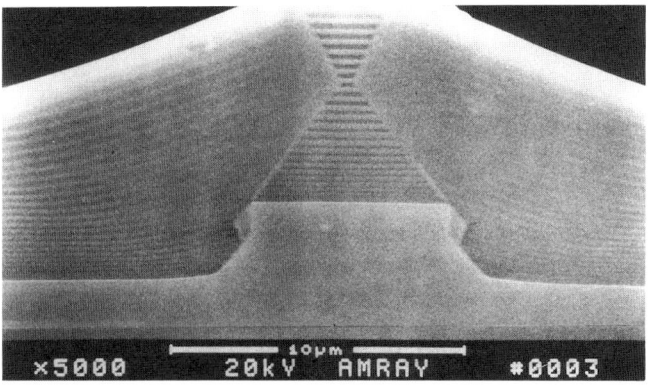

FIG. 16. SEM cross section of a GaAs/AlGaAs multilayer structure grown on [011]-oriented mesa etched on (100) GaAs substrate (courtesy of E. Colas and R. Bhat, Bellcore).

1991, 1992) and by Cox *et al.* (1989). Schematic and TEM cross sections of a GaAs/AlGaAs structure grown on a patterned GaAs substrate oriented 6° off (100) towards (111)A are shown in Fig. 17 (Colas *et al.* 1990a). The [01$\bar{1}$]-oriented grooves were fabricated by wet chemical etching through a 3.5 μm pitch photoresist mask. The $Al_{0.47}Ga_{0.53}As$ cladding layers in this particular structure were grown at 650°C, whereas the (nominally 7 nm thick) GaAs QWL layer was grown at 550°C by repeating 25 times an atomic-layer-epitaxy (ALE) cycle of 0.2 s TMG/0.4 s halt in H_2/1 s AsH_3. The growth was performed in an atmospheric pressure, horizontal OMCVD reactor.

Growth on such patterned vicinal (100) substrates is characterized by formation of periodic *macrosteps* (or *supersteps*), each defined by a reconstructed (100) facet and a high-index facet (close to $\{111\}A$). The (100) planes develop due to their lower surface energy, compared to that of the higher index planes exposed by the corrugated substrate. The periodicity Λ of the macrosteps is identical to that of the etched grating on which the structure is grown. This is different from the case of growth on *nonpatterned* vicinal (100) substrates, where the pitch of the stepped surface is determined solely by the angle of misorientation (Petroff *et al.* 1984). The high-index planes grow at a higher rate than that of the (100) facets, consuming source material supplied via diffusion (mainly in the gas phase). The length of these high-index facets is given approximately by $\Lambda \sin\alpha/\sin\theta$, where α is the vicinal angle, θ is the inclination angle of the high-index plane with respect to (100), and $\alpha \ll \theta$ is assumed. Formation of these macrosteps can also be viewed as "accumulation" of the monolayer steps on the vicinal section within each period of the grating, which forms a superstep whose height is equal to the sum of the monolayer steps (Cox *et al.* 1989). Ideally, subsequent growth on these macrosteps proceeds via lateral flow of the stepped structure while the size and shape of each macrostep is preserved. This self-limiting feature is very attractive for fabrication of vertically stacked quantum structures with uniform cross sections. However, deviation from a perfect "step flow" growth mode due to finite growth rate on the (100) facets tends to smear out the stepped structure, and conformal growth is possible only for a finite number of stacked layers.

In the structure of Fig. 17, the growth ratio between the two facets of the steps is 1:2.5, with a corresponding QWL band-gap modulation of ~ 80 meV. This large growth rate ratio and band-gap modulation amplitude are a direct result of the enhanced surface diffusion made possible during ALE conditions. The diffusion length associated with this migration process is at least as large as the grating pitch (3.5 μm), as suggested by the uniform thickness of the QWL grown on the (100) facet. The thickness variations observable near the concave corner of the step edge are probably due to

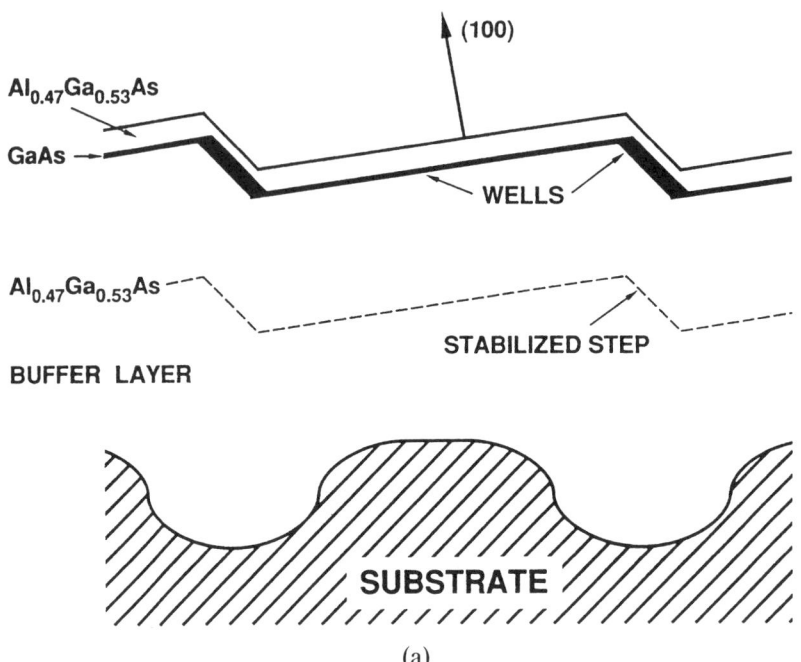

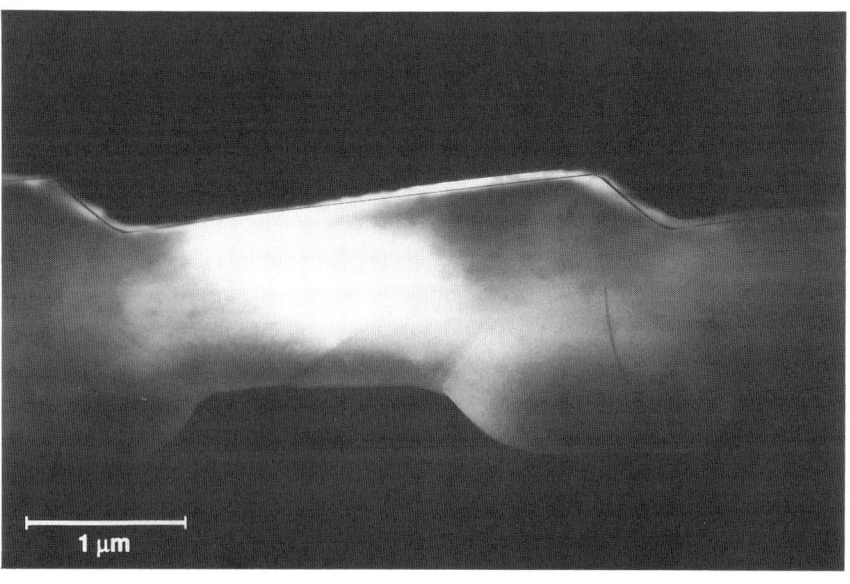

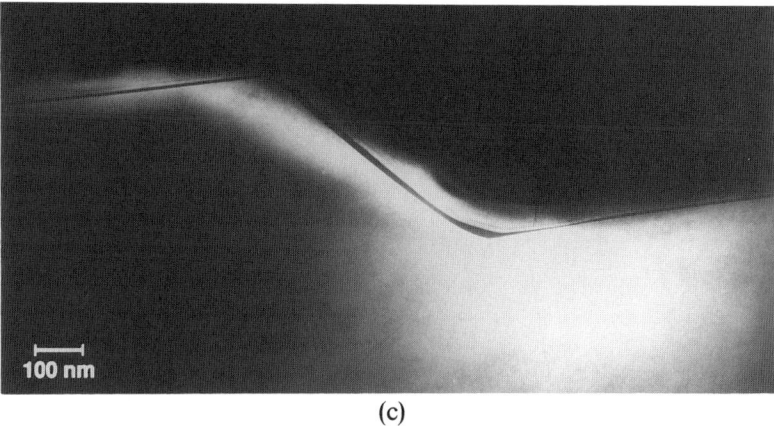

(c)

FIG. 17. (a) Schematic and (b), (c) TEM cross sections of GaAs/$Al_{0.47}Ga_{0.53}$As patterned quantum wells grown by OMCVD on a patterned, off-(100) substrates. Part (c) shows a magnified view of the step edge (Colas et al. 1990a).

surface diffusion with a much shorter (<200 nm) characteristic length. The band-gap modulation is clearly visualized in the CL spectra and the spectrally and spatially resolved images shown in Fig. 18 (Colas et al. 1990a, 1991). Note that the emission emanating from the (100) facet is blue shifted, whereas the emission from the high-index plane is red shifted, with respect to the emission wavelength corresponding to the planar sample. This is a direct consequence of the diffusion effects that reduce and increase the deposition rates at the (100) and high-index facets, respectively. The measured QWL thicknesses at the facets (6 nm and 13–16 nm) yield calculated effective band gaps that are in good agreement with the measured energies (see arrows in Fig. 18(a)). Formation of similar InGaAs/InP wires grown by vapor levitation epitaxy on NP vicinal (100) InP substrates has been reported by Cox et al. (1989).

Growth of InGaAs/InP heterostructures on NP InP substrates has been studied by Galeuchet et al. (1988, 1990, 1991), Scott et al. (1988) and Bhat et al. (1990a). Growth of InP exhibits behavior similar to that of GaAs/AlGaAs, with $\{111\}A$ and $\{111\}B$ facet formation at the sidewalls of [01$\bar{1}$]- and [011]-oriented grooves and ridges, and contiguous growth across the lateral profile. The ternary compounds InGaAs and InAlAs, on the other hand, deposit selectively on the exposed (100) facets, giving rise to completely disconnected, lower band-gap wires that are buried in the InP cladding layers. High surface mobility of the ternary source materials on the $\{111\}B$ facets also leads to enhanced growth on the mesa top, providing a means

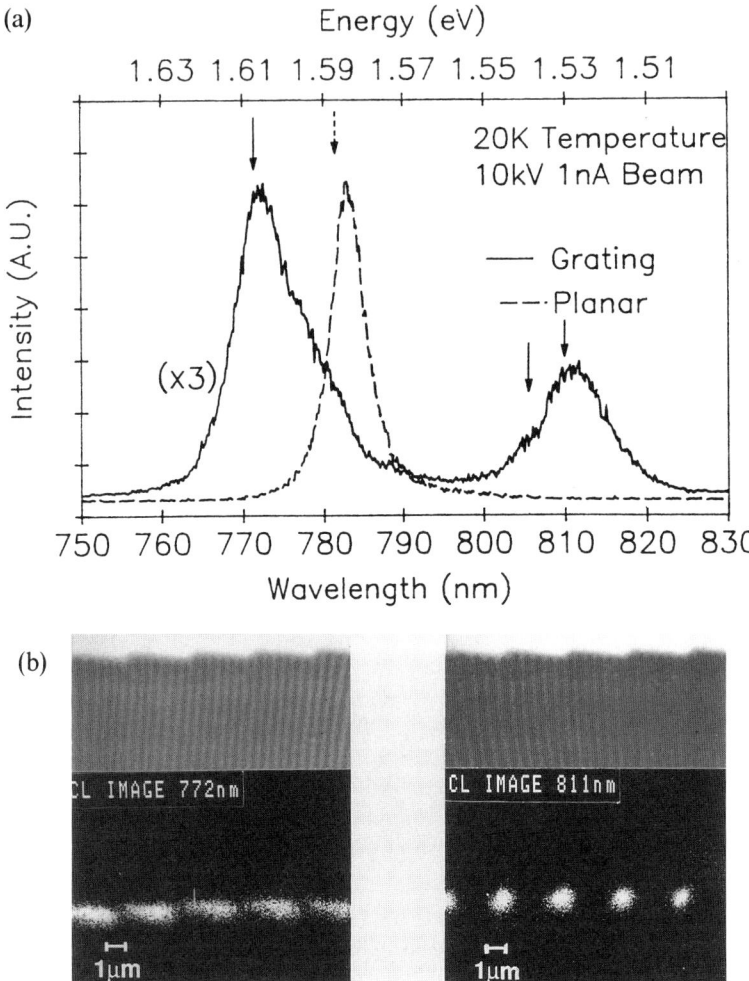

FIG. 18. (a) Low-temperature CL spectra of the patterned QWL structure of Fig. 17 (solid line) and a control, nonpatterned sample grown side by side (dashed line). Arrows indicate the emission energies calculated using the well thicknesses measured from TEM cross sections. (b) Spectrally and spatially resolved CL images recorded at the two CL peaks of the patterned sample; secondary electron cross sectional images are also shown (Colas et al. 1990a).

for controlling the epitaxial layer thickness by the mesa width. This growth behavior has been employed to produce buried heterostructure InGaAs/InP lasers and buried ridge InAlAs/InP optical waveguides via a single epitaxial step (Scott et al. 1988).

The orientation dependence of doping in layers grown by OMCVD on planar substrates of various orientations was investigated in some detail by Bhat et al. (1991) and Kondo et al. (1992). The doping behavior is rather complex and in many cases shows a nonmonotonic dependence on the wafer orientation. Generally, the incorporation rate of donors on the B faces is greater than that on the corresponding A faces. Group-II acceptors, on the other hand, are incorporated preferentially on the A faces as compared with the (100), and in many cases their incorporation on the B faces is lower than on the (100) one.

Lateral patterning of the doping profile in InP layers grown on $[01\bar{1}]$ V-grooves has been demonstrated by Bhat et al. (1990b). A double heterostructure InGaAsP/InP laser in which the upper InP cladding layer was composed of S and Zn doped alternating layers, 55 nm thick each, was grown on the grooved substrate. The stained, cross sectional SEM image of the resulting patterned structure (see Fig. 19) reveals formation of alternating p-n layers on the (100) planes on both sides of the groove, whereas a uniformly p-doped region is obtained inside the groove. This lateral doping profile was attributed by Bhat et al. to the higher effective III/V ratio at the side walls of the groove, caused by the more effective decomposition of PH_3 on the $\{111\}A$ plane. This, in turn, leads to a higher In-vacancy concentration (and lower P-vacancy concentration), which increases the incorporation of Zn and decreases that of S inside the groove. Fast Zn diffusion then converts the S-doped layers into n-type regions inside the groove. Kondo et

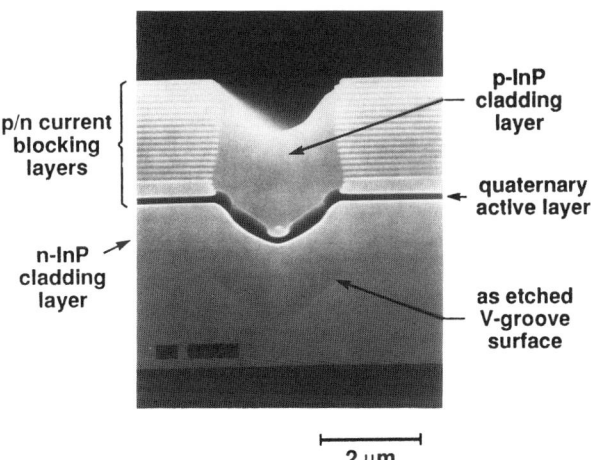

FIG. 19. SEM cross sectional image of a stained double-heterostructure laser grown in a $[01\bar{1}]$-oriented V-groove showing the selective formation of alternating p-n current blocking layers. The staining distorts the shape of the crescent-shaped quaternary active layer at the bottom of the V-groove (Bhat et al. 1990b).

al. (1992) argued that the dopant incorporation is controlled by the nature of the steps on the misoriented facets. The alternating p-n layer structure of Fig. 19 was employed for effective current blocking in these lasers. This doping effect was also used to grow lateral arrays of p-n junctions, by growing InP in [01$\bar{1}$] grooves with both p- and n-dopants present simultaneously (Bhat et al. 1991).

8. Formation of Quantum Wires by Self-Ordering

The growth features of OMCVD on [01$\bar{1}$]-oriented grooves offers a unique possibility for producing uniform arrays of QWRs whose size and shape are independent of the initial profile of the etched surface. The corners between the various planes, which evolve during growth on grooves etched in (100) or vicinal (100) substrates, provide a surface profile determined (in each case) solely by the growth condition and cladding layer composition. This *self-limiting* process is important for realization of quantum structures since it results in surface profiles insensitive to lithography imperfections, thereby reducing possible size fluctuations of the low-dimensional potential wells. Since the radius of curvature at the evolving crystal corners is determined by surface diffusion effects, which typically have a range of a few 10 nm, the lateral dimensions of the resulting quantum structures can be small enough to yield significant lateral quantum size effects.

The evolution of crescent-shaped GaAs QWRs by OMCVD in [01$\bar{1}$]-oriented corrugations is illustrated schematically in Fig. 20. Growth of the AlGaAs layer in the channel results in formation of a sharp corner between two $\{111\}A$ planes due to their faster growth rate. The curvature at this corner, ρ_l, is determined by the surface diffusion length of the group III growth species from the side walls to the bottom of the groove. This radius of curvature depends on the Al mole fraction, the group V/III ratio, and the growth temperature, and can be as small as 10 nm for conventional growth conditions (Bhat et al. 1988; Kapon et al. 1989a). Subsequent growth of a GaAs layer in this sharp V-groove then results in an increase of the radius of curvature to ρ_u, because of the larger effective diffusion length of Ga compared to Al, yielding a *crescent-shaped* wire. Subsequent growth of AlGaAs material leads to resharpening of the groove; in fact, a full recovery of the value of ρ_l can be accomplished with a sufficiently thick AlGaAs layer above the crescent.

The cresecent-shaped wires formed in this way have two remarkable properties. First, the value of the radius of curvature developed during growth of a sufficiently thick AlGaAs barrier, ρ_l, is independent of the initial groove shape and is self-limiting; i.e., it does not vary upon further growth

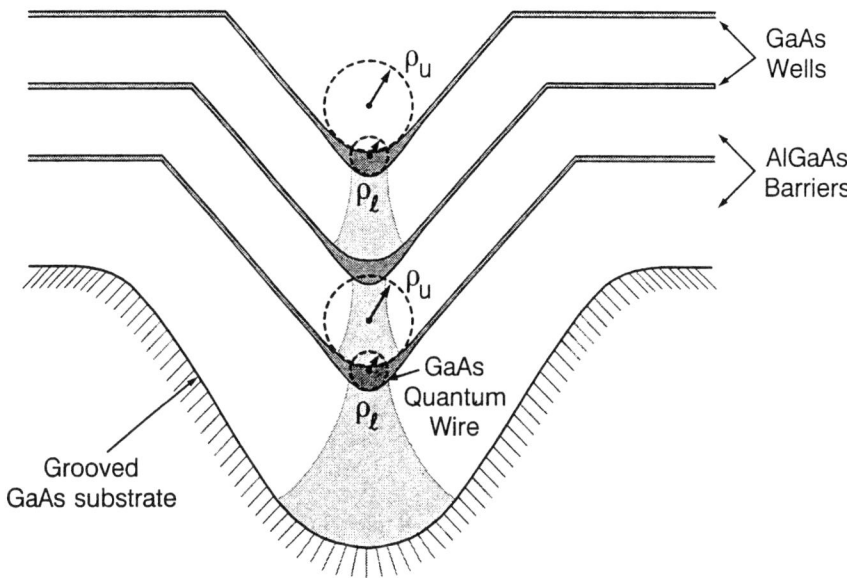

FIG. 20. Schematic cross section of a GaAs/AlGaAs heterostructure grown by OMCVD in a [01$\bar{1}$]-oriented channel etched in a (100) GaAs substrate, showing formation of crescent shaped QWRs. The terms ρ_s, ρ_l and ρ_u indicate the radius of curvature at the bottom of the etched channel, underneath and above the crescent, respectively.

of AlGaAs. This means that the shape and size of the crescent depend only on the growth conditions and GaAs thickness and is thus insensitive to pattern imperfections. Second, complete recovery of the radius of curvature during AlGaAs growth allows stacking of identical crescents in a vertical array configuration. Note also that the effective width of the crescent can be much smaller than its geometrical width due to the quantum size effects discussed in subsection 3.

TEM cross sections of a GaAs/AlGaAs QWR heterostructure grown by OMCVD are shown n Fig. 21 (Kapon et al. 1992c). The substrate was patterned in this case with periodic corrugations of 3.5 µm pitch, oriented along the [01$\bar{1}$] direction. The structure consists of 15X(1.5 nm GaAs/ 1.5 nm $Al_{0.7}Ga_{0.3}As$) buffer SL, 0.5 µm $Al_{0.7}Ga_{0.3}As$, 2 nm GaAs, 40 nm $Al_{0.7}Ga_{0.3}As$ and 30 nm GaAs, and was grown at atmospheric pressure and $T_s = 750°C$. The radius of curvature at the bottom of the grooves reduces during growth from $\rho_s \sim 0.2$ µm to $\rho_l = 23$ nm. The GaAs crescents are 8 nm thick at their center and ~ 50 nm in full width. Thinner QWL layers also form on the $\{111\}A$ side walls as well as on the (100) tops of the grating ridges.

(a)

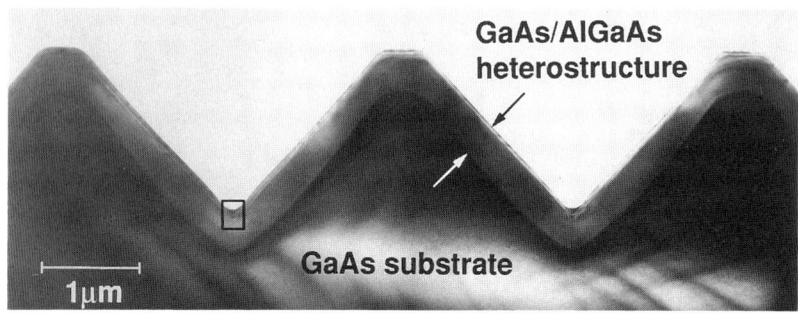

(b)

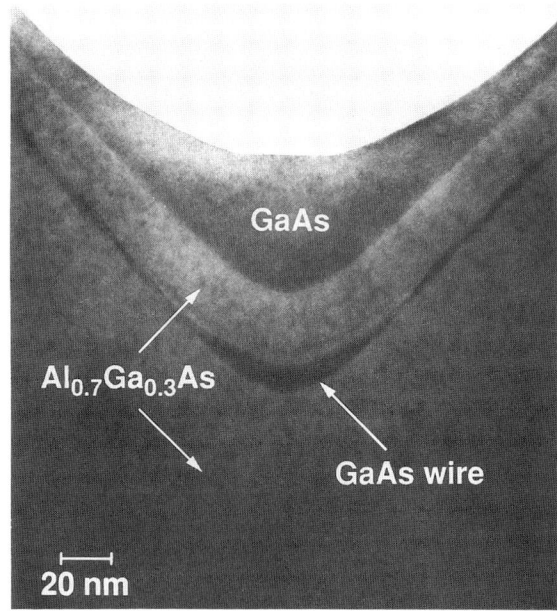

FIG. 21. TEM cross sectional images of GaAs/Al$_{0.7}$Ga$_{0.3}$As QWR heterostructure grown by OMCVD on V-grooved substrate: (a) cross section of the grating and (b) magnified view of the framed area in (a), showing the crescent-shaped GaAs wire (Kapon et al. 1992c).

A vertically stacked array of GaAs/AlGaAs crescent-shaped QWRs grown in a similar way is shown in Fig. 22 (Kapon et al. 1992a; Christen et al. 1992b). This structure was grown at 670°C on a periodically corrugated substrate similar to the one of Fig. 21 and consists of 0.7 μm Al$_{0.45}$Ga$_{0.55}$As

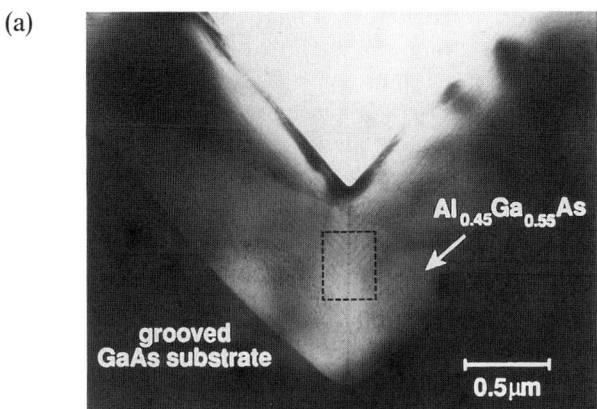

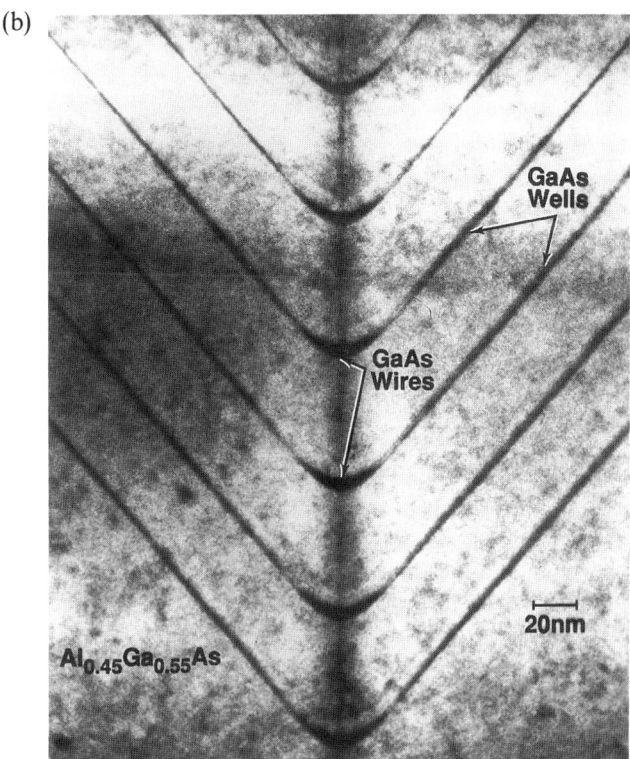

FIG. 22. TEM cross section of a vertical array of GaAs/AlGaAs QWRs formed by self-ordering using OMCVD on V-grooved substrates. Part (b) shows a magnified view of the framed region in (a) (Christen et al. 1992b).

and 10 periods of 2 nm GaAs QWLs separated by 50 nm $Al_{0.45}Ga_{0.55}As$ barriers. The radius of curvature is $\rho_l = 15$ nm at the bottom of the crescents and increases to $\rho_u = 23$ nm after growth of the GaAs. However, the thickness of the AlGaAs barriers is sufficient for complete recovery of ρ, which results in a vertical array of identical, crescent-shaped wires. In this sense, the peculiar features of OMCVD on [01$\bar{1}$]-oriented channels lead to the *self-ordering* of arrays of QWRs whose shape and size depend only on the growth conditions. It is interesting to note that the migration of growth species to the bottom of the groove, which gives rise to the GaAs crescents, also results in formation of a narrow (<20 nm), vertical stripe of Ga-rich AlGaAs region running through the center of the groove. This *vertical* AlGaAs QWL, formed due to the longer diffusion length of Ga compared to Al, plays an important role in the collection of excited charge carriers into the QWR, as discussed in the subsection 9. Similar vertical AlGaAs QWLs have been reported also by Vermeire et al. (1992).

Vertical array of QWRs can also be formed by vertical stacking on the macrosteps that evolve during OMCVD on patterned vicinal (100) substrates. Figure 23(a) shows a SEM cross section of a 30 nm $Al_{0.5}Ga_{0.5}As/10$ nm GaAs SL grown at $T_s = 650°C$ on a GaAs substrate oriented 6° off the (100) plane towards the (111)A orientation and patterned with 0.7 μm-pitch periodic corrugations aligned in the [01$\bar{1}$] direction (Colas et al. 1991). This structure clearly displays the evolution of the macrosteps, which become fully developed after 15 layers. The steps tend to smear out after a few SL periods; however, there exists a "window" of total grown thickness in which a relatively uniform array of steps can develop. The TEM cross section in Fig. 23(b) shows a magnified view of several vertically stacked wires grown at the step edges (Colas et al. 1992).

Growth of InGaAs/GaAs strained-layer QWL heterostructure grown by OMCVD on NP substrates was studied by Zou et al. (1991) and Grodzinski et al. (1991). Crescent-shaped InGaAs wire structures were observed at the bottom of V-grooves oriented along the [01$\bar{1}$] direction. In addition, quasi-periodic InGaAs wire structures formed on the {133}-oriented side walls of the grooves, with periodicities increasing from $\sim$65 to 175 nm for nominal well layer thickness increasing from 8.5 to 23 nm. The origin of these quasi-periodic structures may be related to strain accumulation in the InGaAs layer that results in three-dimensional growth, although the exact mechanism is not clear at present. For growth on [011] mesas, the formation of {111}B no-growth planes allows growth of (100) InGaAs layers of limited lateral extent, which can yield strained layers with thicknesses exceeding the critical values for generation of dislocations on planar substrates. A $\sim$50% increase in critical thickness was inferred from PL studies.

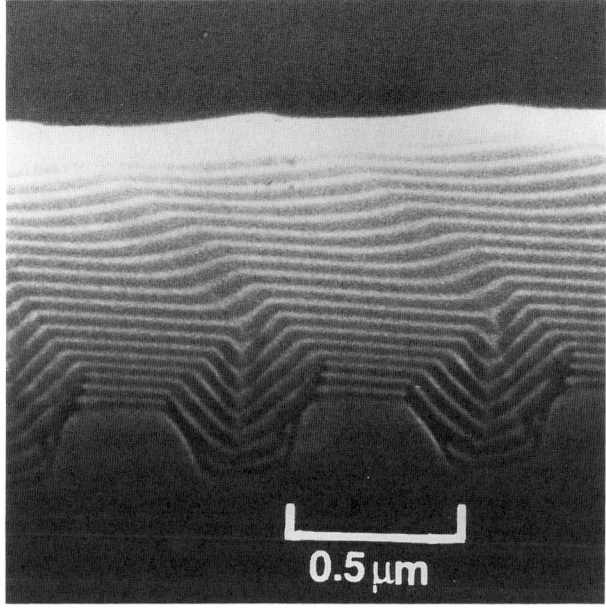

(a)

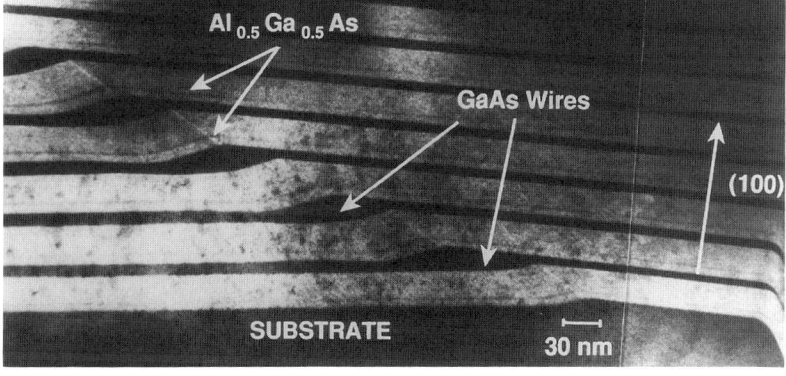

(b)

FIG. 23. (a) SEM cross section of a GaAs/Al$_{0.5}$Ga$_{0.5}$As SL grown on a 6°-off (100) corrugated substrate, showing the evolution of macrosteps on a patterned vicinal substrate (Colas et al. 1991). (b) TEM cross section of a vertically stacked GaAs/Al$_{0.5}$Ga$_{0.5}$As QWR structure grown on a 6°-off-(100) patterned vicinal substrate (Colas et al. 1992). Corrugations are along [01$\bar{1}$] in both cases.

Formation of crescent-shaped InGaAsP/InP QWRs by OMCVD in [01$\bar{1}$]-oriented V-grooves was also demonstrated (Bhat et al. 1990a). The grown heterostructure consisted of eight periods of 4 nm InGaAs/17.5 nm InGaAsP (both lattice matched to InP and the quaternary band gap corresponding to 1.3 μm wavelength, for growth on (100) substrates) sandwiched between InP/InGaAsP SL layers, and was grown at 625°C and 76 Torr. Two types of patterns were studied, 2 μm deep, V-shaped grooves and 1.5 μm deep, flat-bottom grooves. TEM studies revealed a dense network of dislocations on the side walls of the V-groove structures, indicating deviation from lattice-matching conditions in these regions. This probably arises from a difference in the sticking coefficients and migration of the various growth species from the side walls to more favorable deposition sites via surface or gas phase diffusion. However, at the bottom of the grooves, crescent-shaped InGaAs wires were formed in an area that was essentially dislocation free. The observed full width of the crescents was 37.5 nm. The flat-bottom groove structures, on the other hand, were dislocation free and exhibited thicker InGaAs QWL layers at their center. Further work is required in this area to identify growth conditions and pattern features that allow lattice-matched patterned QWL growth.

Dense, *lateral* arrays of crescent-shaped QWRs can be achieved by OMCVD growth on NP substrates patterned with sub-μm-pitch corrugations (Colas et al. 1990b). Figure 24 shows TEM cross sections of two such GaAs/AlGaAs lateral QWR arrays, grown side by side on patterned (100) and 6°-off-(100) (towards (111)A) GaAs substrates at atmospheric pressure. The gratings were fabricated by holographic photolithography and wet chemical etching, and the grating pitch is 0.25 μm ([01$\bar{1}$]-oriented grooves). The grown structure consists of 150 nm $Al_{0.5}Ga_{0.5}As$, 6 nm GaAs and 30 nm $Al_{0.5}Ga_{0.5}As$, all nominally undoped, and the substrate temperature was 650°C. Growth on the (100) patterned substrates results in formation of crescent-shaped GaAs wires, as in the case of the larger grooves described previously. However, since the groove separation in this case is smaller than the surface diffusion length of the growth species, the QWL layers connecting the wires are much thinner (1–2.5 nm) than the nominal QWL thickness, indicating strong migration effects into the concave AlGaAs groove corner. Growth on the vicinal substrate also yields GaAs tapered wire structures, at the corners between the reconstructed (100) planes and the macroscopic step edges. The wires in this case are thinner than those formed on the (100) patterned substrates, and the connecting QWL layers have thickness comparable to that obtained on a flat substrate. The difference in the surface migration effects in the two samples may be due to the fact that the connecting QWL layers are bounded by *two concave* corners in the (100) case, and by *a concave and a convex* corner in the vicinal case. Note also the

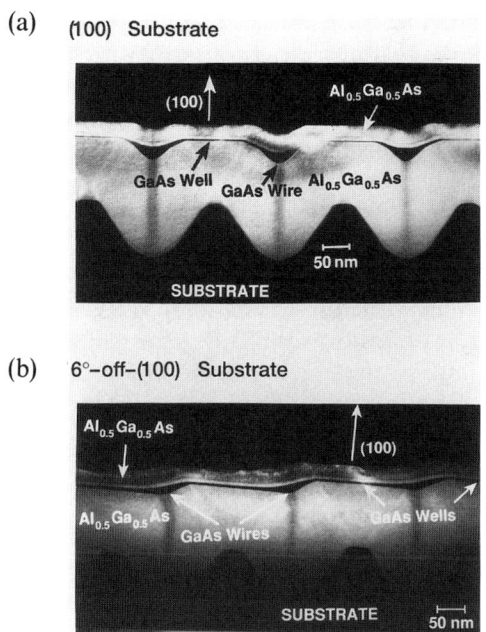

FIG. 24. Lateral arrays of quantum wires grown by OMCVD on sub-μm gratings: (a) (100) substrate, (b) 6°-off (100) substrate. Corrugations are in the [01 $\bar{1}$] direction in both cases (Colas et al. 1990b).

vertical AlGaAs QWL structures which are visible as a darker contrast in both grating structures of Fig. 24. Similar QWR arrays have been grown by Ismail et al. (1991) on 200 nm-pitch gratings fabricated by x-ray lithography and wet chemical etching, and by Vermeire et al. (1992) on 650–765 nm pitch gratings made by holographic photolithography and wet etching. Arrays of strained InGaAs/GaAs QWRs grown on 0.25 μm-pitch gratings were reported by Walther et al. (1992c, 1993). Patterned modulation doped GaAs/AlGaAs heterostructures grown on sub-μm pitch gratings were also reported, by Helm et al. (1991).

The nonplanar nature of selective-area OMCVD on substrates patterned with dielectric masks leads to many growth features that are similar to those of growth on NP substrate, particularly in the case of very small (below 1 μm) lateral dimensions (Lebens et al. 1990; Fukui et al. 1991; Galeuchet et al. 1988, 1990, 1991; Tsukamoto et al. 1992, 1993; Nagamune et al. 1993). The evolution of specific facets during OMCVD on the open areas in the masked substrate establishes a surface profile that can be controlled by the growth parameters, and subsequent deposition of QWL layers results in lateral band-gap patterning as described earlier for NP substrates. The preparation of the NP substrate via the selective-area growth may have some advantages over the preparation by etching, e.g., access to families of crystal facets that are not achievable by common etchants. Tsukamoto et al. (1992) have used this approach to fabricate vertical arrays of crescent-shaped GaAs QWRs, similar to the ones of Fig. 22, grown on 200 nm-pitch SiO_2 gratings made by electron beam lithography and wet etching. Tsukamoto et al. (1993) have also used the same approach to generate crescent-shaped GaAs wires formed near the apex of the ridges between the V-grooves (similar to the structure in Fig. 5(b)). Lebens et al. (1990) and Nagamune et al. (1993) studied the OMCVD growth of GaAs/AlGaAs dots prepared by growth on dot-patterned dielectric masks deposited on (100) GaAs substrates. Rather complex faceting is exhibited by dots grown in this way, and tapered QWL layers bounded by characteristic growth planes have been observed in cross sections of these structures (Nagamune et al. 1993). Tetrahedral GaAs/AlGaAs dot structures have been grown with a similar approach on $\{111\}B$ masked substrates by Fukui et al. (1991). Dot and wire InGaAs/InP structures grown by the selective-area OMCVD technique were also fabricated and studied by Galeuchet et al. (1990, 1991).

9. Optical Properties of Quantum Wire Heterostructures

The subband structure of the $GaAs/Al_{0.7}Ga_{0.3}As$ QWRs of Fig. 21, modeled using the approach discussed in subsection 3, is shown in Fig. 25 (Kapon et al. 1992c). The lateral thickness distributions of two different crescents, corresponding to nominal GaAs thicknesses of 2 nm (sample A) and 3 nm (sample B), were measured from TEM cross sectional images. The corresponding lateral potential wells for each carrier type were fitted to parabolic profiles, with resulting subband separations of 21.7, 3.9 and 16.7 meV (sample A) and 19.1, 2.9 and 14.0 meV (sample B) for electrons, heavy holes and light holes, respectively. The effective width W_{eff} of the wires, for electrons in the ground state of sample A, is 16 nm (see Fig. 25).

Low-temperature CL spectrum and cross sectional images of the QWR heterostructure of Fig. 21 are displayed in Fig. 26 (Kapon et al. 1992c). The spectrum exhibits three luminescence lines, centered at 1.751, 1.671 and 1.606 eV (708, 742 and 772 nm). The CL images demonstrate that the longest wavelength line is due to emission from the QWRs, whereas emission from the QWL layers on the side walls and the top of the corrugations occurs at the two shorter wavelengths. It is interesting to note that the emission patterns from the QWRs are much more uniform than those emanating from the top of the corrugations. This indicates a more uniform well

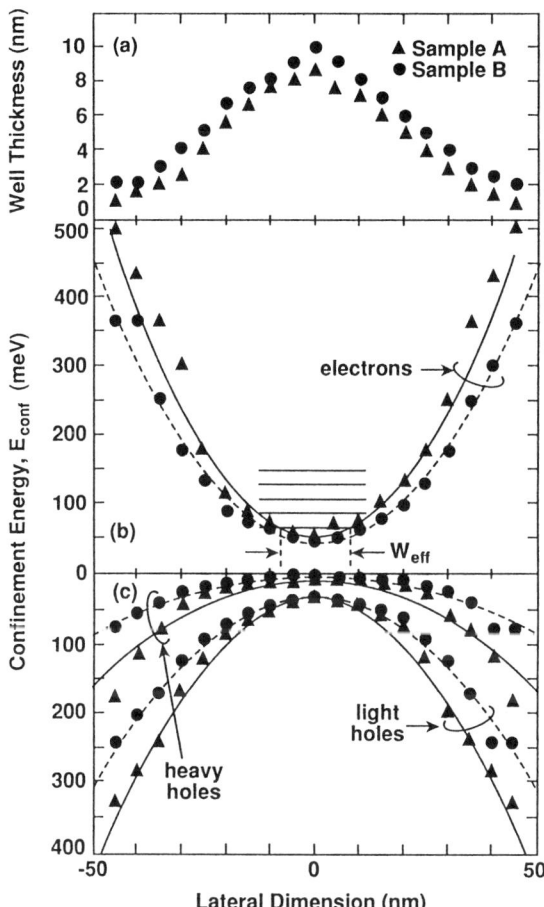

FIG. 25. Model of the subband structure for the GaAs/Al$_{0.7}$Ga$_{0.3}$As crescent-shaped QWRs of Fig. 21: (a) well thickness distribution, (b) lateral distribution of confinement energies for electrons and (c) lateral distribution of confinement energies for holes. Parabolic fits to potential profiles are given by solid (sample A) and dashed (sample B) lines. Horizontal lines indicate electron subbands for structure A. Samples A and B correspond to different nominal GaAs well thickness; sample A is shown in Fig. 21 (Kapon *et al.* 1992c).

thickness along the wires than along the QWL stripes on top of the ridges, which may be due to different surface diffusion processes into the concave and convex features of the surface.

Photoluminescence excitation (PLE) spectra of the same QWR structures, recorded at the QWR emission wavelength for pump beams polarized parallel and perpendicular to the wires, are displayed in Fig. 27 (Kapon *et*

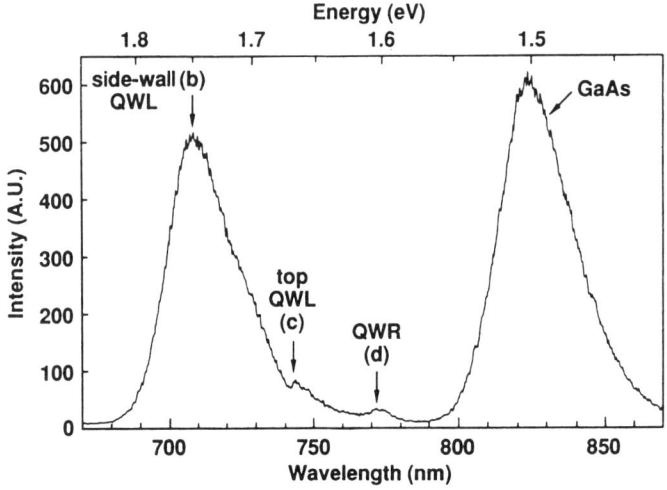
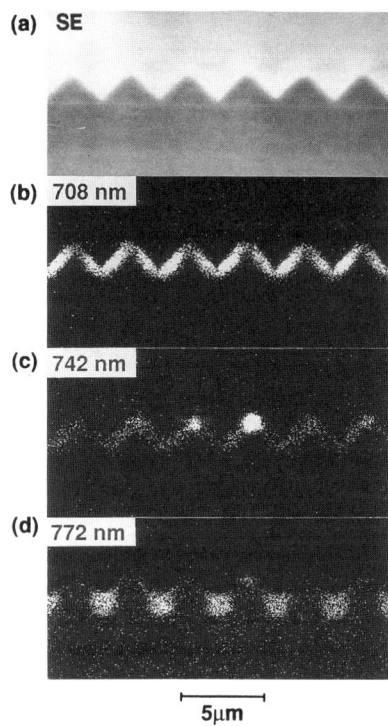

FIG. 26. Cathodoluminescence spectrum (a) and cross sectional images (b) of the QWR heterostructure of Fig. 21. The spectrum was measured at $T = 20$K, with beam voltage of 10 kV and beam current of 2 nA. The top cross section is a secondary electron image, whereas the others are CL images recorded at the designated wavelengths, as also indicated in part (a) (Kapon et al. 1992c).

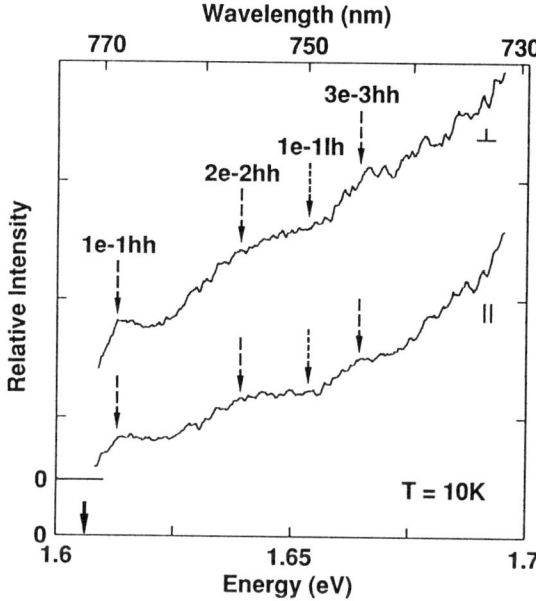

FIG. 27. PLE spectra of the QWR structure of Fig. 21, detected at 772 nm (see solid arrow), for pump beams polarized parallel (⊥) and perpendicular (∥) to the wires. Dashed arrows indicate calculated e–hh and e–lh transition energies with 1e–1hh energy fitted to the lowest-energy peak (Kapon *et al.* 1992c).

al. 1992a). Several peaks are observed in the spectra, representing the enhanced absorption at photon energies corresponding to transitions between the QWR subbands. The separations of these peaks agree well with the calculated ones, as shown by the arrows indicating the positions of the electron–hole transition energies. Thus, these structures exhibit quasi-1D subbands separated by ~22 meV for electrons. The observed Stokes shift of 7 meV is comparable to the half-width of the PL line. Note also that no distinct polarization anisotropy can be observed in the PLE spectra, which is attributed to the nonsymmetric cross section of the crescent-shaped wires (Bockelmann and Bastard 1991).

The side-wall GaAs QWL layers, as well as the vertical AlGaAs regions peculiar to the QWR heterostructures formed by OMCVD on NP substrates, play an important role in the dynamics of carrier collection into the wires. The kinetics of carrier capture, relaxation and recombination in these structures has been investigated using transient CL spectroscopy techniques (Christen *et al.* 1992b). Electron beam pulses of 1.3 ns duration were used to pump the multi-QWR sample shown in Fig. 22. The CL transient (recorded at 675 nm wavelength) as well as spectra measured at eight successive time windows within the light pulse are shown in Fig. 28. All CL spectra show emissions at three different wavelengths, corresponding to recombination at the QWR, the vertical AlGaAs QWL, and the sidewall GaAs QWLs. (The QWL layers on top of the grating ridges were removed by selective etching and hence do not show up in the spectrum). A striking feature of the time-resolved spectra of Fig. 28(b) is the relative increase in emission intensity of the QWR line that occurs during their time evolution. This is due to transport of carriers from the QWLs, where they are first captured, into the lower potential well of the QWRs. The capture process is illustrated schematically in Fig. 29, which depicts the carrier flow from the AlGaAs barrier material into the wells and finally into the lower potential wire. The QWLs connected to the wires thus increase the effective cross section for capture into the QWRs by directing the diffusing carriers into the wires. Note, however, that the carrier feeding into the wires prohibits direct evaluation of the carrier lifetime in the wire using a measurement of the CL transient decay time (as in Fig. 28(a)). Measurement of this parameter is possible in QWR arrays whose spacing is less than the carrier diffusion length, as shown later.

The optical properties of $GaAs/Al_{0.5}Ga_{0.5}As$ QWRs grown on sub-μm-pitch gratings were investigated using PL and PLE by Walther et al. (1992a) and using time-resolved CL by Christen et al. (1992a). The fact that the carrier diffusion length was smaller than the lateral wire spacings (0.24 μm) results in thermalization of the excited carriers in the QWR potential wells, and thus the continuous wave (cw) luminescence spectra of these arrays is dominated by luminescence from the wires. PLE spectra of the structure show enhanced emission at the QWR subbands, with measured subband separations of 34–39 meV, in agreement with model calculations of the energy separations between e–hh subband transitions. Time-resolved CL spectra of the same structure show luminescence from the QWRs even at the earliest time windows employed (limited by the ~ 100 ps time resolution of the measurement system), confirming that the carrier capture in these wires is not limited by carrier diffusion (Christen et al. 1992a). Furthermore, since carrier transport does not affect the measured duration of the luminescence signal in this case, a direct measurement of the carrier lifetime could be obtained, yielding a value of 310 ps for the carrier lifetime. This value is comparable to those measured for high-quality QWL layers of similar thickness (Feldman et al. 1987), demonstrating the absence of significant nonradiative recombination at the interfaces of these wires, which could otherwise severely shorten carrier lifetime. Very efficient carrier

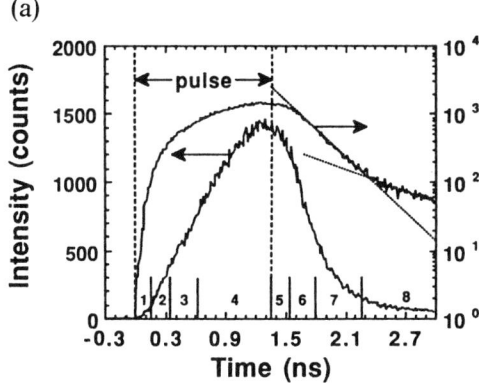

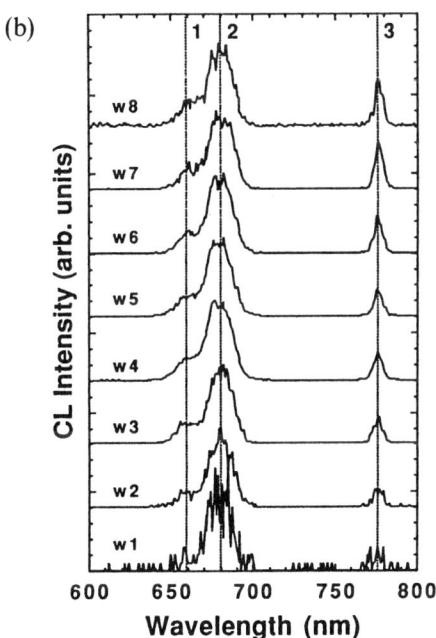

FIG. 28. Time-resolved CL data for the vertically stacked multi-QWR structure of Fig. 22: (a) CL transient recorded at 675 nm and (b) time-delayed spectra at eight time windows depicted in (a), all scaled to the same maximum (Christen et al. 1992b).

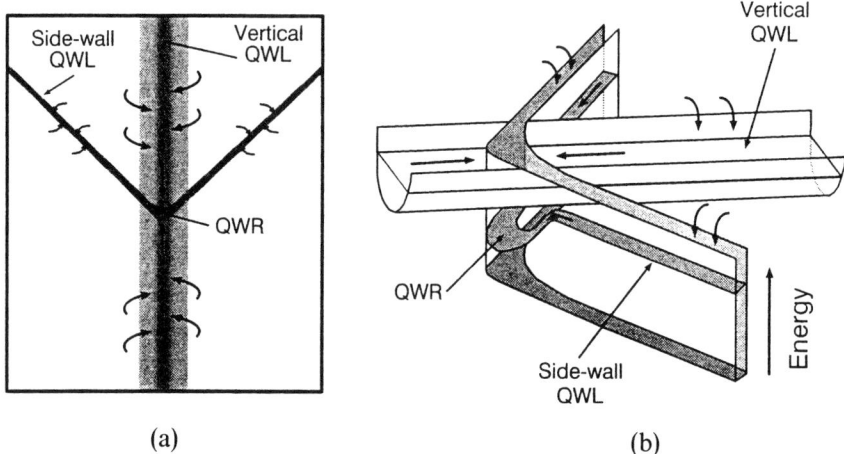

FIG. 29. QWL-assisted carrier capture in a QWR structure grown on a nonplanar substrate: (a) cross section and (b) schematic of the (e.g., conduction) band structure. The outer edges of the structure in (b) represent the band edge of the barrier material, and the arrows indicating the various QWL/QWR regions point to the quantum-confined energy levels.

capture (at low temperatures) was also observed by Walther *et al.* (1993) in strained InGaAs/GaAs QWR laser structures grown on sub-μm gratings.

Luminescence from crescent-shaped GaAs/AlGaAs QWR structures grown on the NP surface that evolves during growth on (100) GaAs substrates masked with patterned dielectric films was reported by Tsukamoto *et al.* (1992, 1993). In this case, however, considerable luminescence from the surrounding barrier material was also observed, in spite of the short periodicity (0.2 μm) of the wire arrays; this may indicate problems with carrier capture or reduced quantum efficiency at the QWRs in these structures. Anisotropy in the PL intensity was interpreted as arising from 2D quantum confinement in the wires. Efficient luminescence was also reported for GaAs/AlGaAs dots (Lebens *et al.* 1990; Nagamune *et al.* 1993), and InGaAs/InP wires and dots (Galeuchet *et al.* 1990, 1991) prepared by selective-area OMCVD.

V. Device Applications

The possibility of lateral band-gap patterning via a single epitaxial step offered by growth of QWL heterostructures on NP substrates is useful for various electronic and optoelectronic devices relying on such patterning.

The approach is particularly attractive for bipolar optoelectronic device applications (Kapon et al. 1992d). Both 1D and 2D laterally patterned structures have found applications in optoelectronics: the former in low-threshold QWL and QWR lasers employing narrow recombination regions, and the latter in integration of lasers with transparent waveguides or optical modulators.

10. Patterned Quantum Well Lasers

Efficient semiconductor diode lasers rely on tight confinement of both the injected charge carriers and the emitted photons to the vicinity of their active region. These confinement mechanisms are required to minimize the volume of the inverted material and to ensure efficient interaction between the photons and the charge carriers, and they can be utilized to yield low threshold currents and high quantum efficiency in these devices. Etching and regrowth techniques have been traditionally used to fabricate buried heterostructure (BH) diode lasers with optical waveguide and recombination regions in the μm range (Tsukada 1974). These structures, however, require multiple epitaxial growth steps to achieve the lateral heterojunctions and refractive index steps. Furthermore, with this approach it is difficult to fabricate laser structures with active region widths in the sub-μm regime, which is necessary for minimizing the threshold current through optimization of the active volume and the optical confinement factor (Kapon 1990). Growth of QWLs on NP substrates can yield patterned QWL (PQW) laser structures providing lateral carrier and optical confinement using a single epitaxial step on the patterned substrate. In addition, the in situ formation of the lateral interfaces results in reduced nonradiative interface recombination compared to that in ex situ formed interfaces, which is particularly important for the narrow active regions required for optimally designed structures.

The lateral potential wells formed at the bottom of grooves regrown with QWL heterostructures by MBE have been utilized to fabricate very narrow ($<1\,\mu$m wide) GaAs/AlGaAs PQW lasers (Kapon et al. 1988a, 1988d, 1990). Figure 30 shows schematic and SEM cross sections of a graded index, separate confinement heterostructure (GRIN-SCH) GaAs/AlGaAs laser grown by MBE in a $[01\bar{1}]$-oriented groove ($T_s = 700°$C, As-rich conditions) (Kapon et al. 1988a). The 7 nm-thick GaAs active layer is sandwiched between two GRIN 0.2 μm thick layers, with x linearly graded from 0.5 to 0.2. The bending and tapering in the active QWL and waveguide layers result in lateral confinement of both the injected carriers and the optical mode to an active region $\sim 0.8\,\mu$m wide. The thinner (4 nm) QWL layers at

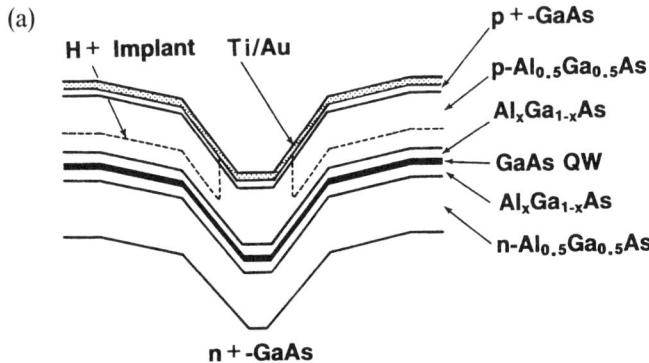

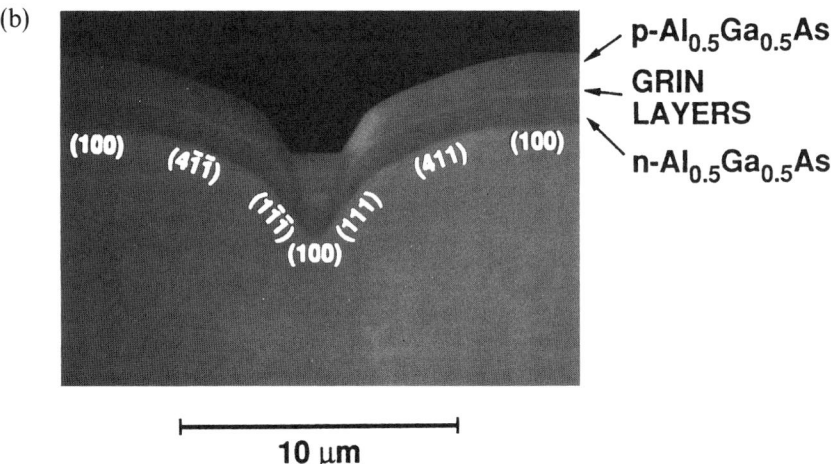

FIG. 30. (a) Schematic and (b) scanning electron microscope cross sections of a patterned QWL GaAs/AlGaAs laser grown by MBE (Kapon et al. 1988a).

the $\{111\}A$ facets exhibit negligible interband absorption at the lasing wavelength of 845 nm, which reduces the threshold current and increases the output efficiency. Growth on the NP substrate thus results in an effectively buried heterostructure laser obtained in a single growth step.

The MBE grown PQW lasers operated with threshold currents as low as 1.8 mA, for $L = 350 \mu m$ cavity length and uncoated facets, with output differential efficiency of 63% (Kapon et al. 1988d). High yield of low threshold devices was obtained because of the simple fabrication process, particularly the single growth step. High reflection ($\sim 98\%$) facet coatings of 125 μm long devices reduced the threshold currents to 0.35 mA under pulsed operation and 0.5 mA for cw operation (see Fig. 31). Threshold currents as low as 100 µA have been predicted for buried heterostructure lasers having optimized QWL widths of ~0.2 µm (Kapon 1990; Kapon *et al.* 1990). Such narrow QWL lasers are difficult to fabricate by conventional etching and regrowth, but could be realized using MBE growth in narrower grooves.

Low-threshold patterned GaAs/AlGaAs QWL lasers grown by MBE were also reported by Marclay *et al.* (1989). Their devices were grown on a p-GaAs substrate grooved along [01$\bar{1}$] and utilized the lateral doping-type variation in the Si-doped layers to achieve current confinement. Lateral optical and carrier confinement were provided by facets lying in the {211}A and {311}A planes. Threshold currents as low as 0.65 mA cw and external efficiency of 0.5 mW/mA were achieved. Output powers (cw) as high as 130 mW in the fundamental waveguide mode with 50% overall (electrical to optical) efficiency have been obtained with similar laser structures having wider (3 µm) active regions (Jaeckel *et al.* 1989). Imanaka *et al.* (1987) have

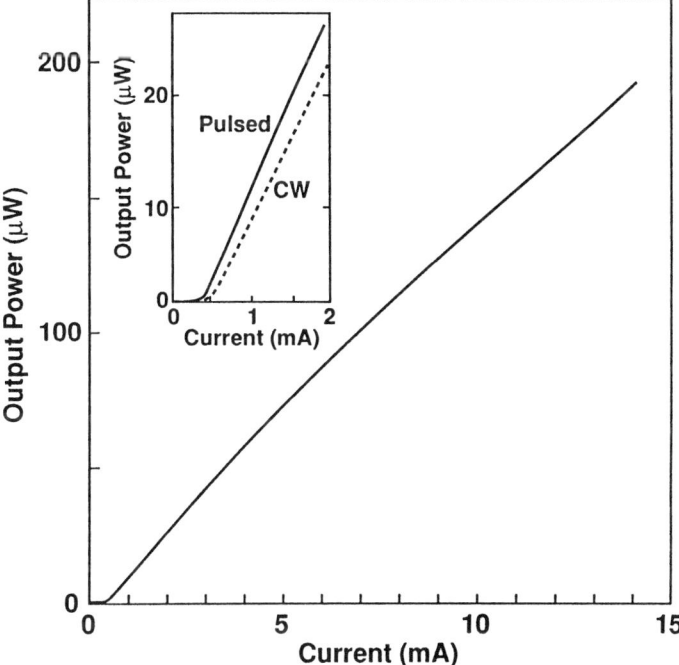

FIG. 31. Light versus current characteristics of a patterned QWL laser grown by MBE. Laser facets are high-reflection coated, and cavity length is 125 µm (room temperature operation) (Kapon *et al.* 1990).

also employed the lateral doping patterning effects to achieve current confinement in double heterostructure GaAs/AlgaAs lasers grown by MBE in [01$\bar{1}$] grooves etched in p-GaAs substrates. Patterned QWL GaAs/ AlGaAs lasers with active region widths in the few-μm range were also grown by MBE (Hong et al. 1987; Meier et al. 1988b; Van Gieson et al. 1989) and OMCVD (Fekete et al. 1987) on substrates patterned with dielectric masks. More recently, very low threshold currents, as low as 203 μA at room temperature, were reported for *uncoated* InGaAs patterned QWL lasers grown by MBE in grooves etched through silicon nitride masks (Tiwari et al. 1992). In this case, the polycrystalline growth on the dielectric mask left on both sides of the groove was very effective in reducing current leakage.

Double heterostructure and multiple-QWL GaAs/AlGaAs lasers grown by MBE in [011]-oriented grooves showing real-index lateral waveguiding were reported by Wu et al. (1984a, 1984b). GaAs/AlGaAs PQW lasers grown on [01$\bar{1}$]-oriented mesas were demonstrated by Mannoh et al. (1985) and Yuasa et al. (1987). These latter multiple-QWL structures utilize the mesa narrowing during growth to reduce the active region width; lasers with active regions as narrow as ~ 1.4 μm have been achieved, with threshold currents as low as 20 mA.

Patterned QWL lasers with narrow (< 1 μm) active regions can also be grown by OMCVD on [011] or [01$\bar{1}$]-oriented mesas. Dzurko et al. (1989a, 1989b) have used the effect of narrowing [01$\bar{1}$]-oriented ridges by growth at elevated substrate temperatures (850°C) to fabricate GaAs/AlGaAs QWL lasers with active regions as narrow as 0.5 μm. Growth of the upper cladding layers of these lasers at lower temperatures (650°C) yielded a wider ridge cap that facilitated electrical contacting. Threshold currents as low as 3.4 mA (pulsed) and 3.8 mA (cw) were obtained at room temperature for uncoated devices. Similar devices incorporating strained InGaAs QWL active layers (1 μm wavelength) with threshold currents as low as 3.0 mA (pulsed) and 3.5 mA (cw) have also been demonstrated (Frateschi et al. 1992). Wider (~ 2 μm) active region multiple QWL GaAs/AlGaAs lasers of similar structure have been reported earlier by Yamada et al. (1986).

Yoshikawa et al. (1987) have utilized the no-growth $\{111\}B$ planes that evolve during OMCVD growth on [011]-oriented mesa to fabricate narrow (2 μm) active region GaAs/AlGaAs lasers. The active region is designed to grow close to the top of the triangular prism bounded by the $\{111\}B$ facets, so that sub-μm active stripe widths can be obtained (see Fig. 16). Growth of subsequent higher band-gap and lower refractive index layers completely embed the active region. Narui et al. (1988, 1990) have also used this technique to grow multiple-QWL GaAs/AlGaAs lasers with active regions <1 μm in width. Room temperature cw threshold currents as low as

0.88 mA were achieved for these lasers *without* facet coatings (Narui et al. 1990).

Low-threshold currents are of particular importance for densely packed arrays of diode lasers to reduce the thermal load. Figure 32 shows cross sections of a GaAs/AlGaAs laser array grown by MBE on a periodically corrugated substrate (Kapon et al. 1989a). The array pitch is 3.5 μm with the grooves oriented in the [01$\bar{1}$] direction. The grown heterostructure is similar to that of the single channel laser of Fig. 30. However, the shallower grooves ($\sim 1\,\mu$m) and their smaller lateral dimensions result in more dominant diffusion effects that tend to planarize the structure. Furthermore, the tone variations in the SEM cross section indicate a higher Al mole fraction between the grooves due the greater diffusion length of Ga adatoms. This effect enhances the lateral confinement of both the carriers and the optical mode at each array channel.

Groups of ~ 14 lasers of these arrays were operated in parallel using

(a)
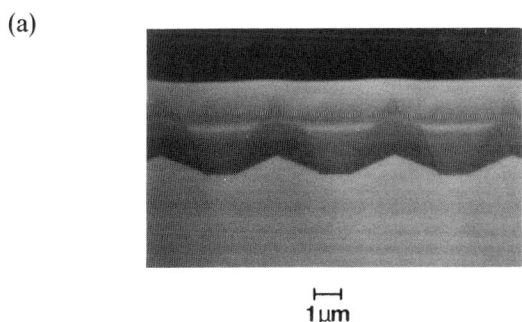
1 μm

(b)
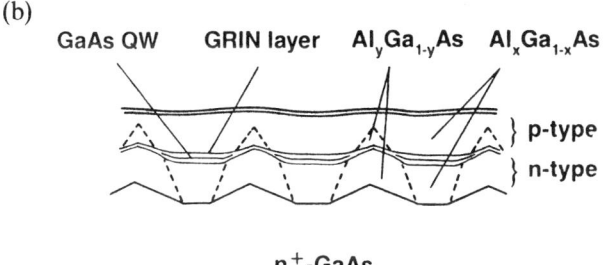
n$^+$-GaAs

FIG. 32. (a) Scanning electron microscope and (b) schematic cross sections of a patterned QWL GaAs/AlGaAs laser array grown by MBE on a periodically corrugated substrate. Regions indicated by Al$_y$Ga$_{1-y}$As have a higher Al mole fraction than those indicated by Al$_x$Ga$_{1-x}$As (Kapon et al. 1989a).

50 μm wide stripe contacts. Pulsed operation with threshold currents of 3.6 mA per laser and output powers up to 375 mW per facet were achieved for uncoated devices. The lateral modulation in waveguide layer thickness and Al mole fraction result in periodic modulation of the waveguide effective index of refraction, giving rise to tight 2D waveguiding at each laser channel. This effect is visualized by the distinct lasing spots observed at the near-field pattern of the array shown in Fig. 33(a). The tight lateral waveguiding also leads to very small overlap between the optical fields of adjacent lasers, which prohibits phase-locking of the laser array. This effect is evidenced by the spectrally resolved near field pattern of the array (Fig. 33(b)), which exhibits uncorrelated laser spectra. Such arrays of optically uncoupled devices are important for applications involving individual addressing of laser arrays. The diffusion effects during MBE help in this case to enhance the lateral refractive index modulation, which allows placing the array elements in close proximity while maintaining negligible optical cross talk.

The strong migration effects during growth of this array structure lead to a GaAs QWL active layer thicker at the laser channels than it is on unpatterned parts of the wafer. The resulting difference in the effective band gap shows up in the different lasing wavelength of the array compared to that of a broad area (50 μm stripe width) unpatterned laser grown on the same wafer, as shown in Fig. 34. The spectrum of the patterned QWL array is red shifted by ∼30 nm, or 55 meV, which corresponds to a QWL thickness difference of ∼3 nm. This illustrates the possibility of achieving controlled variation in lasing wavelength by changing the feature size (e.g., the groove width) in PQW lasers grown on NP substrates. Similar controlled variation in lasing wavelength was demonstrated with InGaAs/AlGaAs QWL lasers grown by MBE on [01$\bar{1}$]-oriented ridges of different widths (Brovelli et al. 1990); wavelength tuning from 980 nm to 1040 nm was obtained by varying the ridge width from 28 to 5 μm. This approach for emission wavelength control should be useful for application in multiple wavelength laser arrays for wavelength division multiplexing (WDM) communications systems.

Large, monolithic arrays of low-threshold multiple-QWL GaAs/AlGaAs lasers grown on NP substrates by OMCVD were reported by Hirata et al. (1991). The arrays were grown on 4.5 μm-pitch periodic corrugations aligned along the [011] direction and utilize the {111}B no-growth planes to completely embed the active region during growth, as for the single emitters described earlier. Arrays of 102 elements were operated with average threshold currents of 1.8 mA per laser, emitting maximum power of 850 mW per facet at 2 A current. The lasing wavelength of the array (861 nm) was longer than that of a single device with nominally the same QWL structure, probably as a result of the QWL thickness enhancement caused by migra-

(a)

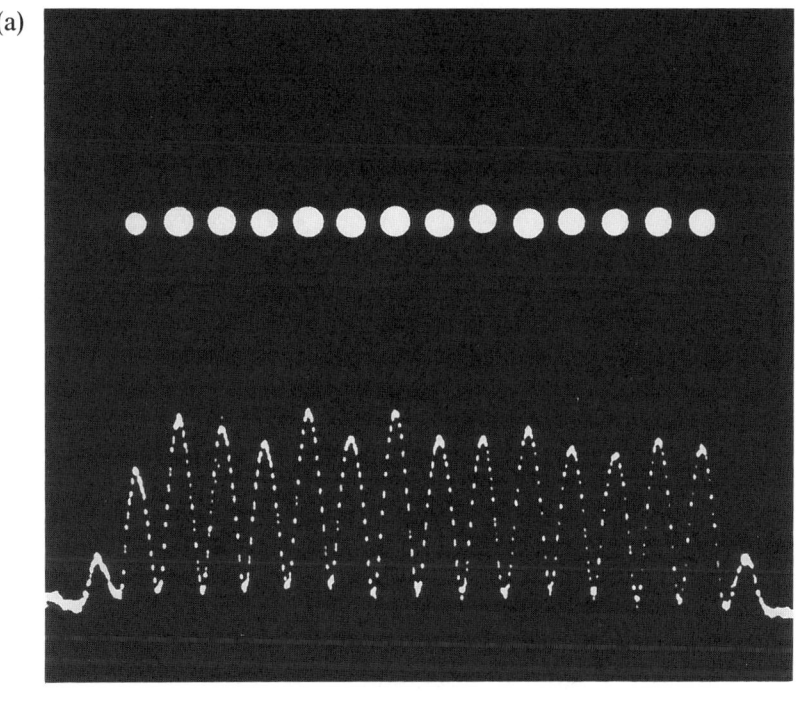

10 µm

(b)

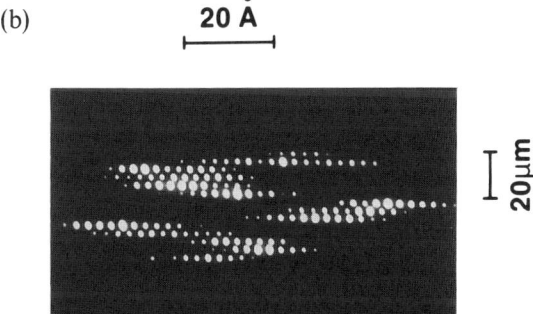

FIG. 33. (a) Near-field and (b) spectrally resolved near-field patterns of the patterned QWL laser array of Fig. 32, operating above threshold. The (vertical) spatial axis in (b) corresponds to the lateral axis in (a) (Kapon et al. 1989a).

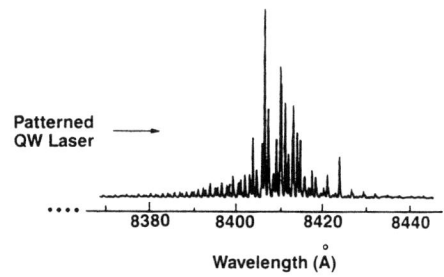

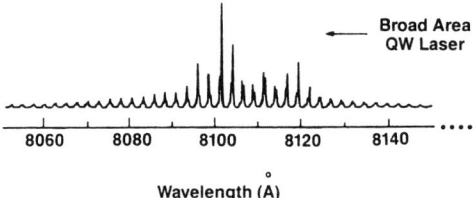

FIG. 34. Spectra of the patterned QWL laser array of Fig. 32 and a broad area laser grown on an unpatterned part of the same wafer (Kapon et al. 1989a).

tion of Ga species during OMCVD from the multiple $\{111\}B$ no-growth planes to the active regions.

11. Integrated Laser Structures

Monolithic integration of diode lasers with other optical devices requires means for tailoring the heterostructure properties in both lateral dimensions. A generic structure in such integration schemes involves a combination of a laser and a "transparent" optical waveguide. The band gap at the transparent waveguide section should usually be higher than that of the laser's active region to minimize absorption of the laser beam at the waveguide region. This is conventionally accomplished by removing the active layer in the passive waveguide section, followed by regrowth of the higher band-gap waveguide structure (Dutta et al. 1986). However, this complicates the fabrication process and makes it difficult to achieve high optical coupling efficiency between the laser and waveguide parts.

A number of patterned heterostructure configurations grown on NP substrates that have been employed for such integration via a single epitaxial growth step are shown in Fig. 35. One scheme (Fig. 35(a)) utilizes the offset in the waveguide layers grown on a stepped substrate to achieve

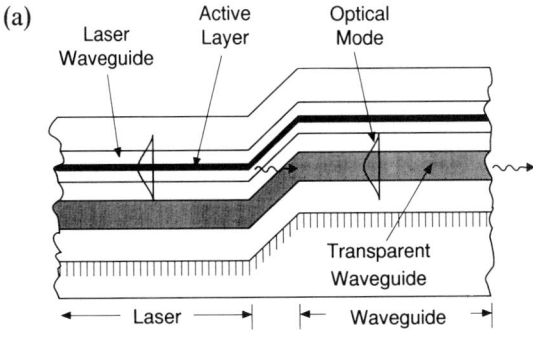

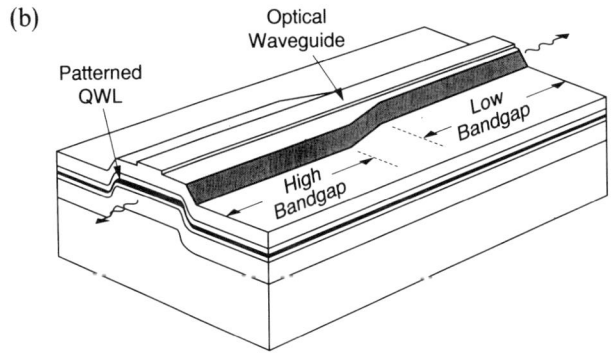

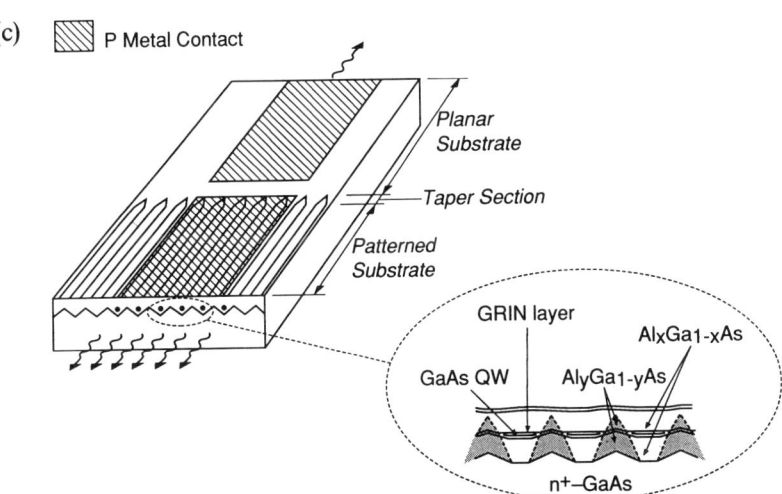

FIG. 35. Schematics of integrated laser structures grown on NP substrates: (a) stepped substrate structure, (b) tapered ridge structure and (c) patterned substrate laser array.

the transition between the active laser waveguide and the transparent waveguide (Azoulay et al. 1989; Bryan et al. 1989). The structure is designed such that light propagating in the active laser waveguide is coupled into higher band-gap optical waveguide layers as it crosses the step edge. An OMCVD grown GaAs/AlGaAs structure of this type showing laser–waveguide coupling efficiencies as high as 70% has been demonstrated (Azoulay et al. 1989). Integrated structures in which the transparent section consists of the higher band-gap cladding layers of the waveguide (i.e., without the additional waveguide structure underneath the laser waveguide layers) are also useful, but with relatively short (<10–$20\,\mu$m) transparent sections to avoid excessive diffraction losses. Such *window* structures can operate to higher output powers since the power limit for catstrophic mirror degradation is increased at the transparent mirror region (Bryan et al. 1989).

The integrated structure of Fig. 35(b) relies on effective band-gap variations along the laser waveguide, achieved by growing the QWL laser structure on a ridge of varying width. Lateral surface diffusion from the sidewalls to the top of the ridge results in a lower band-gap QWL at the narrower ridge section, where the active laser section is placed. Band-gap variations as large as ~ 70 meV in InGaAs/GaAs lasers grown by MBE on ridges of various widths have been observed (Arent et al. 1990a; Brovelli et al. 1990). With a proper design of the tapered QWL structure of Fig. 35(b), and using an additional contact at the waveguide section, the integrated waveguide can function as an external modulator utilizing controlled electroabsorption at the modulator section. Such InGaAs/GaAs integrated laser structures grown by MBE have been demonstrated by Brovelli et al. (1990).

Index-guided GaAs/AlGaAs laser arrays monolithically integrated with transparent optical waveguides have been fabricated and characterized by Chang-Hasnain et al. (1990) using the configuration shown in Fig. 35(c). The MBE grown laser array structure is similar to that described in subsection 10 (Fig. 32). In this case, however, only part of the wafer was grooved, and the grooves were tapered along a section of $50\,\mu$m. The laser p-contact was placed at the patterned, lower band-gap region of the wafer, where the observed emission wavelength was 862 nm (1.439 eV). A second p-contact was fabricated at the broad-area, unpatterned part, where the observed emission wavelength was 837 nm (1.481 eV). The combined structure oscillated at 853 nm (1.45 eV). Propagation and diffraction of the individual laser beams in the transparent waveguide led to coupling of their optical fields, resulting in phase-locked operation of the laser array. In addition, current injection into the broad-area section was employed to achieve ~ 8 nm wavelength variation via tuning of the peak of the spectral gain distribution with increasing carrier density.

12. QUANTUM WIRE LASERS

Two-dimensional quantum confinement of charge carriers is expected to yield significant improvements in laser performance compared to lasers incorporating bulk or QWL material (Arakawa and Sakaki 1982; Asada et al. 1985; Arakawa and Yariv 1986; Kapon 1992). This enhanced performance results from the modified DOS in quasi-1D structures, which assumes sharp peaks at the energies of the QWR subbands. The sharper profile of the DOS, in turn, leads to *spectral* confinement of the injected carriers, allowing achievement of higher carrier densities at the subband energies. The higher optical gain and *differential gain* (i.e., the rate of change of gain with carrier density), combined with the extremely small volume of the QWRs, should yield very low threshold currents, in the μA regime (Asada et al. 1986; Yariv 1988), reduced temperature sensitivity, higher modulation bandwidth and narrower spectral linewidth. Such QWR devices would be useful in applications requiring dense laser arrays or integration of lasers with low-power electronics, e.g., optical computer interconnects, optical signal processing and integrated optoelectronics.

Approaches for realization of QWR lasers (Kapon 1992) have included simulation in high-magnetic fields (Arakawa and Sakaki 1982; Arakawa et al. 1985, 1986) and fabrication by etching-and-regrowth (Cao et al. 1990) or growth on vicinal substrates (Tsuchiya et al. 1989). InGaAs/InP QWR lasers fabricated by etching and regrowth exhibited spectral features attributed to formation of 1D subbands. These devices, however, operated at low temperatures only and suffered from high threshold currents due to defects introduced during the fabrication process (Cao et al. 1990). GaAs/AlGaAs QWR lasers fabricated by growth on vicinal (100) substrates operated at room temperature, but their lasing characteristics are similar to those of conventional QWL lasers, possibly due to nonuniformities in the QWR arrays constituting their active region (Tsuchiya et al. 1989).

Figure 36 shows scehmatic and SEM cross sections of single-QWR GaAs/AlGaAs laser grown by OMCVD in a [01$\bar{1}$]-oriented groove (Kapon et al. 1989b, 1989c). The laser core consisted of a (nominally) 7 nm thick GaAs active layer surrounded by two, 150nm thick GRIN $Al_xGa_{1-x}As$ waveguide layers, and it was grown at 750°C and atmospheric pressure. The Al mole fraction in these layers was graded linearly from 0.7 to 0.2 using a short-period $GaAs/Al_{0.7}Ga_{0.3}As$ SL. Growth of this GRIN-SCH laser structure yields a crescent-shaped GaAs active region surrounded by a 2D optical waveguide formed by the sharp bending and lateral tapering in the GRIN layers. Similar 2D waveguides in double-heterostructure GaAs/AlGaAs lasers grown using the same technique were reported by Mori et al. (1980). The 2D waveguide structure is visible as the lighter contrast layers

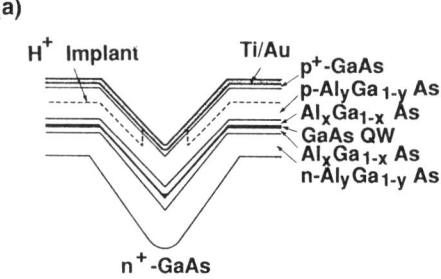

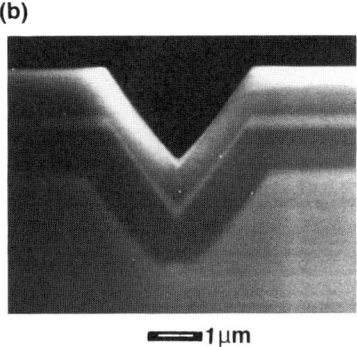

FIG. 36. (a) Schematic and (b) SEM cross sections of a GaAs/AlGaAs QWR laser grown by OMCVD on a V-grooved substrate (Kapon et al. 1989c).

in Fig. 36(b). Such a waveguide is particularly important for achieving low threshold currents in these devices, since the small dimensions of the wire require tight optical confinement to ensure sufficiently high modal gain at low carrier densities. It is equally important to minimize the optical cavity losses in these structures, by using low-loss optical waveguides and high facet reflectivities, in order to comply with the relatively low modal gain. Efficient confinement of the injected carriers into the active region is also required. Confinement of the diode current to the vicinity of the QWR region was accomplished using a narrow ($\sim 1\ \mu$m) p-contact stripe fabricated by proton implantation.

A TEM cross section of the single-QWR laser is displayed in Fig. 37. The crescent-shaped QWR is 12.5 nm thick at its center and ~ 90 nm in full width. It is interesting to note that the use of a SL barrier layer significantly increases the radius of curvature $\rho_l (= 58$ nm) at the bottom of the groove, which leads to wider effective wire widths compared to crescents grown on

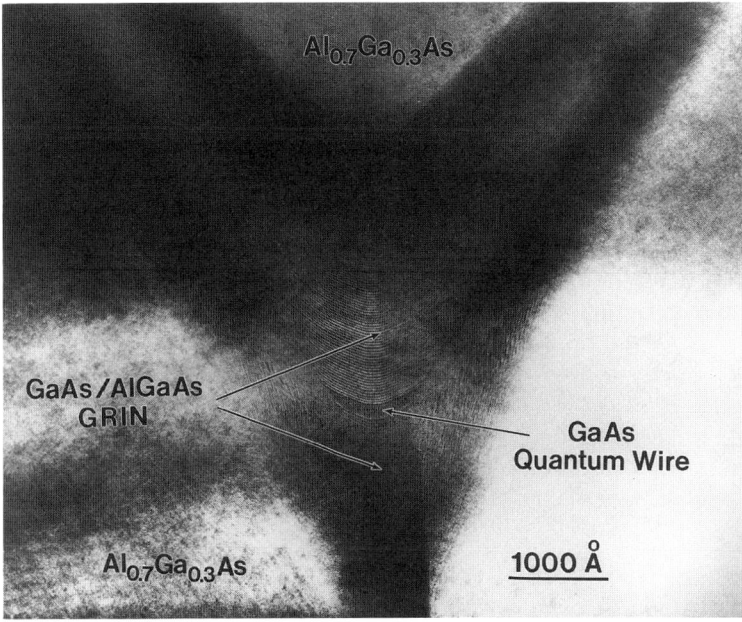

FIG. 37. TEM cross section of a GaAs/AlGaAs QWR laser grown by OMCVD on a V-grooved substrate (Kapon et al. 1989c).

AlGaAs alloys. This is probably due to the larger increase in ρ during growth of the GaAs layers in the SL.

The single-QWR lasers of Fig. 37 operated at room temperature with threshold currents as low as 3.5 mA (for uncoated facets). Their lasing and amplified spontaneous emission (ASE) spectra exhibited enhanced emission at photon energies corresponding to transitions between the QWR subbands, with subband separations of ~ 10 meV (Kapon et al. 1989b, 1989c). Although these threshold currents are comparable to those of conventional QWL lasers, they are still much higher than the expected values for ideal QWR lasers. This is partly due to the low optical confinement factors (on the order of 10^{-3}), which are limited by the size of the wire. Optimal design of QWR lasers, particularly achievement of μA threshold currents, requires the use of multiple-QWR active regions to balance the effects of small active volumes and low confinement factors on the current values.

Vertically stacked multiple-QWR laser structures can be realized using the self-ordering effects discussed in subsection 8 (Simhony et al. 1990, 1991; Kapon et al. 1992b). TEM cross sections of the cores of two 4-QWR laser

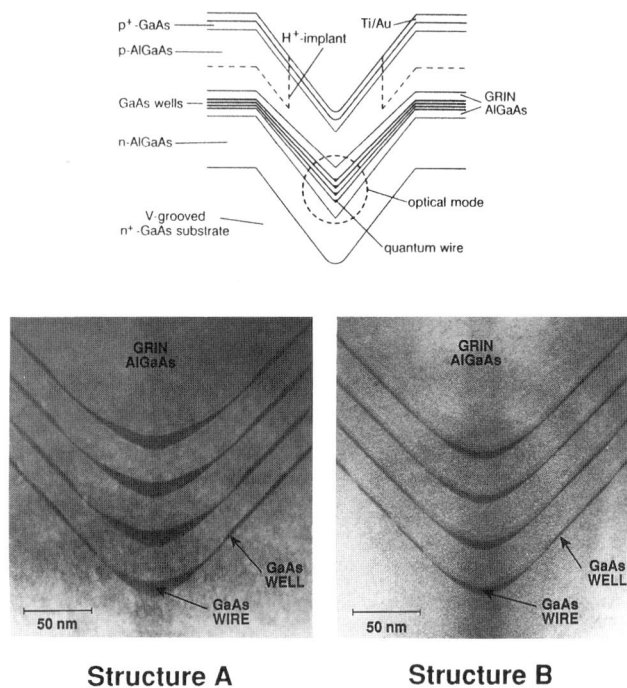

FIG. 38. TEM cross sections of the cores of 4-QWR GaAs/AlGaAs lasers of two different wire sizes (Kapon et al. 1992b).

structures achieved in this way are shown in Fig. 38 (Kapon et al. 1992b). The two structures have different wire size, but are otherwise identical. The GRIN waveguide layers consist of $Al_xGa_{1-x}As$ alloy, with x linearly graded from 0.66 to 0.3 ($x = 0.3$ near the crescents), and the growth temperature was 750°C. The other parts of the laser structure are similar to those of the single-QWR device of Fig. 36. The use of alloys in the QWR barriers in this case leads to a smaller radius of curvature ($\rho_l = 33$ nm) below the crescents, resulting in narrower wires. A model of the QWR subbands in these lasers yields parabolic lateral potential wells, with subband separations of 10.6, 1.7 and 7.6 meV, and 15.9, 2.7 and 12.0 meV for electrons, heavy holes and light holes for the wires of structure A and B, respectively, in Fig. 38.

Emission spectra of the two 4-QWR laser structures, measured at room temperature near lasing threshold, are shown in Fig. 39. The amplified spontaneous emission spectra exhibit sharp peaks due to enhancement of gain at the QWR subbands, and lasing is achieved at one of these subband peaks. The calculated transition energies at the various subbands, assuming

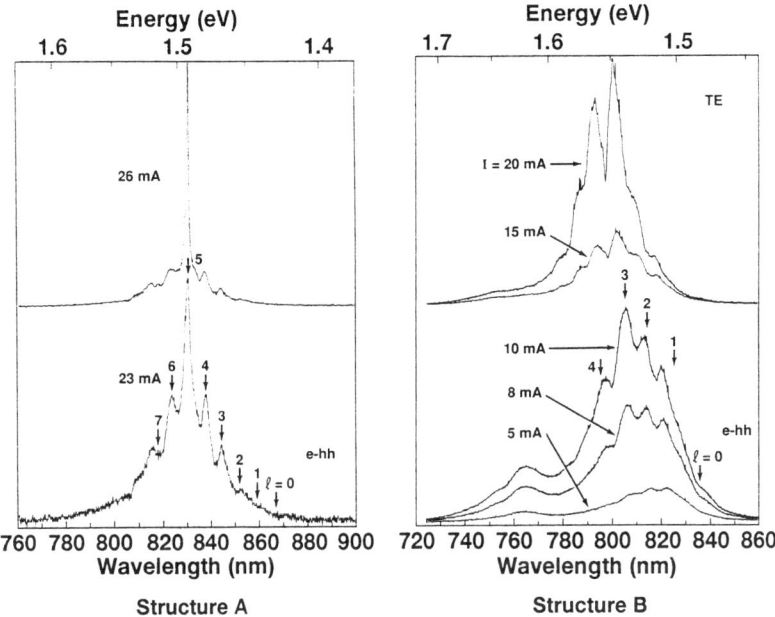

FIG. 39. Emission spectra of the two 4-QWR laser structures of Fig. 38, measured at several currents near threshold. Calculated energies of electron–heavy hole transitions between subbands l are indicated by arrows (Kapon et al. 1992b).

30 meV band-gap shrinkage, are also indicated in the spectra. Very good agreement is obtained between the calculated and measured subband separations, which demonstrates the expected scaling of the quasi-1D subband separation with wire size. The separation of electron subbands in the smaller wires structure is close to $k_B T$ (at room temperature), and lasing in this structure occurs at the fourth QWR subband.

It is interesting to note that in these QWR lasers, lasing at room temperature is due to recombination in the QWRs rather than the surrounding QWL layers, although the optical confinement factor at the wells is much larger. This is a result of thermalization of the injected carriers in the QWR potential wells via the connected QWL layers. Evidence for this process is provided by the temperature dependence of the PL spectra, as shown in Fig. 40 (Walther et al. 1992a). The spectrum at low temperature (80K) shows two peaks, due to recombination at the QWR and at the surrounding QWLs. At higher temperatures, a dramatic increase in the emission intensity of the wires relative to that of the wells is observed. This reflects the increase in carrier diffusion length, which results in a more

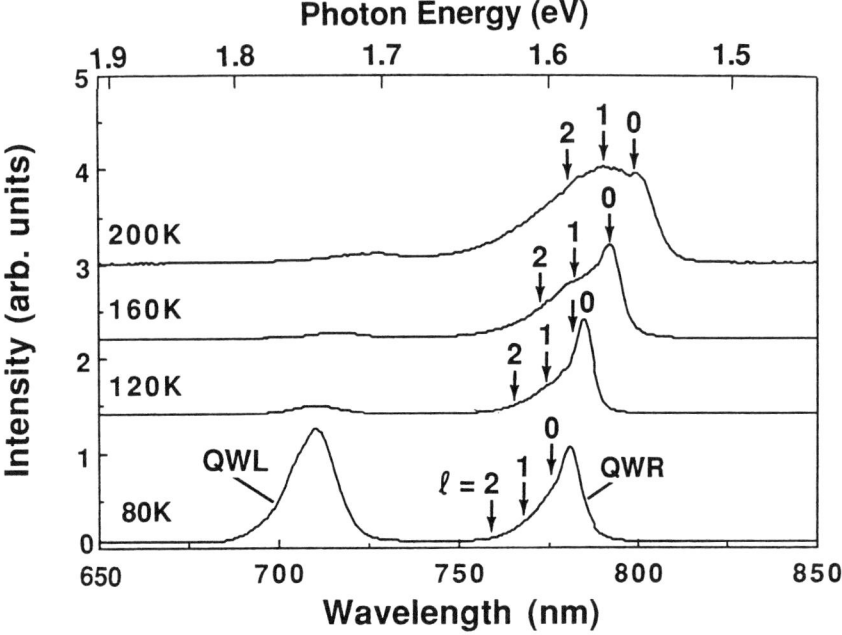

FIG. 40. PL spectra (normalized to the QWR peak) of the 4-QWR laser of Fig. 38(B) at various temperatures. The arrows indicate the calculated energies of the electron–heavy hole transitions between the various QWR subbands $l = 0, 1, 2,\ldots$ assuming $\Delta l = 0$ (Walther et al. 1992a).

efficient carrier collection via the wells. The broadening in the QWR line on the high energy side results from thermal population of higher energy subbands. The measured subband separations agree well with the calculated values (see arrows in Fig. 40). Note that above ~ 100K, the PL spectrum is dominated by emission from the QWRs. The efficient carrier capture and radiative recombination at the wires thus allow lasing with high efficiency from the QWR active regions in these devices.

Reducing the threshold currents of the QWR lasers to the μA range would require optimization of the number of wires and their size, as well as reduction of optical cavity losses to minimize the necessary modal gain. Emission spectra of a high-reflection coated 3-QWR laser showing room temperature threshold current of 0.6 mA (pulsed operation) are displayed in Fig. 41 (Simhony et al. 1991). The cavity length was 135 μm and the mirror reflectivities were $\sim 95\%$. Lasing from two adjacent subbands was observed in this case, with the lowest energy one corresponding to the $\sim$fifth electron–heavy hole subband transition. Further reduction in threshold

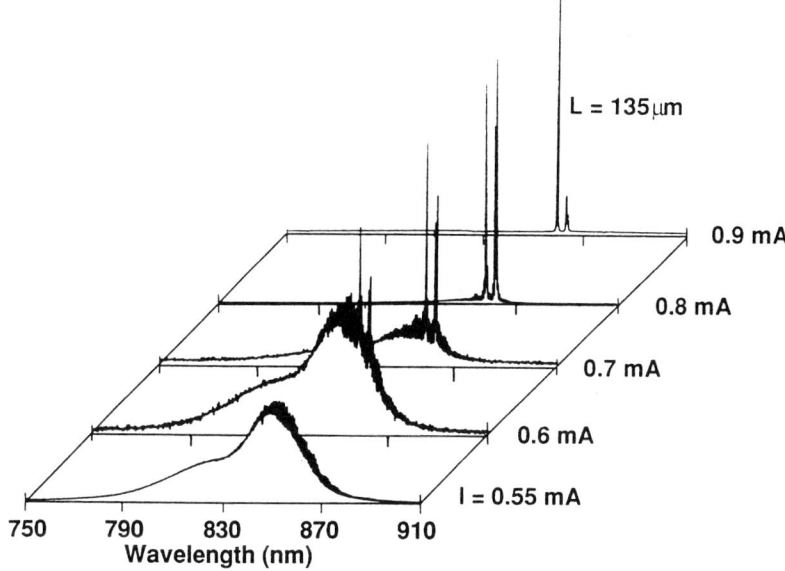

FIG. 41 Spectra of a 3-QWR laser with high-reflection facet coatings. Cavity length $L = 135\,\mu m$, $I_{th} = 0.6\,mA$ (room temperature, pulsed operation) (Simhony et al. 1991).

current would require also minimization of leakage currents in these devices.

One of the important challenges in this area is the achievement of lasing from the ground state of the QWR, since only in that case can full advantage of the quasi-1D carriers be utilized. This requires increasing the modal gain at lower carrier densities. One approach would be to use lateral, dense arrays of QWRs similar to those of Fig. 24. Room temperature operation of such InGaAs/GaAs QWR lasers grown by OMCVD on $0.25\,\mu m$-pitch gratings has recently been demonstrated (Walther et al. 1992c). Increasing the subband separation to a few times $k_B T$ (at room temperature) is also important to minimize gain reduction due to thermal population of higher subbands. As shown earlier, the corresponding range of QWR widths ($\sim 10\,nm$) is readily achievable by OMCVD on corrugated substrates.

VI. Conclusions and Future Directions

Two- and three-dimensional patterning of semiconductor heterostructures via epitaxial growth on NP substrates offers a number of unique features. All interfaces of the resulting patterned structure are formed in situ, thus having the potential of being free from interfacial defects or contamination, a problem often encountered in other fabrication techniques involving growth interruption. The specific crystalline facets that evolve during growth and the different growth and dopant incorporation rates they exhibit provide the means for lateral patterning of the band gap, doping and refractive index. In addition, various types of extremely narrow (~ 10 nm size) lateral potential wells capable of confining both electrons and holes can be formed, allowing realization of structures showing multidimensional quantum confinement. However, since this patterning approach relies on inherent growth evolution of heterostructures on the patterned substrate, detailed understanding of these growth features should be acquired before it can be employed reliably. Investigation of these rich growth features is also useful in shedding light on some aspects of the epitaxial growth kinetics and surface chemistry that are difficult to access in conventional, planar growth.

Considerable insight has been gained into the NP growth characteristics of III-V heterostructures using MBE and OMCVD. The two techniques exhibit significantly different growth morphologies, which can generally be attributed to the difference in both ambience (i.e., vacuum versus gas phase) and surface chemistry. Some clues on the origins of the different growth habits can be obtained from growth by organometallic molecular beam epitaxy (OMMBE), in which the source materials are introduced via the same group III molecular species as in OMCVD while the high vacuum conditions and molecular beam configuration are typical of the MBE process. The OMMBE growth characteristics of GaAs/AlGaAs and InGaAs/AlGaAs QWL heterostructures on (100) substrates corrugated along the [01$\bar{1}$] direction have been investigated by Tamargo et al. (1992). Similar to MBE growth, these patterned structures develop $\{811\}A$, $\{111\}A$ and $\{311\}A$ facets at the bottom and side walls of the grooves, respectively. However, the ratio between the growth rates at the $\{311\}A$ and the $\{111\}A$ faces is higher than for MBE, indicating stronger surface diffusion to the $\{311\}A$ slopes. This leads to sharper corners at the peaks of the corrugations, which are now defined by $\{111\}A$ rather than the $\{311\}A$ planes for sufficiently thick growth. The completely different set of crystal planes evolving in the case of MBE/OMMBE and OMCVD on NP substrates may be due to the different mass transport mechanisms, particularly the long-range gas-phase diffusion in OMCVD, or the different V/III ratio at the surface (Heineke et al. 1992).

One of the remarkable features of NP growth is the ability to generate laterally confined quantum structures with defect-free interfaces. In some cases, the growth dynamics makes possible the self-ordering of arrays of such quantum structures with size and shape determined solely by composition and growth conditions. This is particularly attractive for reducing inhomogeneous broadening effects in the energy spectrum of such quantum structures that can severely degrade their potential advantages. The groove sharpening effects typical of OMCVD growth on [01$\bar{1}$]-oriented grooves allow formation of *nm-size* QWRs starting with *μm-size* patterned features, which can be readily made with conventional photolithography. The curvature recovery observed in these grooved structures is useful for making high density (1D) *uniform* arrays of QWRs with wire separation in the 10 nm range, also without resorting to nanolithography technology. It should be noted that, although the use of facets in this patterning approach somewhat limits its flexibility in terms of designing the quantum structures, adequate flexibility is maintained due to the use of lithography in substrate patterning that allows, e.g., design of the array periodicity and location on the wafer. This is different from the situation in other patterning approaches that avoid using lithography and rely completely on "natural" substrate patterning, e.g., growth on nonpatterned vicinal (100) substrates.

The growth features of QWL layers on NP substrates described in this chapter are useful for applications in a variety of electronic and optoelectronic devices. Quantum wire structures with sufficiently large subband separations (particularly larger than $k_B T$ at room temperature) should be useful for investigation of transport in single quantum channel systems, which is difficult to accomplish with wider wires due to thermal or scattering-induced population of higher energy subbands. This should make possible the investigation of achieving higher mobility in quasi-1D structures due to reduction in scattering effects (Sakaki 1980), as well as various quantum interference devices requiring single-channel conductance.

The high-quality interfaces of patterned QWLs grown on NP substrates is particularly important for optoelectronic device applications, in which nonradiative surface recombination at interface defects can severely reduce carrier lifetime and quantum efficiency. A number of novel optoelectronic devices employing such patterned QWLs have been demonstrated and studied, including record low threshold currents lasers and large laser arrays, laser–waveguide integrated structures and QWR lasers, all fabricated using a simple single-step epitaxial growth. This patterning technique is particularly attractive for generating heterostructure wafers with controlled lateral band-gap variations for integration of photonic circuits comprising of different devices. In this context it is worth mentioning that in most cases the patterned wafers can be completely planarized after a few microns of growth above the patterned surface level, which could be advantageous for some integration schemes.

One of the important possible limitations of this approach is formation of defects in thick layers of compounds that are not lattice-matched due to difference in diffusion lengths and sticking coefficients of the group III and V species over the different facets of the NP substrate. More work is required to identify windows in the growth and composition parameters that allow lattice matching of the patterned structures. In addition, the rapid planarization of NP structures involving feature sizes much smaller than the diffusion length limits the thickness of buffer layers grown underneath the patterned QWLs. This requires optimization of the grown thicknesses to ensure high interface quality with minimal thickness of buffer layer. Alternatively, identifying growth conditions that will allow preservation of the NP profile even after growth of μm-thick layers on sub-μm NP features should also be useful in this context.

Further work is required to gain understanding of the NP growth process on an atomic level. One of the yet unexplained important features of this technique is the microscopic origin of the characteristic sets of crystalline planes that evolve in each case, currently phenomenologically attributed to growth variation with facet angle. Such understanding will be particularly important for the design and growth of quantum structures in which the interface morphology on an atomic scale becomes significant. Although some attempts to model epitaxial growth on NP substrates have been made for both MBE (Ohtsuka and Miyazawa 1988) and OMCVD (Asai 1987; Jones *et al.* 1991; Ratsch and Zangwill 1991; Colas *et al.* 1992), more work is needed for developing detailed growth models that could be used as design tools for generating these complex structures. Extension of growth investigations to other material systems would also extend the impact of this patterning approach. The more complete understanding of growth on NP substrates will increase its usefulness as a unique technique for 3D patterning of semiconductor heterostructures.

Acknowledgments

I am indebted to many of my colleagues at Bellcore for the fruitful collaboration in the past several years on much of the work described in this review. I would like to particularly thank R. Bhat, E. Colas, C. Caneau, L. T. Florez, J. P. Harbison, M. C. Tamargo and F. Turco-Sandroff for OMCVD and MBE growth, and D. M. Hwang for TEM analysis of the structures described. Special thanks are also due to S. Simhony, M. Walther, M. Krahl, J. Christen, M. Grundmann, E. M. Clausen, Jr., K. Kash, N. G. Stoffel, P. S. D. Lin, C. P. Yun, C. Y. Chen, L. M. Schiavone and

B. P. Van der Gaag. I also thank Raj Bhat, David Hwang, Jim Harbison and Kathy Kash (Bellcore), and Richard Mirin (UCSB) for critically reading the manuscript.

REFERENCES

Arakawa, Y., and Sakaki, H. (1982). *Appl. Phys. Lett.* **40**, 939.
Arakawa, Y., and Yariv, A. (1986). *IEEE J. Quantum Electron.* **QE-22**, 1887.
Arakawa, Y., Vahala, K., Yariv, A., and Lau, K. (1985). *Appl. Phys. Lett.* **47**, 1142.
Arakawa, Y., Vahala, K., Yariv, A., and Lau, K. (1986). *Appl. Phys. Lett.* **48**, 384.
Arent, D. J., Nilsson, S., Galeuchet, Y. D., and Meier, H. P. (1989). *Appl. Phys. Lett.* **55**, 2611.
Arent, D. J., Brovelli, L., Jackel, H., Marclay, E., and Meier, H. P. (1990a). *Appl. Phys. Lett.* **56**, 1939.
Arent, D. J., Galeuchet, Y. D., Nilsson, S., and Meier, H. P. (1990b). *J. Vac. Sci. Technol. B* **8**, 145.
Asada, M., Miyamoto, Y., and Suematsu, Y. (1985). *Jpn. J. Appl. Phys.* **24**, L95.
Asada, M., Miyamoto, Y., and Suematsu, Y. (1986). *IEEE J. Quantum Electron.* **QE-22**, 1915.
Asai, H. (1987). *J. Crystal Growth* **80**, 425.
Azoulay, R., Remiens, D., Menigaux, L., and Dugrand, L. (1989). *Appl. Phys. Lett.* **54**, 1857.
Ballingall, J. M., and Wood, C. E. C. (1982). *Appl. Phys. Lett.* **41**, 947.
Banyai, L., Galbraith, I., Ell, C., and Haung, H. (1987). *Phys. Rev. B* **36**, 6099.
Beenakker, C. W. J., and van Houten, H. (1991). *Solid State Physics*, **44** H. Eherenreich, and D. Turnbull, eds.), p. 1. Academic Press, Boston.
Bhat, R., Kapon, E., Hwang, D. M., Koza, M. A., and Yun, C. P. (1988). *J. Crystal Growth* **93**, 850.
Bhat, R., Kapon, E., Werner, J., Hwang, D. M. Stoffel, N. G., and Koza, M. A. (1990a). *Appl. Phys. Lett.* **56**, 863.
Bhat, R., Zah, C. E., Caneau, C., Koza, M. A., Menocal, S. G., Schwarz, S. A., and Faivre, F. J. (1990b). *Appl. Phys. Lett.* **56**, 1691.
Bhat, R., Caneau, C., Zah, C. E., Koza, M., Bonner, W. A., Hwang, D. M., Schwarz, S. A., Menocal, S. G., and Faivre, F. J. (1991). *J. Crystal Growth* **107**, 772.
Bockelmann, U., and Bastard, G. (1991). *Europhys. Lett.* **15**, 215.
Brandt, O., Tapfer, L., Ploog, K., Bierwolf, R., Hohenstein, M., Phillipp, F., Lage, H., and Heberle, A. (1991). *Phys. Rev. B* **44**, 8043.
Brovelli, L., Arent, D. J., Jackel, H., and Meier, H. P. (1990). 12th IEEE International Semiconductor Laser Conference, Davos, Switzerland, September, paper D-5. *Conference Digest*, p. 50.
Bryan, R. P., Miller, L. M., Cockerill, T. M., and Coleman, J. J. (1989). *Appl. Phys. Lett.* **54**, 1634.
Burnham, R. D., and Scifres, D. R. (1975). *Appl. Phys. Lett.* **27**, 510.
Cao, M., Miyake, Y., Tamura, Hirayama, H., Arai, S., Suematsu, Y., and Miyamoto, Y. (1990). *Transactions of the IEICE E* **73**, 63.
Chang-Hasnain, C. J., Kapon, E., Harbison, J. P., and Florez, L. T. (1990). *Appl. Phys. Lett.* **56**, 429.
Cho, A. Y., and Ballamy, W. C. (1975). *J. Appl. Phys.* **46**, 783.
Christen, J., Grundmann, M., Kapon, E., Colas, E., Hwang, D. M., and Bimberg, D. (1992a). *Appl. Phys. Lett.* **61**, 67.
Christen, J., Kapon, E., Colas, E., Hwang, D. M., and Schiavone, L. M. (1992b). *Surf. Sci.* **267**, 257.

Cibert, J., Petroff, P. M., Dolan, G. J., Pearton, S. J., Gossard, A. C., and English, J. H. (1986). *Appl. Phys. Lett.* **49**, 1275.
Clausen, E. M., Jr., Craighead, H. G., Worlock, J. M., Harbison, J. P., Schiavone, L. M., Florez, L., and van der Gaag, B. (1989). *Appl. Phys. Lett.* **55**, 1427.
Clausen, E. M., Jr., Kapon, E., Tamargo, M. C., and Hwang, D. M. (1990). *Appl. Phys. Lett.* **56**, 776.
Colas, E., Kapon, E., Simhony, S., Cox, H. M., Bhat, R., Kash, K., and Lin, P. S. D. (1989). *Appl. Phys. Lett.* **55**, 867.
Colas, E., Clausen, E. M., Jr., Kapon, E., Hwang, D. M., and Simhony, S. (1990a). *Appl. Phys. Lett.* **57**, 2472.
Colas, E., Simhony, S., Kapon, E., Bhat, R., Hwang, D. M., and Lin, P. S. D. (1990b). *Appl. Phys. Lett.* **57**, 914.
Colas, E., Clausen, E. M., Jr., Kapon, E., Hwang, D. M., Simhony, S., Bhat, R., Chen, C. Y., Lin, P. S. D., Schiavone, L., and Van der Gaag, B. (1991). *J. Crystal Growth* **107**, 243.
Colas, E., Nihous, G. C., and Hwang, D. M. (1992). *J. Vac. Sci. Technol. A* **40**, 691.
Cox, H. M., Lin, P. S., Yi-Yan, A., Kash, K., Seto, M., and Bastos, P. (1989). *Appl. Phys. Lett.* **55**, 472.
Crawford, D. L., Nagarajan, R. L., and Bowers, J. E. (1991). *Appl. Phys. Lett.* **58**, 1629.
Demeester, P., Van Daele, P., and Baets, R. (1988). *J. Appl. Phys.* **63**, 2284.
Dutta, N. K., Cella, T., Picirilli, A. B., and Brown, R. L. (1986). *Appl. Phys. Lett.* **49**, 1227.
Dzurko, K. M., Menu, E. P., Beyler, C. A., Osinski, J. S., and Dapkus, P. D. (1989a). *Appl. Phys. Lett.* **54**, 105.
Dzurko, K. M., Menu, E. P., Beyler, C. A., Osinski, J. S., and Dapkus, P. D. (1989b). *IEEE J. Quantum Electron.* **QE-25**, 1450.
Fekete, D., Burnham, R. D., Scifres, D. R., Streifer, W., and Yingling, R. D. (1981). *Appl. Phys. Lett.* **38**, 607.
Fekete, D., Bour, D., Ballantyne, J. M., and Eastman, L. F. (1987). *Appl. Phys. Lett.* **50**, 635.
Feldman, J., Peter, G., Goebel, E. O., Dawson, P., Moore, K., and Foxon, C. (1987). *Phys. Rev. Lett.* **59**, 2337.
Ford, C. J. B., Thornton, T. J., Newbury, R., Pepper, M., Ahmed, H., Foxon, C. T., Harris, J. J., and Roberts, C. (1988). *J. Phys. C* **21**, L325.
Frateschi, N. C., Osinski, J. S., Beyler, C. A., and Dapkus, P. D. (1992). *Photon. Technol. Lett.* **4**, 209.
Fujii, M., Yamamoto, T., Shigeta, M., Takebe, T., Kobayashi, K., Hiyamizu, S., and Fujimoto, I. (1992). *Surf. Sci.* **267**, 26.
Fukui, T., and Saito, H. (1988). *J. Vac. Sci. Technol. B* **6**, 1373.
Fukui, T., Ando, S., Kokura, Y., and Toriyama, T. (1991). *Appl. Phys. Lett.* **58**, 2018.
Galeuchet, Y. G., and Roentgen, P. (1991). *J. Crystal Growth* **107**, 147.
Galeuchet, Y. G., Roentgen, P., and Graf, V. (1988). *Appl. Phys. Lett.* **53**, 2638.
Galeuchet, Y. G., Roentgen, P., and Graf, V. (1990). *J. Appl. Phys.* **68**, 560.
Galeuchet, Y. G., Rothuizen, H., and Roentgen, P. (1991). *Appl. Phys. Lett.* **58**, 2423.
Grodzinski, P. Zou, Y., Osinski, J. S., and Dapkus, P. D. (1991). *J. Crystal Growth* **107**, 583.
Guha, S., Madhukar, A., Kaviani, K., Chen, L., Kuchibhotla, R., Kapre, R., Hyugaji, M., and Xie, Z. (1990). *Proc. MRS Symp. III–V Heterostructures for Electronic/Photonic Devices* **145**, 27.
Hashimoto, A., Fukunaga, T., and Watanabe, N. (1989). *Appl. Phys. Lett.* **54**, 998.
Hata, M., Isu, T., Watanabe, A., and Katayama, Y. (1990). *J. Vac. Sci. Technol.* **8**, 692.
Heineke, H., Baur, B., Hoger, R., Jobst, B., and Miklis, A. (1992). *J. Crystal Growth* **124**, 170.
Helm, M., Colas, E., Hayes, J. R., Van der Gaag, B. P., Schiavone, L. M., and Hwang, D. M. (1991). *Appl. Phys. Lett.* **58**, 1320.
Henry, C. H. (1982). *IEEE J. Quantum Electron.* **QE-18**, 259.

Hersee, S. D., Barbier, E., and Blondeau, R. (1986). *J. Cryst. Growth* **77**, 319.
Herring, C. (1951). *Phys. Rev.* **82**, 87.
Hirata, S., Narui, H., and Mori, Y. (1991). *Appl. Phys. Lett.* **58**, 319.
Hoenk, H., and Vahala, K. J. (1989a). *Appl. Phys. Lett.* **54**, 1347.
Hoenk, M. E., Nieh, C. W., Chen, H. Z., and Vahala, K. J. (1989b). *Appl. Phys. Lett.* **55**, 53.
Holonyak, N., Jr., Kolbas, R. M., Dupuis, R. D., and Dapkus, P. D. (1980). *IEEE J. Quantum Electron.* **QE-16**, 170.
Hong, J. M., Wu, M. C., Wang, S., Wang, W. I., and Chang, L. L. (1987). *Appl. Phys. Lett.* **51**, 886.
Imanaka, K., Imamoto, H., Sato, F., Asai, M., and Shimura, M. (1987). *Electron. Lett.* **23**, 209.
Ismail, K., Burkhardt, M., Smith, H. I., Karam, N. H., and Sekula-Moise, P. A. (1991). *Appl. Phys. Lett.* **58**, 1539.
Izrael, A., Sermage, B., Marzin, J. Y., Ougazzaden, A., Azoulay, R., Etrillard, J., Theirry-Mieg, V., and Henry, L. (1990). *Appl. Phys. Lett.* **56**, 830.
Jaeckel, H., Meier, H. P., Bona, G. L., Walter, W., Webb, D. J., and Van Gieson, E. (1989). *Appl. Phys. Lett.* **55**, 1059.
Jones, S. H., Seidel, L. K., Lau, K. M., and Harold, M. (1991). *J. Crystal Growth* **108**, 73.
Kamon, K., Takagishi, S., and Mori, H. (1985). *J. Crystal Growth* **73**, 73.
Kapon, E. (1985). Unpublished.
Kapon, E. (1990). *Opt. Lett.* **15**, 801.
Kapon, E. (1992). *Proc. IEEE* **80**, 398.
Kapon, E., Tamargo, M. C., and Hwang, D. M. (1987), *Appl. Phys. Lett.* **50**, 347.
Kapon, E., Harbison, J. P., Yun, C. P., and Stoffel, N. G. (1988a). *Appl. Phys. Lett.* **52**, 607.
Kapon, E., Hwang, D. M., Bhat, R., and Tamargo, M. C. (1988b). *Superlattices and Microstructures* **4**, 297.
Kapon, E., Stoffel, N. G., Dobisz, E. A., and Bhat, R. (1988c). *Appl. Phys. Lett.* **52**, 351.
Kapon, E., Yun, C. P., Harbison, J. P., Florez, L. T. and Stoffel, N. G. (1988d). *Electron. Lett.* **24**, 985.
Kapon, E., Harbison, J. P., Yun, C. P., and Florez, L. T. (1989a). *Appl. Phys. Lett.* **54**, 304.
Kapon, E., Hwang, D. M., and Bhat, R. (1989b). *Phys. Rev. Lett.* **63**, 430.
Kapon, E., Simhony, S., Bhat, R., and Hwang, D. M. (1989c). *Appl. Phys. Lett.* **55**, 2715.
Kapon, E., Simhony, S., Harbison, J. P., Florez, L. T., and Worland, P. (1990). *Appl. Phys. Lett.* **56**, 1825.
Kapon, E., Christen, J., Colas, E., Bhat, R., Hwang, D. M., and Schiavone, L. M. (1992a). *Proc. Int. Symp. Nanostructure and Mesoscopic Systems* (W. P. Kirk and M. A. Reed, eds.), p. 63. Academic Press, Boston.
Kapon, E., Hwang, D. M., Walther, M., Bhat, R., and Stoffel, N. G. (1992b). *Surf. Sci.* **267**, 593.
Kapon, E., Kash, K., Clausen, E. M., Jr., Hwang, D. M., and Colas, E. (1992c). *Appl. Phys. Lett.* **60**, 477.
Kapon, E., Walther, M., Christen, J., Grundmann, M., Caneau, C., Hwang, D. M., Colas, E., Bhat, R., Song, G. H., and Bimberg, D. (1992d). *Superlattices and Microstructures* **12**, 491.
Kapon, E., Walther, M., Christen, J., Grundmann, M., Colas, E., and Bimberg, D. (1993). *Proc. 7th International Winter School on New Developments in Solid State Physics* (G. Bauer, ed.), p. 300. Springer-Verlag, Berlin.
Kash, K., Scherer, A., Worlock, J. M., Craighead, H. G., and Tamargo, M. C. (1986). *Appl. Phys. Lett.* **49**, 1043.
Kash, K., Bhat, R., Mahoney, D. D., Lin, P. S. D., Scherer, A., Worlock, J. M., Van der Gaag, B. P., Koza, M., and Grabbe, P. (1989). *Appl. Phys. Lett.* **55**, 681.
Kash, K., Van der Gaag, B. P., Mahoney, D. D., Gozdz, A. S., Florez, L. T., Harbison, J. P., and Sturge, M. D. (1991). *Phys. Rev. Lett.* **67**, 1326.
Kondo, M., Anayama, C., Tanahashi, T., and Yamazaki, S. (1992). *J. Crystal Growth* **124**, 449.

Kojima, K., Mitsunaga, K., and Kyuma, K. (1989). *Appl. Phys. Lett.* **55**, 882.
Kojima, K., Mitsunaga, K., and Kyuma, K. (1990). *Appl. Phys. Lett.* **56**, 154.
Krahl, M., Kapon, E., Schiavone, L. M., Van der Gaag, B. P., Harbison, J. P., and Florez, L. T. (1992). *Appl. Phys. Lett.* **61**, 813.
Laidig, W. D., Holonyak, N., Jr., Camras, M. D., Hess, K., Coleman, J. J., Dapkus, D. P., and Bardeen, J. (1981). *Appl. Phys. Lett.* **38**, 776.
Lebens, J. A., Tsai, C. S., Vahala, K. J., and Kuech, T. F. (1990). *Appl. Phys. Lett.* **56**, 2642.
Mannoh, M., Yuasa, T., Naritsuka, S., Shinozaki, K., and Ishii, M. (1985). *Appl. Phys. Lett.* **47**, 728.
Marclay, E., Arent, D. J., Harder, C., Meier, H. P., Walter, W., and Webb, D. J. (1989). *Electron. Lett.* **25**, 892.
Marti, U., Proctor, M., Monnard, D., Martin, D., Morier-Genoud, F., Reinhart, F. K., Widmer, R., and Lehmann, H. (1991). Advanced Processing and Characterization Technologies (APCT '91), Clearwater Beach, FL.
Marzin, J. Y., Izrael, A., Birotheau, L., Sermage, B., Roy, N., Azoulay, R., Robein, D., Benchimol, J.-L., Henry, L., Theirry-Mieg, V., Ladan, F. R., and Taylor, L. (1992). *Surf. Sci.* **267**, 253.
Matsubara, K., Ravikumar, K. G., Asada, M., and Suematsu, Y. (1989). *Trans. IEICE* **E72**, 1179.
Meier, H. P., Broom, R. F., Epperlein, P. W., Van Gieson, E., Harder, C., Jackel, H., Walter, W., and Webb, D. J. (1988a). *J. Vac. Sci. Technol. B* **6**, 692.
Meier, H. P., Van Gieson, E., Walter, W., Harder, Buchmann, C., Webb, D., and Moser, A. (1988b). *Electron. Lett.* **24**, 1123.
Meier, H. P., Van Gieson, E., Epperlein, P. W., Harder, C., Walter, W., Krahl, M., and Bimberg, D. (1989a). *J. Crystal Growth* **95**, 66.
Meier, H. P., Van Gieson, E., Walter, W., and Harder, C. (1989b). *Appl. Phys. Lett.* **54**, 433.
Miller, D. A. B., Chemla, D. S., and Schmitt-Rink, S. (1988). *Appl. Phys. Lett.* **52**, 2154.
Miller, D. L. (1985). *Appl. Phys. Lett.* **47**, 1309.
Miller, D. L., and Asbeck, P. M. (1987). *J. Crystal Growth* **81**, 368.
Miller, M. S., Weman, H., Pryor, C. E., Krishnamurthy, M., Petroff, P. M., Kroemer, H., and Merz, J. L. (1992). *Phys. Rev. Lett.* **68**, 3464.
Mimura, T., Hiyamizu, S., Fujii, T., and Nanbu, K. (1980). *Jpn. J. Appl. Phys.* **19**, L225.
Mirin, R. P., Tan, I.-H., Weman, H., Leonard, M., Yasuda, T., Bowers, J. E., and Hu, E. L. (1992). *J. Vac. Sci. Technol. A* **10**, 697.
Mori, Y., Matsuda, O., Morizane, K., and Watanabe, N. (1980). *Electron. Lett.* **16**, 785.
Nagamune, Y., Tsukamoto, S., Nishioka, M., and Arakawa, Y. (1993). *J. Crystal Growth* **126**, 707.
Nagata, S., and Tanaka, T. (1977). *J. Appl. Phys.* **48**, 940.
Nagata, S., Tanaka, T., and Fukai, M. (1977). *Appl. Phys. Lett.* **30**, 505.
Narui, H., Ohta, T., and Mori, Y. (1988). *Electron. Lett.* **24**, 1249.
Narui, H., Hirata, S., and Mori, Y. (1990). 12th Int. Semicond. *Laser Conference, Davos, Switzerland*, paper F1. Conference Digest, p. 78.
Neave, J. H., Dobson, P. J., Joyce, B. A., and Zhang, J. (1985). *Appl. Phys. Lett.* **41**, 100.
Nilsson, S., Van Gieson, E., Arent, D. J., Meier, H. P., Walter, W., and Forster, T. (1989). *Appl. Phys. Lett.* **55**, 972.
Ohtsuka, M., and Miyazawa, S. (1988). *J. Appl. Phys.* **64**, 3522.
Okano, Y., Seto, H., Katahama, H., Nishine, S., Fujimoto, I., and Suzuki, T. (1989). *Japan. J. Appl. Phys.* **28**, L 151.
Osorio, F. A. P., Degani, M. H., and Hipolito, O. (1987). *Phys. Rev. B* **37**, 1402.
Petroff, P. M., Gossard, A. C., and Wiegmann, W. (1984). *Appl. Phys. Lett.* **45**, 620.

Pfeiffer, L., West, K. W., Stormer, H. L., Eisenstein, J. P., Baldwin, K. W., Gershoni, D., and Spector, J. (1990). *Appl. Phys. Lett.* **56**, 1697.
Ratsch, C., and Zangwill, A. (1991). *Appl. Phys. Lett.* **58**, 403.
Sakaki, H. (1980). *Jpn. J. Appl. Phys.* **19**, L735.
Sanada, T., Yamakoshi, S., Wada, O., Fujii, T., Sakurai, T., and Sasaki, M. (1984). *Appl. Phys. Lett.* **44**, 325.
Schmitt-Rink, S., Chemla, D. S., and Miller, D. A. B. (1989). *Advances in Physics* **38**, 89.
Scott, M. D., Riffat, J. R., Griffith, I., Davies, J. L., and Marshall, A. C. (1988). *J. Crystal Growth* **93**, 820.
Shiegeta, M., Okano, Y., Seto, H., Katahama, H., Nishine, S., Kobayashi, K., and Fujimoto, I. (1991). *J. Crystal Growth* **111**, 284.
Shen, X. Q., Tanaka, M., and Nishinaga, T. (1992). Eleventh Record of Alloy Semiconductor Physics and Electronics Symposium, July 8–10, Kyto, Japan, p. 333.
Simhony, S., Kapon, E., Colas, E., Bhat, R., Stoffel, N. G., and Hwang, D. M. (1990). *Photon. Technol. Lett.* **2**, 305.
Simhony, S., Kapon, E., Colas, E., Hwang, D. M., Stoffel, N. G., and Worland, P. (1991). *Appl. Phys. Lett.* **59**, 2225.
Smith, J. S., Derry, P. L., Margalit, S., and Yariv, A. (1985). *Appl. Phys. Lett.* **47**, 712.
Sollner, T. C. L. G., Goodhue, W. D., Tannenwald, P. E., Parker, C. D., and Peck, D. D. (1983). *Appl. Phys. Lett.* **43**, 588.
Subbanna, S., Kroemer, and Merz, J. L. (1986). *J. Appl. Phys.* **59**, 488.
Tamargo, M. C., Quinn, W. E., Hwang, D. M., Chen, C. Y., Kapon, E., Schiavone, L. M., Brasil, M. J. S., and Nahory, R. E. (1992). *J. Vac. Sci. Technol. B* **10**, 982.
Tate, A. Ohmori, Y., and Kobayashi, M. (198). *J. Cryst. Growth* **89**, 360.
Tiwari, S., Petit, G. D., Milkove, K. R., Davis, R. J., Woodal, J. M., and Legoues, F. (1992). *Proc. Int. Electron Devices Mtg.*, San Francisco, December 13–16; paper 34.2.1.
Tsang, W. T., and Cho, A. Y. (1977). *Appl. Phys. Lett.* **30**, 293.
Tsang, W. T., and Wang, S. (1976). *Appl. Phys. Lett.* **28**, 44.
Tsuchiya, M., Petroff, P. M., and Coldren, L. A. (1989). *IEEE Trans. Electron Devices*, **TED-36**, 2612.
Tsukada, T. (1974). *J. Appl. Phys.* **45**, 4899.
Tsukamoto, S., Nagamune, Y., Nishioka, M., and Arakawa, Y. (1992). *J. Appl. Phys.* **71**, 533.
Tsukamoto, S., Nagamune, Y., Nishioka, M., and Arakawa, Y. (1993). *Appl. Phys. Lett.* **62**, 49.
Turco, F. S., Simhony, S., Kash, K., Hwang, D. M., Ravi, T. S., Kapon, E., and Tamargo, M. C. (1990a). *J. Crystal Growth* **104**, 766.
Turco, F. S., Tamargo, M. C., Hwang, D. M., Nahory, R. E., Werner, J., Kash, K., and Kapon, E. (1990b). *Appl. Phys. Lett.* **56**, 72.
Van Gieson, E., Meier, H. P., Harder, C., Buchmann, P., Webb, D., and Walter, W. (1989). *J. Vac. Sci. Technol. B* **7**, 405.
Vermeire, G., Yu, Z. Q., Vermaerke, F., Buydens, L., Van Daele, P., and Demeester, P. (1992). *J. Crystal Growth* **124**, 513.
Walther, M., Kapon, E., Christen, J., Hwang, D. M., and Bhat, R. (1992a). *Appl. Phys. Lett.* **60**, 521.
Walther, M., Kapon, E., Hwang, D. M., Colas, E., and Nunes, L. (1992b). *Phys. Rev. B* **45** (Rapid Commun.), 6333.
Walther, M., Kapon, E., Schiavone, L. M., Hwang, D. M., Caneau, C., Scherer, A., and Stoffel, N. G. (1992c). Conference on *Lasers and Electr-Optics (CLEO '92), Anaheim, CA, May. Conference Digest*, p. 378.
Walther, M., Kapon, E., Caneau, C., Hwang, D. M., and Schiavone, L. M. (1993). *Appl. Phys. Lett.*, in print.

Wang, W. I., Mendez, E. E., Khan, T. S., and Esaki, L. (1985). *Appl. Phys. Lett.* **47**, 826.
Weisbuch, C. (1987). *Semiconductors and Semimetals*, **24**, *Applications of Multiquantum Wells, Selective Doping and Superlattices* (R. Dingle, ed.), p. 1. Academic Press, New York.
Wu, Y. H., Werner, M., Chen, K. L., and Wang, S. (1984a). *Appl. Phys. Lett.* **44**, 834.
Wu, Y. H., Werner, M., and Wang, S. (1984b). *Appl. Phys. Lett.* **45**, 606.
Wulff, B. (1901). *Z. Krist.* **34**, 449.
Yablonovitch, E. (1987). *Phys. Rev. Lett.* **58**, 2059.
Yamada, T., Yuasa, T., Kamon, K., Shimazu, M., and Ishii, M. (1986). *Electron. Lett.* **22**, 1165.
Yariv, A. (1988). *Appl. Phys. Lett.* **53**, 1033.
Yu, Y. H., Werner, M., Chen, K. L., and Wang, S. (1984). *Appl. Phys. Lett.* **44**, 834.
Yoshikawa, A., Yamamoto, A., Hirose, M., Sugino, T., Kano, G., and Teramoto, I. (1987). *IEEE J. Quantum Electron.* **QE-23**, 725.
Yuasa, T., Mannhoh, M., Yamada, T., Naritsuka, S., Shinozaki, K., and Ishii, M. (1987). *J. Appl. Phys.* **62**, 764.
Zarem, H. A., Sercel, P. C., Hoenk, M. E., Lebens, J. A., and Vahala, K. J. (1989a). *Appl. Phys. Lett.* **54**, 2692.
Zarem, H., Vahala, K., and Yariv, A. (1989b). *IEEE J. Quantum Electron.* **QE-25**, 705.
Zinke-Allmang, M., Feldman, L. C., and Nakahara, S. (1988). *Appl. Phys. Lett.* **52**, 144.
Zmudzinski, C. A., Givens, M. E., Bryan, R. P., and Coleman, J. J. (1989). *IEEE J. Quantum Electron.* **QE-25**, 1539.
Zou, Y., Grodzinski, P., Osinski, J. S., and Dapkus, P. D. (1991). *Appl. Phys. Lett.* **58**, 717.

CHAPTER 5

Optical Properties of $Ga_{1-x}In_xAs/InP$ Quantum Wells

H. Temkin
ELECTRICAL ENGINEERING DEPARTMENT, COLORADO STATE UNIVERSITY, FORT COLLINS

D. Gershoni
DEPARTMENT OF PHYSICS, TECHNION, HAIFA, ISRAEL

M. B. Panish
AT&T BELL LABORATORIES, MURRAY HILL, NEW JERSEY

I.	INTRODUCTION.	338
II.	COMPUTATION OF ENERGY LEVELS IN QUANTUM WELLS.	340
III.	OPTICAL PROPERTIES OF QUANTUM WELLS AND SUPERLATTICES.	345
IV.	DYNAMIC EFFECTS.	361
V.	QUANTUM WIRES AND BOXES.	363
VI.	ELECTRIC FIELD EFFECTS.	376
VII.	STRAINED LAYER SUPERLATTICES.	385
	1. *Critical Layer Thickness.*	385
	2. *Optical Properties and Electronic Energy Levels.*	390
	3. *Higher Order Optical Transitions.*	402
VIII.	THERMAL STABILITY AND IMPURITY INDUCED DISORDERING.	404
IX.	SUMMARY.	415
	ACKNOWLEDGMENTS.	415
	REFERENCES	415

Abstract

In the past few years, high quality quantum wells of $Ga_{1-x}In_xAs/InP$ have been grown and structurally characterized by precision X-ray diffraction and transmission electron microscopy. Such well-characterized heterostructures, both in intentionally strained and in unstrained versions, have been studied by photoluminescence, absorption, and photoluminescence and photocurrent excitation techniques. This chapter reviews the optical properties of these quantum wells, reduced dimensionality structures (quantum wires and boxes) derived from them, and changes induced by applied electric field, lattice mismatch strain, and thermal annealing. The combination of precise structural characterization and optical studies allows for comparison with the simple one-band envelope function model.

I. Introduction

In the last few years high-quality quantum wells of $Ga_xIn_{1-x}As_{1-y}P_y/$ InP have become available. The epitaxial methods used for growth of these structures are listed in Fig. 1. They range from conventional molecular beam epitaxy (MBE) (Claxton et al. 1987), through several intermediate methods that we call *gas source molecular beam epitaxy* (GSMBE), to metalorganic chemical vapor deposition (MOCVD). The latter two techniques in particular have achieved the reproducibility and precision of multilayer epitaxy previously reserved to the $GaAs/AlGa_{1-x}As$ system. There are two major subdivisions of GSMBE method, hydride source MBE (HSMBE) (Panish 1980, 1987; Panish and Sumski 1984; Panish et al. 1985) and metalorganic MBE (MOMBE) (Tsang et al. 1986; Tsang and Miller 1986). In the discussion of optical properties of $Ga_xIn_{1-x}As_{1-y}P$ quantum wells we shall rely primarily on the results obtained on the structures grown by HSMBE, with which we are most familiar. However, notable examples of the work carried out on structures grown by MOMBE and MOCVD (Razeghi et al. 1983; Miller et al. 1986a; Siednich et al. 1986) will also be discussed.

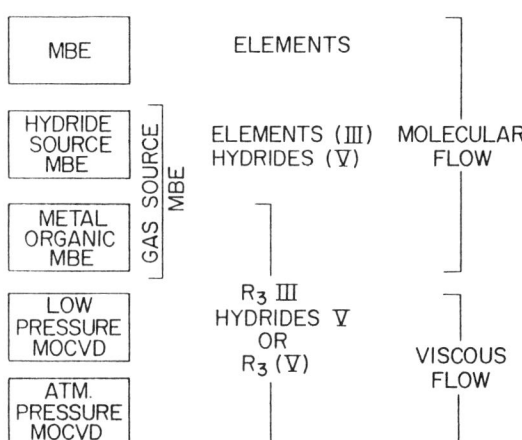

FIG. 1. Block diagram illustrating the range of epitaxial techniques used for the preparation of GaInAs and GaInAsP quantum wells.

5. Optical Properties of $Ga_{1-x}In_xAs/InP$ Quantum Wells

The interpretation of the optical experiments relies quite heavily on the structural information available for the sample under investigation. Two techniques, transmission electron microscopy (TEM) and high-resolution X-ray diffraction, were found particularly effective in studies of single quantum wells and superlattices. An example of the TEM cross section obtained on a HSMBE grown lattice matched structure consisting of 50 ternary quantum wells, each ~ 32 Å thick, and separated by 103 Å thick InP barriers is shown in Fig. 2. The ($\bar{2}00$) direction cross section shows the well and barrier thicknesses and provides some measure of sample uniformity. The well–barrier interfaces are very flat and well defined, at least on the scale of the micrograph, and the sample is free of defects.

High-resolution X-ray diffraction studies provide very detailed structural information averaged over a sample area of several mm². These studies, which were initiated by Vandenberg et al. (1986, 1987), use four crystal geometry X-ray diffractometer as proposed by Bartels (1973). The diffractometer produces a very well-collimated monochromatic X-ray beam for high-resolution rocking curves in and out of the growth direction. The X-ray rocking curves can be computer simulated to a remarkable precision with a kinematic step model of Segmuller and Blakeslee (1973). The model considers the superlattice to be a series of layers with abrupt interfaces. The

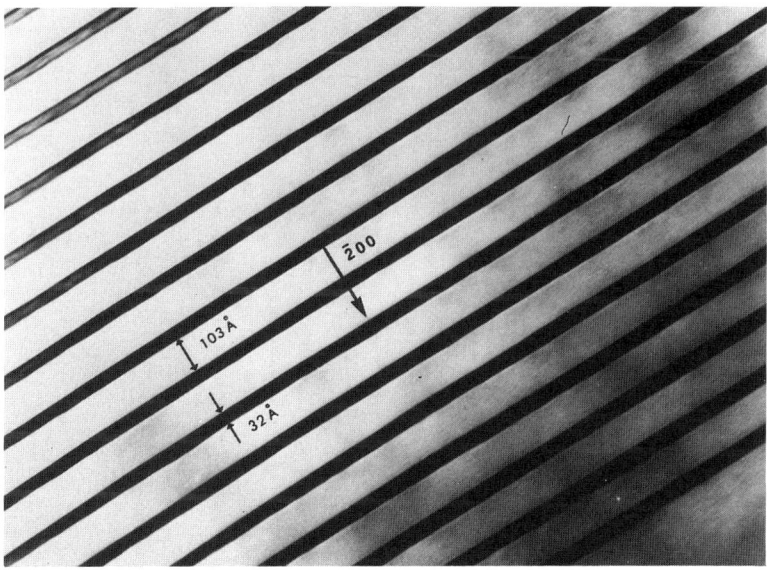

FIG. 2. Transmission electron micrograph illustrating a cross section of a GaInAs/InP superlattice grown by gas source molecular beam epitaxy.

(400) X-ray rocking curve for a superlattice that was grown with 100 periods, each nominally composed of a 20 Å layer of $Ga_{0.47}In_{0.53}As$ and a 100 Å thick layer of InP, is shown in the upper trace of Fig. 3. The lower trace shows the rocking curve simulated for this structure. Although excellent agreement is obtained between the experimental and calculated rocking curves, the experimental data shows in addition broadening and splitting of individual superlattice overtones. This behavior is indicative of period variations of ± 1 monolayer in some of the layers. This degree of growth precision is typical of GSMBE samples.

II. Computation of Energy Levels in Quantum Wells

The optical properties of semiconductor quantum structures are dramatically different from those of the corresponding bulk layers. These differences are attributed to the confinement of carriers to the thin epitaxial layers. Although a complete solution of the quantum mechanical problem involved

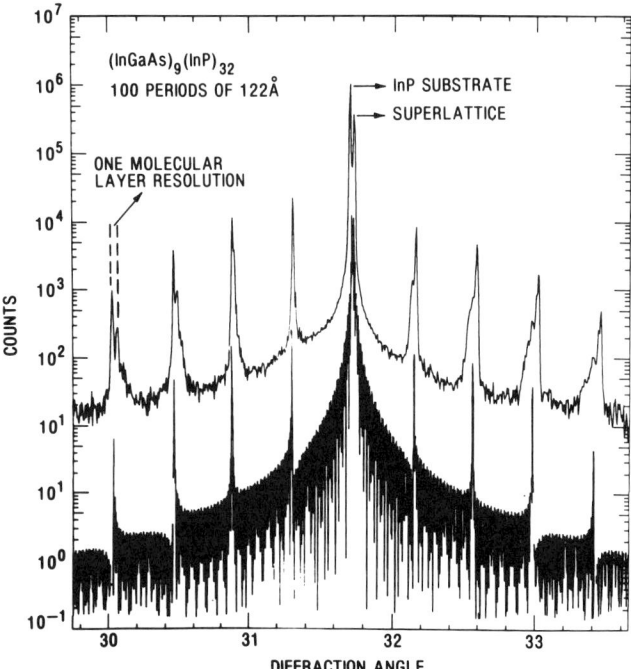

FIG. 3. High-resolution, four crystals, X-ray diffraction spectrum of a 100 well superlattice. Lower curve shows a fit calculated with a kinematic step model.

5. OPTICAL PROPERTIES OF $Ga_{1-x}In_xAs/InP$ QUANTUM WELLS

is by no means simple or straightforward we have found that a one-band effective mass model is usually sufficient to explain most of the experimental observations. A more accurate treatment, as well as a current review of the relevant literature, can be found in Baraff and Gershoni (1991).

The energy levels of confined particle states can be obtained by solving the one-dimensional single-particle Schrödinger equation (Ikonic et al. 1988), which we write within the envelope function and effective mass approximation as

$$\left[\frac{-\hbar^2}{2m_0}\left(\frac{\partial}{\partial z}\frac{1}{m^*(z)}\frac{\partial}{\partial z}\right) + V_0(z)\right]\Psi(z) = E\Psi(z) \quad (1)$$

where $m^*(z)$ is the layer, i.e. material dependent carrier effective mass, and $V_0(z)$ is the effective potential energy that is, in the case of a square well, the height of the potential step. The standard boundary conditions of continuity of the wave function and the current probability over each internal interface (Bastard 1981b) are automatically satisfied by the Schrödinger equation (Baraff and Gershoni 1991). The solution for a square potential well can be then written in a well-known implicit form as

$$\tan(kL/2) = m_w^*\alpha/m_b^*k \quad \text{for even } n \quad (2)$$

and

$$\tan(kL/2) = -m_b^*k/m_w^*\alpha \quad \text{for odd } n \quad (3)$$

where $\alpha^2 = 2m_b^*E/\hbar^2$ and $k^2 = (2m_w^*(E - V_0(z))/\hbar^2$. For computational simplicity Eqs. (2) and (3) can be transformed to a combination of sine and cosine.

All that is then needed to obtain the energy levels of confined particle states in a quantum well are the electron and hole effective masses in the well and barrier materials, the conduction and valence band discontinuities, and the well dimensions. A set of such calculated levels is shown in a scale drawing of the 60 Å thick GaInAs/InP quantum well of Fig. 4. The oscillator strength of the optical transitions between conduction and valence band sublevels is proportional in this approximation to a product of the dipole moment between the Bloch states and the envelope wave function overlap integral between the carrier envelope wave functions (Miller et al. 1985):

$$M_{op} \propto \langle U_v|\vec{p}|U_c\rangle \int_v \Psi_{env}^{v*}\Psi_{env}^c d^3\vec{r} \quad (4)$$

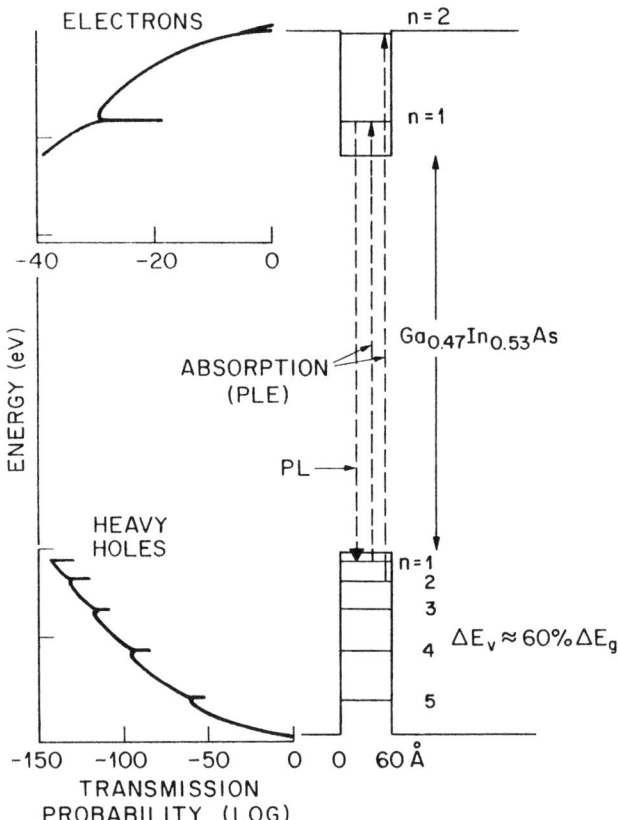

FIG. 4. Schematic drawing of the band-gap of a GaInAs/InP double heterostructure. Notice that about 60% of the band-gap discontinuity lies in the valence band. Confined particle energy levels are drawn to scale. Electron and hole transmission probability curves through a single 60 Å thick quantum well are calculated by resonant tunneling method.

The resulting allowed optical transitions are mainly between sublevels of equal quantum number ($\Delta n = 0$) since the overlap integral is vanishingly small for the others. The polarization selection rules follows mostly from the dipole moment between the Bloch states. The heavy and light hole zone center valence band Bloch states (U_v) are characterized by the respective total angular momentum states of $|3/2; \pm 3/2\rangle$ and $|3/2; \pm 1/2\rangle$. The zone center conduction band edge electron state is characterized by the total angular momentum state of $|1/2; \pm 1/2\rangle$. Neglecting any mixing mechanisms, the excitonic transition probabilities between the zone center conduction band electron and hole states are then given by the square of the respective Clebsh–Gordan coefficients, which relate the direction of the electric field associated with the electromagnetic radiation ($|1; \pm 1\rangle$, TE polarization, with respect to the growth direction, and $|1;0\rangle$, TM polarization) to the composite electron–hole total angular momentum numbers. That is electron–heavy hole = 3/4, electron–light hole = 1/4 for the TE polarization and electron–heavy hole = 0 and electron–light hole = 1 for the TM polarization. The heavy hole transition is thus strictly forbidden in the TM polarization, while the TE polarization favors it by a 3:1 ratio over the light hole transition.

The implicit formulae (2) and (3) are valid only for the simple case of a potential square well. It is often much easier to use the resonant tunneling calculation, which is computationally simple and can be applied to wells of arbitrarily complex shapes (Miller et al. 1985). The principle of such calculation is shown in Fig. 5, which represents an arbitrarily shaped one-dimensional potential well. The barrier and well are approximated by a series of constant width potential steps. The number of steps is adjusted to produce a solution independent of the step size. The wave function in each potential section s is given by

$$\psi_s = A_s e^{ik_s z} + B_s e^{-ik_s z} \tag{5}$$

where k_s is the particle momentum in the sth section:

$$k_s^2 = 2m_s^*(E_i - V_s)/\hbar^2 \tag{6}$$

and E_i and V_s are the incident energy and potential step height of that section, and m_s^* is the carrier's effective mass. We assume here parabolic bands, however, nonparabolicity effects can be introduced through the

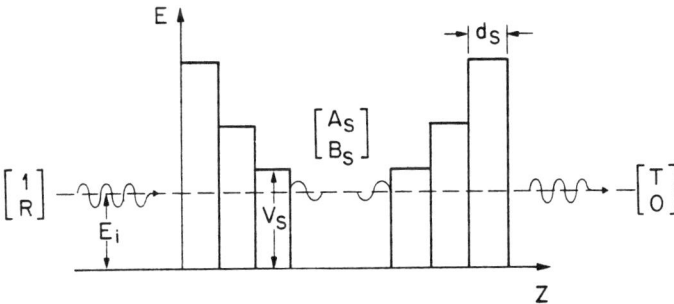

FIG. 5. Arbitrarily shaped potential well used in resonant tunneling calculation of energy levels.

energy dependence of the effective mass (Gershoni et al. 1988b). The coefficients A_s and B_s describe the amplitudes of the forward and backward going wave packets, respectively. The requirements of the wave function continuity and current probability conservation when applied at each step of the potential result in a transfer matrix that relates amplitudes in the section s to those in section $s + 1$:

$$\begin{bmatrix} A_{s+1} \\ B_{s+1} \end{bmatrix} = \begin{bmatrix} M^s_{11} M^s_{12} \\ M^s_{21} M^s_{22} \end{bmatrix} \begin{bmatrix} A_s \\ B_s \end{bmatrix} \qquad (7)$$

where

$$M^s_{11} = \frac{1}{2}\left(1 + \frac{m^*_{s+1} k_s}{m^*_s k_{s+1}}\right) \exp[i(k_s - k_{s+1})d_s] \quad \text{and} \quad m^s_{22} = (m^s_{11})^* \qquad (8)$$

$$M^s_{12} = \frac{1}{2}\left(1 - \frac{m^*_{s+1} k_s}{m^*_s k_{s+1}}\right) \exp[-i(k_s + k_{s+1})d_s] \quad \text{and} \quad m^s_{21} = (m^s_{12})^*$$

and d_s is the width of the sth potential section. The transfer matrix for the entire potential is obtained by the left multiplication of the subsequent transfer matrices. If the incident amplitude is equal to 1, the transmitted amplitude is given by the following equation:

$$\begin{bmatrix} T \\ 0 \end{bmatrix} = \prod_{s=1}^{n-1} \begin{bmatrix} M^s_{11} M^s_{12} \\ M^s_{21} M^s_{22} \end{bmatrix} \begin{bmatrix} 1 \\ R \end{bmatrix} \qquad (9)$$

and the transmission probability is given by $|T|^2$. By calculating $|T|^2$ as a function of the incident energy E one can locate resonances in the transmission probability. These correspond exactly to bound states in the well (Landau and Lifshitz 1977). The wave packet amplitudes are readily available in the course of this procedure and thus allow for reconstruction of the eigenfunctions corresponding to the bound states eigenvalues. The resulting electron and heavy hole transmission probability curves for a 60 Å wide $Ga_{1-x}In_xAs$ well are illustrated in Fig. 4.

In this section we have described a method of calculating sublevel energies and their envelope wave functions for semiconductor quantum wells. The method is very simple and requires only minimal computing resources. It provides, however, as will be made clear later, an extremely useful tool to analyze the optical properties of semiconductor quantum structures.

III. Optical Properties of Quantum Wells and Superlattices

Photoluminescence (PL) and photoluminescence excitation (PLE) studies of lattice matched heterostructures of $Ga_{0.47}In_{0.53}As/InP$ and $Ga_xIn_{1-x}As_{1-y}P_y$ grown in the configuration illustrated in the TEM micrograph of Fig. 6, or in very similar configurations, have been done by Temkin et al. (1985b) and Panish et al. (1986). The low-temperature (6K) PL spectrum of one such structure is shown together with a corresponding atomic resolution TEM micrograph in Fig. 6. The particular HSMBE grown structure contains quantum wells 6, 9, 12, and 24 Å thick, all to within ± 2 Å, and a 600 Å thick $Ga_{0.47}In_{0.53}As$ reference layer (not shown in the micrograph). The purpose of the reference layer is to provide an independent check of the ternary composition used in the growth of the quantum wells. The excitation source was a low-power HeNe laser emitting at a wavelength of 632.8 nm. Its photon energy is larger than the band gap of the InP barrier. The PL emission of all four wells, characterized by excellent efficiency and very large blue shifts due to the quantum size effect,

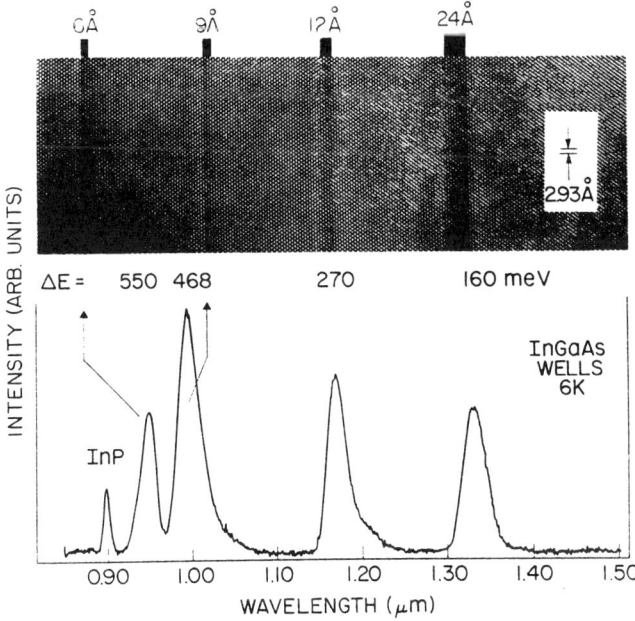

FIG. 6. Low-temperature photoluminescence spectrum of a sample consisting of four GaInAs quantum wells of decreasing thickness. Inset shows the atomic resolution TEM cross section of the same sample.

is ascribed to the recombination of electron and heavy hole from the $n = 1$ subband level (1H). The shift exhibited by the narrowest well, measured from the PL peak of the ternary reference layer, is as large as 534 meV. This constitutes approximately 85% of the band-gap discontinuity. The large energy shifts for the thin wells are consistent with their size measured by TEM. The InP emission at ~ 1.4 eV delineates the maximum energy shift possible in this material system. Although the electron–hole pairs are generated in the InP layers, the PL signal originating in these layers is very weak and most of the recombination occurs in the wells. Thus the quantum wells show excellent carrier capture efficiency, consistent with a low density of interfacial traps. The results of Fig. 6 are independent of the experimental conditions. Neither the energies nor line shapes are significantly affected by a two orders of magnitude decrease in the incident power. Qualitatively similar PL spectra have been also observed on a number of samples grown in the temperature range 450–550°C. Virtually identical spectra were obtained by probing various area of the sample at 5 mm intervals, as shown in Fig. 7. Thus, these structures are spatially uniform.

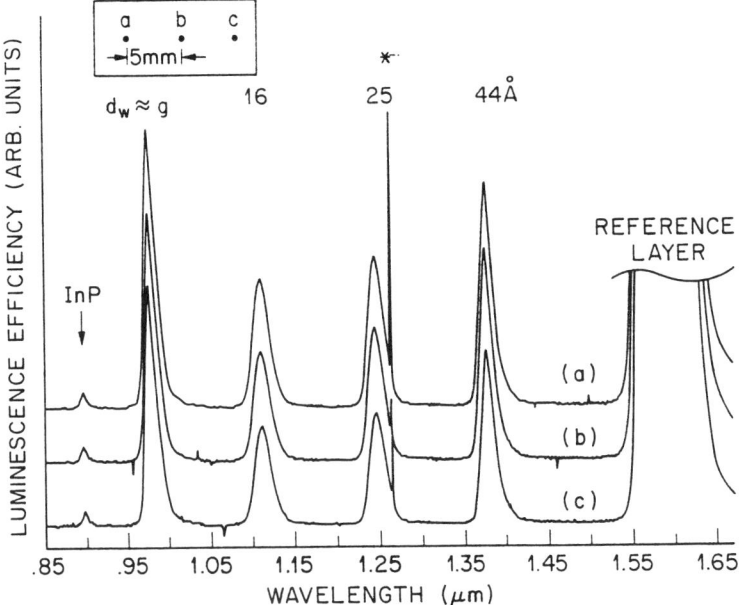

FIG. 7. Compositional and structural uniformity of GaInAs wells as demonstrated by the superposition of three PL spectra obtained from a large sample at 5 mm intervals. The sample consists of four ternary quantum wells and a 1000 Å thick GaInAs reference layer. Star denotes HeNe laser line used for excitation.

5. OPTICAL PROPERTIES OF $Ga_{1-x}In_xAs/InP$ QUANTUM WELLS 347

Low-temperature PL spectra of a sample containing four single wells of GaInAsP, all with a composition adjusted to give a nominal room temperature bulk band gap of 1.55 μm, are shown in Fig. 8(a). The well thicknesses were again determined by the TEM cross sections. Similarly to the ternary well sample, the PL emission of all the wells is quite intense and the large quantum size shifts are of the magnitude consistent with well thicknesses.

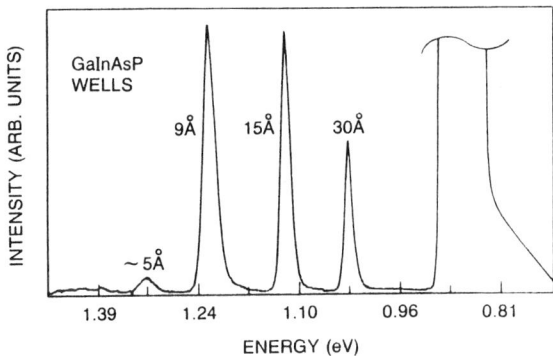

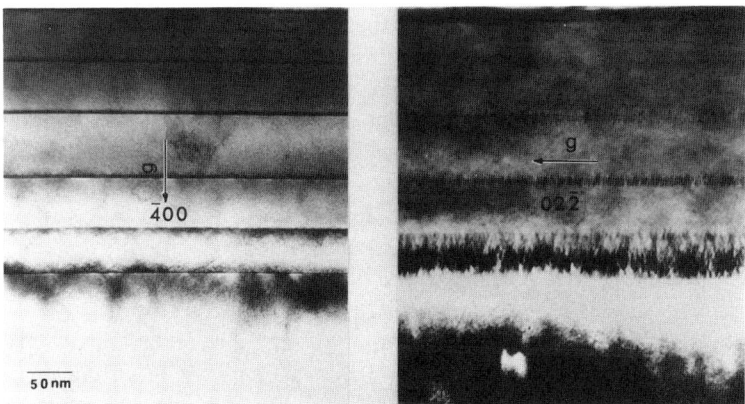

FIG. 8. (a) Low-temperature PL spectrum of a sample consisting of our narrow GaInAsP quantum wells. (b) TEM cross sections of this sample using (400) and (02$\bar{2}$) reflections. The latter is sensitive to strain and reveals phase separation contrast.

The sample uniformity measured by the PL spectra obtained from various surface points is similar to that illustrated in Fig. 7 for the ternary structures. Furthermore, the exciton linewidth is only slightly wider than that of the ternary samples. Fig. 8(b) shows the TEM image in the $(0\bar{2}2)$ direction, which is sensitive to stress. The faint, and somewhat random, contrast modulation on a ~ 100 Å scale visible even in the thinnest of the four wells is attributed to compositional modulation (DeCremoux et al. 1981; Mahajan et al. 1984). The problem of sample homogeneity on the scale of the well size and the associated question of alloy potential fluctuation has thus far received little attention. As is discussed later in this section, the effect of potential fluctuations on optical properties of quantum wells is quite considerable, particularly in this material system with the ternary mixed crystal constituting the quantum well.

The GaInAs wells discussed previously consist of a ternary alloy in which the constituent group III elements are randomly distributed in their sublattice. An interesting attempt to prepare an ordered alloy has been reported by Razeghi et al. (1987b). Low-pressure MOCVD was used to grow quantum wells of $(GaAs)_2(InAs)_2/InP$. The crystal growth deposits two monolayers of either GaAs or InAs at a time. The procedure is quite complicated and slow, and a long waiting time appears necessary after each two monolayer steps to prevent contamination of the subsequent layer. During the 1 min wait AsH_3 flow was used to prevent surface decomposition. The experimental structure consisted of four wells, 12, 25, 50 and 100 Å thick separated from each other by 500 Å of InP. The layer thicknesses were estimated from the growth parameters. Fig. 9 shows the low temperature PL spectrum of the sample obtained with the HeNe (excitation above the InP barrier band gap) and Nd:Yag (below band-gap excitation) lasers. Four distinct peaks are observed in the PL spectrum, each identifiable with the 1H exciton of the individual well. The linewidth is unusually large, indicating most likely some degree of intermixing and compositional fluctuation in the well that results in the inhomogeneous broadening of the PL lines. The difference between the two excitation energies, not seen in random alloy samples, also suggest some degree of interdiffusion with the barrier. The notable feature of the spectrum is the high spatial uniformity, claimed to be far better than that of standard MOCVD samples.

The 1H exciton linewidths for layers grown by hydride source MBE, as well as other epitaxial techniques, are in most samples less than would be calculated for the difference in energy of excitons sampling atomically smooth regions differing in thickness by one monolayer. Given that the diameter of an exciton confined in a quantum well is expected to equal about a hundred atomic sites, it is presumed that the luminescence line broadening that is observed in most ternary quantum wells results from the

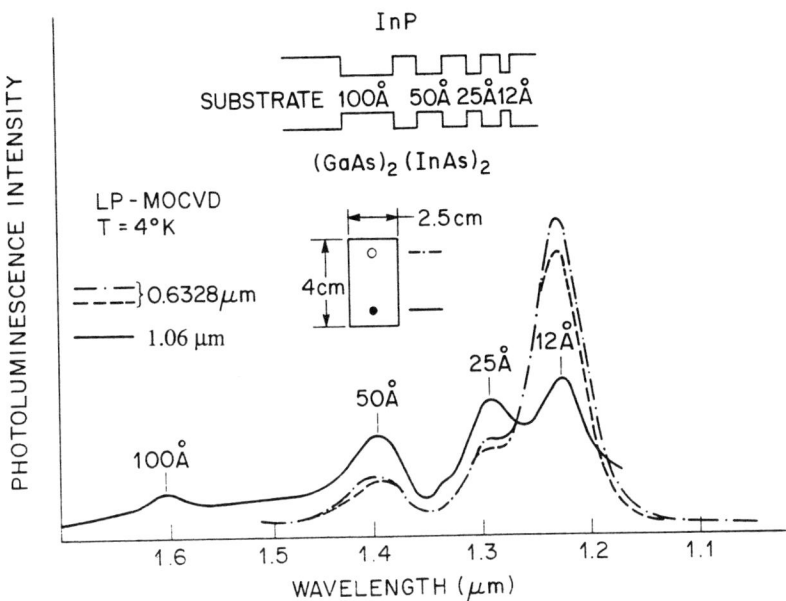

FIG. 9. Low-temperature PL spectra of a sample consisting of the ordered $(GaAs)_2(InAs)_2$ wells. The two spectra shown for the HeNe laser excitation were obtained from different areas of the wafer (After Razeghi et al. 1987).

interface roughness extending on average less than monolayer in depth and laterally on a scale smaller than the exciton size. Figure 10 shows a plot of the PL linewidth as a function of the well thickness. The solid line shows the linewidth expected with a monolayer interfacial step height. While some scattering is seen in the data, the linewidth generally decreases with increasing well thickness, as expected from the interface roughness model. The linewidths reported by MBE techniques are the lowest, also consistent with a better control over the interface sharpness. The contribution of interface roughness to the linewidths was convincingly demonstrated by the studies of very smooth GaAs/GaAlAs quantum wells by Tu et al. (1987) who showed line separations expected from excitons confined to domains that differ in thickness by one monolayer. Such line splitting is not observed in the $Ga_{0.47}In_{0.53}As$ wells. Although some of the $Ga_{0.47}In_{0.53}As$ exciton linewidth could be also attributed to potential fluctuations of the ternary well material, it is interesting to note that the excitonic transitions of the quaternary wells grown by HSMBE are also quite narrow. We would like, however, to caution against giving too much importance to the PL linewidth as a measure of the sample quality since it is often influenced by

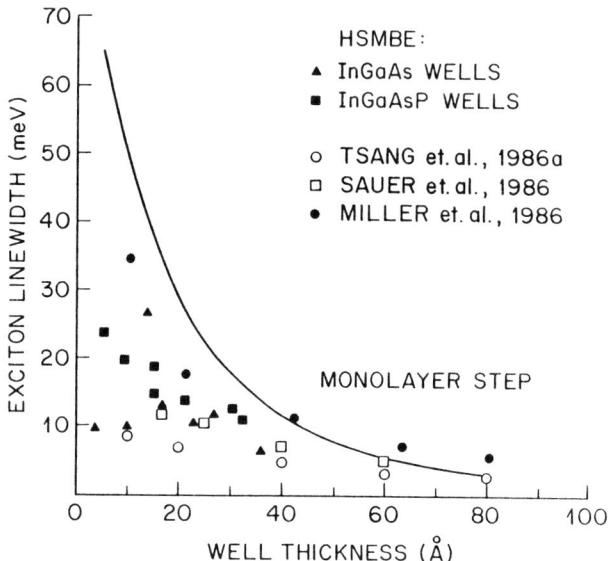

FIG. 10. Low-temperature PL linewidth of the $n = 1$ exciton is plotted versus the well width. The calculated width corresponding to a single monolayer thickness fluctuations is shown in a solid line. The HSMBE data are taken from Panish et al. (1986). Data points labeled ○, □, and ● refer to wells of InGaAs.

extrinsic impurity effects (Bhat and Miller 1987). Furthermore, as discussed later, the absorption or excitation spectra, which are more representative of the intrinsic processes, typically show considerably larger excitonic linewidths.

The energy shift due to the quantum size effect in GaInAs/InP heterostructures has been observed in photoluminescence by a number of workers. A summary of such data is presented in Fig. 11. There is rather poor agreement between the experiment and theory, and the data among various workers does not appear to be internally consistent. Many attempts have been made to reconcile the PL results. These included reinterpretation of the conduction–valence band partition of the band gap, in contradiction to the results of most transport experiments (Sauer et al. 1986), and an ad-hoc inclusion of lattice mismatch strain (Skolnick et al. 1987). In our opinion this discrepancy is caused by the sensitivity of PL measurements to impurity and potential fluctuation related recombination, in which the emission energy is lower than that of the free exciton. We cannot also rule out errors in layer thickness determination and for the thinnest of the layers, to compositional errors that may result from a small amount of interdiffusion with the barrier.

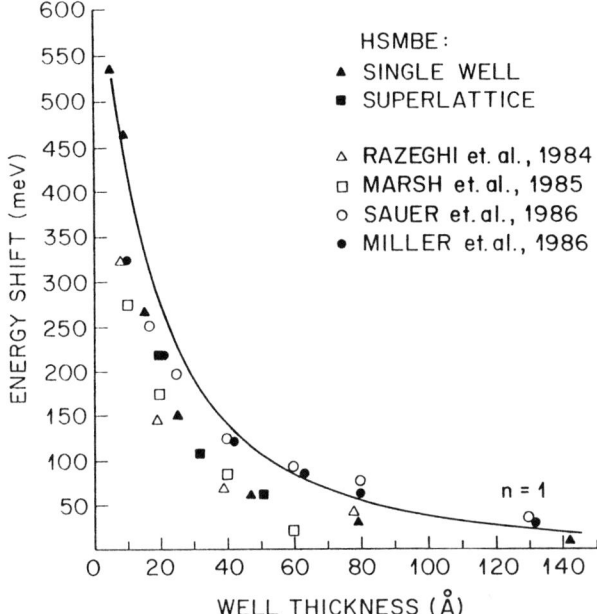

FIG. 11. Confined particle energy shift measured by low temperature PL are plotted as a function of the ternary well width. The solid line is calculated assuming 60% of the band-gap discontinuity lies in the valence band. Data points labeled ▲ and ■ are taken from Panish *et al.* (1986) and Gershoni and Temkin (1989).

A more fruitful approach to the examination of the confined particle energy levels is the study of absorption, either directly or by PLE or photocurrent excitation (PCE) spectroscopy.

The low-temperature (~ 10K) PLE spectra of the four quantum well sample previously discussed are shown in Fig. 12. The PL spectrum of this sample, obtained with a HeNe laser, was shown in Fig. 6. The PLE experiments were carried out by Gershoni *et al.* (1988b) using tungsten lamp light dispersed through a 0.64 m spectrometer as a continuous excitation source. The PL spectrum of the sample, obtained here with the lamp source and shown in the lower trace of Fig. 12, is of course dominated by $n = 1$ exciton recombination. The PLE spectra of individual quantum wells are shown displaced one above the other. Excellent luminescence efficiency of the material permits observation of the PLE spectrum even from a 9 Å thick well. In addition to the 1H exciton, all the wells also show well resolved 1L exciton transitions. The 1L level of the narrowest well is, however, quite weak. The 1H exciton energies measured by PLE are all at higher energies

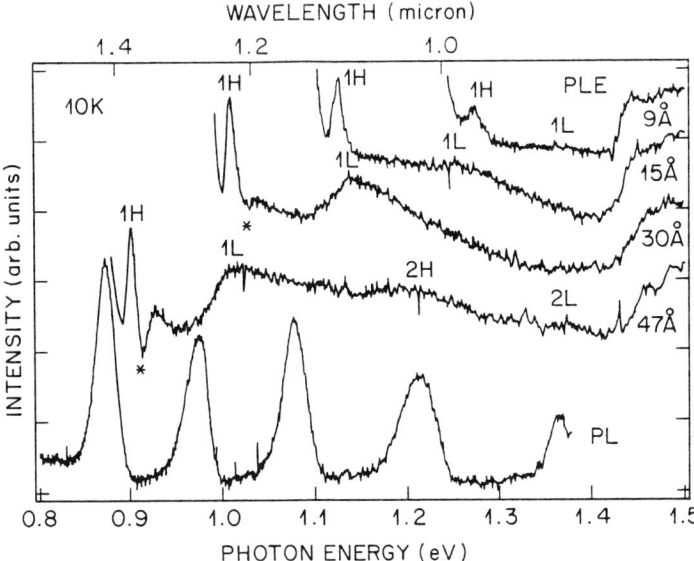

FIG. 12. PL and PLE spectra of a sample consisting of four single quantum wells. The narrowest well was 9 Å wide. Stars indicate dips discussed in the text.

than the corresponding PL peaks. The blue shift of the PLE levels increases with decreasing well thickness. This resembles the behavior of excitons bound to impurities in quantum wells, as was previously seen by Miller et al. (1982) in GaAs wells and calculated by Bastard (1981a). Following the 1H exciton of the 47 Å wide well there is a dip in the PLE response. This dip is up-shifted by ~10 meV from the exciton line. A similar feature appears to be developing above the 1H transition of the 30 Å well. We interpret the dips as the onset of the continuum for a hydrogenlike 2D heavy hole exciton. Following Miller et al. (1981) an exciton binding energy of ~10 meV can be estimated from this shift. This observation agrees with calculations that predict maximum binding energy for these quantum well dimensions. Smaller binding energies (~6 meV) make it difficult to resolve this feature in narrower wells.

Before the discussion of optical properties of multi-quantum well structures the question of coupling between the neighboring quantum wells should be addressed. Experimentally, this problem has been studied by Sauer et al. (1988). Each of their MOMBE grown samples consisted of two GaInAs quantum wells, 60 and 100 Å thick, separated by a thin InP barrier. At room temperature, exciton emission characteristic of individual quantum wells could be observed even in a sample with a barrier as thin as 20 Å. As the ambient temperature was lowered to the cryogenic level the PL signal originating in a narrower well gradually weakened in the 60 Å thick barrier. This was interpreted in terms of depopulation of the narrower well through the exciton tunneling into the wider adjacent well, as schematically shown in Fig. 13. A more detailed study of the temperature dependence of the PL intensity ratio revealed an Arrhenius-type behavior with an activation energy of $\delta E_{th} = 30.0$ meV. This activation energy is not consistent with an independent electron or hole tunneling between wells. Instead, good agreement with the experimental value is obtained assuming a combined electron–hole, or exciton, tunneling. A multi-quantum well structure with the barriers in this thickness range could be then defined as an optical superlattice in which excitons are not strictly bound to individual wells. For a similar two-well sample with a 100 Å barrier the PL intensity ratio was found to vary less than a factor of two in the temperature range of 300K to 4.2K. The wells separated by barriers with a thickness on the order of 100 Å can therefore be considered uncoupled or independent of each other. This is due to vastly different electron and hole emission rates. While such a

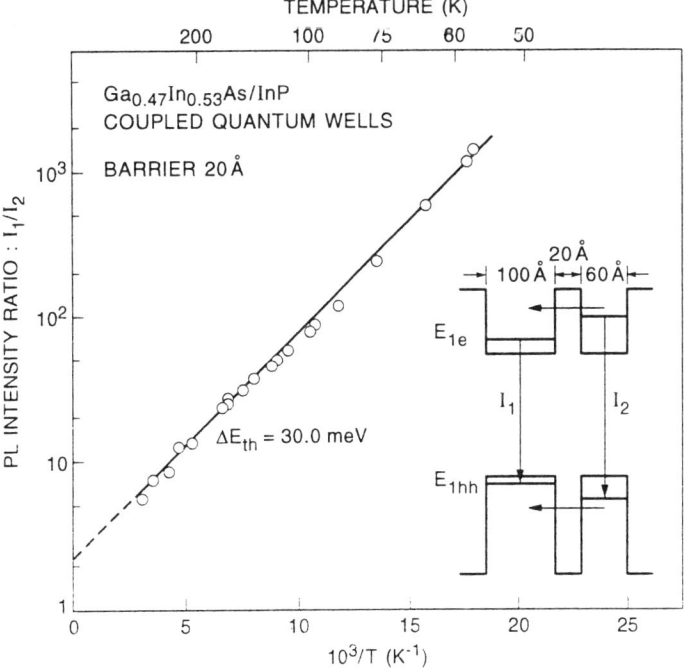

FIG. 13. Temperature dependence of the relative PL efficiency collected from a two-well sample with a 20 Å thick barrier (after Sauer et al. 1988).

collection of wells introduces a new spatial periodicity, its optical properties are essentially identical to those of a single well. Nevertheless, such structures are also commonly referred to as superlattices, even in the context of optical properties.

The typical results of PL and absorption measurements of a multi-quantum well structures are illustrated in Fig. 14. The heterostructure had 20 periods, with each period consisting of an 80 Å quantum well of $Ga_xIn_{1-x}As_{1-y}P_y$, with x and y adjusted to yield bulk material with a room temperature band gap of 0.8 eV and a 150 Å InP barrier. This was, in fact,

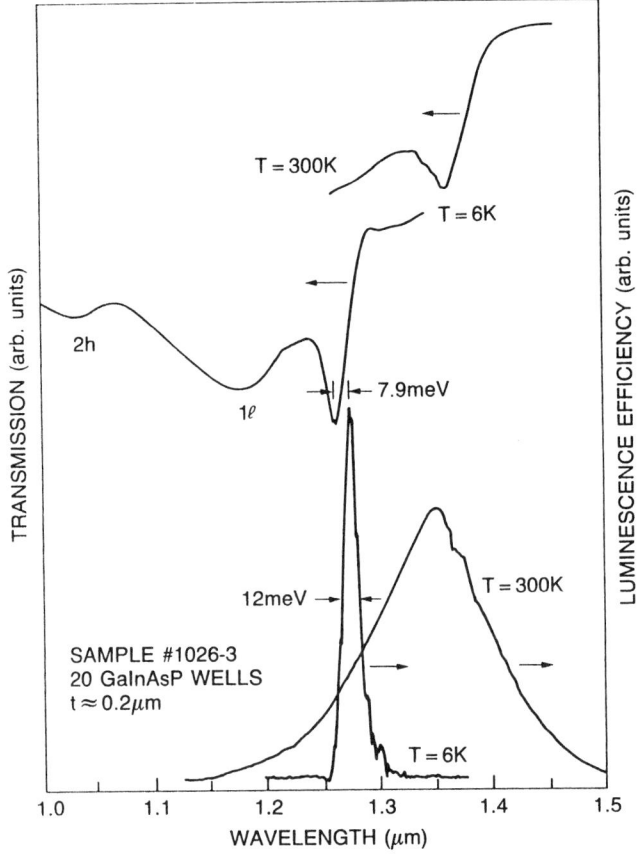

FIG. 14. Photoluminescence and absorption spectra obtained on a 20-well quaternary superlattice at 6K and 300K. Composition of the quantum well corresponds to a room temperature band gap of 0.79 eV. Notice the slight shift between the $n = 1$ exciton energies measured by PL and absorption.

the first superlattice grown by GSMBE (Temkin et al. 1985b) and the first superlattice reported for the GaInAsP system. The 6K PL shows the usual $n = 1$ electron–heavy hole transition. The PL is very intense and comparable in efficiency to ternary structures. The exciton linewidth of 12 meV is equal to that observed in single quantum wells of similar thickness grown using the same procedures. This indicates excellent uniformity of dimensions and well composition in each period. This result is typical of the luminescence from quaternary and ternary superlattice heterostructures grown by HSMBE. Although broadened of course, the luminescence is readily observed at and well above room temperature. Similarly, the sharp excitonic absorption edge is also readily observable at temperatures of 100°C and beyond. Higher order exciton levels are also clearly resolved in the absorption spectrum of this sample. The 8 meV red shift of the PL peak with respect to the absorption measurement, indicative of the nonintrinsic contribution to the exciton emission, is similar to that found in the single well ternary samples discussed previously.

The optical properties of a ternary structure consisting of 10 GaInAs wells, each 79 Å thick, separated by 460 Å InP layers are shown in Fig. 15. The quantum wells are confined within the intrinsic layer of a *p-i-n* structure

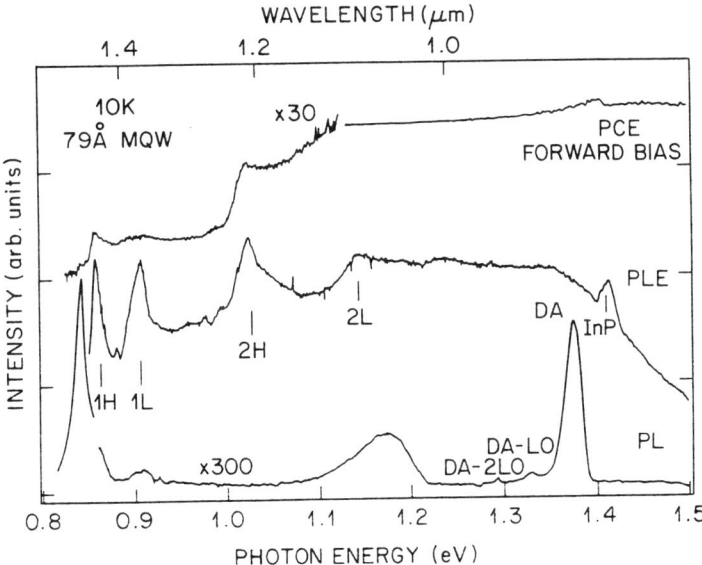

FIG. 15. Low-temperature PL, PLE, and PCE spectra of a 10-well GaInAs superlattice grown in a *p-i-n* configuration. Large Stokes shift is observed in the PL spectrum of the $n = 1$ exciton.

to facilitate studies of PL, PLE, and PCE as a function of bias. The layer dimensions were determined to within a monolayer using high-resolution X-ray diffraction. The PL spectrum is dominated by a single line at 0.843 eV, which is due to the excitonic recombination of electron and heavy hole from the $n = 1$ subband level (1H). Another line, of less than three orders of magnitude in intensity, is seen 63 meV higher in energy. This line is due to the recombination of $n = 1$ electron with a light hole (1L). The intensity ratio between the two luminescence lines reflects the ratio of the lifetime of the carriers, approximately 2 nsec (Feldman et al. 1987), to the thermalization time of holes to the lowest energy state. This latter process is thus estimated to occur in about 1 psec. The additional PL features, a broad line between 1.1 and 1.2 eV and a triplet of sharp lines at 1.373, 1.335, and 1.297 eV originate in InP. The broad line is one of many such features in this spectral region and associated with complexes of stoichiometric defects and transition-metal impurities (Temkin et al. 1981a). The three high energy lines are due to donor-acceptor (Be) pairs and their two LO-phonon replicas. The PLE spectrum shows very well resolved lines due to 1H, 1L, 2H, and 2L transitions at energies of 0.858, 0.905, 1.028, and 1.134 eV, respectively. These peaks are also observed in the PCE spectrum, which was taken under a slight forward bias to eliminate Stark shifts of the exciton peaks. The 1H line in the PL spectrum is also red shifted, here by ~ 15 meV, relative to the transition energy observed in the PLE and PCE experiments. In contrast, the 1L line exhibits no shift. We argue that the red shift of the 1H level is due to recombination processes that involve exciton binding by potential fluctuations in the quantum well. These may arise either from composition fluctuations or interface roughness. At low temperatures excitons are rapidly thermalized to localized states of lower energy associated with these potential fluctuations. The absence of such a red shift in the 1L transition indicates that the time for exciton thermalization to a state associated with the potential fluctuation is longer than the hole thermalization time to the lowest energy state. This is in agreement with the measured localized exciton hopping time of 16 psec (Tai et al. 1988). Radiative recombination of the 1H excitons bound to those final states gives rise both to the observed red shift and to the narrower PL linewidth (12 meV) compared to the width of the same line measured in absorption (14 meV). A simplified diagram of these transitions and proposed recombination lifetimes are shown in Fig. 16.

The linewidth obtained in absorption measurements is indicative of the real density of the exciton states, and it is a better measure of the sample quality as pointed out previously by Miller and Bhat (1988). The absorption linewidth at low temperatures (in the absence of phonon broadening) is mainly due to inhomogeneous broadening due to the spatial potential

5. OPTICAL PROPERTIES OF $Ga_{1-x}In_xAs/InP$ QUANTUM WELLS

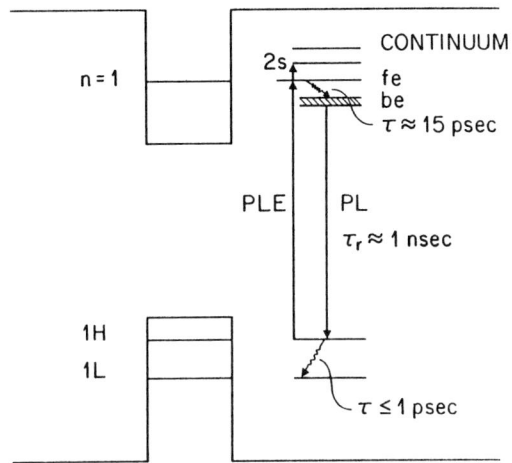

FIG. 16. A simplified diagram of the optical transitions and recombination lifetimes in a GaInAs well.

fluctuations. Excitons bound to different potential fluctuations where the radiative recombination takes place are the reason for the red shift of the PL line, similar in magnitude to the absorption linewidth. The shift is more sensitive to the well dimensions than in the case of exciton bound to impurities. This is because the exciton radius decreases with the quantum well width, which in turn reduces the volume it samples. The average over a smaller volume results in deeper potential fluctuations and thus a larger binding energy. As discussed later, the 1H energies measured by PLE are in good agreement, in contrast to those measured by PL, with the calculated values. The role of potential fluctuations in determining the optical properties of GaInAs/InP heterostructures should be studied further in the ordered alloy quantum well, once their structural properties improve.

The optical transition energies measured at low temperatures by absorption techniques in the samples shown in Figs. 12 and 15, together with the data obtained on two other samples with wider quantum wells (142 Å and 189 Å), are plotted in Fig. 17 as a function of the quantum well width. The transition energies were calculated by numerically finding the roots of Eqs. (2) and (3) (Gershoni et al. 1988b). The dispersion relations that correlate the wavevector k to the particle energy E are then

$$E = \hbar^2 k_w^2 / 2m_w^*(E) \qquad (10)$$

and

$$E - V_0 = \hbar^2 k_b^2 / 2m_b^*(E) \qquad (11)$$

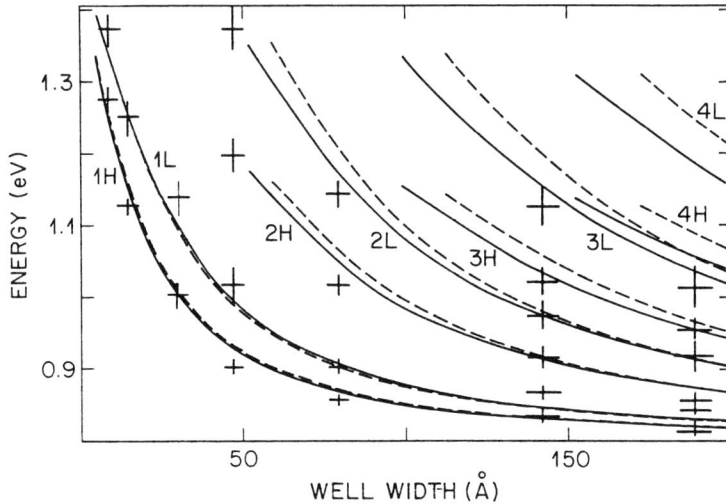

FIG. 17. A summary of low-temperature exciton energies plotted for GaInAs wells with the thicknesses ranging from 9 to 200 Å. Solid lines were calculated taking into account band nonparabolicity, dashed lines neglect this effect.

where V_0 is the barrier potential, or band discontinuity. The effects of band nonparabolicity were introduced through the energy-dependent effective masses of electrons and light holes:

$$m(E) = \hbar^2 k^2(E)/2E \qquad (12)$$

where $k(E)$ is given implicitly by the Kane (1957) model dispersion relation:

$$(E')(E' - E_g)(E' + \Delta_{so}) - k^2 p^2(E' + \tfrac{2}{3}\Delta_{so}) = 0 \qquad (13)$$

Here $E' = E + E_B - \hbar^2 k^2/2m_0$, where E_B is the relevant band energy, i.e., $E_g(0)$ for electron (light hole), E_g and Δ_{so} are the material band gap and split off energies, and P is the $k \cdot p$ matrix element. The last was chosen so that $m(E = 0)$ is given by the known band edge effective masses listed in Table I. Different P values are used for the electron and light hole (Blakemore 1982). In the barriers the energy E is replaced by $E - V_b$. Thus, the nonparabolicity is calculated from the $k \cdot p$ theory and is not a free parameter. The dashed lines in Fig. 17 represent calculated transition energies neglecting the nonparabolicity effects. These are included in the calculation represented by the solid lines. The inclusion of nonparabolicity has a negligible effect on the calculated energy of the lowest order excitonic

TABLE I

Name	InP	GaAs	InAs	$Ga_{0.47}In_{0.53}As$
E_g(eV)	1.424[a]	1.519[a]	0.418[a]	0.812[d]
Δ_{so}(eV)	0.11[a]	0.34[a]	0.37[a]	0.356[e]
m^{el}(emu)	0.079[a]	0.067[a]	0.023[a]	0.041[f,e]
m_z^{lh}(emu)	0.121[b]	0.094[e]	0.027[b]	0.056[f,e]
m_z^{hh}(emu)	0.606[b]	0.341[b]	0.4[b]	0.377[f,e]
Valence band offset relative to InP (eV)	0	0.34[g]	0.41[g]	0.37[h,e]

[a]Landolt-Bornstein (1982).

[b]We use $m_z^{lh} = \dfrac{1}{\gamma 1 + 2\gamma 2}$ and $m_z^{lh} = \dfrac{1}{\gamma 1 - 2\gamma 2}$ and Luttinger parameters $\gamma 1$, $\gamma 2$ from note a.

[c]Nelson et al. (1987).
[d]Nicholas et al. (1979); Pearsall (1982).
[e]Linearly interpolated from the binary compounds data.
[f]Alavi and Aggrawal (1980), in very close agreement with note e.
[g]Bauer and Margaritondo (1987).
[h]Forrest et al. (1984).

transitions. However, a considerable effect is seen in the higher order transitions. For these, the inclusion of nonparabolicity results in a better agreement with the data. The calculation also includes the exciton binding energy as estimated earlier. However, other small effects, such as the background doping in the wells (Moss–Brustein shift) and the binding energy dependence on well size are neglected. From an experimental point of view these corrections are at most comparable to the linewidth of the observed optical transitions in this material system.

The GaInAs and GaInAsP quantum wells are used in many optoelectronic devices that depend on excitonic transitions for their optical characteristics. It is, therefore, important to consider room temperature optical properties of quantum well based structures. Figure 18 shows the room temperature photoresponse spectra of avalanche photodiodes grown by HSMBE (Temkin et al. 1985a, 1986) in which the absorbing layer consists of multi-quantum well structures. The three curves were obtained for diodes having $Ga_{0.47}In_{0.53}As$ wells 51, 32, and 20 Å wide. The spectra were obtained under a low reverse bias of -1 V. Transitions between the electron and heavy and light hole levels are clearly resolved for all three structures. Similar quality spectra can be obtained at elevated temperatures, at least up to 100°C. Excellent optical properties of GaInAs superlattices grown by

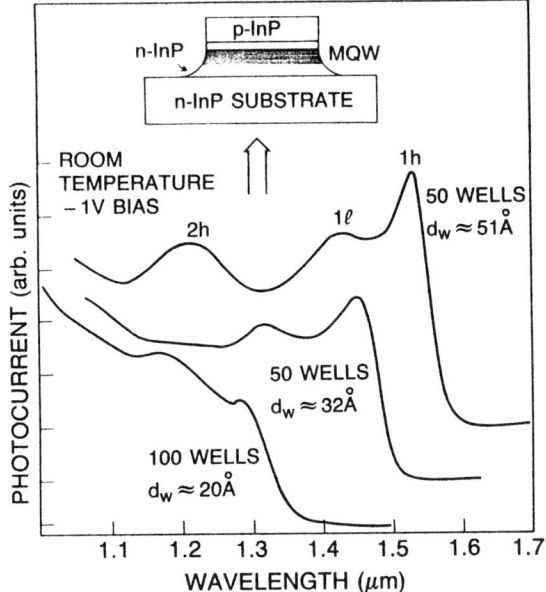

FIG. 18. Room temperature photocurrent spectra of three superlattices with different well thicknesses (in different samples). Well resolved exciton transitions remain visible even at temperatures in excess of 100°C.

Tsang et al. (1987) by MOMBE were also reported well above room temperature.

The transition energies for the lowest, $n = 1$ and 2, order excitons are plotted in Fig. 19 as a function of temperature. The data were obtained by measuring the absorption spectrum of a 50-well sample. Each $Ga_{0.47}In_{0.53}As$ well was approximately 75 Å thick. In the temperature range of 6–400K the energy shift can be described by the Varshni equation:

$$E = E_0 - \alpha T^2/(\beta + T) \qquad (14)$$

where E_0 is the low temperature value of the band gap and the coefficients $\alpha = 4.9 \times 10^{-4}$ eV/K and $\beta = 327$K. The calculated temperature shift is plotted in Fig. 19 as solid lines. It is important to note that the excitonic energy levels exhibit a temperature dependence identical to that of the bulk band gap of InP and $Ga_{0.47}In_{0.53}As$ (Temkin et al. 1981b). Similar results were also reported by Kawaguchi and Asahi (1987).

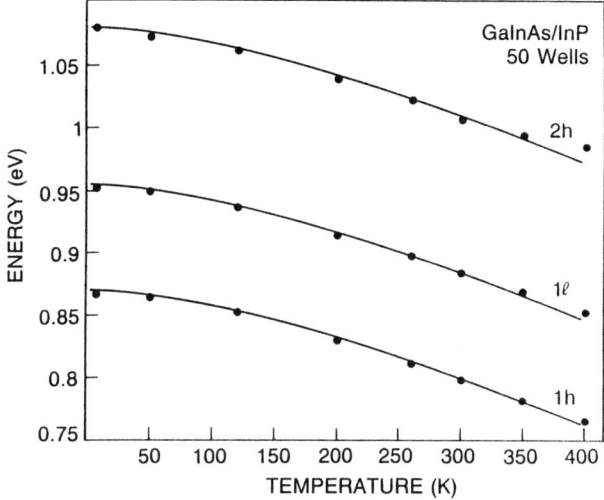

FIG. 19. Energy shift with temperature for the three lowest order excitons confined in GaInAs wells. Solid lines are calculated from the known temperature dependence of the GaInAs and InP band gap.

IV. Dynamic Effects

A deeper understanding of optical properties of quantum wells is obtained when the time evolution of the luminescence and absorption spectra is considered. The detailed knowledge of carrier lifetime is important since it permits quantitative comparisons of samples grown under different conditions and by different techniques. The dynamics of carrier recombination in GaAs/GaAlAs quantum wells has been studied extensively (Hegarty and Sturge 1985). Theoretical models relate radiative lifetime, and the related oscillator strength, to the width of the quantum well (Feldman et al. 1987). It is then possible to study 2D excitons in a systematic fashion. The investigation of the carrier lifetime dependence on the well width in GaInAs/InP wells was reported (Cebulla et al. 1989; Ritter et al. 1990; Brenner et al. 1991).

The time-resolved luminescence spectra obtained on a sample containing wells of GaInAs with the widths of 10, 40, 80, and 150 Å are shown in Fig. 20. The wells were separated by the 400 Å thick barriers of InP. Experimental traces are shifted vertically to provide clear display of the data. The luminescence intensity was found not to depend on the well width, in agreement with the data of Figs. 6 and 12. All the wells showed exponential

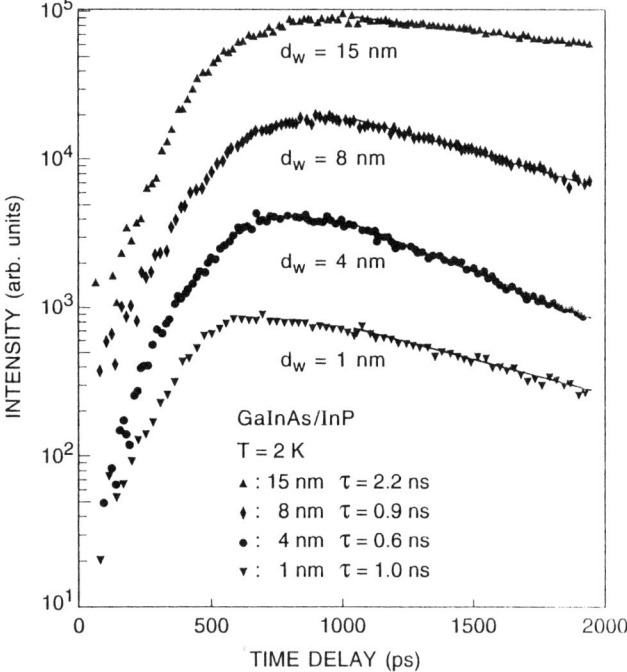

FIG. 20. Time resolved PL measurements carried out on quantum wells of different widths (After Cebulla et al. 1989).

decay of the luminescence intensity characterized by a single time constant. Lifetimes of 1.0, 0.6, 0.9, and 2.2 nsec were measured for increasing well width. The data of Cebulla et al. (1989), obtained on a set of wells with a wide range of thicknesses, showed a systematic increase in the exciton lifetime with the well width. The exciton lifetimes in two- and three-dimensional systems can be written (Feldman et al. 1987), respectively, as

$$\tau^{2D} \propto \frac{1}{M_{op}^2} \tag{15}$$

and

$$\tau^{3D} \propto (V/V_w) \frac{1}{M_{op}^2} \tag{16}$$

where M_{op} is the oscillator strength defined in Eq. (4), V is the exciton volume, and V_w is the volume of the well. The transition from a 3D to 2D system, for well thicknesses progressively smaller than the exciton radius, thus results in shorter exciton lifetime in narrower wells. This is in good agreement with the experimental results shown in Fig. 20. This data set also suggests an increased lifetime in very narrow wells of less than 10 Å.

In a purely 2D system the lifetime is inversely proportional to the square of the overlap integral, which decreases rapidly for thinner wells. This effect might lead to longer lifetime in narrow wells of GaInAs, in which the conduction band offset is relatively shallow and the electron wave function spreads into the barrier for well thicknesses below 50 Å. The effective width of the well, the width of the electron wave function, is then much larger than the physical width. For the same range of well thicknesses the hole wave functions remain largely confined to the valence band wells. This is a result of larger valence band offset and larger effective mass of the holes. A detailed theory was discussed by Feldman et al. (1987), who reported a systematic increase in the exciton lifetime in wells of GaAs/GaAlAs. Similar experiments on GaInAs/InP wells were carried out by Ritter et al. (1990). Lifetimes of 4.1, 3.6, 4.0, and 2.6 nsec were measured for well widths of 14, 27, 40, and 60 Å, respectively. In this set of samples, the lifetimes of narrower wells were largely independent of the well thickness, as expected for the wells in this width range.

V. Quantum Wires and Boxes

The physics of spatially quantized systems has been the subject of intense interest since the preparation of atomically sharp heterojunction interfaces was made possible by the development of molecular beam epitaxy. Optical properties of the two-dimensional quantum wells have been studied since the first work by Dingle et al. (1974) two decades ago. The high-quality quantum wells and superlattices made available by advances in the precision epitaxial growth have sparked the more recent studies of structures of even lower dimensionality, such as quantum wires and boxes. In principle, systems of lower dimensionality can be formed from such two-dimensional structures by lithographically patterning the layer to transverse dimensions comparable to the carrier wavelength. This dimension is typically $\sim 0.02\,\mu m$, which is near the present limits of electron beam lithography. However, the patterning must not degrade the layer's quality. Such degradation may occur as a result of processing induced defects and enhanced recombination rate expected at layer side walls. In very small structures any increase in the surface recombination rate will dominate the structure's behavior since the surface to volume ratio becomes very large (Temkin et al. 1987a). In this respect GaInAs/InP is a much better material system than GaAs/AlGaAs since its surface recombination velocity is about two orders of magnitude lower (Casey and Buehler 1977).

The physical reason for changes in optical and electrical characteristics of low dimensional structures lies in a markedly different nature of their density of states. It is well known that upon a reduction from three to two dimensions (2D) the density of states $\rho(E)$ of carriers with parabolic dispersion changes from a continuous $\rho(E) \sim E^{1/2}$ to a steplike dependence, shown in Fig. 21. This direct consequence of the quantization of motion in one direction gives rise to a characteristic steplike absorption and a greatly enhanced exciton intensity in quantum well superlattices. Similar considerations show that in a 1D system (quantum wire) the density of states is characterized by single steps followed by a $\rho(E) \sim E^{-1/2}$ dependence. In a 0D system (quantum box or quantum dot), where the particle motion is fully quantized, $\rho(E)$ is reduced to a set of delta functions. The optical properties of lower dimensionality systems have been first studied experimentally in CuCl and CdS microcrystals suspended in glass matrices by Ekimov and Onushchenko (1982) and Brus (1986). They have observed large increases in the absorption edge energy due to three-dimensional confinement. Early transport studies have been also made on very small Si structures by

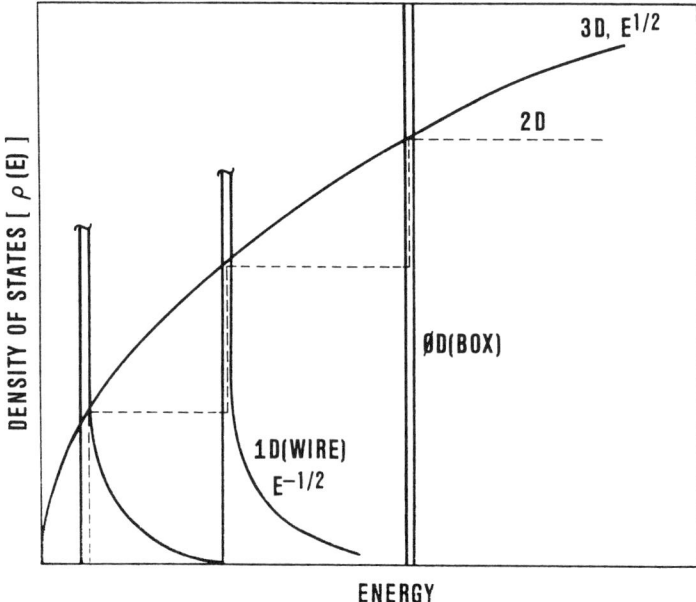

FIG. 21. Schematic representation of the density of states for quantum confined structures with reduced dimensionality.

5. OPTICAL PROPERTIES OF $Ga_{1-x}In_xAs/InP$ QUANTUM WELLS

Skocpol et al. (1982). Significant improvement in the performance of quantum wire semiconductor lasers have been predicted by Asada et al. (1985).

With III–V compounds, research in this field has concentrated mainly on the GaAs/AlGaAs material system. Quantum wires and boxes with dimensions upward of 500 Å have been fabricated by a number of techniques and studied by spatially and spectrally resolved low-temperature cathodoluminescence by Petroff et al. (1982) and Cibert et al. (1986). PL studies of quantum boxes were carried out by Reed et al. (1986). As mentioned previously the GaAs system suffers from a relatively large surface recombination velocity and in order to fabricate structures with efficient PL, very sophisticated processes must be employed. Cibert et al. (1986) observed 2D and 3D confinement in structures prepared by quantum well disordering and hence without the vacuum interface.

A number of specialized crystal growth processes are being investigated specifically for the preparation of very small structures. These include tilted and serpentine superlattices (Tsuchiya et al. 1989b; Miller et al. 1992) formed by the deposition of a fraction of a monolayer of GaAs/AlGaAs at a time and the growth of quantum wells on patterned substrates (Kapon et al. 1989). These novel approaches have already produced high-quality arrays of very fine quantum wires with interesting optical properties (Colas et al. 1990; Pryor 1991) and quantum wire based semiconductor lasers (Tsuchiya et al. 1989a; Kapon et al. 1989). Quantum confinement to lateral dimensions as narrow as 70 Å, which resulted in very pronounced energy shifts of excitonic transitions and drastic changes of the polarization selection rules, were reported recently by Gershoni et al. (1990b). In these studies the lateral confinement was achieved by strain modulation in the quantum well plane. The strain modulation was achieved by the growth of GaAs/AlGAs quantum wells on the cleaved edge of an InGaAs/GaAs strained layer superlattice. In-situ processing methods combining epitaxy of GaInAs/InP and vacuum patterning techniques have produced buried quantum wires as narrow as 600 Å (Harriott et al. 1991). Another possibility of forming completely buried quantum wires, i.e., wires with a negligibly low surface recombination velocity, is offered by selective area epitaxy. This technique has produced GaInAs wires as narrow as 1000 Å with excellent optical properties at room temperature (Galeuchet et al. 1991; Cotta et al. 1992).

The most detailed investigations of the optical properties of quantum wires and boxes based on GaInAs/InP were performed on structures prepared by conventional techniques of planar crystal growth followed by patterning and etching. Two types of structures were grown for this purpose by HSMBE (Temkin et al. 1987; Gershoni et al. 1988b). The first one, most suitable for experiments on quantum boxes, consisted of a single 50 Å thick

$Ga_{0.47}In_{0.53}As$ quantum well grown on a $\sim 1\,\mu m$ thick InP buffer layer and capped with a 100 Å thick InP layer. The active layer of the second structure contained four $Ga_{0.47}In_{0.53}As$ quantum wells, each 50 Å thick, separated by 100 Å thick InP barriers. The quantum wells were grown on a 0.5 μm thick InP buffer layer, and a 1500 Å thick GaInAsP guide layer with a band gap corresponding to a wavelength of $\sim 1.3\,\mu m$ and capped with a 300 Å thick InP cladding layer. This asymmetric quaternary waveguide structure is designed to guide the light generated in the quantum wire. The layers were not intentionally doped, resulting in a n-type background carrier concentration at the mid $-10^{15}\,cm^{-3}$ level. Schematic cross sections of the structures used in the fabrication of these quantum wires and boxes are shown in Fig. 22.

Wires and round boxes were carved from the quantum wells using electron beam lithography and argon ion milling. The JEOL JBX-5D electron beam writing instrument used for this purpose is capable of creating openings as small as $0.02\,\mu m$ (200 Å). For such a pattern, a composite polymethyl methacrylate/metal film stencil (Dolan and Fulton 1983) was made to form metal wires and dots by the "lift-off" technique. The

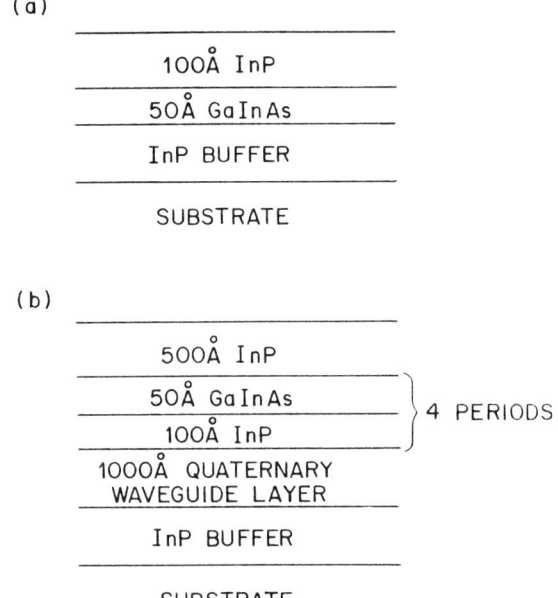

FIG. 22. Cross section of the quantum well structures used in the fabrication of quantum wires and boxes.

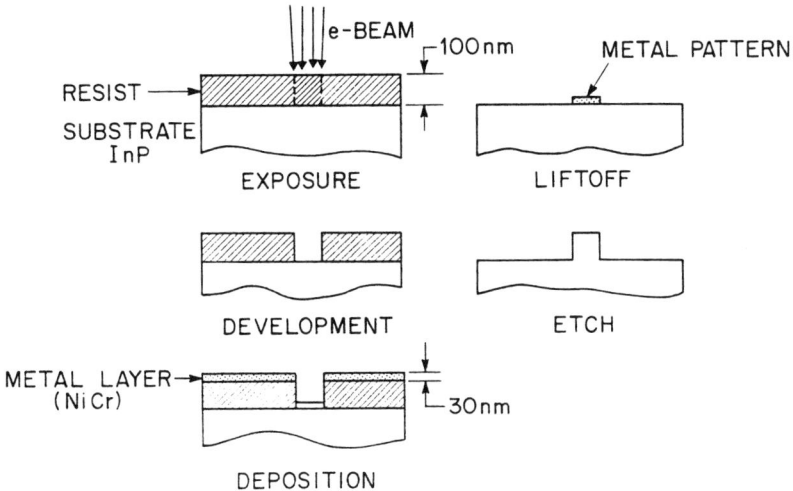

FIG. 23. Simplified diagram of the lift-off process used in the fabrication of nanostructures.

fabrication sequence is illustrated in Fig. 23; this process has been also discussed in detail by Scherer and Craighead (1986). It should be pointed out that the lift-off process works best when the thickness of the metal mask is approximately equal to or smaller than the transverse feature dimensions. Since the metal mask tends to be eroded in the subsequent transfer by ion milling the etched features are also limited, to within factors of two, to a similar height.

The metal patterns were then used as masks for ion milling in a mixture of argon and oxygen. The ion milling removed ~ 30–90 nm from unmasked regions of the sample leaving the desired lines and dots of material. We have used Cr masks with a high energy, ~ 500 V, milling process on single quantum well samples. These metal masks are believed to be eroded by the ion milling and cannot be completely removed after the fabrication process. Another variant of the dry etching process employed a NiCr metal mask. In this sample the mask pattern was transferred to the quantum well structure by milling with a 100 V Ar ion beam. The low energy ion beam does not appear to attack the metal mask, which is subsequently dissolved by wet etching. The size and shape of the islands of material finally obtained depend on all of the fabrication elements, including the resist openings, the grain structure of the masking metal, and the erosive properties of the ion milling process with the mask used. All of these factors involve uncertainties and nonuniformities at the 50–100 Å level so that the size of the boxes, for example, is expected to vary by such amounts. Optical properties of these samples, as discussed later, are affected by details of the fabrication process, much in the same way properties of the quantum well change with the subtle alterations of the crystal growth.

The TEM image of quantum boxes fabricated by the high-energy milling process is shown in Fig. 24. The micrograph shows a top view of the (100) surface with the rows of boxes aligned parallel to the cleavage directions. Under high magnification the boxes appear as round spots of somewhat different shapes with the average diameter of 300 ($\pm$ 50) Å. Since the ion milling transfer does not produce undercutting, this image reveals the true size of the quantum boxes. To put these sizes in perspective, each box is estimated to contain only 1.4×10^5 atoms. With the average box separation of ~ 2000 Å about 1.7% of the surface area (filling factor) is taken by the boxes. Quantum wires of similar dimensions, down to a width of 300 Å, have been fabricated with a filling factor of 15%.

Low-temperature PL experiments were carried out with a low power, <1 mW, unpolarized HeNe laser operating at 632.8 nm. The exciting photon energy is larger than the band gap of InP and PL is expected to result from the capture of photocarriers generated in InP by the smaller band-gap GaInAs wells. This capture process is very efficient and luminescence from the InP barriers is not observed, as already discussed. The

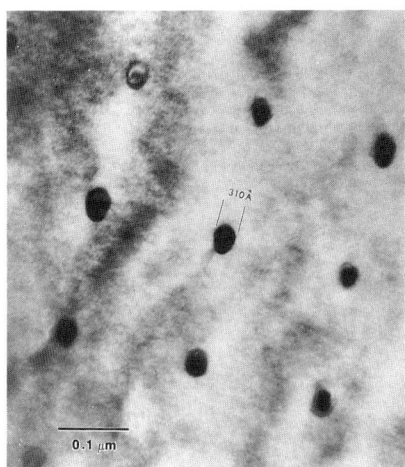
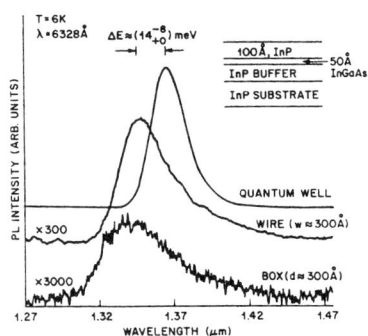

FIG. 24. Transmission electron micrograph of the GaInAs/InP quantum boxes and low-temperature PL spectra of 300 Å wide boxes and wires. PL spectrum of the control sample is shown for comparison.

sample areas covered with quantum wires, which form efficient diffraction gratings, were found by observing white light diffraction.

The resulting PL spectra are shown in Fig. 24. The upper trace shows a spectrum typical of the unprocessed quantum well structure. No changes were observed in the part of the control sample previously covered by the Cr mask. The lower two traces show the spectra of wires and boxes. All three traces were obtained at the same incident power and angle of incidence. The PL efficiency is high and the spectra of wires and dots could be observed at temperatures up to 60K. The exciton wavelength of 1.37 μm is consistent with the 50 Å well width. The PL spectra of the wires and dots are found shifted to shorter energies by as much as 14 meV. The linewidth increase, particularly in the case of quantum boxes, is expected in view of a fairly large distribution of sizes shown in Fig. 24, although we cannot exclude other causes. Prior to processing, the PL response of the quantum well sample was carefully mapped. Neither the exciton wavelength nor its linewidth and intensity were found to vary from position to position over the 1 cm^2 sample. Significant changes in the PL spectrum were found only as a function of the incident power, as expected from band-filling effects. Attenuation of the incident power by a factor of 4×10^3 results in the exciton peak shifting towards longer wavelength by nearly 8 meV. Thus as much as 8 meV of the observed energy up-shift may have to be attributed, in this sample, to effects other than the low dimensional confinement. A similar blue shift of about 11 meV, relative to the PL energy of the control sample, is observed in the quantum wires fabricated from the asymmetric waveguide structure.(The shift is consistent with transverse dimension of the quantum wire. In this sample wavelength of the PL emission did not change even when the excitation power density was varied by five orders of magnitude, thus precluding band-filling effects. The 2D confinement in disordered GaAs/AlGaAs wells is obtained with a lateral confinement energy of only 60 meV. This leads to a large number of very closely spaced exciton levels (Cibert et al. 1986). The lateral confinement in the present structure is formed by an infinitely high potential barrier with a much larger level spacing. Thus the higher order levels are effectively depopulated and only a single exciton peak is expected in the PL spectrum. Using the InGaAs effective mass of $m_e^* = 0.041\, m_e$ (Nicolas et al. 1979) an additional energy shift of a 10 meV due to lateral confinement is expected for the 300 Å wide quantum wire. In view of the experimental uncertainties discussed earlier, the agreement with the observed exciton shift is excellent.

Another feature of the PL spectra of control sample, quantum wires, and dots fabricated with a high energy Ar beam, shown in Fig. 24, is the luminescence efficiency that does not scale with the surface filling factors. This could be attributed either to nonradiative recombination at the sidewall of small structures fabricated by ion beam milling or to shadowing effects of the incompletely removed metal mask used in the fabrication process. The two effects can be distinguished by the measurement of the luminescence efficiency as a function of the incident laser power, shown in Fig. 25. Reduction of the incident intensity by as much as a factor of 4000, from the full power of only ~1 mW, results in a smooth decrease in the integrated luminescence efficiency of the quantum well sample. In particular the saturation of the PL efficiency due to nonradiative centers was not observed down to the lowest laser intensities. The PL efficiency of the wires, for which fairly complete data could be obtained, was found to track the data obtained on the unprocessed control sample. The intensity ratio for wires compared to dots is furthermore close to that implied by the filling factors, but does not agree with the filling factor when compared to the control sample, suggesting shadowing effects of the metal mask. This is confirmed by the results obtained on quantum wires fabricated by a low-energy ion transfer process, which show the PL intensity ratio of about 1:10, approximately equal to the area ratio. This confirms our conjecture of negligibly low density of nonradiative recombination centers in the side

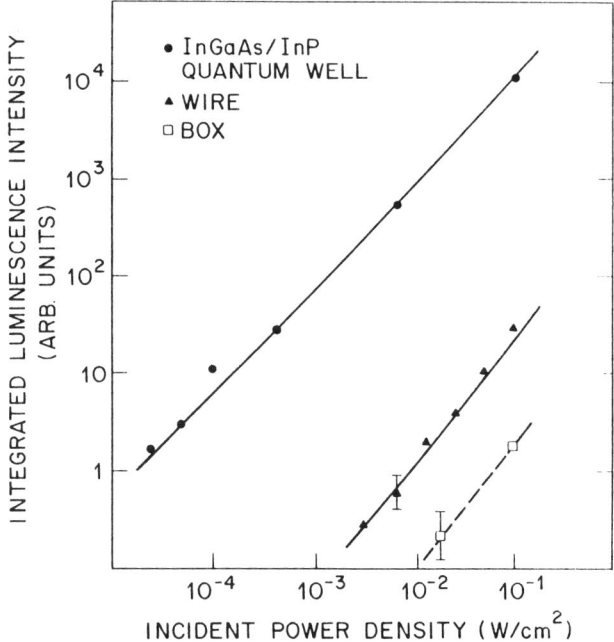

FIG. 25. PL efficiency of the quantum wires and dots, and the control sample, measured as a function of the incident power.

walls of Ar-milled GaInAs/InP quantum wires and boxes (Temkin et al. 1987).

In addition to the PL energy shift, strong evidence for the lateral confinement is provided by the PLE spectrum of the quantum wire sample. This experiment carried out by Gershoni et al. (1988b) used dispersed light of a tungsten lamp as the continuously tunable excitation source. The weakness of the excitation source required samples with excellent optical properties. The PLE spectrum of the asymmetric waveguide control sample shown in Fig. 26 displays well defined $n = 1$ electron–heavy and light hole transitions, in excellent agreement with the energies calculated for a 50 Å wide ternary quantum well. The InP band edge is also clearly seen at energy of 1.415 eV. The PLE spectrum of the quantum wire, fabricated by a low energy Ar-ion milling, clearly shows two additional excited states labeled 1,2H and 1,3H at 0.934 eV and 0.963 eV, respectively. These can be assigned to the excitonic transitions associated with the lateral confinement in the y-direction, $n_y = 2$ and $n_y = 3$. Both transitions belong to the single $n_z = 1$ electron state confined in the original quantum well. The PL peak that resulted from recombination of carriers from the $n_y = n_z = 1$ sublevels is red shifted (Stokes shift) by ~12 meV, similar to the shift in the control sample.

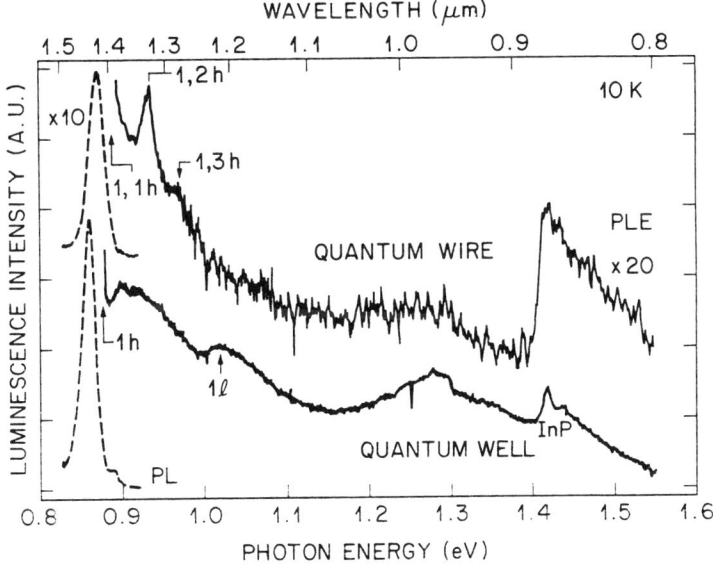

FIG. 26. Low-temperature PL (dashed lines) and PLE (solid lines) obtained on ~350 Å wide quantum wire structure fabricated by electron beam writing and low-energy Ar-ion milling.

To compare our experimental results with the calculated energies of excitons confined to the quantum wires, we have developed a new method of solving the two-dimensional (2D) Schrödinger equation. The solution is based on a mathematically similar problem of the 2D optical waveguide solved by Henry and Verbeek (1988). The method provides for a very efficient solution of 2D potential well problems of arbitrary shape and yields their energies as well as their eigenfunctions. In the absence of a general 2D calculation the standard practice of this field was to approximate the solution by successively solving the 1D problem in both directions (Cibert et al. 1986; Bayant 1987; Hirayama et al. 1988). This is acceptable only in few separable cases, such as the infinitely deep 2D well problem.

In the following, we describe the solution to a one-band 2D problem. Baraff and Gershoni (1991) have recently used a very similar method to solve a multiband problem as well. The single particle 2D effective mass problem is given by the Schrödinger equation, which we write in the following form:

$$\left[\frac{-\hbar^2}{2m_0}\left(\vec{\nabla}\frac{1}{m^*(z,y)}\vec{\nabla}\right) + V(z,y)\right]\Psi(z,y) = E\Psi(z,y) \quad (17)$$

$\Psi(z,y)$ is then expanded in terms of a complete orthonormal set of functions:

$$\Psi(z,y) = \sum_l a_l \Phi_l(z,y) \quad (18)$$

where $\Phi_l(z,y)$ are products of sine and cosine functions that go to zero at the planes $z = \pm L_z/2$ and $y = \pm L_y/2$. L_z and L_y can be chosen at will. Care should be taken to move the boundaries away from the quantum wire so that the eigenvalues are essentially independent of their choice. A typical example can be found in Fig. 27.

The function $\Phi_l(z,y)$ can be written as

$$\Phi_l(z,y) = u_m(z)u_n(y) \quad (19)$$

where

$$u_m(t) = \sqrt{\frac{2}{L_t}}\begin{cases} \sin\dfrac{m\pi t}{L_t}, & \text{for } m = 2, 4, 6,\ldots \\ \cos\dfrac{m\pi t}{L_t}, & \text{for } m = 1, 3, 5, 7,\ldots \end{cases} \quad (20)$$

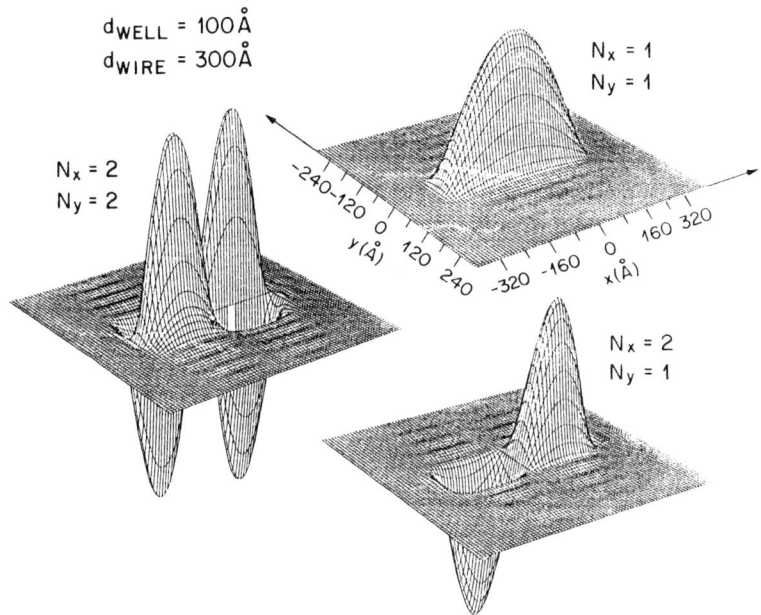

FIG. 27. Wave functions of the lowest energy states calculated wire based on a 100 Å thick ternary well with a transverse dimension of 300 Å.

and a simple one-to-one mapping between the subscripts (m, n) and l is applied. Substitution into Eq. (17), left multiplication by $\Phi_l(z, y)^*$, and integration, using the orthonormality condition, $\int \Phi_l^* \Phi_l \, dz \, dy = \delta_{ll'}$, allows for converting the partial differential equation to a matrix eigenvalue equation:

$$[M_{ll'} - E\delta_{ll'}]a_l = 0 \qquad (21)$$

where the matrix $M_{ll'}$ is given by

$$M_{ll'} = \frac{-\hbar^2}{2m_0} \left[\int \Phi_l \vec{\nabla} \left(\frac{1}{m^*}\right) \nabla \Phi_{l'} \, dz \, dy + \int \Phi_l \left(\frac{1}{m^*}\right) \nabla^2 \Phi_{l'} \, dz \, dy + \int \Phi_l V \Phi_{l'} \, dz \, dy \right]$$

(22)

The advantage of this approach is that if $V(z, y)$ and $m^*(z, y)$ have stairlike behavior (or can be approximated by such functions) all the integrals in eq.

(22) can be easily performed analytically and Eq. (17) is then solved using well-developed standard software tools such as EISPACK (Smith et al. 1976).

The method can be readily applied to the 3D confinement, quantum box problems as well as to the "conventional" 1D confined quantum well problem: the latter can be used as a check of the program and indeed complete agreement is obtained with the resonant tunneling method.

The three wave functions of a 300 Å wide quantum wire formed from a 100 Å single $Ga_{0.47}In_{0.53}As/InP$ quantum well are displayed in Fig. 27. The two lowest energy states are associated with the $n_z = 1$ quantum number and the $n_y = 1$ and 2. Another higher energy state associated with the $n_z = 2$, $n_y = 2$ quantum numbers is also shown. The problem was solved using 15 waves in each direction and taking advantage of the symmetry of the problem (thus eliminating the vanishing wave terms). The 225×225 matrix has thus 50625 elements to be computed. The solution of the problem consumes ~ 150 sec on a medium-speed workstation, considerably less than the time it takes to plot the wave functions! Another advantage of the method is the simple way it handles calculations of overlap integrals between electronic and hole states. One has only to solve the two single particle problems, using the same orthonormal basis set (i.e., using the same L_z, L_y, and number of waves) and the overlap integral is then given by

$$\langle \Psi_{el} | \Psi_h^* \rangle = \int dz\, dy \sum_l a_l^{el} \Phi_l \cdot \sum_{l'} a_{l'}^h \Phi_{l'} = \sum_l a_l^{el} a_l^h \qquad (23)$$

In Fig. 28, we show the calculated electronic energies of a 50 Å thick $Ga_{0.47}In_{0.53}As/InP$ quantum well as a function of the transverse dimension. For this case, there is only one bound state in the growth direction $n_z = 1$, the six first eigenenergies associated with the quantum numbers $n_y = 1, 2, \ldots, 6$ are shown together with the experimental energies extracted from Fig. 26. Excellent agreement is obtained for the first three el → hh transitions. The absence of the corresponding el → lh transitions is not clear at this point, it can be partially understood by the 3:1 ratio of the dipole transitions for this polarization. Furthermore, similar absence of this transition was observed in luminescence from a much wider quantum wire by Reed et al. (1986). They attributed this effect to the singularity in the density of hh states, which depletes higher energy hole states. This, however, does not hold for absorption, as is the case of our experiment, and we believe that stronger mixing between hh and lh states due to the additional lateral confinement may be the right explanation for both experiments.

Significantly narrower quantum wires were prepared recently by Notomi

5. OPTICAL PROPERTIES OF $Ga_{1-x}In_xAs/InP$ QUANTUM WELLS

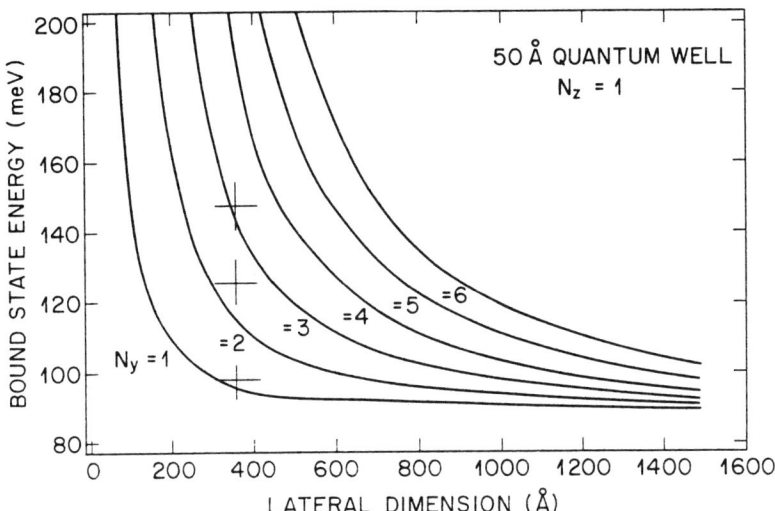

FIG. 28. Calculated energies of the $N_z = 1, 2, 3,...$ sublevels of the $N_y = 1$ state plotted as a function of the wire width. Experimental points are extracted from the PLE data.

et al. (1991). The sample was grown by HSMBE and consisted of a single ternary well, 50 Å thick, covered with a layer of InP, 800 Å thick. The structure was terminated with a 30 Å thick layer of GaInAs. The resist pattern was exposed by electron beam lithography and consisted of 1500 Å wide line-and-space pattern oriented in the (110) direction. The resist mask was used only for the etching of the thin top ternary layer. The subsequent etching of the underlying InP and the quantum well was carried out with wet selective etchants using the lines formed in the top ternary layer. The etched structure formed undercut (reverse) mesas with a transverse dimension of about 100 Å. Fluctuations in wire dimension were minimized by the use of anisotropic etchants that have higher etch rates along selected crystallographic directions. Finally, the etched structure was overgrown with InP by MOCVD.

The PL spectra obtained on these narrow and completely buried wires are shown in Fig. 29 (Notomi et al. 1991). A remarkable blue shift could be seen very clearly with decreasing wire width. The 100 Å wide wire showed a blue shift of over 35 meV with respect to the reference well. The shoulder visible on the high-energy side of the PL spectrum of a 200 Å wide wire was assigned to a transition induced by the lateral confinement in the well's plane, i.e., the 1,2H level. Its energy is in good agreement with the calculated splitting of the lowest bound level.

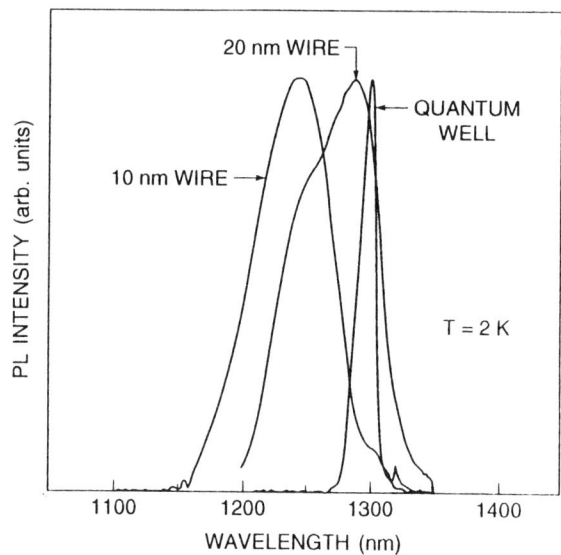

FIG. 29. Low-temperature PL spectra obtained on buried quantum wires as narrow as 100 Å (after Notomi et al. 1991).

The effects of transverse confinement are clearly observed in the optical spectra of GaInAs/InP nanostructures. These are the blue shift of the $n = 1$ exciton as well as the lateral confinement induced line splitting. The experimental data is in good agreement with a one-band solution of the two-dimensional effective mass Schrödinger equation. The integrity of the original material appears to be maintained in the patterning of InP heterostructures, making it a particularly favorable material system for a wide variety of experiments on ultra-small samples.

VI. Electric Field Effects

It is well known that the application of an electric field in the direction perpendicular to the semiconductor heterostructure layers results in an energy shift of the optical absorption edge. In quantum well structures, this effect, which is commonly referred to as the *quantum confined Stark effect* (QCSE), produces a remarkably large shift without appreciable smearing of the sharp excitonic features (Miller et al. 1985). The effect is schematically illustrated in Fig. 30. Applied electric field changes the shape of the potential that the carriers are subjected to resulting in a reduction of the electron–

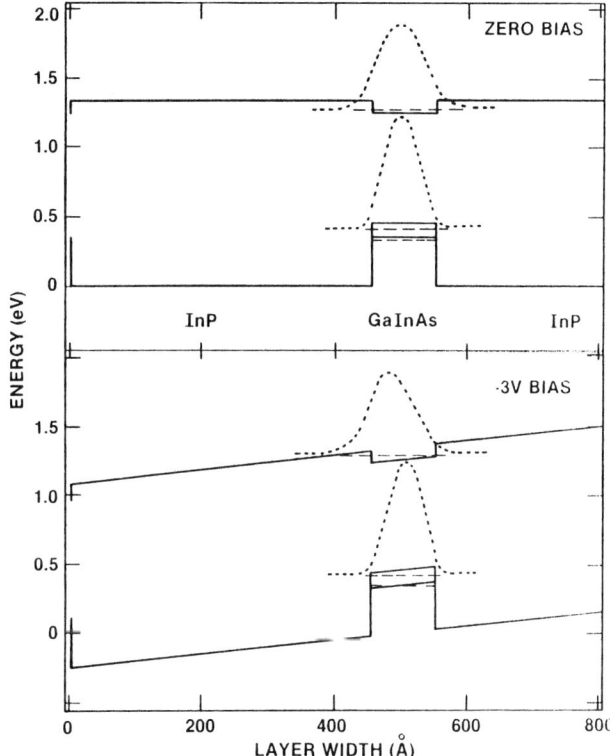

FIG. 30. Calculated energy levels and wave functions of a quantum well without (a) and with (b) an applied electric field.

hole transition energy. This red shift of the absorption edge has been demonstrated well above room temperature. Thus, it gives rise to many interesting optical device possibilities. QCSE has been analyzed in great detail in GaAs/GaAlAs and very good agreement between the theory end experiments was obtained (Miller et al. 1985). Such experiments have been carried out on the GaInAs/InP structures (Bar-Joseph et al. 1987; Temkin et al. 1987). The electro-absorptive modulator based on QCSE in GaInAs heterostructures has potential for applications in fiber communications in the 1.3–1.55 μm range.

Room temperature absorption spectra obtained in a 100-period GaInAs/InP superlattice at 0 and −30 V bias are shown in Fig. 31. The experiment of Bar-Joseph et al. (1987) uses light absorption in the direction perpendicular to the superlattice and the modulator structure is illustrated schematically in the inset. The quantum wells and barriers, grown by atmospheric

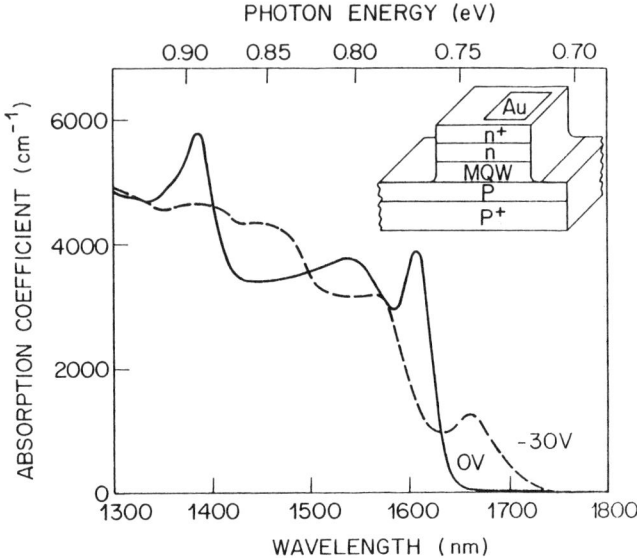

FIG. 31. Room temperature absorption spectra of a GaInAs superlattice consisting of 100 wells, each 100 Å thick. The spectra were measured with and without a reverse bias. Device structure is shown in the inset (after Bar-Joseph et al. 1987).

pressure MOCVD (Miller et al. 1986a), are each approximately 100 Å thick. The background carrier concentration in the superlattice, an important parameter for a modulator, is estimated at 2–3×10^{15} cm^{-3}. The zero-bias spectrum shows well resolved 2H and 1L transitions, as well as a strong 2H exciton. The slight red energy shift of the exciton features, compared to the calculated energies, was attributed to a small deviation from the exact lattice matched composition. The 1H exciton shifts dramatically to lower energies with the applied voltage. The new peak observed at a bias of -30 V at 0.84 eV is located close to the forbidden $n = 1$ electron to $n = 3$ hh transition energy. The energy shift of the 1H exciton has been studied as function of bias and good agreement with calculations has been obtained up to a bias of 20 V, above which the observed QCSE was found smaller than expected.

A more detailed spectral response of the 1H and 1L excitons as a function of applied electric field is illustrated in Fig. 32. These polarized photocurrent spectra were obtained with a rib waveguide modulator grown by HSMBE (Temkin et al. 1987c). The superlattice consisted of 50 nonintentionally doped GaInAs wells, each 100 Å thick, separated by 120 Å thick InP barriers. The background carrier concentration was measured by the capcaitance–voltage profiling to be $\sim 1 \times 10^{16}$ cm^{-3}. At this background level

5. OPTICAL PROPERTIES OF $Ga_{1-x}In_xAs/InP$ QUANTUM WELLS

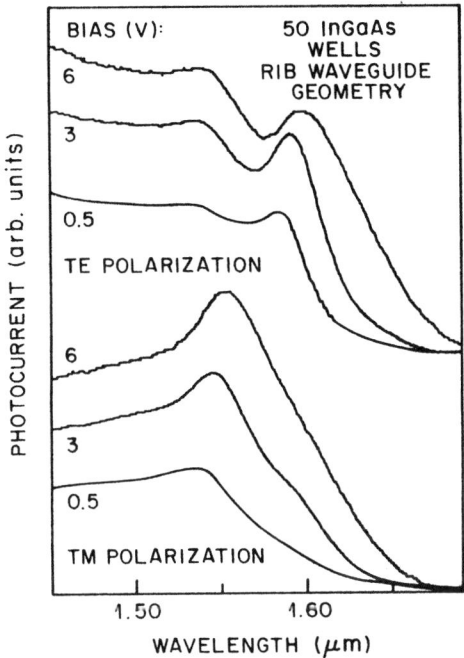

FIG. 32. Room temperature polarized photocurrent response of a GaInAs modulator as a function of the applied bias. Quantum wells of this sample were 100 Å thick.

the depletion layer width is estimated to reach 1 μm at a reverse bias of 15V. Thus the total thickness of this superlattice, chosen mainly for convenience of the optical evaluation, may exceed the depletion layer width at low bias levels. In addition, similar p-i-n structures with 100 Å thick GaInAsP wells of 1.3 μm and 1.55 μm compositions were also evaluated. These contained only 20 quantum wells each and the p-n junction was displaced by 0.15 μm into the top InP cap layer. After the growth, the wafers were processed into 250 μm long rib waveguide devices using standard techniques. The current–voltage characteristics of these waveguides showed the reverse breakdown voltage of up to 25 V and a leakage current below 1 μA up to the breakdown. The photocurrent response shown in Fig. 32 is separated into two polarization components. The spectra are obtained with the incident light polarized either parallel (TE polarization) or normal (TM polarization) to the superlattice layers. The TE polarization, in which the electric field vector is parallel to the superlattice plane, is shown in the upper part of Fig. 32. In agreement with the polarization selection rules discussed in Section I, two peaks are observed in the TE spectrum, at 1.58 μm and 1.54 μm, respectively. The energies of these exciton levels agree with those calculated for a 100 Å thick quantum well of lattice matched GaInAs. In the TM polarization, shown in the lower part of Fig. 32, only the light hole transition is observed in the photocurrent spectrum, again in agreement with the dipole selection rules discussed in Section I. With the applied reverse bias above 0.5 V, the photocurrent spectra shift to longer wavelength. With a bias of 6 V the exciton peaks exhibit a shift by approximately 250 Å from their zero-bias positions. The shift continues for another 100 Å with a bias increasing to 10 V, however a rapid smearing of the excitonic peak is observed above 6 V reflecting the decreasing overlap between electron and hole envelope wave functions. The excitons seen in the two polarizations shift with bias at approximately the same rate.

The QCSE can be used to modulate the waveguide transmission at wavelengths close to the exciton edge. Using a semiconductor laser emitting in a single longitudinal mode tuned to a wavelength of 1.63 μm, we have been able to obtain a modulation depth of 35% at a bias of 6 V. The extinction ratio was maintained up to the modulation frequency of 500 MHz. At this wavelength the modulator acted only in the TE polarization, as expected from its spectral response shown in Fig. 32. More recently, more sophisticated waveguide modulators have been reported by Koren et al. (1987). These devices, operating at a wavelength of 1.67 μm, showed an on–off ratio of 47:1 and a bandwidth of 3 GHz.

The structure with 100 Å thick GaInAs quantum wells produces the $n = 1$ exciton at the wavelength of 1.58 μm, which limits the modulator operation to even longer wavelengths. It would appear that a natural extension of the operating range of such devices to the technologically important shorter wavelengths of 1.3–1.55 μm could be accomplished simply by decreasing the well thickness. Indeed, we have shown in Section II that excellent room temperature response could be obtained in GaInAs superlattices with wells as thin as 20 Å, in which the exciton edge was shifted up to 1.28 μm. However, in the case of electro-absorption modulator, the usefulness of this approach is limited by two effects. First, as the well thickness decreases, the carrier density in the well and the related interfacial charge density increase. The electric field of these charges screens the externally applied bias thus reducing the exciton shift. This effect is extrinsic and can be reduced by improving the material purity. The second effect is the intrinsic dependence of the QCSE on the well size. The exciton shift occurs because of a field induced change in the well shape, from rectangular to triangular. With decreasing well thickness, exciton levels move away from the well's bottom and their energies thus become less sensitive to the shape of the well. Qualitatively a simple second-order perturbation calculation yields the following expression for the ground state energy shift of an infinitely deep well:

$$\Delta E \sim m^* e^2 F^2 L^4 \qquad (24)$$

where e is the carrier charge, m^* its effective mass, F the external field, and L is the well thickness. Thus the energy level shift decreases, for a constant field, as a fourth power of the well thickness. In quantitative terms this dependence is illustrated in Fig. 33, where a more accurate model, which takes into account the finite depth of the potential well, was used for the calculations of the $n = 1$ excitonic transition energy as a function of the applied electric field. The calculations are based on the resonance tunneling (RT) method as explained in Section II. The parameters used for these calculations are listed in Table I. The applied voltage is assumed to drop uniformly across the superlattice waveguide layer. A shift of $\sim 10\,\text{meV}$ demonstrated under a low bias in Fig. 32 is in good agreement with the calculated curves. Similarly, good agreement is obtained with the data of Bar-Joseph discussed previously. Based on the results of Fig. 33, we estimate that, in the absence of exciton broadening, a bias of 15 V should be sufficient for a shift of $\sim 40\,\text{meV}$ in a structure with 100 Å wells. Such a shift of the

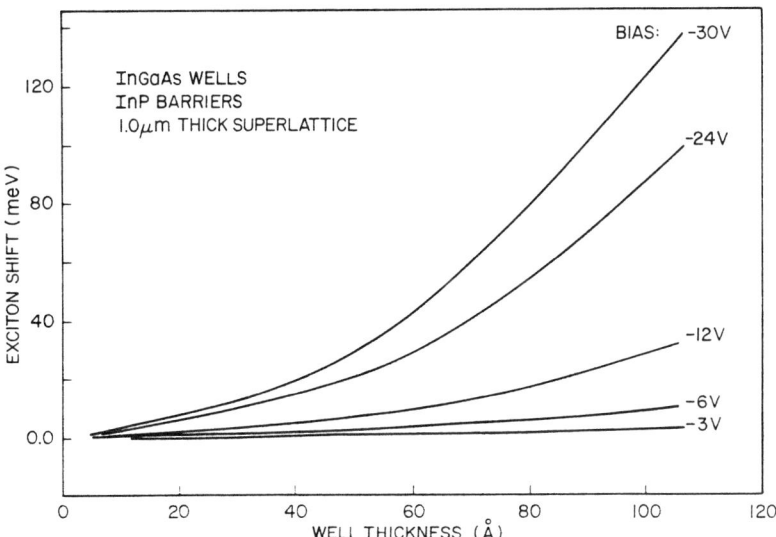

FIG. 33. Calculated quantum confined Stark plotted as a function of the well thickness for a GaInAs/InP superlattice.

exciton from the operating wavelength should result in low loss operation and an on–off ratio on the order of 50:1 (Koren et al. 1987; Walcita et al. 1986). In comparison, a bias of nearly 20 V would be needed to obtain a similar shift in a structure with 85 Å wells of InGaAs needed for 1.55 μm operation. An exciton shift of this magnitude in a modulator operating at 1.3 μm, for which the well thickness would have to be reduced to ~25 Å, would require unreasonably large bias levels.

These design constraints can be removed by replacing the ternary well material with various compositions of GaInAsP. This allows for an independent adjustment of the well size and the 1H exciton wavelength. Calculations carried out for the 1.3 μm well composition indicate a rate of the exciton shift with bias similar to that found in the ternary wells of equal dimensions. To demonstrate this effect we have studied the photocurrent spectra and the quantum confined Stark effect in two superlattices with 100 Å wide wells with 1.55 and 1.3 μm compositions of InGaAsP. The resulting waveguide response, in the TE polarization, is shown in Fig. 34. The wide spectral range photocurrent response of the waveguide device with GaInAs wells is included for comparison. The spectra of quaternary superlattices also show a well-resolved series of the exciton peaks, which can be assigned to the $n = 1$ electron to heavy and light hole transitions, as indicated by arrows in Fig. 34. The edge exciton in the quaternary wells, and especially in the 1.3 μm sample, is somewhat less pronounced and the slope of the exciton edge less steep than in the ternary sample. On the other hand, the ternary structures exhibit a bias dependence strikingly similar to that shown in Figs. 31 and 32. Above the bias of 5 V the edge exciton becomes visibly broader and is completely suppressed at 10 V. This demonstrates the quantum confined Stark effect in quaternary superlattices, in qualitative agreement with the calculated results.

A very different design of an electroabsorption modulator based on strained quantum wells was proposed by Gershoni et al. (1990a). The band structure and the strain profile are shown in Fig. 35. The structure uses strained layers, and the corresponding changes in the electronic band structures as discussed in Section VII, to minimize the hole confinement in the wells. The two wells are compressively strained, and the barrier between them was grown under a tensile strain to balance the overall strain in the structure, thus increasing the critical layer thickness. Small changes in the applied electric field would then introduce large changes in the oscillator strength of the band-to-band optical transitions. The oscillator strength is proportional to the overlap integral between the carrier's envelope wave functions as given by the Eq. (4).

In the structure shown in Fig. 35, and without the applied electric field, the only allowed transitions are those between the same symmetry states, for

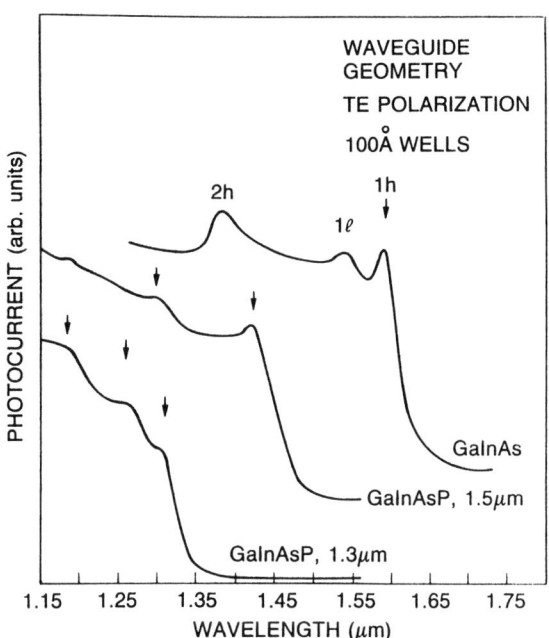

FIG. 34. Photocurrent response spectra of GaInAsP based waveguide modulators with 100 Å thick wells. The spectrum of a GaInAs device is shown for comparison.

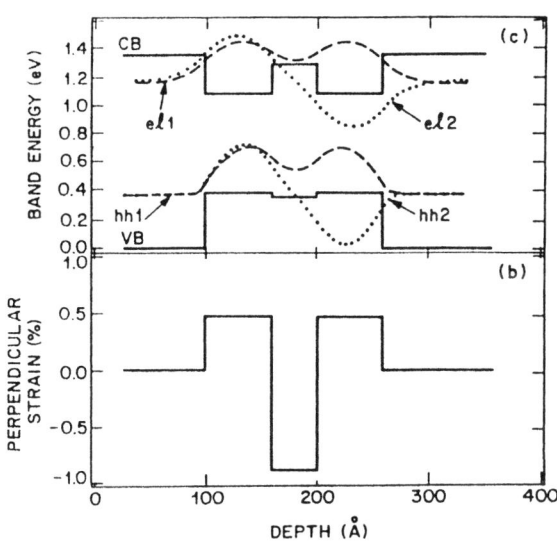

FIG. 35. Band-gap structure of an individual quantum well in a strained well modulator. Lower panel plots the lattice mismatch strain profile of a quantum well (after Gershoni et al. 1990a).

383

instance 11h = el1–hh1 or 22h = el2–hh2. Transitions between the symmetric and antisymmetric state, such as 12h = el1–hh2, are spatially forbidden as can be seen from the wave functions plotted in Fig. 35. The electric field applied in the direction perpendicular to the wells reduces the strength of the normally allowed transitions and enhances the forbidden ones. The ratio of the oscillator strengths can even be reversed under moderate electric fields with the forbidden transitions becoming dominant.

The photocurrent spectra of a sample containing ten wells, each consisting of three compositions of GaInAs, are shown in Fig. 36 as a function of bias. The 12h excitonic transition shows a blue shift of about 10 meV with the applied bias and its intensity peaks at a field of about 28 kV/cm, in good agreement with the calculation (Gershoni *et al.* 1990a). The built-in field caused by the *p-i-n* junction is enough to make the asymmetric transitions dominant in oscillator strength. Since the HH2 state moves up in energy faster than the el1 state moves down (with applied electric field), the exciton peaks associated with the optical transition el1-hh2 is blue shifted. This behavior is in marked contrast to the conventional QCSE and has potential for applications in bistable devices such as the SEED (Miller *et al.* 1982).

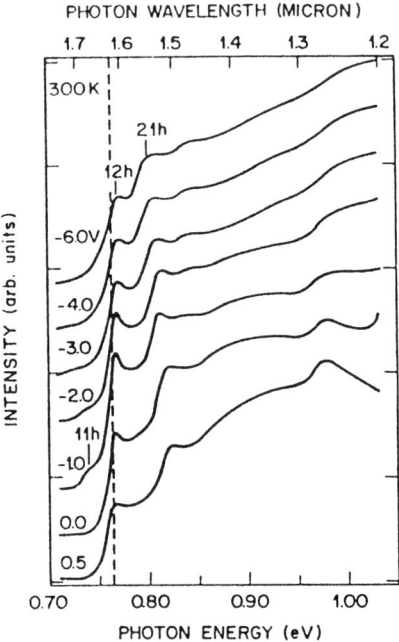

FIG. 36. Room temperature photocurrent response spectra of a strained layer modulator as a function of bias. Notice the blue shift of the 12 h exciton with applied bias.

VII. Strained Layer Superlattices

In preceding sections we have discussed the properties of the ternary and quaternary quantum wells carefully lattice matched to InP. In recent years much attention has been given to strained layer superlattices (SLS), in which multilayer structures are grown lattice mismatched to the substrate but with the layer thickness small enough to accommodate the mismatch strain coherently rather than by misfit dislocation (Osbourn 1984). The elastic layer strain gives rise to a number of interesting properties and SLS can be conveniently used to study strain effects. However, in most lattice mismatched systems the sign of strain is difficult to vary. In contrast, $Ga_{1-x}In_xAs/InP$ SLS structures can be either positively or negatively strained, from $x = 0$ (GaAs) to $x = 1$ (InAs), to explore the relationship between strain and electronic levels of quantum wells (People 1987; Gershoni et al. 1987, 1988c). The lattice matched and strain free $Ga_{0.47}In_{0.53}As$ can then be used as a convenient reference point at which the strain and quantum size effects can be independently controlled, and a direct comparison to the nonstrained, lattice matched system can be made.

In this section we describe the use of room and low temperature PL and polarized PCE spectroscopy to identify the electron to light and electron to heavy hole transitions as a function of compositional strain (Gershoni et al. 1987; Temkin et al. 1989; Gershoni and Temkin 1989). Most of the SLS structures grown for this purpose were coherently strained and the critical layer thickness was established as a function of composition for the entire range of In concentrations. The superlattice dimensions, as well as the strain, were measured by high-resolution X-ray diffraction and transmission electron microscopy. The samples were grown in a p-i-n configuration that permits the application of an electric field to the superlattice.

11. Critical Layer Thickness

The experiments designed to determine the critical layer thickness were done on two sets of SLS structures embedded within the intrinsic region of p-i-n diodes (Gershoni and Temkin 1989; Gershoni et al. 1989a; Temkin et al. 1989). The first set consisted of five identical quantum wells separated by 300 Å thick barriers. The well composition of $Ga_{1-x}In_xAs$ was kept at $x = 0.37$ and $x = 0.75$, in different samples. The well widths were 40, 60, 100, and 160 Å, again in different samples, spanning the range from below to well over the critical layer thickness at each composition. In the second set of fourteen samples the superlattice region consisted of ten $Ga_{1-x}In_xAs$ wells, with the thicknesses close to the critical limit estimated from the force-balance model of Matthews and Blakeslee (1974). The number of wells, their widths and the widths of InP barriers in these samples were chosen to assure that the critical thickness of the entire superlattice remained below the critical limit. The concentration of In in these structures was varied from $x = 0.0$ to 1.0, resulting in four samples with lattice constant larger than that of the substrate, i.e., subjected to compressive biaxial stress, and the reamining eight under tension. The strained wells were separated by InP barriers ranging in thickness from 250 to 400 Å, in different samples. In all the samples the superlattice regions were capped with 1000 Å thick undoped n-InP layer, followed by a 3000 Å thick layer of InP doped p-type with Be. While the junction was misplaced in each case into InP, the SLS structure was well within the depletion region even at very low reverse bias voltages. Electrical measurements were carried out on 100 µm diameter mesa diodes defined by wet etching.

The well dimensions and compositions were determined from TEM cross sections and HRXRD measurements. The TEM cross sections showed well-defined quantum wells for the entire range of compositions. Defects could be seen only in the 160 Å thick SLS samples. These resulted in slip planes and stacking faults along the {111} planes, propagating through the entire structure. Their density was estimated at less than $4 \times 10^8 \text{ cm}^{-2}$. It is well known that TEM, which images only very small areas, is not sensitive to low defect densities. For instance, the lattice mismatch strain in the 100 Å well, $x = 0.37$, sample was apparently just large enough to produce misfit dislocations. A low density (less than $30–50 \text{ cm}^{-1}$) of misfit lines lying only in one of the (110) directions could be seen in Nomarski contrast optical microscope. These defects were not found in the TEM cross sections.

The strain component parallel to the growth direction was determined by fitting the HRXRD spectra with a kinematic diffraction model described by Vandenberg et al. (1986, 1987). This procedure allows for the determination of the Ga and In concentrations with precision better than 1%. Most of the SLS structures show very well resolved (400) spectra, as illustrated in Fig. 37. The spectrum of a control, lattice matched, superlattice shows at least 11 orders of satellite reflections, indicative of the structural perfection of the sample. The satellite reflections show a symmetric intensity distribution around the (400) InP substrate peak, as expected from strain free layers. In comparison the asymmetric shape of the traces obtained on the strained structures is indicative of an in-plane strain resulting in lattice compression or tension in the growth direction, which is monitored by the (400) diffraction scans. However, despite the lattice mismatch the strained samples retain their structural integrity, as judged by the sharpness and intensity of the satellite reflections. Similar quality spectra were obtained from SLS samples spanning the entire range of well thicknesses studied, at a fixed

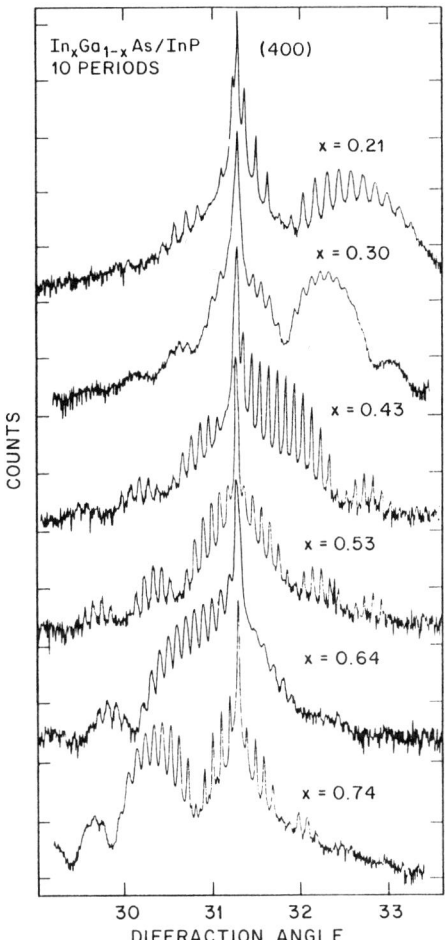

FIG. 37. High-resolution X-ray diffraction scans of lattice matched and strained GaInAs/InP superlattices.

composition, as well as most of the samples spanning the range of In concentrations. The only samples for which superlattice diffraction satellite spectra could not be obtained had $x < 0.1$. In this sample strain induced slip line density, as judged from TEM cross sections, exceeded $6 \times 10^9 \, \text{cm}^{-2}$.

Room temperature PL spectra of a set of SLS structures with $x = 0.37$ and $x = 0.75$ are shown in Fig. 38. The spectra covered a range from 1.3 to nearly 2 μm. The PL linewidth did not vary with composition when plotted on the energy scale. This indicates a high degree of compositional uniformity. The high luminescence efficiency did not degrade significantly with

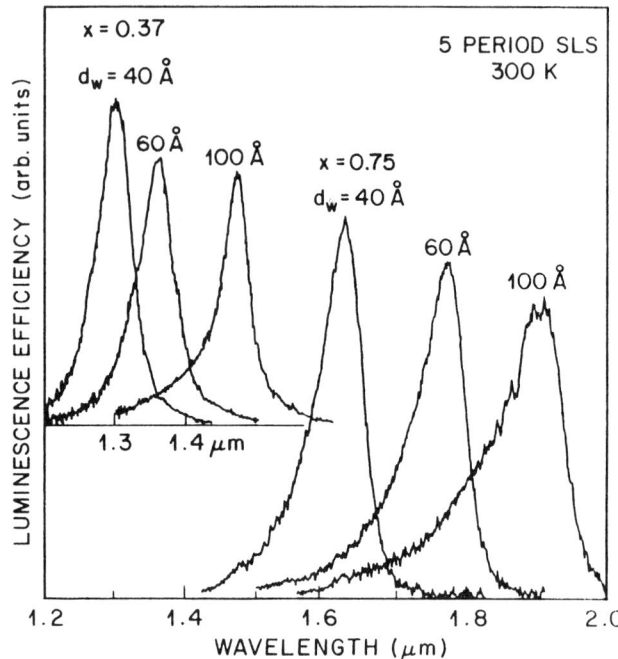

FIG. 38. Room temperature PL spectra of SLS samples with $x = 0.37$ and $x = 0.75$ plotted as a function of the well width.

increasing strain. The changes in the band gap with composition and strain resulted in large shifts from the lattice matched wavelength of 1.66 μm. In addition, the quantum size effect resulted in a considerable blue shift of the PL peak with decreasing well thickness. For instance, the luminescence of the In-rich SLS shifted from $\sim 2\,\mu\text{m}$ to 1.62 μm as the well width was reduced from 100 to 40 Å. The exciton energies seen in the PL spectra were identified as the $n = 1$ electron–light hole (for the $x = 0.37$ set) and electron–heavy hole (for $x = 0.75$) transitions. The energies of light hole excitons of the $x = 0.37$ set were found in excellent agreement, to within 5 meV, with the calculated values (Temkin et al. 1989). This included the sample with 160 Å wide wells which is not illustrated in Fig. 38. The energies of In-rich SLS samples in this set were typically ~ 20–25 meV lower than the calculated values. This is a Stokes shift that was also seen in lattice matched wells.

The good overall agreement between the measured and calculated excitonic transition energies would seem to imply that strain in the wells is accommodated elastically, and all of these structures are commensurate.

However, such a conclusion is not justified since an appreciable amount of strain relief can take place before it is detected by optical measurements. For instance, the linear density of dislocations induced in the thicker well samples, on the order of $2 \times 10^4 \text{cm}^{-1}$, relaxes the elastic strain by no more than 4% (Chu et al. 1985). This is not sufficient to affect the optical measurements.

A better approach for the determination of the onset of relaxation is to study the effect of a relatively small density of dislocations due to relaxation, on the behavior of p-i-n diodes containing the strained region (Temkin et al. 1989). Diodes fabricated from the SLS samples with the well dimensions of 40 and 60 Å showed very well-behaved current–voltage (I–V) characteristics. The revese currents saturated at very low levels, less than 1 nA up to 10 V, and were very reproducible. These I–V characteristics were also indistinguishable from those of the control samples grown without a superlattice in the i-region. The I–V characteristics of the 100 Å well sample were much worse. The leakage currents increased by as much as four orders of magnitude at 10 V and the results were not reproducible. The breakdown voltage decreased from 30–35 to ~ 15 V. The two sets of diodes fabricated from the 160 Å SLS wafers exhibited even poorer characteristics. In general, as established by similar experiments on diodes with a wide range of x, the I–V characteristics do not degrade for SLS thicknesses below and up to the critical thickness. In contrast, the I–V characteristics degrade sharply for larger SLS dimensions. Clearly, the misfit dislocations induced by relaxation as the result of both negative and positive strain are significantly more electrically active than the already existing threading dislocations (at concentrations of 10^3 to 10^4cm^{-2}) that are incorporated into the growing layers from the substrate.

It should be stressed that the preceding comments apply specifically to p-i-n diodes in which the depletion front moves through the SLS region under reverse bias. It is possible to mask the effect of strain induced defects by doping the SLS much higher than the adjacent, and oppositely doped, bulk layer. In such an abrupt diode, an example of which could be the base–collector junction of a bipolar transistor, the depletion layer will be constrained mostly to the defect-free, low-doped, cladding layer. Experiments carried out on such diodes in the Ge_xSi_{1-x}/Si system show high junction quality for SLS thicknesses as much as a factor of two larger than the critical layer thickness (King et al. 1988). Clearly, the use of such a device provides a poor test of the onset of the elastic strain relaxation.

Figure 39 shows a plot of the critical layer thickness of $Ga_xIn_{1-x}As$ as a function of In concentration. Commensurate samples, i.e., those largely free of strain induced defects, are shown as open circles. Samples that have relaxed, as judged by the p-n junction quality, are labeled by filled circles.

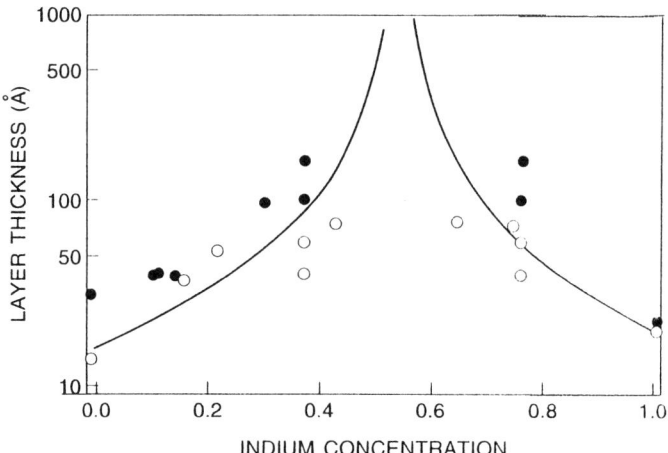

FIG. 39. Critical layer thickness of $Ga_{1-x}In_xAs/InP$ plotted as a function of In concentration x.

The solid curve represent the critical thicknesses calculated on the basis of force balance (FB) model proposed by Matthews and Blakeslee (1974). The elastic parameters used in the calculation were obtained from the compilations of Adachi (1982). The application of the FB model to III–V alloy systems such as $Ga_xIn_{1-x}As/GaAs$ now appears to be well recognized (Fritz et al. 1987), despite occasional challenges based on low-resolution X-ray diffraction and PL measurements (Reithmaier et al. 1989). The data of Fig. 39 convincingly show the applicability of the FB model to $Ga_xIn_{1-x}Aa/InP$. It does not appear easy to exceed the critical thickness and maintain high electrical quality of the SLS contained in the depletion region.

2. OPTICAL PROPERTIES AND ELECTRONIC ENERGY LEVELS

Figure 40 shows the low-temperature PL spectra of representative SLS samples as a function of In concentration x. The spectral peaks shown are due to radiative recombination of the 1s exciton within the $Ga_xIn_{1-x}As$ quantum wells. It is noteworthy that, while x varies considerably from the lattice matched composition and the exciton energy shifts by as much as ~ 230 meV, corresponding to a wavelength shift of over 4000 Å, no significant degradation of the luminescence intensity is observed from samples with $x > 0.2$. This is consistent with the absence of nonradiative defects, such as misfit dislocations, in the strained samples. A slight broadening of

5. OPTICAL PROPERTIES OF Ga$_{1-x}$In$_x$As/InP QUANTUM WELLS

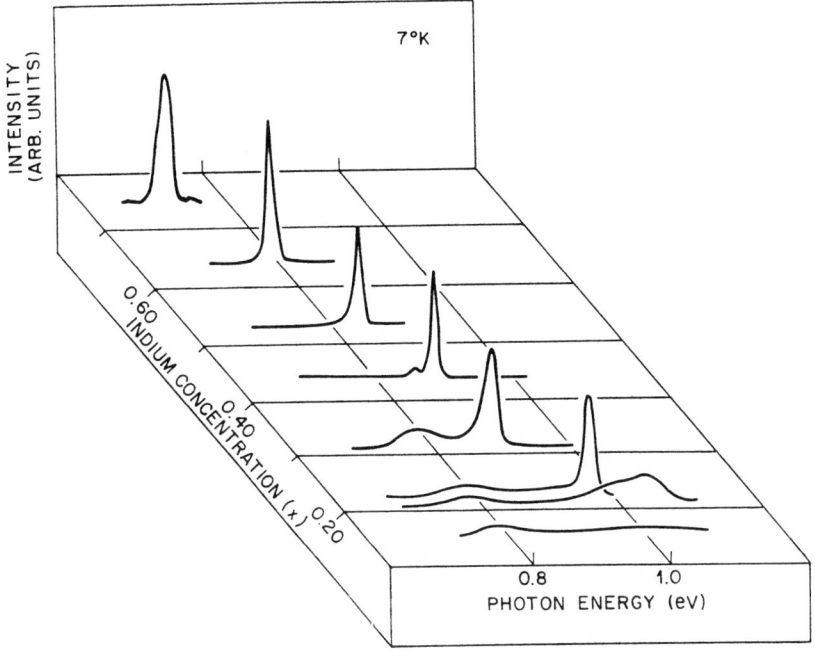

FIG. 40. Low-temperature photoluminescence spectra of strained GaInAs superlattices with different In concentrations.

the exciton peak was observed only in the two samples furthest away from the lattice matched composition. This is attributed to the presence of a few monolayers of the quaternary InGaAsP at the well to barrier interface (Vandenberg et al. 1986). The excellent quality of the optical properties of these samples is maintained up to room temperature. The abrupt quenching of the luminescence in samples of composition close to GaAs will be discussed later in terms of a transition from a type I to a type II superlattice. It is not indicative of a lower sample quality, as was verified by TEM and HRXRD.

The polarized room temperature photocurrent response spectra of rib-loaded waveguides prepared from the SLS structures are used to discriminate between the heavy and light hole transitions. In these experiments linearly polarized light from a tungsten lamp dispersed by a monochromator was coupled into the superlattice the cleaved faccts of the waveguide.

Figure 41 presents a series of room temperature photocurrent spectra obtained on six SLS samples with $0 < x < 0.64$. For $x > 0.2$ each shows

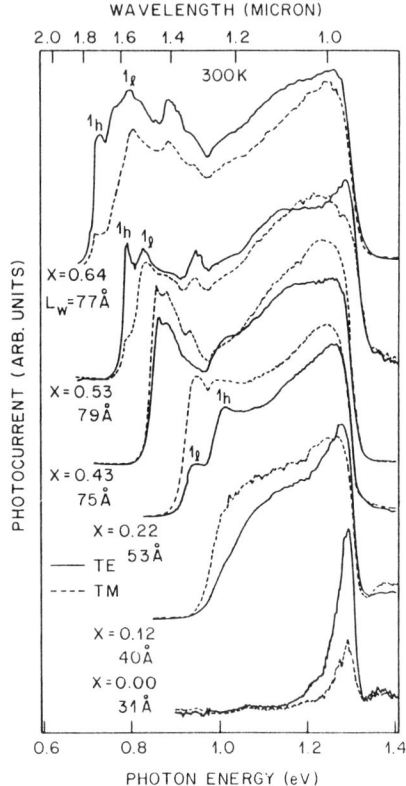

FIG. 41. Room temperature polarized photocurrent spectra of strained layer samples. Heavy and light hole excitonic transitions are marked.

readily identifiable 1H and 1L transitions, as well as a number of higher order exciton peaks. Considerable changes are observed in the energy difference of the two $n = 1$ (1H and 1L) transitions as a function of In concentration. In samples with $x > 0.53$, the lattice matched composition, the light hole level is found, as expected, above the heavy hole level, i.e., at higher energies. In the sample with $x = 0.43$ the energies of the two transitions, as detected by the polarized response spectra, are found to overlap. At even lower In concentration of $x = 0.30$ the exciton levels switch positions so that the light hole becomes the lower energy state. A similar observation was reported by Schuber et al. (1985) who directly measured the light hole masses in $In_{0.2}Ga_{0.8}As/GaAs$ SLS's and by Nishi et al. (1986) who assigned the light hole–electron transition to their lowest energy transmission peaks observed in GaInAs/InAlAs SLS.

5. OPTICAL PROPERTIES OF $Ga_{1-x}In_xAs/InP$ QUANTUM WELLS

Another striking phenomenon in the series of photocurrent spectra shown in Fig. 41 is the vanishing of the exciton structure for In concentration lower than $x \sim 0.2$, and a complete disappearance of the photocurrent response for the $x = 0$ sample. As mentioned earlier these samples exhibit no PL response. The spectra of samples with $x < 0.2$ furthermore exhibit a markedly different behavior under the electric field bias, as plotted in Fig. 42. For the SLS with $x = 0.3$ there is almost no change in the shape of the photocurrent spectrum as the bias is increased from 0 to 6 V, corresponding to a field of 10^5 V/cm applied across the superlattice. This behavior is typical of all the samples with $x > 0.20$, where the only noticeable spectral change with bias was a slight down-shift in the $n = 1$ exciton energy caused by the

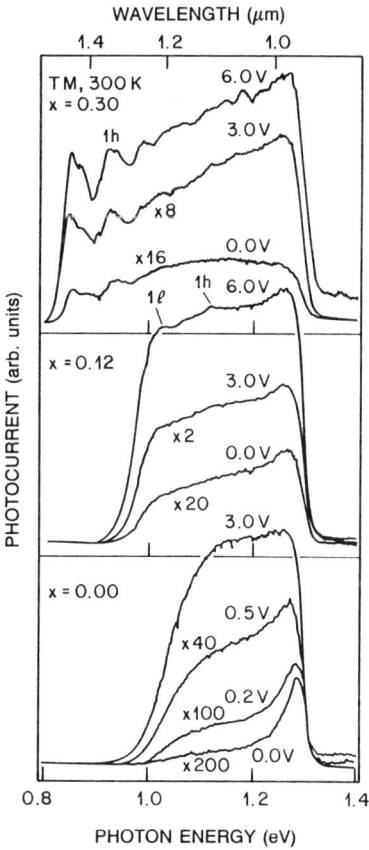

FIG. 42. Applied field dependence of the photocurrent response spectra for three samples with low In concentration.

quantum confined Stark effect. In contrast, in samples with $x < 0.2$ a dramatic increase in the photocurrent response is induced even at a low bias level. At higher bias the response begins to resemble that of the samples with much larger In concentrations and, in the sample with $x = 0.12$, excitonlike structure appears at a bias of 6 V.

These changes cannot be explained by the increased defect density in the very low x samples. First, the absence of the PL response and a different photocurrent response are observed in the sample with $x = 0.12$, which we consider, on the basis of X-ray diffraction and TEM, to be of very high quality. Second, a normal photocurrent response is observed in a sample with $x = 1$, which exhibits a defect density similar to the $x = 0$ sample. The effects are explained later by the transition from type I (quantum well both in the valence and conduction bands) to type II (quantum well in the valence band and a potential barrier in the conduction band) superlattice at In concentrations below $x \sim 0.2$.

The energy levels of the light (full circles) and heavy (open circles) hole excitons deduced from the spectra of Fig. 41 are plotted in Fig. 43 versus In concentration. This figure also shows the calculated energies of both the 1s light and heavy hole-electron transitions for 75 Å thick quantum wells.

Commensurate growth of strained $Ga_xIn_{1-x}As$ on (100) oriented InP

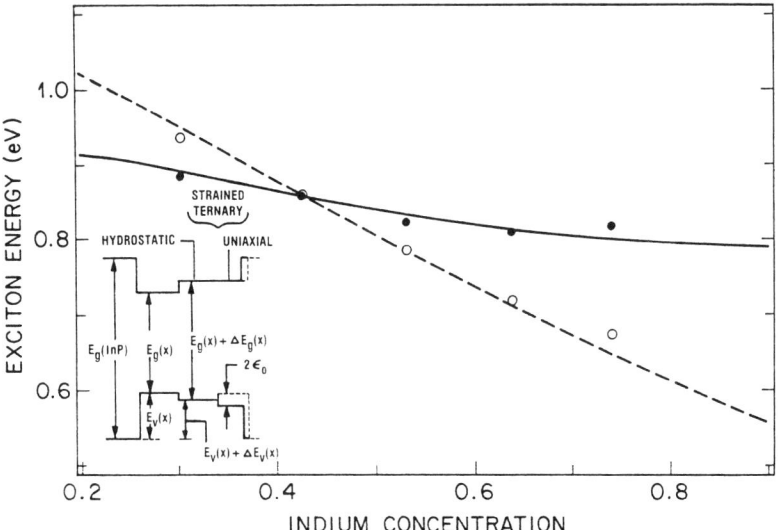

FIG. 43. Excitonic transition energies plotted versus In concentration x. Open and filled circles denote heavy and light hole transitions, rspectively. Lines indicate calculated energies for a 75 Å thick well. Inset shows band-gap changes with biaxial strain.

substrate subjects these layers to a biaxial in-plane strain. The strained epitaxial layers experience a tetragonal distortion, resulting in a very simple form of the strain tensor e_{ij}. If we define $z \parallel [001]$ (i.e., along the growth direction) it follows that e_{ij} has only diagonal components with $e_{xx} = e_{yy} = e_\parallel$ and $e_{zz} = -(2C_{12}/C_{11})e_\parallel = e_\perp$ (Osbourn 1984), where $e_\parallel$ ($e_\perp$) are the diagonal strain tensor components parallel (perpendicular) to the growth plane, and C_{ij} are the components of the elastic stiffness tensor (Hensel and Feher 1963; Bir and Pikus 1974). Knowing the elements of the strain tensor one can readily apply the phenomenological deformation potential theory (Bir and Pikus 1974; Osbourn 1984) to calculate the effect of strain on the electronic states of the system under study. The calculation of the energy levels was carried out in the following way. First, commensurate growth was assumed, so that the in-plane lattice constant of the superlattice is always equal to that of the InP substrate and the in-plane strain components are given by $e_\parallel = [a_{InP} - a(x)]/a(x)$, where $a_{InP} = 5.8688$ Å is the lattice constant of InP, $a(x)$ is the lattice constant of the relaxed $In_xGa_{1-x}As$, and x is the In concentration as extracted from X-ray diffraction measurements. Lattice constants of various ternary compositions can be calculated from the known lattice constants of InAs and GaAs using Vegard's law (Adachi 1982). The same interpolation procedure was used to obtain all other material parameters of ternary alloys, such as the stiffness coefficient tensor $C_{ij}(x)$, the electron and light and heavy hole effective masses, the valence band deformation potentials $a_v(x)$ and $b_v(x)$ (notation of Pikus and Bir 1974) and the energy band-gap dependence on hydrostatic pressure $dEg/dp(x)$ taken from Adachi (1982). The band offsets were estimated in a similar way from the synchrotron radiation data for the parent binary compounds reported by Bauer and Margaritondo (1987). For the lattice matched composition we calculate the valence band offset ΔE_v to be equal to 60% of the total bandgap discontinuity, in agreement with capacitance-voltage measurements of Forrest et al. (1984) and Cavicchi et al. (1989). We note that the linear interpolation of the relative valence band energies of the unstrained binary compounds compiled by Bauer and Margaritondo (1987) ($\Delta E_v^{InAs/InP} = 0.41$ eV, $\Delta E_v^{GaAs/InP} = 0.34$ eV) ignores the effect of strain on the relative valence band energies, which, according to mid-gap theories (Tersoff 1986; Cardona and Christenson 1987), should arise from the strain induced shift of the mid-gap point. Finally, parabolic interpolation, which takes into account the bandgap bowing, was used to calculate the energy gap dependence on the In concentration.

With the calculated strain tensor components, the strain induced change in the ternary band gap could be calculated according to

$$\Delta E_g(x) = [-dE_g(x)/dp] \left\{ \tfrac{1}{3}[C_{11}(x) + 2C_{12}(x)] \right\} [2e_\parallel + e_\perp] \quad (25)$$

where $1/3(C_{11}(x) + 2C_{12}(x))$ is the bulk modulus of the ternary layer. The expressions in the square and curly brackets describe the band-gap hydrostatic deformation potential. The change in the valence band offset with the hydrostatic strain was calculated, similarly to Eq. (25), according to

$$\Delta E_v(x) = a_v(x)(2e_\parallel + e_\perp) \qquad (26)$$

where $a_v(x)$ is the ternary valence band hydrostatic deformation potential. The effects of hydrostatic strain on the conduction band offset can be calculated simply by subtracting Eq. (26) from Eq. (25). The hydrostatic part of the strain changes the direct energy gap, and since the change is not equally distributed between the conduction and valence bands, it directly influences the band offsets. The uniaxial part of the strain results in removing the valence band degeneracy at the Brillouin zone center. This effect has been studied experimentally in the bulk GaAs (Chandrasecar and Pollack 1977) and InAs (Yu et al. 1971), parent binaries of the ternary material. The valence band splitting due to the uniaxial strain was calculated, according to

$$\varepsilon_0 = b_v(x)(e_\perp - e_\parallel) \qquad (27)$$

where ε_0 is half the energy splitting between the (3/2, 3/2) and (3/2, 1/2) valence band sublevels and $b_v(x)$ is the ternary shear deformation potential for the [100] strain. These effects are illustrated schematically in the inset of Fig. 43. The quantum size effect, which like the uniaxial strain determines a preferred spatial direction (again the growth direction $\parallel z$), also removes the valence band degeneracy by creating the heavy and light hole states (Bastard and Brum 1986). The two effects add in case of compressive strain and subtract in case of tensile strain. The quantum size effect was calculated for 75 Å wells, separately for electrons and heavy and light holes. The resulting transition energies of the light (solid line) and heavy (dashed line) holes are plotted as a function of composition in Fig. 43. The exciton binding energy that amounts to ~ 10 meV (see Section III), was neglected. Despite that, the agreement with the experimental data is quite satisfactory. This is especially so in view of the fact that the material parameters are interpolated from the binary values and not measured directly. It should be noted that the energy of the 1s exciton is not very sensitive to the conduction/valence band offset ratio. A systematic 15% variation in the valence band offset did not cause any appreciable change in the calculated curves. The higher order excitonic transitions are much more sensitive to that parameter, as discussed later. On the other hand, to obtain the agreement shown in Fig. 43 one must conclude that most of the band-gap changes induced by variation in the In concentration occur in the conduction band offset.

The conduction and valence band edges of the strained $Ga_{1-x}In_xAs$ calculated by Gershoni et al. (1988b) using the phenomenological deformation potential theory are plotted versus x in Fig. 44. The calculated lines represent bulk alloy energies and do not include quantum size shifts. The energy scale is referenced to the InP valence band and the arrows indicate excitonic transitions experimentally determined from the data of Fig. 41, except for the two points with the lowest x values, which have been obtained under bias. The transition energies are also referenced to the valence band calculated for a given In concentration and corrected for the appropriate quantum size effect using the parameters of Table I. The bandgap of InP is indicated by a straight horizontal line. Remarkably good agreement is obtained between the calculated energy bands and the experimental data points for the entire range of $0.2 < x < 0.74$. In the $x = 1.0$ sample we observe misfit dislocations indicating some strain relaxation resulting in reduced VB(3/2, 3/2)–VB(3/2, 1/2) splitting.

The conduction band discontinuities of a number of SLS samples were measured directly by Cavicchi et al. (1989) using admittance spectroscopy.

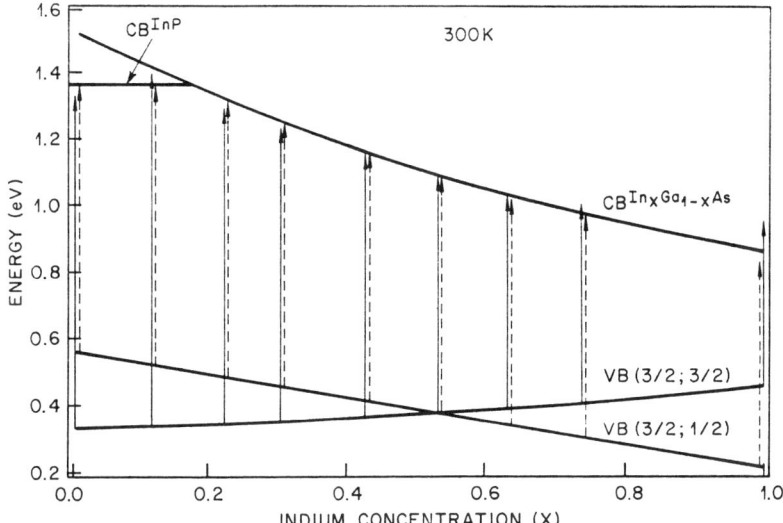

FIG. 44. Calculated band structure of the strained GaInAs. Energy is referenced to the InP valence band. Arrows indicate measured transition energies corrected for the quantum size effect. Samples with the lowest values of x were measured under applied electric field.

The samples used were grown in a *p-i-n* configuration by HSMBE. Each contained a four-period superlattice with approximately 70 Å thick wells, separated by 180 Å thick InP barriers. Capacitance and conductance measurements were performed as a function of temperature to obtain thermal activation energies. These were 0.09 and 0.21 eV for the In concentrations in the wells of $x = 0.37$ and 0.69, respectively. From these raw values conduction band offsets of 175 meV and 315 meV (± 25 meV) were obtained. The values of ΔE_c measured by admittance spectroscopy on the strained layer and control lattice matched samples are in excellent agreement with the calculated energy bands of GaInAs shown in Fig. 44. Excellent agreement is thus obtained between the calculated band structure and the measured band offsets.

At the very low In concentrations, $x < 0.2$, the data of Figs. 40–42 cannot be interpreted in terms of transitions between the conduction band electrons and valence band light and heavy hole of the strained layer quantum well. These data fit only transitions between the unstrained conduction band of the InP barrier and the strained layer valence bands. The spectral changes illustrated in Figs. 41 and 42 can be explained semi-quantitatively in terms of changing overlap between the hole and electron wave functions for $n = 1$ states of a type II superlattice. The wave functions were calculated using the resonant tunneling method. The oscillator strength of the excitonic transition as given by Eq. (4) is proportional to the square of their overlap integral. For $x > 0.2$ both electrons and holes are confined in the strained layers of $Ga_{1-x}In_xAs$, forming a structure known as a type I superlattice. The effect of the external electric field on the probability distribution of these particle states, shown in Fig. 30, is the decreased overlap as the electron and hole distributions are pulled apart and a slight energy shift due to the quantum confined Stark effect. This last effect is very small, less than 10 meV (Temkin et al. 1987). The photocurrent spectrum obtained under a bias of 3 V, does not show significant changes as the square of the overlap integral (calculated for the $x = 0.30$ sample) decreases from 0.94 to 0.80. This minor change is not detectable and the measured photocurrent actually increases due to a larger depletion width. The electric field effect is dramatically different in a type II superlattice, as shown in Fig. 45. The normalized $n = 1$ sublevel probability distributions for electrons and light holes are drawn to scale, together with the quantum well energy structure. In the absence of applied field the probability distributions of electron and holes are spatially separated with the resulting overlap of less than 0.009. The electrons are confined to the 280 Å wide InP wells with the calculated depth of 172 meV below the strained GaAs barriers. The light holes are still confined in the strained layer. As discussed previously, the transitions between these states are not observed either in the photocurrent or luminescence spectra. The

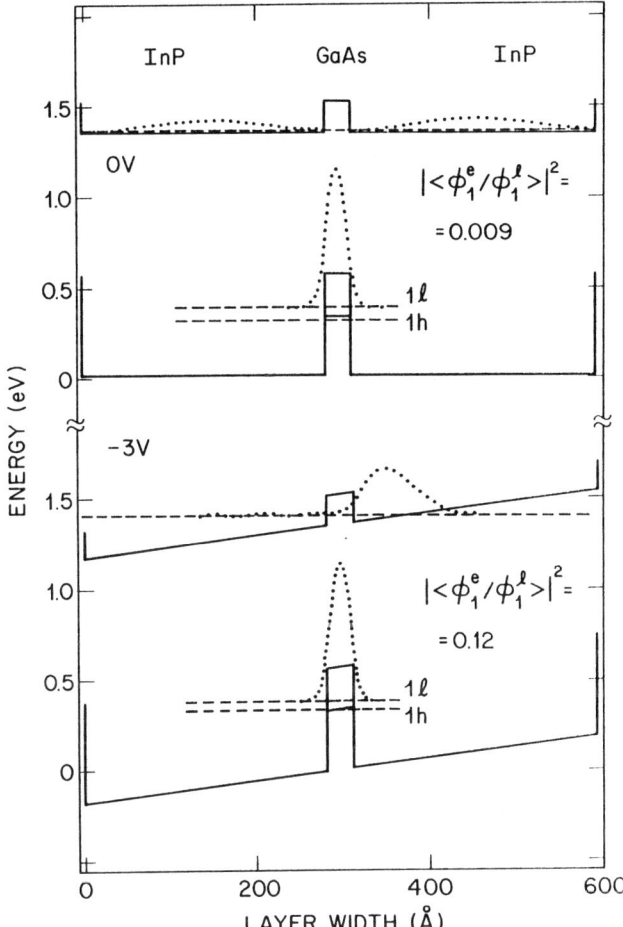

FIG. 45. Calculated quantum well energy levels and wave functions of the electrons and light holes in type II superlattice ($x = 0.0$ sample) with and without the applied electric field.

application of the electric field to this type of superlattice results in a dramatic increase of the calculated overlap, to 0.12, which manifests itself in a large increase of the photocurrent response. However, to account quantitatively for the observed spectral change, higher order exciton states, as well as continuum states, should be included in the calculations (Miller et al. 1986b). With increased bias the wave function of electron quasi-confined to InP penetrates further into the strained layer barrier which gives rise to discernible exciton resonances. This effect is more pronounced for the sample with $x = 0.12$, in which the GaInAs potential barrier is lower.

The In concentration at which the superlattice switches from type I to type II is a sensitive function of the relative (unstrained) GaAs/InP valence band energy. For instance, a 100 meV change in the value of this parameter, from 0.34 eV to 0.24 eV, results in the critical In concentration shifting from $x_c = 0.195$ to $x_c = 0.088$. A change of 170 meV, in the same direction, would eliminate the transition. Experimentally, the transition point can be determined to lie in the range of $0.15 < x < 0.22$, corresponding to ± 60 meV error in the relative valence band energy. The agreement suggests that the effect of strain on the valence band offset $\Delta E_v^{\text{GaAs/InP}}$ is vanishingly small, as suggested by the mid-gap theories.

Additional evidence for the superlattice type transition was provided by the experiments of Gerling et al. (1991), who measured the changes induced in the PL spectrum of strained layers of GaAs grown on InP as a function of hydrostatic pressure. Hydrostatic pressure, which causes changes in the sample volume, can be used to induce direct-to-indirect bandgap transitions in semiconductors (Paul 1961).

Two samples, each containing a single well of GaAs were used. The well thicknesses were approximately 18 and 25 Å. While these dimensions place the GaAs wells somewhat above the critical layer thickness, examination by cathodoluminescence imaging did not show any misfit dislocations. The samples were thus considered to be completely pseudomorphic. Low-temperature (2K) PL and PLE measurements without the applied pressure showed recombination between the electrons confined to InP and holes confined to the strained GaAs well, as expected from type II superlattices. The pressure dependence of the PL emission of the two samples is shown in Fig. 46. At low pressures the PL energy shifted at the rate of approximately 1 meV/kbar. The same rate was measured for the band-edge recombination of InP. The smooth pressure dependence disappeared at the pressures of 64 kbar and 71 kbar for the wider and narrower wells, respectively. The PL signal from the wells decreased abruptly at higher pressures. The change in slope and intensity was found to be completely reversible. These changes could not be associated with the direct–indirect transitions of bulk InP or GaAs, which occur at considerably higher pressures (Tozer et al. 1988).

The effect described in Fig. 46 was attributed to a hydrostatic pressure induced transition from type II to type I superlattice. This transition is similar, but not identical, to that discussed earlier in biaxially strained samples. In the absence of hydrostatic pressure the GaAs/InP structures are type II, as discussed by Gershoni et al. (1988c). The applied hydrostatic pressure results in the increased energy of the Γ point and decreased energy of X point. At a sufficiently high pressure the energy of the X band of the strained GaAs well becomes lower than the energy of the Γ band of the InP

5. OPTICAL PROPERTIES OF Ga$_{1-x}$In$_x$As/InP QUANTUM WELLS

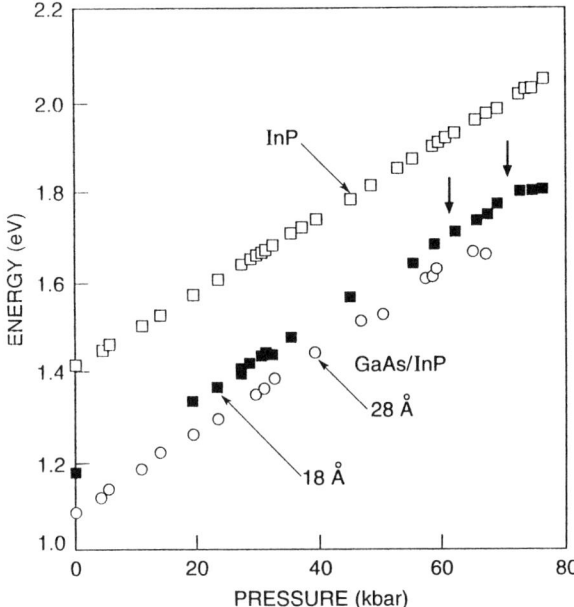

FIG. 46. A plot of the PL energies measured on the GaAs wells grown on InP as a function of the hydrostatic pressure. The open squares illustrate the pressure induced change in the band gap of InP.

barrier, resulting in a type I structure. The conduction band of the GaAs well in that structure is indirect. Gerling et al. (1991) simulated the effect of pressure using a deformation potential model similar to the one discussed in detail previously. Good agreement with the measured energy shift was obtained. The decrease in the PL intensity observed at higher pressures is difficult to model. The effect was attributed to the smallness of the overlap integral between the envelope functions of the Γ point of InP and the X point of GaAs.

The transition from type I to type II superlattice was invoked recently to explain the high performance of the diode lasers in which the active layer structure is based on quantum wells under a tensile strain (Starck et al. 1992). Such structures contain GaInAsP wells of the bulk composition corresponding to the band gap at about 1.5 μm and barriers and waveguides of GaInAsP with a band gap close to 1.3 μm. Quantum well lasers grown under a compressive or highly tensile strain are characterized by very low threshold current density and high slope efficiency (Temkin et al. 1990; Zah et al. 1991; Thijs et al. 1992).

3. Higher Order Optical Transitions

Until now we discussed only the lowest order optical transitions of strained quantum wells. We have shown that these transitions can be explained in terms of a simple one-band model. The only effect of strain in this model is the energy shift of the Brillouin zone center band edges. These shifts are calculated by the phenomenological deformation potential theory. However, since higher order terms in the crystal momentum and strain are no longer negligible away from the zone center, higher order transitions are not expected to follow such a simple model. A multiband model for calculating electronic states of strained heterostructures was developed recently by Baraff and Gershoni (1991) and Gershoni et al. (1992). Their eight-band $k \cdot p$ model incorporates the deformation potential theory and thus consistently contains the higher order terms. For the sake of simplicity, however, we shall continue to use the one-band model in this section.

Within the framework of a one-band effective mass model two important strain induced corrections, band nonparabolicity and nonlinear strain effects, should be considered (Gershoni et al. 1989a). This is because the uniaxial strain mixes the light hole ($|3/2, \pm 1/2\rangle$) valence band and the split-off ($|1/2; \pm 1/2\rangle$) valence band. This mixing results in a correction, nonlinear in strain, to the valence band edge:

$$\Delta E_{NL}^{\pm} = \pm \tfrac{1}{2}\Delta_0(x)[1 + \tfrac{1}{2}X - (1 + X + \tfrac{9}{4}X^2)^{1/2}] \tag{28}$$

where the $+(-)$ signs refer to the $3/2 \pm 1/2$ ($1/2 \pm 1/2$) valence band. The spin-orbit split-off energy of the unstrained ternary is given as $\Delta_0(x)$ and $X = \varepsilon_0/\Delta_0$. The band mixing introduces a correction to the band-edge effective mass of the light holes:

$$\Delta\left[\frac{1}{m_{LH,z}^*(x)}\right] = \frac{1}{4}\left[\frac{1}{m_{LH,x}^*(x)} - \frac{1}{m_{HH,z}^*(x)}\right]\left[\frac{1 + 9X}{(1 + 2X + 9X^2)^{1/2}} - 1\right] \tag{29}$$

where $m_{LH,z}^*(x)$ and $m_{HH,z}^*(x)$ are the light and heavy hole effective masses, in the growth direction z, of the unstrained ternary, respectively. The corrections introduced by Eqs. (28) and (29) are negligibly small for the lowest order ($n = 1$) transitions. They amount to no more than 20 meV even in the most strained samples and are thus comparable to the exciton linewidth. The corrections are much more important for the higher order excitons. For example, a shift of 90 meV is calculated for the $n = 2$ transition of the highly strained sample with $x = 0.73$. The confined carrier energies can be now calculated using Eqs. (2) and (3) or (7)–(9) or in which band nonparabolicity is included according to the dispersion relation given in Eq. (13).

The PCE spectra obtained from a set of HSMBE grown SLS samples are shown in Fig. 47. The set includes samples grown under tensile and compressive strain. Most of the excitonic features are clearly visible even at room temperature. The calculated transition energies are marked with vertical bars. Good agreement for $n = 1$ and $n = 2$ transitions is obtained for the entire range of compositions.

The PL and PCE data presented here, as well as additional PLE data discussed by Gershoni and Temkin (1989), provide a wealth of information on the exciton energies in the ternary SLS structures covering a wide range of strain and well thicknesses. Experimental results for several transitions are summarized in Fig. 48 and compared with the model discussed previously. Calculated optical transition energies are shown as a function of the indium concentration x for 55, 75, and 95 Å thick wells for which the

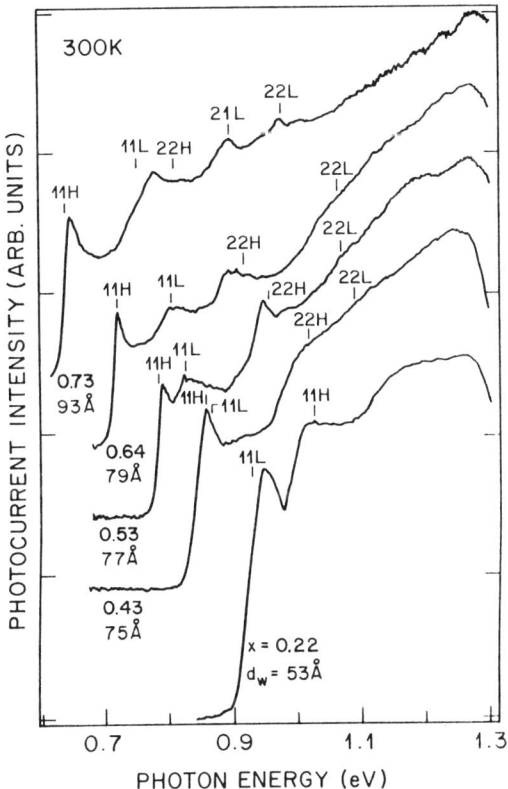

FIG. 47. Room temperature PCE spectra of five strained layer GaInAs superlattice samples. Calculated excitonic transitions are indicated with vertical bars.

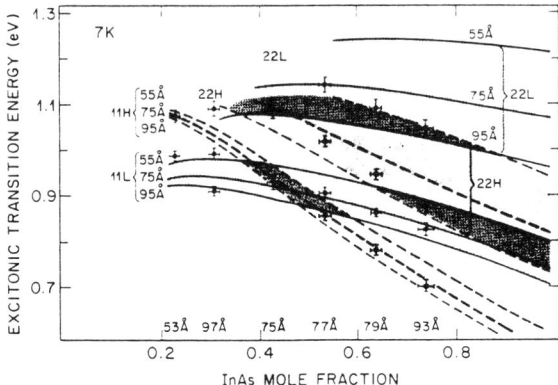

FIG. 48. Calculated $\Delta n = 0$ excitonic transition energies as a function of the In concentration for strained wells of three different thicknesses. The measured energies are indicated for comparison.

thicknesses and compositions were known precisely from the growth data, TEM cross sections, and high-resolution X-ray diffraction measurements. Also indicated in Fig. 48 are the measured $\Delta n = 0$ excitonic transitions. The quantum well widths for the different data points are indicated below them. This figure summarizes the changes in the optical properties of the GaInAs/InP material system as a function of x (and thus of the strain). It shows the absence of excitonic transitions for $x < 0.2$, where the system is a type II superlattice, and the change-over from the light hole to the heavy hole exciton as the lowest energy optical transition (absorption edge) above $x \sim 0.43$. Very good agreement is obtained between all the observed transitions and the calculation for the entire range of strain values.

VII. Thermal Stability and Impurity Induced Disordering

The structural characteristics of single quantum wells and superlattices can be greatly altered by the exposure to high temperatures, well in excess of the growth temperature, and impurity indiffusion. The question of structural stability is very important in all superlattice based devices, particularly so in the case of strained layer superlattices. Some of the changes induced by the high-temperature treatment appear to be unique to quantum well structures, such as the well-studied localized disordering of GaAs/GaAlAs structures upon diffusion or implantation of acceptors (Laidig et al. 1981). Aside from a more fundamental interest, the controlled

5. OPTICAL PROPERTIES OF $Ga_{1-x}In_xAs/InP$ QUANTUM WELLS

disordering of a superlattice can be exploited in many device applications; for instance, in semiconductor lasers (Fukuzawa et al. 1984). In this section we discuss thermal annealing and impurity induced disordering in lattice matched and strained GaInAs/InP quantum wells.

The changes induced by the high-temperature exposure were monitored by low-temperature photoluminescence. In addition, TEM cross sections were used to obtain independent information about interdiffusion depth as a function of temperature. The samples were grown by HSMBE and consisted of four GaInAs quantum wells with the as-grown thicknesses of 9, 16, 30, and 47 Å. These were separated by InP barriers approximately 100 Å thick. A 600 Å thick GaInAs reference layer was included in these samples. The samples were cleaved into a number of small pieces and heat treated in a carbon crucible under an InP cover in a flowing H_2 ambient. The heating cycle consisted in each case of a 20 min ramp, a 5 min exposure at a constant temperature, and a 10 min cooling cycle. Under these conditions no luminescence degradation was observed up to the temperature of 825°C.

The PL spectra illustrating the changes induced by these brief high-temperature anneals are shown in Fig. 49. The top trace shows the 6K spectrum of the untreated control sample, which is discussed in detail in Section II. Prior to high temperature experiments the entire sample was mapped for PL uniformity and the spectra obtained from different parts of the sample did not show variations larger than ± 10 Å in the exciton wavelength and linewidth. The question of sample uniformity was also discussed in Section II. We have not been able to observe any changes in the PL spectra of the quantum well samples as a function of the growth temperature, from 425°C to 550°C, or postgrowth heat treatment of up to 620°C (Temkin et al. 1985b). Using the annealing procedure just described, changes become visible only upon raising the annealing temperature above 700°C. As shown in Fig. 49, increasing the temperature of the heat treatment results in considerable shift of the exciton peaks towards higher energies. The degree of shift depends strongly on the well thickness. Thus the exciton peak corresponding to the largest, 47 Å wide well, shifts from 1.375 μm to 1.10 μm after the exposure to a temperature as high as 842°C, an increase in energy of 225 meV. In the same temperature range, the thinnest well, initially 9 Å wide, exhibited a shift of ~ 85 meV. This is consistent, as will be discussed later, with changes in the well composition being larger than the quantum size effects due to changes in the well sizes. While the high-temperature treatment does result in a visible roughening and localized surface decomposition of the InP cap layer, the intensity of the well luminescence is remarkably constant up to ~ 820°C. An indication of unchanged high quality of the wells and well–barrier interfaces is the relative weakness of the InP emission. No changes were found in the PL

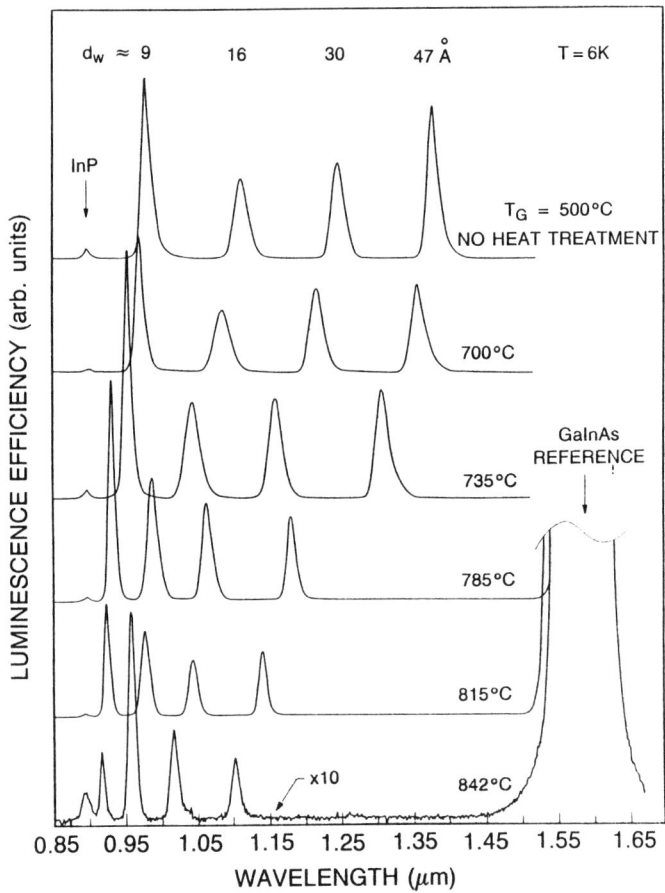

FIG. 49. Low-temperature PL spectra obtained on lattice matched GaInAs quantum wells subjected to brief (5 min) high-temperature anneals.

spectra of the bulk, 600 Å wide, GaInAs reference layer. Degradation of the PL was observed above $\sim 820°C$, as indicated in the lowest race of Fig. 49, which had to be expanded by about a factor of 10. This degradation is most likely due to decomposition of the sample as the result of loss of phosphorus.

A commonly used indicator of the sample quality, and specifically that of the well–barrier interface smoothness, is the PL linewidth. The linewidths shown by the control sample could be interpreted, as discussed in Section II, as due to the well thickness fluctuations of less than a monolayer over the exciton diameter. After the high-temperature annealing the PL linewidths of all of the excitonic transitions were considerably reduced. For instance, the linewidth of the PL from the 9 Å wide well was reduced from 20 meV to 9.5 meV after exposure to the highest temperature. Any conclusion as to the improved interface smoothness, however, must await the precise knowledge of the well dimensions after this heat treatment.

The TEM of the control sample is compared in Fig. 50 with that obtained on a sample treated at 785°C. These micrographs reveal two striking features of the thermally treated GaInAs/InP quantum wells. First, the well dimensions change significantly even after a brief, 5 min, exposure to high temperatures. For instance, the thinnest well, initially 9 Å wide, increases to 24 Å. The widening continues at higher temperatures and the same 9 Å well reaches 33 Å after annealing at 815°C. The increase in the well thickness is expected since any in-diffusion of phosphorus, believed to be the element with the highest diffusion rate, must be accompanied by the out-diffusion of As and Ga. A second, more surprising observation, is the sharpness of the well–barrier interfaces after the heat treatment. Normal diffusion processes should result in gradual changes in composition, which in GaAs/GaAlAs can be approximated by a parabolic potential well (Schlesinger and Kuech 1986). In the case of GaInAs/InP the interfacial contrast remains quite sharp up to the very highest temperature applied. This observation does not seem to be an artifact of the TEM contrast mechanism. The pictures of Fig. 50 were taken using a (200) reflection, which is normally forbidden in a diamond structure (Hirsh *et al.* 1977). In the zinc-blend lattice this reflection is allowed, due to chemical differences between the alternating (100) atomic planes. Thus the sharp TEM contrast between the wells and barriers, shown in Fig. 50, cannot be the result of graded interfaces. Quantitatively, the (200)

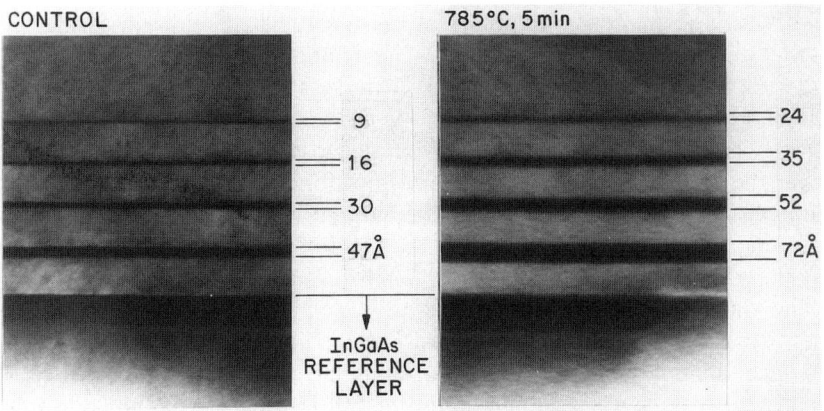

FIG. 50. TEM micrographs obtained on (a) as-grown control sample and (b) same sample annealed at 785°C for 5 min.

structure factors ($F(200)$) for InP and $Ga_{0.47}In_{0.53}As$ are calculated to be 9.16 and 1.83, respectively. Intermediate layers should have intermediate structure factors and thus be readily observed as having contrast intermediate between that of InP and $Ga_{0.47}In_{0.53}As$. Compositionally graded layers should thus be seen with TEM. The fact that they are not observed suggests that the broadened quantum well is compositionally uniform. This speculation is supported by high resolution X-ray studies, such as described in the first part of this chapter, of $Ga_{0.47}In_{0.53}As$/InP superlattices annealed in the same way as the single quantum well structures. Those studies show that the annealing does not cause significant added strain. It appears that the net effect of the widening of the well is to uniformly distribute the InP and $Ga_{0.47}In_{0.53}As$ that formerly occupied the volume of the new well. Thus, if the new well has twice the thickness of the original its composition would be $Ga_{0.23}In_{0.77}As_{0.5}P_{0.5}$, a composition lattice matched to InP. In fact, this mode of well broadening will always lead to a lattice matched structure. We believe that under heat treatment the four elements redistribute themselves to reach the energy minimum. Since the addition of strain is energetically too costly, the structure tends to stay lattice matched.

The shift in the PL energies from the various layers after the heat treatment is a result of a combination of compositional and dimensional changes. The compositional changes tend to increase the optical transition energy since the lattice matched quaternary has a higher band gap. The dimensional changes have an opposite effect, since the energy associated with the quantum confinement is reduced. We estimate the composition of the quaternary alloy constituting the diffused quantum well from its PL, after subtracting the energy associated with the quantum confinement. For example, the width of the largest QW has increased after the heat treatment to 104 Å. The quantum confinement shift associated with this width is estimated to be less than 50 meV. The changes in width should have therefore reduced the exciton shift by 70 meV. This gives a room temperature band gap of approximately 1 eV for the quaternary in the well, which agrees well with the lattice matched composition of $Ga_{0.23}In_{0.77}As_{0.5}P_{0.5}$, as can be seen in the simplified composition band-gap diagram of Fig. 51 (Ludowise 1977). This is exactly the composition obtained previously, assuming complete intermixing of barrier and well materials when the well thickness is doubled.

The increase in the well thicknesses explains the decrease in the PL linewidths of the heat-treated samples. The PL linewidth, which reflects the interface roughness, depends very strongly on the well size. It increases rapidly for narrow wells, as plotted in Fig. 10. The linewidths shown in Fig. 49 can be attributed to an average surface step size of about 1/3–1/2

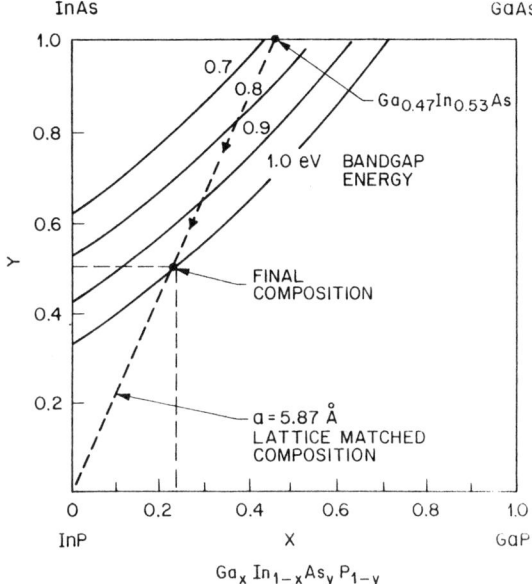

FIG. 51. Change in the well composition resulting from the high-temperature treatment.

monolayer. The linewidth of the heat-treated wells, while decreased in absolute terms, corresponds to a similar surface roughness in considerably wider wells. However, a possibility of smoother interfaces, as a result of diffusion along the interface boundary, cannot be excluded.

The temperature dependence of the PL emission and well thickness is difficult to model. The diffusion sources are very limited, only a few monolayers thick in case of As, and finite source effects must be considered. Vacancy interdiffusion most likely can be ruled out by the extremely high concentrations of the atomic species involved. The interface sharpness observed in interdiffused samples rules out the simple Fick's rule behavior (Crank 1975). Indeed, using the diffusion coefficient $D_0 = 7 \times 10^{10}$ cm^2/s and an activation energy $E_a = 5.65$ eV obtained for P in InP by Goldstein (1961), an interface displacement of 75 Å is calculated for 5 min at 815°C, resulting in the largest well becoming ~ 200 Å thick. This is a factor of two larger than the TEM measurement. It was shown by Weisberg and Blanc (1963) that much sharper diffusion profiles can be produced by concentration dependent diffusion, such as Zn in GaAs. In the quaternary system lattice matched to InP the driving force for interdiffusion, the compositional gradient, will be opposed by the lattice mismatch strain. Inclusion of such strain results in diffusivity varying as the second power of concentration. The results obtained with such a simple model are shown in Fig. 52, which plots diffusion depth as a function of temperature. The solid line is obtained by assuming a diffusion coefficient of the form $D = D_0(C/C_s)^2$, where C_s is the surface concentration. The calculation assumes the diffusion coefficient equal to that of P in InP and only slightly larger activation energy $Q = 5.8\,\text{eV}$. As seen in Fig. 52 excellent agreement can be obtained with the temperature dependence of the well thickness. These results suggests that the diffusion rate is adjusted by strain to maintain a roughly lattice matched quantum well.

The values reported for the activation energy of the well–barrier interdiffusion process vary widely and appear to be strongly dependent on the structure of the sample, method of preparation, etc. An activation energy of $E_A = 2.6\,\text{eV}$ was reported for a covered GaInAs well (Oshinowo et al. 1992b). Their well was grown on an InP buffer and was capped with a 200 Å thick barrier layer of InP. The structure of the sample was thus similar, but not identical, to that shown in Fig. 50. When the top InP layer was etched off, the activation energy dropped to $E_A = 1.3\,\text{eV}$. The bare well sample also showed a shift in the PL energy for the annealing temperature as low as 550°C, considerably lower than that needed to affect buried wells. The difference in the activation energies was attributed to larger point defect density in the bare samples.

We have similarly studied the thermal stability of the SLS structures. The

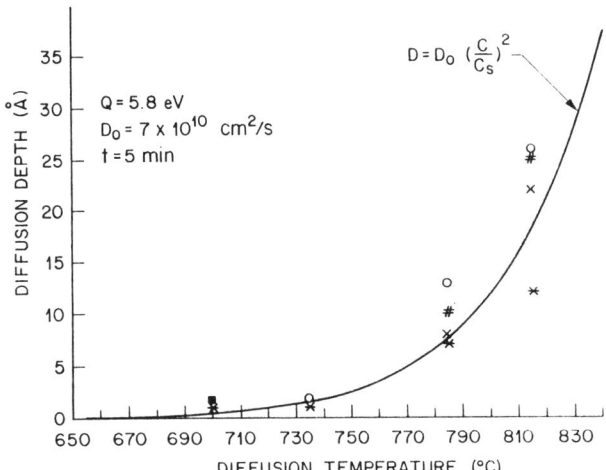

FIG. 52. Increase in the well size calculated on the basis of concentration dependent diffusivity model. Points indicate experimental data for various well thicknesses.

changes in the energy of the $n = 1$ excitonic transitions in two strained superlatices, with $x = 0.43$ (In-poor) and $x = 0.74$ (In-rich), subjected to similar high-temperature heat treatments were also studied by low temperature PL. The SLS structures of these samples consisted of 10 wells, each approximately 75 Å wide. These structures were carefully characterized, as discussed in Section VII and are considered to be coherently strained. Each wafer section was heat treated, measured, and heat treated again. Because of the very high activation energy of the P diffusion only the highest temperature alloying is expected to be significant. The PL results are shown in Fig. 53. Similar to the case of lattice matched wells we do not observe any significant degradation in the low-temperature PL efficiency, at least up to the annealing temperature of $\sim 800°C$. This indicates, as is also confirmed by TEM studies, that the SLS structures remain structurally stable after heat treatment at temperatures considerably higher than the growth temperature of 500°C. There is also a striking difference in the PL line line shift between the In-rich and -poor samples. The PL from the $x = 0.74$ SLS shows an increased energy shift with temperature, very similar to that observed in the lattice matched quantum wells. In contrast, the In-poor sample with $x = 0.43$ shows an initial shift to lower energies, after heat treatment at 730°C. The shift changes course towards higher energies only after heat treatment at higher temperatures. Similar behavior was reported by Razeghi

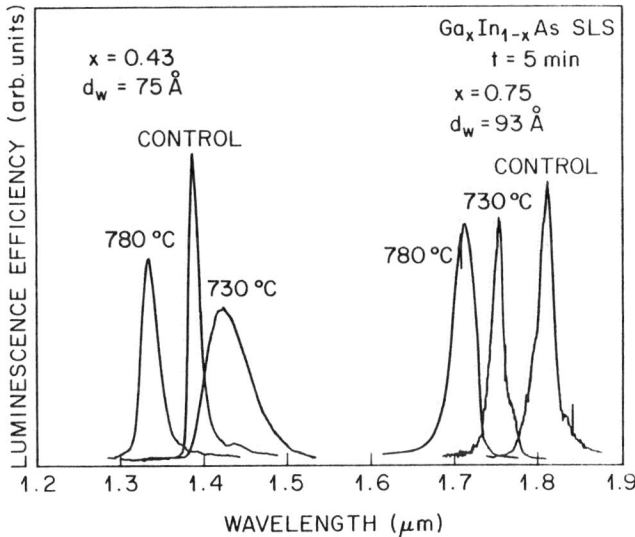

FIG. 53. Low-temperature PL spectra of lattice mismatched superlattices, with $x = 0.43$ and $x = 0.75$, after brief high-temperature anneals.

et al. (1987a). We interpret the initial red shift as a result of the increased In concentration and the decreased Ga concentration in the well as it interdiffuses with the InP barrier. The mixing results in a roughly lattice matched composition, in perfect agreement with our studies of lattice matched superlattices.

The heat treatment induced well–barrier interdiffusion can be greatly accelerated by indiffusion or implantation of impurities. Localized intermixing can be obtained by patterning the wafer with a suitable mask. The Zn-diffusion process has been investigated by Nakashima *et al.* (1988). Conclusive disordering by Si implantation have been achieved by Tell *et al.* (1988). The 25-period superlattice of nominally 50 Å GaInAs and equal thickness InP barriers was capped with a 1000 Å InP. The structure was grown at 625°C by atmospheric pressure MOCVD. Figure 54 shows a TEM cross section of a GaInAs/InP superlattice after a patterned Si implantation followed by 2 h annealing at 650°C. The implanation dose was $5 \times 10^{14}\,\mathrm{cm}^{-2}$ at the energy of 200 keV. As expected the relatively low annealing temperature did not affect the protected, not implanted, areas of the superlattice. The implanted area shows the original structure to be completely homogenized. Only residual implantation damage is visible in this area. The region between the two areas, under a tapered part of the

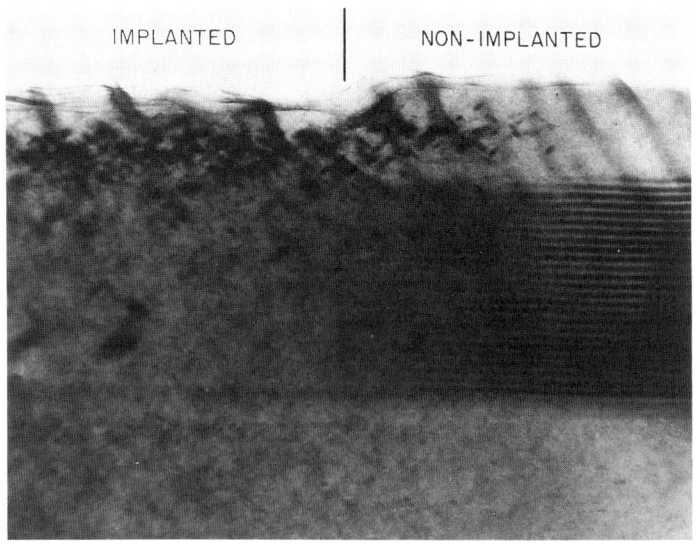

FIG. 54. A TEM micrograph of a superlattice sample subjected to a patterned implantation with Si and a 2 hr anneal at 650°C (after Tell *et al.* 1988).

resist mask, shows some intermixing and significantly lower residual damage. Noticeable grading of the interfaces could be observed after anneals as short as 20 min.

Low-temperature PL spectra of as-grown, annealed, and Si implanted and annealed samples are shown in Fig. 55. These data were obtained on samples implanted with a dose of 2.5×10^{14} and annealed for 20 min at 650°C. The as-grown sample exhibits a single narrow PL peak near 1.5 μm, consistent with the well dimensions. The annealed sample shows a blue shift of the magnitude comparable to that found previously in the single quantum wells. The PL spectrum of the implanted sample shows a considerable change, with a large shift toward higher energies as well as greatly increased linewidth. A significant PL response can be observed in this sample at ~1.2 μm, which corresponds to a 180 meV increase over the un-annealed sample. This was interpreted as a combined effect of intermixing, which can

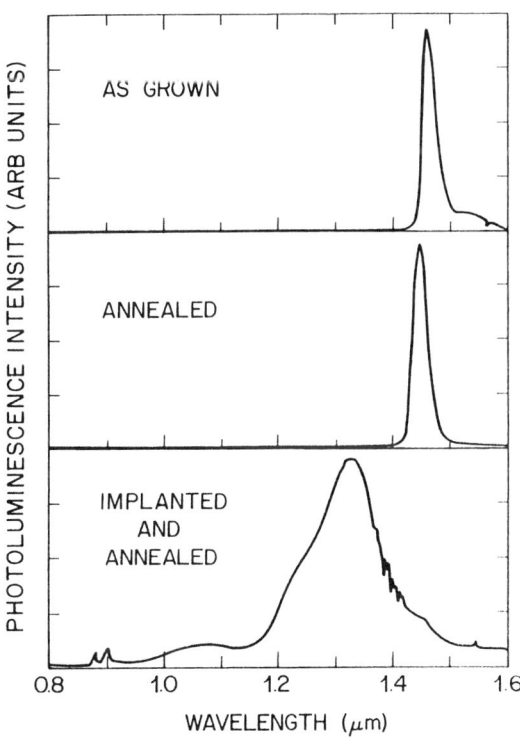

FIG. 55. The PL spectra of (a) as-grown, (b) annealed, and (c) Si implanted and annealed sample (after Tell et al. 1998).

increase the band gap by as much as 250 meV, and band filling effect of increased electron concentration in Si implanted material.

Oshinowo et al. (1992a) used the quantum well disordering accelerated by Ar^+-ion implantation to form quantum wires of GaInAs/InP as narrow as 750 Å. The process is, in principle, similar to that described previously by Cibert et al. (1986) for GaAs/AlGaAs.

The wire pattern was defined by electron beam lithography on the surface of a sample containing a single well of GaInAs, 50 Å thick, sandwiched between 200 Å thick barriers of InP. The resist pattern was used as an implantation mask. The samples were implanted with 30 keV Ar^+-ions with a range of doses from 10^{11} to 10^{14} cm^{-2}. For the annealing temperatures ranging from 600°C to 800°C the PL energy shift was consistently larger in the Ar^+ implanted samples. This implies accelerated interdiffusion in these samples.

The PL spectra of the narrow wire structures are shown in Fig. 56 (Oshinowo et al. 1992a). The large energy difference of 90 meV between the wire and the barrier represents the potential barrier of the lateral confinement. It is interesting to note that he PL efficiency of the wire is much stronger than that of the barrier, indicative of the high quality of the wire material.

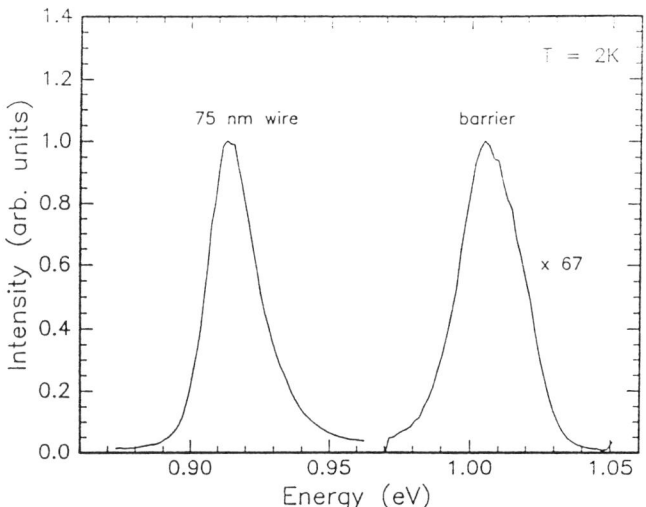

FIG. 56. Low-temperature PL spectra of narrow quantum wires and barriers formed by the process of quantum well disordering accelerated by Ar^+-ion implantation (after Oshinowo et al. 1992b.)

IX. Summary

In summary, we present a review of optical properties of $Ga_{1-x}In_xAs/InP$ quantum wells. A wide variety of lattice matched and strained structures were studied. Their optical properties were compared with a simple theory. Good agreement between the measured and calculated transition energies was obtained for the entire range of $Ga_{1-x}In_xAs$ compositions. Effects of lattice mismatch strain and the application of transverse electric field have been analyzed in detail. Fabrication of ultra-small structures exhibiting the effects of lowered dimensionality was presented and compared with theoretical predictions. Finally studies of the influence of heat treatment induced interdiffusion and ion induced disordering were presented and explained semi-quantitatively.

Acknowledgments

This work has been made possible by a combined effort of many friends and collaborators generous with their time, energy, and ideas. In particular, we would like to thank R. A. Hamm, S. Sumski, J. M. Vandenberg, D. Ritter, S. N. G. Chu, G. J. Dolan, and R. A. Logan.

References

Adachi, S. (1982). *J. Appl. Phys.* **53**, 8775.
Alavi, K., and Aggrawal, R. L. (1980). *Phys. Rev. B* **21**, 1311.
Asada, M., Miyamoto, Y., and Suematsu, M. (1985). *Japn. J. Appl. Phys.* **24**, L95.
Baraff, G. A., and Gershoni, D. B. (1991). *Phys. Rev. B* **43**, 4011.
Bar-Joseph, I., Klingshirn, C., Miller, D. A. B., Chemla, D. S., Koren, U., and Miller, B. I. (1987). *Appl. Phys. Lett.* **50**, 1010.
Bartels, W. J. (1973). *J. Vac. Sci. Technol. B* **1**, 338.
Bastard, G. (1981a). *Phys. Rev. B* **24**, 4714.
Bastard, G. (1981b). *Phys. Rev. B* **24**, 5693.
Bastard, G., and Brum, J. B. (1986). *IEEE J. Quant. Electron.* **QE-22**, 1625.
Bauer, R. S., and Margaritondo, G. (1987). *Phys. Today* **40**, 34.
Bhat, R., and Miller, R. C. (1987). Unpublished.
Bir, G. L., and Pikus, G. E. (1974). *Symmetry and Strain Induced Effects in Semiconductors.* Halsted, New York.
Blakemore, J. S. (1982). *J. Appl. Phys.* **53**, R123.
Brayant, G. H. (1987). *Phys. Rev. Lett.* **59**, 1140.
Brenner, I., Gershoni, D., Ritter, D., Panish, M. B., and Hamm, R. A. (1991). *Appl. Phys. Lett.* **58**, 965.
Brus, L. (1986). *IEEE J. Quant. Electronics* **QE-22**, 1909.

Cardona, M., and Christensen, N. E. (1987). *Phys. Rev. B* **35**, 6182.
Casey, H. C., and Buehler, E. (1977). *Appl. Phys. Lett.* **30**, 247.
Cavicchi, R. E., Lang, D. V., Gershoni, D., Sergent, A. M., Vandenberg, J., Chu, S. N. G., and Panish, M. B. (1989). *Appl. Phys. Lett.* **54**, 739.
Cebulla, U., Bacher, G., Mayer, G., Forchel, A., Tsang, W. T., and Razeghi, M. (1989). *Superlattices and Microstructures* **5**, 227.
Chandrasecar, M., and Pollack, F. H. (1977). *Phys. Rev. B* **15**, 2127.
Chu, S. N. G., Macrander, A. T., Strege, K., and Johnston, W. D., Jr. (1985). *J. Appl. Phys.* **57**, 249.
Cibert, J., Petroff, P. M., Dolan, G. J., Pearton, S. J., Gossard, A. C. and English, J. H. (1986). *Appl. Phys. Lett.* **49**, 223.
Claxton, P. A., Roberts, J. S., David, J. P. R., Sotomayor-Torres, C. M., Skolnick, M. S., Tapster, P. R., and Nash, K. J. (1987). *J. Cryst. Growth* **81**, 288.
Colas, E. Simhony, S., Kapon, E., Bhat, R., Hwang, D. M., and Lin, S. D. (1990). *Appl. Phys. Lett.* **57**, 914.
Cotta, M., Harriott, L. R., Hamm, R. A., and Temkin, H. (1992). To be published.
Crank, J. (1975). *The Mathematics of Diffusion*. Oxford Science Publications, Oxford.
DeCremoux, B., Hirth, P., and Ricciardi, J. (1981). *Inst. Phys. Conf. Ser.* **56**, 115.
Dingle, R., Wiegmann, W., and Henry, C. H. (1974). *Phys. Rev. Lett.* **33**, 827.
Ekimov, A. I., and Onuschenko, A. A. (1982). *Sov. Phys. Semicond.* **16**, 775.
Feldman, J., Peter, G., Gobel, E. O., Dawson, P., Moore, K., Foxon, C., and Elliott, R. J. (1987). *Phys. Rev. Lett.* **59**, 2337.
Forrest, S. R., Schmidt, P. H., Wilson, R. B., and Kaplan, M. L. (1984). *Appl. Phys. Lett.* **45**, 1199.
Fritz, J. J., Gourley, P. L., and Dawson, L. R. (1987). *Appl. Phys. Lett.* **51**, 1004.
Fukuzawa, T., Semura, S., Saito, H., Ohta, T., Uchida, Y., and Nakashima, H. (1984). *Appl. Phys. Lett.* **45**, 1.
Fultin, T. A., and Dolan, G. J. (1983). *Appl. Phys. Lett.* **42**, 752.
Galeuchet, Y. D., Rothuizen, H., and Roentgen, P. (1991). *Appl. Phys. Lett.* **58**, 2423.
Gerling, M., Pistol, M.-E., Samuelson, L., Seifert, W., Fornell, J.-O., and Ledebo, L. (1991). *Appl. Phys. Lett.* **59**, 806.
Gershoni, D., and Temkin, H. (1989). *J. Luminescence* **44**, 381.
Gershoni, D., Vandenberg, J. M., Hamm, R. A., Temkin, H., and Panish, M. B. (1987). *Phys. Rev. B* **36**, 1320.
Gershoni, D., Temkin, H., Dolan, G. J., Dunsmuir, J., Chu, S. N. G., and Panish, M. B. (1988a). *Appl. Phys. Lett.* **53**, 995.
Gershoni, D., Temkin, H., and Panish, M. B. (1988b). *Phys. Rev. B* **38**, 7870.
Gershoni, D., Temkin, H., Vandenberg, J. M., Chu, S. N. G., Hamm, R. A., and Panish, M. B. (1988c). *Phys. Rev. Lett.* **60**, 448.
Gershoni, D., Temkin, H., Panish, M. B., and Hamm, R. A. (1989a). *Phys. Rev. B* **39**, 5531.
Gershoni, D., Temkin, H., Panish, M. B., and Hamm, R. A. (1989b). *Phys. Rev. B* **40**, 1329.
Gershoni, D., Hamm, R. A., Panish, M. B., and Humphrey, D. A. (1990a). *Appl. Phys. Lett.* **56**, 1347.
Gershoni, D., Weiner, J. S., Chu, S. N. G., Baraff, G. A., Vandenberg, J. M. Pfeifer, L. N., West, K., and Chand, N. (1990b). *Phys. Rev. Lett.* **65**, 1631.
Gershoni, D., Henry, C. H., and Baraff, G. A. (1992). To be published.
Goldstein, B. (1961). *Phys. Rev.* **121**, 1305.
Harriott, L. R., Temkin, H., Chu, S. N. G., Wang, Y. L., Hsieh, Y. F., Hamm, R. A., Panish, M. B., and Wade, H. H. (1991). *Proceedings SPIE* **57**, 1565.
Hegarty, J., and Sturge, M. D. (1985). *J. Opt. Soc. Am. B* **2**, 1143.

Hensel, J. C., and Feher, G. (1963). *Phys. Rev.* **129**, 1401.
Henry, C. H., and Verbeek, B. H. (1988). *J. Lightwave Techn.*, to be published.
Hirayama, Y., Taracha, S., Suzuki, Y., and Okamoto, H. (1988). *Phys. Rev. B* **37**, 2774.
Hirsh, P., Howie, A., Nicholson, R. B., Pashley, D. W., and Wheelan, M. J. (1977). *Electron Microscopy in Thin Crystals*. Krieger, New York.
Ikonic, Z., Milanovic, V., Tjapkin, D., and Pajevic, S. (1988). *Phys. Rev. B* **37**, 2097.
Kane, E. O. (1957). *J. Phys. Chem. Solids* **1**, 249.
Kapon, E., Simhony, S., Bhat, R., and Hwang, D. M. (1989). *Appl. Phys. Lett.* **55**, 2715.
Kawaguchi, Y., and Asahi, H. (1987). *Appl. Phys. Lett.* **50**, 1243.
King, C. A., Hoyt, J. L., Gronet, C. M., Gibbons, J. F., and Wilson, S. W. (1988). *IEEE Electron Dev. Lett.* **9**, 229.
Koren, U., Koch, T. L., Presting, H., and Miller, B. I., (1987). *Appl. Phys. Lett.*
Laidig, W. D., Holonyak, N., Jr., Camras, M. D., Hess, K., Coleman, J. J., Dapkus, P., and Bardeen, J. (1981). *Appl. Phys. Lett.* **38**, 776.
Landau, L. D., and Lifshitz, E. M. (1977). *Quantum Mechanics–Nonreslativistic theory*. Pergamon Press, Oxford.
Ludowise, M. J. (1977). Ph.D. thesis, University of Illinois at Champaign-Urbana.
Mahajan, S., Dutt, B. V., Temkin, H., Cava, R. J., and Bonner, W. A. (1984). *J. Cryst. Growth* **68**, 589.
Marsh, J. H., Roberts, J. S., and Claxton, P. A. (1985). *Appl. Phys. Lett.* **46**, 1161.
Matthews, J. W., and Blakeslee, A. E. (1974). *J. Cryst. Growth* **27**, 118.
Miller, R. C., and Bhat., R. (1988).
Miller, R. C., Kleinman, D. A., Tsang, W. T., and Gossard, A. C. (1981). *Phys. Rev. B* **24**, 1134.
Miller, R. C., Gossard, A. C., Tsang, W. T., and Munteanu, O. (1982). *Phys. Rev. B* **25**, 3871.
Miller, D. A. B., Chemla, D. S., Damen, T. C., Gossard, A. C., Wiegman, W., Wood, T. H., and Burrus, C. A. (1985). *Phys. Rev. B* **32**, 1043.
Miller, B. I., Schubert, E. F., Koren, U., Ourmazd, A., Dayem, A. H., and Capik, R. J. (1986a). *Appl. Phys. Lett.* **49**, 1384.
Miller, D. A. B., Weiner, J. S., Chemla, D. S. (1986b). *IEEE J. Quant. Electron.* **QE-22**, 1816.
Miller, M. S., Weman, J., Pryor, C. E., Krishnamurty, M., and Petroff, P. M. (1992), *Phys. Rev. Lett.* **68**, 3464.
Nakashima, K., Kawaguchi, Y., Kawamura, Y., Imamura, Y., and Asahi, H. (1988). *Appl. Phys. Lett.* **52**, 1383.
Nelson, D. F., Miller, R. C., and Kleinman, D. A. (1987). *Phys. Rev. B* **35**, 7770.
Nicholas, R. J., Portal, J. C., Houlbert, C., Perrier, P., and Pearsall, T. P. (1979). *Appl. Phys. Lett.* **34**, 492.
Nishi, K., Hirose, K., and Mizatani, T. (1986). *Appl. Phys. Lett.* **49**, 194.
Notomi, M., Naganuma, M., Nishida, T., Tamamura, T., Iwamura, H., Nojima, S., and Okamoto, M. (1991). *Appl. Phys. Lett.* **58**, 720.
Osbourn, G. C. (1984). *Mater. Res. Soc. Symp. Proc.* **25**, 455.
Oshinowo, J., Dreybrodt, J., Forchel, A., Emmerling, M., Gyuro, I., Speier, P., and Zelinski, E. (1992a). Proceedings, Int. Conf. InP and Rel. Mat., p. 63.
Oshinowo, J., Forchel, A., Grutzmacher, D., Stollenwerk, M., Heuken, M., and Heime, K., (1992b). *Appl. Phys. Lett.*, **60**, 2260.
Panish, M. B. (1980). *J. Electrochem. Soc.* **127**, 2729.
Panish, M. B. (1987). *J. Cryst. Growth*. **81**, 249.
Panish, M. B., and Sumski, S. (1984). *J. Appl. Phys.* **55**, 3571.
Panish, M. B., Temkin, H., and Sumski, S. (1985). *J. Vac. Sci. and Techn.* **3**, 657.
Panish, M. B., Temkin, H., Hamm, R. A., and Chu, S. N. G. (1986). *Appl. Phys. Lett.* **49**, 164.
Paul, W. (1961). *J. Appl. Phys. Suppl.* **32**, 2082.

Pearsall, T. P. (1982). *GaInAsP Alloy Semiconductors*. Wiley, New York.
People, R. (1987). *Appl. Phys. Lett.* **50**, 1604.
People, R., and Bean, J. C. (1985). *Appl. Phys. Lett.* **47**, 322.
Petroff, P. M., Gossard, A. C. Logan, R. A., and Wiegemann, W. (1982). *Appl. Phys. Lett.* **41**, 635.
Pryor, C. (1991). *Phys. Rev. B* **44**, 12912.
Razeghi, M., and Duchemin, J. P. (1984). *J. Cryst. Growth* **70**, 145.
Razeghi, M., Hirtz, J. P., Ziemelis, U. O., Delalande, C., Etiene, B., and Voos, M. (1983). *Appl. Phys. Lett.* **43**, 585.
Razeghi, M., Archer, O., and Launay, F. (1987a). *Semicond. Sci. Technol.* **2**, 793.
Razeghi, M., Maurel, P., Omnes, F., and Nagle, J. (1987b). *Appl. Phys. Lett.* **51**, 2216.
Reed, M. A., Bate, R. T., Bradshaw, K., Duncan, W. M., Frensley, W. R., Lee, J. W., and Shih, H. D. (1986). *J. Vac. Sci. Technol. B* **4**, 358.
Reithmaier, J. P., Cerva, H., and Losch, R. (1989). *Appl. Phys. Lett.* **54**, 48.
Ritter, D., Panish, M. B., Hamm, R. A., Gershoni, D., and Brenner, I. (1990). *Appl. Phys. Lett.* **56**, 1448.
Sauer, R., Harris, T. D., and Tsang, W. T. (1986). *Phys. Rev. B* **34**, 9023.
Sauer, R., Harris, T. D., and Tsang, W. T., (1988). *Phys. Rev. B*, to be published.
Segmuller, A., jand Blakeslee, A. E. (1973). *J. Appl. Cryst.* **6**, 19.
Scherer, A., and Craighead, H. G. (1986). *Appl. Phys. Lett.* **49**, 1284.
Schlesinger, T. E., and Kuech, T. (1986). *Appl. Phys. Lett.* **49**, 519.
Schuber, J. E., Fritz, I. J., and Dawson, L. R. (1985). *Appl. Phys. Lett.* **46**, 182.
Skocpol, W. J., Jackel, L. D., Hu, E. L., Howard, R. E., and Fetter, L. A. (1982). *Phys. Rev. Lett.* **49**, 951.
Skolnick, M. S., Tapster, P. R., Bass, S. J., Apsley, N., Pitt, A. D., Chew, N. G., Cullis, A. G., Aldred, S. P., and Warwick, C. A. (1986). *Appl, Phys. Lett.* **48**, 1455.
Skolnick, M. S., Taylor, L. L., Bass, S. J., Pitt, A. D., Mowbray, D. J., Cullis, A. G., and Chew, N. G. (1987). *Appl. Phys. Lett.* **51**, 24.
Smith, B. T., Boyle, J. M., Dongarra, J. J., Garbow, B. S., Ibebe, Y., Klema, V. C., and Moler, C. B. (1976). *Matrix Eigensystem Routines-EISPACK Guide*. Springer-Verlag, New York.
Starck, C., Emergy, J-Y., Simes, R. J., Pagnod-Rossiaux, P., Gaborit, F., Matabon, M., Pommereau, F., Goldstein, L., and Barrau, J. (1992). Proceedings, Int. Conf. InP and Related Mat., 0. 457.
Tai, K., Tsang, W. T., and Hegarty, J. (1988). Proc. Int. Conf. Quant. Electronics (IQEC), July 18–21, Tokyo.
Tell, B., Johnson, B. C., Zyskind, Brown, J. M., Sulhoff, J. W., Brown-Goebeler, K. F., Miller, B. I., and Koren, U. (1988). *Appl. Phys. Lett.* **52**, 1428.
Temkin, H., Dutt, B. V., and Bonner, W. B. (1981a). *Appl. Phys. Lett.* **38**, 431.
Temkin, H., Keramidas, V. G., Pollack, M. A., and Wagner, W. R. (1981b). *J. Appl. Phys.* **52**, 1574.
Temkin, H., Panish, M. B., and Logan, R. A. (1985a). *Appl. Phys. Lett.* **47**, 978.
Temkin, H., Panish, M. B., Petroff, P. M., Hamm, R. A., Vandenberg, J. M., and Sumski, S. (1985b). *Appl. Phys. Lett.* **47**, 394.
Temkin, H., Panish, M. B., and Chu, S. N. G. (1986). *Appl. Phys. Lett.* **49**, 859.
Temkin, H., Chu, S. N. G., Panish, M. B. and Logan, R. A. (1987a). *Appl. Phys. Lett.* **50**, 956.
Temkin, H., Dolan, G. J., Panish, M. B., and Chu, S. N. G. (1987b). *Appl. Phys. Lett.* **50**, 413.
Temkin, H., Gershoni, D., and Panish, M. B. (1987c). *Appl. Phys. Lett.* **50**, 1776.
Temkin, H., Gershoni, D., Chu, S. N. G., Vandenberg,J., Hamm, R. A., and Panish, M. B. (1989). *Appl. Phys. Lett.* **55**, 1668.
Temkin, H., Tanbun-Ek, T., and Logan, R. A. (1990). *Appl. Phys. Lett.* **56**, 1210.

Tersoff, J. (1986). *Phys. Rev. Lett.* **12**, 2755.
Thijs, P. J. A., Binsma, J. J., Tiemeijer, L. F., and van Dongen, T. (1992). *Electron. Lett.* **28**, 829.
Tozer, S. W., Wolford, D. J., Bradley, J. A., Bour, D., and Stringfellow, G. B. (1988). *Proc. 19th Int. Conf. Phys. Semicond.*, (W. Zawadzki, ed.), p. 881. Warsaw.
Tsang, W. T., and Schubert, E. F. (1986a). *Appl. Phys. Lett.* **49**, 220.
Tsang, W. T., and Miller, R. C. (1986). *J. Cryst. Growth* **77**, 55.
Tsang, W. T., Schubert, E. F., Chu, S. N. G., Tai, K., and Sauer, R. (1987). *Appl. Phys. Lett.* **50**, 540.
Tsuchiya, M., Gaines, J. M., Yan, R. H., Simes, R. J., Holtz, P. O., Coldren, L. A., and Petroff, P. M. (1989a). *Phys. Rev. Lett.* **62**, 466.
Tsuchiya, M., Petroff, P. M., and Coldren, L. (1989b). *Appl. Phys. Lett.* **54**, 1690.
Tu, C. W., Miller, R. C., Wilson, B. A., Petroff, P. M., Harris, T. D., Kopf, R. F., Sputz, R. F., and Lamont, M. G. (1987). *J. Cryst. Growth* **81**, 159.
Vandenberg, J. M., Chu, S. N. G., Hamm, R. A., Panish, M. B., and Temkin, H. (1986). *Appl. Phys. Lett.* **49**, 1305.
Vandenberg, J. M., Hamm, R. A., Panish, M. B., and Temkin, H. (1987). *J. Appl. Phys.* **62**, 1278.
Wakita, K., Kawamura, Y., Yoshikuni, Y., and Asahi, H. (1986). *Electron. Lett.* **22**, 907.
Weisberg, L. R., and Blanc, J. (1963). *Phys. Rev.* **131**, 1548.
Yu, P. Y., Cardona, M., and Pollack, F. H. (1971). *Phys. Rev. B* **3**, 340.
Zah, C. E., Bhat, R., Pathak, B., Caneau, C., Favire, F. J., Andreadakis, N. C., Hwang, D. M., Koza, M. A., Chen, C. Y., and Lee, T. P. (1991). *Electron. Lett.* **27**, 1414.

Index

A

Absorption, 350, 354, 356–357, 360, 364, 374, 377
Absorption coefficient, 125
Absorption measurement, 111
Absorption spectrum, 111
Activation energy, 52, 55
Admittance spectroscopy, 397
Airy differential equation, 11
Airy functions, 9, 11
Alloy, 322, 357
AlSb, 232, 252
$Al_{1-x}Ga_xAs$, 231–233
Analytical solutions, 9
Annealing, 56
As-stabilized surface, 6
Asymmetric potential wells, 172
Asymmetric sawtooth structure, 22
Auger electron spectroscopy, 43
Autocompensation, 5, 78

B

Band discontinuity, 397
Band filling, 369, 414
Band gap
  effective, 267
  patterning, 260
    lateral, 261, 284, 286, 294
Band mixing, 402
Be
  in δ-doped GaAs, 58
  in GaAs, 57–58
Bloch-type eigenfunction, 9

Boundary conditions, 12, 341
Brownian motion, 51
Bulk modulus, 396

C

Carbon, 58
Carrier Hall concentration, 79
Centroid of charge, 32
Charge carriers
  capture, 310, 326
  diffusion, 308
  lateral confinement, 267
  recombination, 308
Composition
  change, 382, 407–408
  modulation, 348
  uniformity, 340, 346, 355, 357
Compositional grading, 6, 23, 154–158
  applications, 154, 158
  band edge profiles, 155
Compositional superlattices, 106
Conduction band offset, 397–398
Confined energy levels, 341, 350, 358, 369, 371, 374–375, 394
Contact resistance, 94
Continuum, 352
Coulomb correlation effect, 1, 70, 74
Coulomb repulsion, 71–72
Crystal growth, 1, 3
Cubic lattice, *see* Simple cubic lattice
Current-voltage characteristic, 128
CV-concentration, 30
CV-measurements, 38
CV-profile, 33, 53

D

de Broglie wavelength, 9, 17
Debye-Hueckel screening length, *see* Screening radius
Deep donor, 91
Defects, 356, 363, 386, 390
Deformation potential, 395–396
Degenerately doped, 83
δ-doped barriers, 131
δ-doped doping superlattice, 1, 106, 108, 122
δ-doped FET, 93
  heterostructure, 1
δ-doped semiconductor, 3, 27, 80
delta-doping, 1
  definition, 27
Density fluctuations, doping atoms, 21
Density of states, 364
  inhomogeneous broadening, 270
  low dimensional, 270
Depletion-mode δ-doped FET, 100
Depth, 30
Diamagnetic shift, 251
Diffusion, 1, 51
  activation energy, 409–410
  concentration dependence, 410
  length, 52, 55
Diffusion coefficient, 6, 52–53, 55–56, 67, 409
  Be in GaAs, 60
  C-acceptors in GaAs, 58
  C in GaAs, 61
  Si
    in $Al_xGa_{1-x}As$, 58
    in GaAs, 56
Digital alloy, 156–165
  design, 165
  effective bandgap, mass, 163–165
  energy levels, 164–165
  flux transients, 160–161, 165
  shutter motion, 160–161, 165–166
Dirac's delta function, 7
Discreteness of charge, 21
Distribution, dopants, 1
Donor, *see* Shallow donor
Doping, lateral patterning, 269, 283, 295
Doping superlattice, 106–107, 113
Doping superlattice laser, 1, 135, 139
  current injection, 135
Drift mobility, 72
Dry etching, 367

DX-center, 78
Dynamic effects, 357, 361

E

Effective mass, 25, 266
Effective masses, 341, 358–359, 402
Eigenstate energies, 9, 12
Einstein relation, 67, 72
Elastic scattering, 21
Elastic tensor, 395
Electric field, 376, 393
Electro-absorption, 377, 382
Electroluminescence, 133
Electronic properties, 1, 80, 85
Electronic structure, 6
Energy shift, 345, 351, 376, 388, 391–392, 401, 404, 406
Enhancement-mode operation, 99
Envelope function, 10
Evaporation, dopants, 3
Even symmetry of states, 12
Exact solution, V-shaped well, 10
Exchange reactions at surfaces, 244–245
Exciton linewidth, 348, 350, 355, 402, 406
Exiton, 111

F

Face-centered cubic lattice, 76
Faceting, 240
Fermi level pining induced segregation, 69
  surface, 63
Fermi's golden rule, 124
Fermi velocity, 21
Fickian dopant diffusion, 52
Field-driven redistribution, 69
Field-effect transistor, 1, 91
Fractional monolayer superlattice, 235–236, 243
Free carrier concentration, 77
Free hole concentration, 80

G

GaAs, 231, 233
GaSb, 232, 252
Gaussian distribution, 52

Gaussian function, 6
Gaussian wave function, 35
Generation rate, 122
Graded well growth, 155–168
   chemical vapor deposition (CVD), 158–159
   curvature control, 167
   molecular beam epitaxy (MBE), 155–156, 158–168
      analog alloy, 156, 159–161
      analog versus digital alloy, 166–167
      digital alloy, 156–165
Growth interruption, 5
Growth kinetics
   on {100} GaAs surfaces, 221
   on {110} GaAs surfaces, 238

H

Half parabolic potential wells, 172–175
Hall carrier densities, 77
Hall factor, 77
Hall measurement, 77, 83
Hall mobility, 90
Heterostructure, 1, 85, 102, 104, 261
   FET, 101
   patterned, nonplanar, 264–266
Higher order transitions, 357, 402
Holes, 342, 356, 358–359, 371, 394
   level crossing, 394
   masses, 359
Homostructure, 80
Hydrostatic pressure, 395

I

Impurity disordering, 404, 412, 414
InAs, 232
Inelastic scattering, 21
Interdiffusion, 407
Interface states, 63
Ion-gauge flux measurement, 161, 166–167
Ionized-impurity mobility, 83, 90
Ionized-impurity scattering, 83, 90

K

Kronig-Penney model, 25

L

Lateral superlattices, 219, 235
Lifetime, 356, 362–363
Lifetime, carriers, 122
Light-emitting diode (LED), 1, 132
Localization, Si in GaAs, 37
Long-period doping superlattices, 114–115
Loss, 140

M

Madelung energy, 75
Magneto optical properties, quantum wire superlattices, 251
Magnetoresistance, 104
Matthews and Blakesle, 390
MBE, see Molecular beam epitaxy
Metal-semiconductor FET, 1
Mid gap theory, 395
Migration enhanced epitaxy, 233
Miniband width, 128
Mixing effect, 45
Mobility, 79, 82
   enhancement, 83–84
Modulator, 378, 380
Modulators, 1, 143
Molecular beam epitaxy, 4, 220, 230
   doping, 269
   nonplanar, 274
Monolayer, 5
Multisubband transition, 117

N

Negative differential conductivity (NDC), 130
Nonalloyed ohmic contacts, 94–95
Nondegenerately doped, 83
Non-parabolicity, 343, 358, 402

O

Odd symmetry, 12
Optoelectronic devices, 132
Ordered alloy, 348, 357
Organometallic chemical vapor deposition
   doping, 270
   nonplanar, 288, 296

Organometallic vapor phase deposition, quantum wire superlattices, 240
Oscillator strength, 119, 341, 361–362, 382, 398
Overlap integral, 341, 398, 401

P

Parabolic potential wells, 153–217
 applications, 158
 characterization, 169–176, 181–189
  capacitance voltage (CV), magneto-capacitance voltage (MCV), 156, 176, 181–184
  inelastic (Raman) light scattering, 156, 171–173
  photoluminescence excitation (PLE), 156, 169–172, 205–207
  photoluminescence (PL), 156, 169–172
  resonant tunneling (RT), 156, 172–176
 electron density profile, 176–184
  experiment, 181–184
  theory, 176–181
 electronic structure, 169, 180–181, 194–195
  band offsets, 169, 171, 175
  calculations, 169, 180–181
  magnetic field, 194–195
   charge density wave, 194
   spin density wave, 194–195
   Wigner crystal, 194
  interband optical transitions, 158, 204–207
   inelastic (Raman) light scattering, 156, 171–173
   photoluminescence excitation (PLE), 156, 169–172, 205–207
   photoluminescence (PL), 156, 169–172
   selection rules, 169–172, 206
 magnetotransport, 189–195
  fractional quantum Hall effect (FQHE), 158, 192
  quantum Hall effect (FQHE), 158, 189–192
  tilted magnetic field, 192–193
 modulation doping, 176, 184–189
  mobility, 184–189

persistent photoconductivity (PPC), 185–186
perturbations, 202–211
 superlattice, 207–211
 sinusoidal, 157, 211
 square, 157, 207–211
 unintentional, 167, 202–204
plasma excitations, 195–204, 211–214
 generalized Kohn theorem, 196, 201
 radiation coupling schemes, 196–204
 resonances, 196–204, 211–214
  coupled cyclotron plasmon, 158, 196, 198–204
  cyclotron, 158, 200–201
  non-linear response, 158
  plasmon, 158, 196–204, 211–214
  surface plasmon, 201
 scattering, 158, 186–189
  alloy disorder, 186–187
  background impurity, 187, 189
  interface roughness, 189
  intersubband, 187
  size effect, 187–189
 subband filling effects, 183–184
 subband tunability, 157
Perpendicular transport, 126
Phase coherence, 21
Phase separation, 332
Photocurrent excitation, 355, 360, 392–393, 403
Photoluminescence, 1, 116, 121
 photoluminescence excitation spectroscopy, 249
 polarization, 248–249
 quantum wire superlattices, 247–248
Photoluminescence energy, 119
Photoluminescence line shape, 117
Photoluminescence spectroscopy, 113
Photonic switching, 144
Pinch-off voltage, 98
p-i-n diodes, 355, 359, 379, 382, 384, 391
Poisson's equation, 8
Polarization selection rules, 343
Position expectation value, 28
Post-growth annealing, 51
Potential fluctuation, 86, 348, 350, 356
Potential wells
 asymmetric, 172

half parabolic, 172–175
parabolic, 153–217
square, 154, 167–169
Pressure dependence, 400
Proximity annealing, 51

Q

Quantum box, 364, 368
Quantum capacitance, 32
Quantum-confined absorption, 1, 109
Quantum-confined interband transitions, 107, 123
Quantum-confined transitions, 119
Quantum confinement, 265
  energy, 268
  multi dimensional, 260, 270, 272, 305
Quantum dot, 270, 287
Quantum well
  patterned, 266, 271
    integrated laser, 318
  laser, 311
  subbands, 266, 272
Quantum wire, 364, 371, 374–375
  crescent, 273, 297
  heterostructure, 263, 270
  laser, 321
  subbands, 305, 307, 325
Quantum wire superlattices, 242
  optical properties, 242, 247–248
Quaternary wells, 347, 354, 382

R

Radiative lifetime, 119
Random distribution, 76
Recombination rate, 119
Reflection high-energy electron diffraction (RHEED), 4, 222, 224, 228
Repulsive Coulomb interaction, 70–71
Resolution
  CV profile, 35
  CV technique, 36
  SIMS, 45
Resonant tunneling, 128
Roughening, 45

S

Saturation, free carrier concentration, 1, 77
Sawtooth doping superlattice, 22
Sawtooth modulator, 145
Sawtooth structure, *see* Symmetric sawtooth structure
Sawtooth superlattice, 109, 111
Schottky barrier height, 41
Schottky contact, 20, 29, 37
Schroedinger equation, 341, 372
Screening, 120
Screening radius, 30, 89
Secondary ion mass spectrometry (SIMS), 62, 161, 167
  doping profile, 45
Segregation, 1, 62, 70
  of dopants, 60
  velocity, 66–67
Selection rules, 111
Selectively δ-doped heterostructure (SDH), *see* Heterostructure
Selectively doped heterostructure, *see* Heterostructure
Selectively doped heterostructure transistors (SDHT), 101
Self-alignment, 96
Self-consistent solutions, 1, 18
Self ordering, 264, 269, 280, 296
Serpentine superlattices, 241–242, 247
Shallow donor, 91
Shockley diode, 131
Short-channel effect, 93
Short-period doping superlattice, 114–115
Si, in $Al_xGa_{1-x}As$, 67
Si-air, 78
Simple cubic lattice, 76
Singular surface, 220
Spatial distribution, 27
Spatial extent of wave function, 35
Spatial localization, 1, 27
Spatial separation, 84
Square potential wells, 154, 167–169
Standard deviation, 6
Stark capacitance, 32
Stark effect, 376
Stark ladder energy, 128
Step bunching on surfaces, 239–240
Stimulated emission, 138–139

Stokes shift, 350, 355–356, 388
Strain, 385
  band mixing, 402
  band offsets, 395
  critical thickness, 385
  effective mass dependence, 395
  modulator, 382
  non-linear effects, 402
  tensor, 395
  thermal stability, 411
Submonolayer deposition, 236, 243
Superkinks, 246
Superlattice, 1, 22, 336, 354, 359, 378, 386, 411
  lateral, 263
  in parabolic wells, 207–214
    capacitance voltage, 210–211
    electronic structure, 208–209
    magnetoplasmon dispersion, 211–214
    optical properties, calculation, 211
    transport, 207, 210–211
  type I to type II transition, 398, 400
Surface
  diffusion, 265, 277, 293, 302
  energy, 265
  reconstruction, 6, 222, 226
  segregation, 63, 65
    process, 65
    Sn in GaAs, 60
  states, 41, 63
Suspension of epitaxial crystal growth, 3
Symmetric sawtooth structure, 22

T

Temperature dependence, 353, 360, 405, 407, 410–411
Theory of capacitance-voltage (CV) profiling, 29
Thermal stability, 404
Thomas-Fermi screening length, 30
Tilted superlattices, 236–237
Transfer matrix, 344
Transistor, see Selectively doped heterostructure transistor
Transition energies, 123
Transmission, 1
Transmission electron microscopy, 339, 345, 347, 386, 407, 412
Transmission probability, 342–343
Transmission spectroscopy, 122
Transmission spectrum, 123
Transport, 1
Triangular well, 15
Tunability, 115
Tunable spontaneous emission, 134
Two dimensional electron gas, 104
Two-mask process, 95
Two-region model, 93

U

Ultra-high vacuum, 4

V

Valence band offset, 341–342, 359, 395
Variational method, 9
Variational solutions, 13
Varshni equation, 360
Vegard's law, 395
Velocity-field characteristic, 93
Vicinal surface, 224, 229
V-shaped-quantum-well, 1, 6, 9–10, 13, 18
V-shaped well, 15

W

Wave function, 8
Wenzel-Kramers-Brillouin (WKB) approximation, 15
WKB approximation, 15
WKB method, 9
WKB solutions, 15
Work function, 63

X

X-ray diffraction, 340, 387

Z

Zero-order approximation, 17

Contents of Volumes in this Series

Volume 1 Physics of III–V Compounds

C. Hilsum, Some Key Features of III–V Compounds
Franco Bassani, Methods of Band Calculations Applicable to III–V Compounds
E. O. Kane, The k-p Method
V. L. Bonch-Bruevich, Effect of Heavy Doping on the Semiconductor Band Structure
Donald Long, Energy Band Structures of Mixed Crystals of III–V Compounds
Laura M. Roth and Petros N. Argyres, Magnetic Quantum Effects
S. M. Puri and T. H. Geballe, Thermomagnetic Effects in the Quantum Region
W. M. Becker, Band Characteristics near Principal Minima from Magnetoresistance
E. H. Putley, Freeze-Out Effects, Hot Electron Effects, and Submillimeter Photoconductivity in InSb
H. Weiss, Magnetoresistance
Betsy Ancker-Johnson, Plasma in Semiconductors and Semimetals

Volume 2 Physics of III–V Compounds

M. G. Holland, Thermal Conductivity
S. I. Novkova, Thermal Expansion
U. Piesbergen, Heat Capacity and Debye Temperatures
G. Giesecke, Lattice Constants
J. R. Drabble, Elastic Properties
A. U. Mac Rae and G. W. Gobeli, Low Energy Electron Diffraction Studies
Robert Lee Mieher, Nuclear Magnetic Resonance
Bernard Goldstein, Electron Paramagnetic Resonance
T. S. Moss, Photoconduction in III–V Compounds
E. Antončik and J. Tauc, Quantum Efficiency of the Internal Photoelectric Effect in InSb
G. W. Gobeli and F. G. Allen, Photoelectric Threshold and Work Function
P. S. Pershan, Nonlinear Optics in III–V Compounds
M. Gershenzon, Radiative Recombination in the III–V Compounds
Frank Stern, Stimulated Emission in Semiconductors

Volume 3 Optical of Properties III–V Compounds

Marvin Hass, Lattice Reflection
William G. Spitzer, Multiphonon Lattice Absorption
D. L. Stierwalt and R. F. Potter, Emittance Studies
H. R. Philipp and H. Ehrenveich, Ultraviolet Optical Properties
Manuel Cardona, Optical Absorption above the Fundamental Edge
Earnest J. Johnson, Absorption near the Fundamental Edge
John O. Dimmock, Introduction to the Theory of Exciton States in Semiconductors
B. Lax and J. G. Mavroides, Interband Magnetooptical Effects

CONTENTS OF VOLUMES IN THIS SERIES

H. Y. Fan, Effects of Free Carries on Optical Properties
Edward D. Palik and George B. Wright, Free-Carrier Magnetooptical Effects
Richard H. Bube, Photoelectronic Analysis
B. O. Seraphin and H. E. Bennett, Optical Constants

Volume 4   Physics of III–V Compounds

N. A. Goryunova, A. S. Borschevskii, and D. N. Tretiakov, Hardness
N. N. Sirota, Heats of Formation and Temperatures and Heats of Fusion of Compounds $A^{III}B^{V}$
Don L. Kendall, Diffusion
A. G. Chynoweth, Charge Multiplication Phenomena
Robert W. Keyes, The Effects of Hydrostatic Pressure on the Properties of III–V Semiconductors
L. W. Aukerman, Radiation Effects
N. A. Goryunova, F. P. Kesamanly, and D. N. Nasledov, Phenomena in Solid Solutions
R. T. Bate, Electrical Properties of Nonuniform Crystals

Volume 5   Infrared Detectors

Henry Levinstein, Characterization of Infrared Detectors
Paul W. Kruse, Indium Antimonide Photoconductive and Photoelectromagnetic Detectors
M. B. Prince, Narrowband Self-Filtering Detectors
Ivars Melngalis and T. C. Harman, Single-Crystal Lead-Tin Chalcogenides
Donald Long and Joseph L. Schmidt, Mercury-Cadmium Telluride and Closely Related Alloys
E. H. Putley, The Pyroelectric Detector
Norman B. Stevens, Radiation Thermopiles
R. J. Keyes and T. M. Quist, Low Level Coherent and Incoherent Detection in the Infrared
M. C. Teich, Coherent Detection in the Infrared
F. R. Arams, E. W. Sard, B. J. Peyton, and F. P. Pace, Infrared Heterodyne Detection with Gigahertz IF Response
H. S. Sommers, Jr., Macrowave-Based Photoconductive Detector
Robert Sehr and Rainer Zuleeg, Imaging and Display

Volume 6   Injection Phenomena

Murray A. Lampert and Ronald B. Schilling, Current Injection in Solids: The Regional Approximation Method
Richard Willliams, Injection by Internal Photoemission
Allen M. Barnett, Current Filament Formation
R. Baron and J. W. Mayer, Double Injection in Semiconductors
W. Ruppel, The Photoconductor-Metal Contact

CONTENTS OF VOLUMES IN THIS SERIES

Volume 7 Application and Devices
PART A

John A. Copeland and Stephen Knight, Applications Utilizing Bulk Negative Resistance
F. A. Padovani, The Voltage-Current Characteristics of Metal-Semiconductor Contacts
P. L. Hower, W. W. Hooper, B. R. Cairns, R. D. Fairman, and D. A. Tremere, The GaAs Field-Effect Transistor
Marvin H. White, MOS Transistors
G. R. Antell, Gallium Arsenide Transistors
T. L. Tansley, Heterojunction Properties

PART B

T. Misawa, IMPATT Diodes
H. C. Okean, Tunnel Diodes
Robert B. Campbell and Hung-Chi Chang, Silicon Carbide Junction Devices
R. E. Enstrom, H. Kressel, and L. Krassner, High-Temperature Power Rectifiers of $GaAs_{1-x}P_x$

Volume 8 Transport and Optical Phenomena

Richard J. Stirn, Band Structure and Galvanomagnetic Effects in III–V Compounds with Indirect Band Gaps
Roland W. Ure, Jr., Thermoelectric Effects in III–V Compounds
Herbert Piller, Faraday Rotation
H. Barry Bebb and E. W. Williams, Photoluminescence 1: Theory
E. W. Williams and H. Barry Bebb, Photoluminescence II: Gallium Arsenide

Volume 9 Modulation Techniques

B. O. Seraphin, Electroreflectance
R. L. Aggarwal, Modulated Interband Magnetooptics
Daniel F. Blossey and Paul Handler, Electroabsorption
Bruno Batz, Thermal and Wavelength Modulation Spectroscopy
Ivar Balslev, Piezopptical Effects
D. E. Aspnes and N. Bottka, Electric-Field Effects on the Dielectric Function of Semiconductors and Insulators

Volume 10 Transport Phenomena

R. L. Rode, Low-Field Electron Transport
J. D. Wiley, Mobility of Holes in III–V Compounds
C. M. Wolfe and G. E. Stillman, Apparent Mobility Enhancement in Inhomogeneous Crystals
Robert L. Petersen, The Magnetophonon Effect

CONTENTS OF VOLUMES IN THIS SERIES

Volume 11 Solar Cells

Harold J. Hovel, Introduction; Carrier Collection, Spectral Response, and Photocurrent; Solar Cell Electrical Characteristics; Efficiency; Thickness; Other Solar Cell Devices; Radiation Effects; Temperature and Intensity; Solar Cell Technology

Volume 12 Infrared Detectors (II)

W. L. Eiseman, J. D. Merriam, and R. F. Potter, Operational Characteristics of Infrared Photodetectors
Peter R. Bratt, Impurity Germanium and Silicon Infrared Detectors
E. H. Putley, InSb Submillimeter Photoconductive Detectors
G. E. Stillman, C. M. Wolfe, and J. O. Dimmock, Far-Infrared Photoconductivity in High Purity GaAs
G. E. Stillman and C. M. Wolfe, Avalanche Photodiodes
P. L. Richards, The Josephson Junction as a Detector of Microwave and Far-Infrared Radiation
E. H. Putley, The Pyroelectric Detector–An Update

Volume 13 Cadmium Telluride

Kenneth Zanio, Materials Preparation; Physics; Defects; Applications

Volume 14 Lasers, Junctions, Transport

N. Holonyak, Jr. and M. H. Lee, Photopumped III–V Semiconductor Lasers
Henry Kressel and Jerome K. Butler, Heterojunction Laser Diodes
A. Van der Ziel, Space-Charge-Limited Solid-State Diodes
Peter J. Price, Monte Carlo Calculation of Electron Transport in Solids

Volume 15 Contacts, Junctions, Emitters

B. L. Sharma, Ohmic Contacts to III–V Compound Semiconductors
Allen Nussbaum, The Theory of Semiconducting Junctions
John S. Escher, NEA Semiconductor Photoemitters

Volume 16 Defects, (HgCd)Se, (HgCd)Te

Henry Kressel, The Effect of Crystal Defects on Optoelectronic Devices
C. R. Whitsett, J. G. Broerman, and C. J. Summers, Crystal Growth and Properties of $Hg_{1-x}Cd_x Se$ alloys
M. H. Weiler, Magnetooptical Properties of $Hg_{1-x}Cd_x Te$ Alloys
Paul W. Kruse and John G. Ready, Nonlinear Optical Effects in $Hg_{1-x}Cd_x Te$

CONTENTS OF VOLUMES IN THIS SERIES

Volume 17 CW Processing of Silicon and Other Semiconductors

James F. Gibbons, Beam Processing of Silicon
Arto Lietoila, Richard B. Gold, James F. Gibbons, and Lee A. Christel, Temperature Distributions and Solid Phase Reaction Rates Produced by Scanning CW Beams
Arto Leitoila and James F. Gibbons, Applications of CW Beam Processing to Ion Implanted Crystalline Silicon
N. M. Johnson, Electronic Defects in CW Transient Thermal Processed Silicon
K. F. Lee, T. J. Stultz, and James F. Gibbons, Beam Recrystallized Polycrystalline Silicon: Properties, Applications, and Techniques
T. Shibata, A. Wakita, T. W. Sigmon, and James F. Gibbons, Metal-Silicon Reactions and Silicide
Yves I. Nissim and James F. Gibbons, CW Beam Processing of Gallium Arsenide

Volume 18 Mercury Cadmium Telluride

Paul W. Kruse, The Emergence of $(Hg_{1-x}Cd_x)Te$ as a Modern Infrared Sensitive Material
H. E. Hirsch, S. C. Liang, and A. G. White, Preparation of High-Purity Cadmium, Mercury, and Tellurium
W. F. H. Micklethwaite, The Crystal Growth of Cadmium Mercury Telluride
Paul E. Petersen, Auger Recombination in Mercury Cadmium Telluride
R. M. Broudy and V. J. Mazurczyck, (HgCd)Te Photoconductive Detectors
M. B. Reine, A. K. Sood, and T. J. Tredwell, Photovoltaic Infrared Detectors
M. A. Kinch, Metal-Insulator-Semiconductor Infrared Detectors

Volume 19 Deep Levels, GaAs, Alloys, Photochemistry

G. F. Neumark and K. Kosai, Deep Levels in Wide Band-Gap III–V Semiconductors
David C. Look, The Electrical and Photoelectronic Properties of Semi-Insulating GaAs
R. F. Brebrick, Ching-Hua Su, and Pok-Kai Liao, Associated Solution Model for Ga–In–Sb and Hg–Cd–Te
Yu. Ya. Gurevich and Yu. V. Pleskon, Photoelectrochemistry of Semiconductors

Volume 20 Semi-Insulating GaAs

R. N. Thomas, H. M. Hobgood, G. W. Eldridge, D. L. Barrett, T. T. Braggins, L. B. Ta, and S. K. Wang, High-Purity LEC Growth and Direct Implantation of GaAs for Monolithic Microwave Circuits
C. A. Stolte, Ion Implantation and Materials for GaAs Integrated Circuits
C. G. Kirkpatrick, R. T. Chen, D. E. Holmes, P. M. Asbeck, K. R. Elliott, R. D. Fairman, and J. R. Oliver, LEC GaAs for Integrated Circuit Applications
J. S. Blakemore and S. Rahimi, Models for Mid-Gap Centers in Gallium Arsenide

Volume 21 Hydrogenated Amorphous Silicon
Part A

Jacques I. Pankove Introduction
Masataka Hirose, Glow Discharge; Chemical Vapor Deposition

Yoshiyuki Uchida, dc Glow Discharge
T. D. Moustakas, Sputtering
Isao Yamada, Ionized-Cluster Beam Deposition
Bruce A. Scott, Homogeneous Chemical Vapor Deposition
Frank J. Kampas, Chemical Reactions in Plasma Deposition
Paul A. Longeway, Plasma Kinetics
Herbert A. Weakliem, Diagnostics of Silane Glow Discharges Using Probes and Mass Spectroscopy
Lester Guttman, Relation between the Atomic and the Electronic Structures
A. Chenevas-Paule, Experiment Determination of Structure
S. Minomura, Pressure Effects on the Local Atomic Structure
David Adler, Defects and Density of Localized States

Part B

Jacques I. Pankove, Introduction
G. D. Cody, The Optical Absorption Edge of a-Si: H
Nabil M. Amer and Warren B. Jackson, Optical Properties of Defect States in a-Si: H
P. J. Zanzucchi, The Vibrational Spectra of a-Si: H
Yoshihiro Hamakawa, Electroreflectance and Electroabsorption
Jeffrey S. Lannin, Raman Scattering of Amorphous Si, Ge, and Their Alloys
R. A. Street, Luminescence in a-Si: H
Richard S. Crandall, Photoconductivity
J. Tauc, Time-Resolved Spectroscopy of Electronic Relaxation Processes
P. E. Vanier, IR-Induced Quenching and Enhancement of Photoconductivity and Photoluminescence
H. Schade, Irradiation-Induced Metastable Effects
L. Ley, Photoelectron Emission Studies

Part C

Jacques I. Pankove, Introduction
J. David Cohen, Density of States from Junction Measurements in Hydrogenated Amorphous Silicon
P. C. Taylor, Magnetic Resonance Measurements in a-Si: H
K. Morigaki, Optically Detected Magnetic Resonance
J. Dresner, Carrier Mobility in a-Si: H
T. Tiedje, information about band-Tail States from Time-of-Flight Experiments
Arnold R. Moore, Diffusion Length in Undoped a-Si: H
W. Beyer and J. Overhof, Doping Effects in a-Si: H
H. Fritzche, Electronic Properties of Surfaces in a-Si: H
C. R. Wronski, The Staebler-Wronski Effect
R. J. Nemanich, Schottky Barriers on a-Si: H
B. Abeles and T. Tiedje, Amorphous Semiconductor Superlattices

Part D

Jacques I. Pankove, Introduction
D. E. Carlson, Solar Cells

G. A. Swartz, Closed-Form Solution of I–V Characteristic for a-Si: H Solar Cells
Isamu Shimizu, Electrophotography
Sachio Ishioka, Image Pickup Tubes
P. G. LeComber and W. E. Spear, The Development of the a-Si: H Field-Effect Transistor and Its Possible Applications
D. G. Ast, a-Si: H FET-Addressed LCD Panel
S. Kaneko, Solid-State Image Sensor
Masakiyo Matsumura, Charge-Coupled Devices
M. A. Bosch, Optical Recording
A. D'Amico and G. Fortunato, Ambient Sensors
Hiroshi Kukimoto, Amorphous Light-Emitting Devices
Robert J. Phelan, Jr., Fast Detectors and Modulators
Jacques I. Pankove, Hybrid Structures
P. G. LeComber, A. E. Owen, W. E. Spear, J. Hajto, and W. K. Choi, Electronic Switching in Amorphous Siliocn Junction Devices

Volume 22 Lightwave Communications Technology
Part A

Kazuo Nakajima, The Liquid-Phase Epitaxial Growth of IngaAsp
W. T. Tsang, Molecular Beam Epitaxy for III–V Compound Semiconductors
G. B. Stringfellow, Organometallic Vapor-Phase Epitaxial Growth of III–V Semiconductors
G. Beuchet, Halide and Chloride Transport Vapor-Phase Deposition of InGaAsP and GaAs
Manijeh Razeghi, Low-Pressure Metallo-Organic Chemical Vapor Deposition of $Ga_xIn_{1-x}AsP_{1-y}$ Alloys
P. M. Petroff, Defects in III–V Compound Semiconductors

Part B

J. P. van der Ziel, Mode Locking of Semiconductor Lasers
Kam Y. Lau and Ammon Yariv, High-Frequency Current Modulation of Semiconductor Injection Lasers
Charles H. Henry, Spectral Properties of Semiconductor Lasers
Yasuharu Suematsu, Katsumi Kishino, Shigehisa Arai, and Fumio Koyama, Dynamic Single-Mode Semiconductor Lasers with a Distributed Reflector
W. T. Tsang, The Cleaved-Coupled-Cavity (C^3) Laser

Part C

R. J. Nelson and N. K. Dutta, Review of InGaAsP InP laser Structures and Comparison of Their Performance
N. Chinone and M. Nakamura, Mode-Stabilized Semiconductor Lasers for 0.7–0.8- and 1.1–1.6-μm Regions
Yoshiji Horikoshi, Semiconductor Lasers with Wavelengths Exceeding 2 μm
B. A. Dean and M. Dixon, The Functional Reliability of Semiconductor Lasers as Optical Transmitters

CONTENTS OF VOLUMES IN THIS SERIES

R. H. Saul, T. P. Lee, and C. A. Burus, Light-Emitting Device Design
C. L. Zipfel, Light-Emitting Diode-Reliability
Tien Pei Lee and Tingye Li, LED-Based Multimode Lightwave Systems
Kinichiro Ogawa, Semiconductor Noise-Mode Partition Noise

Part D

Federico Capasso, The Physics of Avalanche Photodiodes
T. P. Pearsall and M. A. Pollack, Compound Semiconductor Photodiodes
Takao Kaneda, Silicon and Germanium Avalanche Photodiodes
S. R. Forrest, Sensitivity of Avalanche Photodetector Receivers for High-Bit-Rate Long-Wavelength Optical Communication Systems
J. C. Campbell, Phototransistors for Lightwave Communications

Part E

Shyh Wang, Principles and Characteristics of Integratable Active and Passive Optical Devices
Shlomo Margalit and Amnon Yariv, Integrated Electronic and Photonic Devices
Takaoki Mukai, Yoshihisa Yamamoto, and Tatsuya Kimura, Optical Amplification by Semi-conductor Lasers

Volume 23 Pulsed Laser Processing of Semiconductors

R. F. Wood, C. W. White, and R. T. Young, Laser Processing of Semiconductors: An Overview
C. W. White, Segregation, Solute Trapping, and Supersaturated Alloys
G. E. Jellison, Jr., Optical and Electrical Properties of Pulsed Laser-Annealed Silicon
R. F. Wood and G. E. Jellison, Jr., Melting Model of Pulsed Laser Processing
R. F. Wood and F. W. Young, Jr., Nonequilibrium Solidification Following Pulsed Laser Melting
D. H. Lowndes and G. E. Jellison, Jr., Time-Resolved Measurements During Pulsed Laser Irradiation of Silicon
D. M. Zebner, Surface Studies of Pulsed Laser Irradiated Semiconductors
D. H. Lowndes, Pulsed Beam Processing of Gallium Arsenide
R. B. James, Pulsed CO_2 Laser Annealing of Semiconductors
R. T. Young and R. F. Wood, Applications of Pulsed Laser Processing

Volume 24 Applications of Multiquantum Wells, Selective Doping, and Superlattices

C. Weisbuch, Fundamental Properties of III–V Semiconductor Two-Dimensional Quantized Structures: The Basis for Optical and Electronic Device Applications
H. Morkoc and H. Unlu, Factors Affecting the Performance of (Al, Ga)As/GaAs and (Al, Ga)As/InGaAs Modulation-Doped Field-Effect Transistors: Microwave and Digital Applications

N. T. Linh, Two-Dimensional Electron Gas FETs: Microwave Applications
M. Abe et al, Ultra-High-Speed HEMT Integrated Circuits
D. S. Chemla, D. A. B. Miller, and P. W. Smith, Nonlinear Optical Properties of Multiple Quantum Well Structures for Optical Signal Processing
F. Capasso, Graded-Gap and Superlattice Devices by Band-Gap Engineering
W. T. Tsang, Quantum Confinement Heterostructure Semiconductor Lasers
G. C. Osbourn et al., Principles and Applications of Semiconductor Strained-Layer Superlattices

Volume 25 Diluted Magnetic Semiconductors

W. Giriat and J. K. Furdyna, Crystal Structure, Composition, and Materials Preparation of Diluted Magnetic Semiconductors
W. M. Becker, Band Structure and Optical Properties of Wide-Gap $A^{II}_{1-x}Mn_xB^{VI}$ Alloys at Zero Magnetic Field
Saul Oseroff and Pieter H. Keesom, Magnetic Properties: Macroscopic Studies
T. Giebultowicz and T. M. Holden, Neutron Scattering Studies of the Magnetic Structure and Dynamics of Diluted Magnetic Semiconductors
J. Kossut, Band Structure and Quantum Transport Phenomena in Narrow-Gap Diluted Magnetic Semiconductors
C. Riquaux, Magnetooptical Properties of Large-Gap Diluted Magnetic Semiconductors
J. A. Gaj, Magnetooptical Properties of Large-Gap Diluted Magnetic Semiconductors
J. Mycielski, Shallow Acceptors in Diluted Magnetic Semiconductors. Splitting, Boil-off, Giant Negative Magnetoressitance
A. K. Ramdas and R. Rodriquez, Raman Scattering in Diluted Magnetic Semiconductors
P. A. Wolff. Theory of Bound Magnetic Polarons in Semimagnetic Semiconductors

Volume 26 III–V Compound Semiconductors and Semiconductor Properties of Superionic Materials

Zou Yuanxi, III–V Compounds
H. V. Winston, A. T. Hunter, H. Kimura, and R. E. Lee, InAs-Alloyed GaAs Substrates for Direct Implantation
P. K. Bhattachary and S. Dhar, Deep Levels in III–V Compound Semiconductors Grown by MBE
Yu. Yu. Gurevich and A. K. Ivanov-Shits, Semiconductor Properties of Superionic Materials

Volume 27 High Conducting Quasi-One-Dimensional Organic Crystals

E. M. Conwell, Introduction to Highly Conducting Quasi-One-Dimensional Organic Crystals
I. A. Howard, A Reference Guide to the Conducting Quasi-One-Dimensional Organic Molecular Crystals
J. P. Pouquet, Structural Instabilities
E. M. Conwell, Transport Properties

C. S. Jacobsen, Optical Properties
J. C. Scott, Magnetic Properties
L. Zuppiroli, Irradiation Effects: Perfect Crystals and Real Crystals

Volume 28 Measurement of High-Speed Signals in Solid State Devices

J. Frey and D. Ioannou, Materials and Devices for High-Speed and Optoelectronic Applications
H. Schumacher and E. Strid, Electronic Wafer Probing Techniques
D. H. Auston, Picosecond Photoconductivity: High-Speed Measurements of Devices and Materials
J. A. Valdmanis, Electro-Optic Measurement Techniques for Picosecond Materials, Devices, and Integrated Circuits
J. M. Wiesenfeld and R. K. Jain, Direct Optical Probing of Integrated Circuits and High-Speed Devices
G. Plows, Electron-Beam Probing
A. M. Weiner and R. B. Marcus, Photoemissive Probing

Volume 29 Very High Speed Integrated Circuits: Gallium Arsenide LSI

M. Kuzuhara and T. Nazaki, Active Layer Formation by Ion Implantation
H. Hasimoto, Focused Ion Beam Implantation Technology
T. Nozaki and A. Higashisaka, Device Fabrication Process Technology
M. Ino and T. Takada, GaAs LSI Circuit Design
M. Hirayama, M. Ohmori, and K. Yamasaki, GaAs LSI Fabrication and Performance

Volume 30 Very High Speed Integrated Circuits: Heterostructure

H. Watanabe, T. Mizutani, and A. Usui, Fundamentals of Epitaxial Growth and Atomic Layer Epitaxy
S. Hiyamizu, Characteristics of Two-Dimensional Electron Gas in III–V Compound Heterostructures Grown by MBE
T. Nakanisi, Metalorganic Vapor Phase Epitaxy for High-Quality Active Layers
T. Nimura, High Electron Mobility Transistor and LSI Applications
T. Sugeta and T. Ishibashi, Hetero-Bipolar Transistor and Its LSI Application
H. Matsueda, T. Tanaka, and M. Nakamura, Optoelectronic Integrated Circuits

Volume 31 Indium Phosphide: Crystal Growth and Characterization

J. P. Farges, Growth of Discoloration-free InP
M. J. McCollum and G. E. Stillman, High Purity InP Grown by Hydride Vapor Phase Epitaxy
T. Inada and T. Fukuda, Direct Synthesis and Growth of Indium Phosphide by the Liquid Phosphorous Encapsulated Czochralski Method
O. Oda, K. Katagiri, K. Shinohara, S. Katsura, Y. Takahashi, K. Kainosho, K. Kohiro, and R. Hirano, InP Crystal Growth, Substrate Preparation and Evaluation

K. Tada, M. Tatsumi, M. Morioka, T. Araki, and T. Kawase, InP Substrates: Production and Quality Control
M. Razeghi, LP-MOCVD Growth, Characterization, and Application of InP Material
T. A. Kennedy and P. J. Lin-Chung, Stoichiometric Defects in InP

Volume 32 Strained-Layer Superlattices: Physics

T. P. Pearsall, Strained-Layer Superlattices
Fred H. Pollack, Effects of Homogeneous Strain on the Electronic and Vibrational Levels in Semiconductors
J. Y. Marzin, J. M. Gerárd, P. Voisin, and J. A. Brum, Optical Studies of Strained III–V Heterolayers
R. People and S. A. Jackson, Structurally Induced States from Strain and Confinement
M. Jaros, Microscopic Phenomena in Ordered Superlattices

Volume 33 Strained-Layer Superlattices: Materials Science and Technology

R. Hull and J. C. Bean, Principles and Concepts of Strained-Layer Epitaxy
William J. Schaff, Paul J. Tasker, Mark C. Foisy, and Lester F. Eastman, Device Applications of Strained-Layer Epitaxy
S. T. Picraux, B. L. Doyle, and J. Y. Tsao, Structure and Characterization of Strained-Layer Superlattices
E. Kasper and F. Schaffler, Group IV Compounds
Dale L. Martin, Molecular Beam Epitaxy of IV–VI Compound Heterojunction
Robert L. Gunshor, Leslie A. Kolodziejski, Arto V. Nurmikko, and Nobuo Otsuka, Molecular Beam Epitaxy of II–VI Semiconductor Microstructures

Volume 34 Hydrogen in Semiconductors

J. I. Pankove and N. M. Johnson, Introduction to Hydrogen in Semiconductors
C. H. Seager, Hydrogenation Methods
J. I. Pankove, Hydrogenation of Defects in Crystalline Silicon
J. W. Corbett, P. Deák, U. V. Desnica, and S. J. Pearton, Hydrogen Passivation of Damage Centers in Semiconductors
S. J. Pearton, Neutralization of Deep Levels in Silicon
J. I. Pankove, Neutralization of Shallow Acceptors in Silicon
N. M. Johnson, Neutralization of Donor Dopants and Formation of Hydrogen-Induced Defects in n-Type Silicon
M. Stavola and S. J. Pearton, Vibrational Spectroscopy of Hydrogen-Related Defects in Silicon
A. D. Marwick, Hydrogen in Semiconductors: Ion Beam Techniques
C. Herring and N. M. Johnson, Hydrogen Migration and Solubility in Silicon
E. E. Haller, Hydrogen-Related Phenomena in Crystalline Germanium
J. Kakalios, Hydrogen Diffusion in Amorphous Silicon
J. Chevalier, B. Clerjaud, and B. Pajot, Neutralization of Defects and Dopants in III–V Semiconductors

G. G. DeLeo and W. B. Fowler, Computational Studies of Hydrogen-Containing Complexes in Semiconductors
R. F. Kiefl and T. L. Estle, Muonium in Semiconductors
C. G. Van de Walle, Theory of Isolated Interstitial Hydrogen and Muonium in Crystalline Semiconductors

Volume 35 Nanostructured Systems

Mark Reed, Introduction
H. van Houten, C. W. J. Beenakker, and B. J. van Wees, Quantum Point Contacts
G. Timp, When Does a Wire Become an Electron Waveguide?
M. Büttiker, The Quantum Hall Effect in Open Conductors
W. Hansen, J. P. Kotthaus, and U. Merkt, Electrons in Laterally Periodic Nanostructures

Volume 36 The Spectroscopy of Semiconductors

D. Heiman, Laser Spectroscopy of Semiconductors at Low Temperatures and High Magnetic Fields
Arto V. Nurmikko, Transient Spectroscopy by Ultrashort Laser Pulse Techniques
A. K. Ramdas and S. Rodriguez, Piezospectroscopy of Semiconductors
Orest J. Glembocki and Benjamin V. Shanabrook, Photoreflectance Spectroscopy of Microstructures
David G. Seiler, Christopher L. Littler, and Margaret H. Weiler, One- and Two-Photon Magneto-Optical Spectroscopy of InSb and $Hg_{1-x}CD_xTe$

Volume 37 The Mechanical Properties of Semiconductors

A.-B. Chen, Arden Sher and W. T. Yost, Elastic Constants and Related Properties of Semiconductor Compounds and Their Alloys
David R. Clarke, Fracture of Silicon and Other Semiconductors
Hans Siethoff, The Plasticity of Elemental and Compound Semiconductors
Sivaraman Guruswamy, Katherine T. Faber and John P. Hirth, Mechanical Behavior of Compound Semiconductors
Subhanh Mahajan, Deformation Behavior of Compound Semiconductors
John P. Hirth, Injection of Dislocations into Strained Multilayer Structures
Don Kendall, Charles B. Fleddermann, and Kevin J. Malloy, Critical Technologies for the Micromachining of Silicon
Ikuo Matsuba and Kinji Mokuya, Processing and Semiconductor Thermoelastic Behavior

Volume 38 Imperfections in III/V Materials

Udo Scherz and Matthias Scheffler, Density-Functional Theory of sp-Bonded Defects in III/V Semiconductors
Maria Kaminska and Eicke R. Weber, EL2 Defect in GaAs
David C. Look, Defects Relevant for Compensation in Semi-Insulating GaAs

R. C. Newman, Local Vibrational Mode Spectroscopy of Defects in III/V Compounds
Andrzej M. Hennel, Transition Metals in III/V Compounds
Kevin J. Malloy and Ken Khachaturyan, DX and Related Defects in Semiconductors
V. Swaminathan and Andrew S. Jordan, Dislocations in III/V Compounds
Krzysztof W. Nauka, Deep Level Defects in the Epitaxial III/V Materials

Volume 39 Minority Carriers in III–V Semiconductors: Physics and Applications

Niloy K. Dutta, Radiative Transitions in GaAs and Other III–V Compounds
Richard K. Ahrenkiel, Minority-Carrier Lifetime in III–V Semiconductors
Tomofumi Furuta, High Field Minority Electron Transport in p-GaAs
Mark S. Lundstrom, Minority-Carrier Transport in III–V Semiconductors
Richard A. Abram, Effects of Heavy Doping and High Excitation on the Band Structure of GaAs
David Yevick and Witold Bardyszewski, An Introduction to Non-Equilibrium Many-Body Analyses of Optical Processes in III–V Semiconductors

Volume 40 Epitaxial Microstructures

E. F. Schubert, Delta-Doping of Semiconductors: Electronic, Optical, and Structural Properties of Materials and Devices
A. Gossard, M. Sundaram, and P. Hopkins, Wide Graded Potential Wells
P. Petroff, Direct Growth of Nanometer-Size Quantum Wire Superlattices
E. Kapon, Lateral Patterning of Quantum Well Heterostructures by Growth of Nonplanar Substrates
H. Temkin, D. Gershoni, and M. Panish, Optical Properties of Ga$_{1-x}$In$_x$As/InP Quantum Wells

Volume 41 High Speed Heterostructure Devices

F. Capasso, F. Beltram, S. Sen, A. Pahlevi, and A. Y. Cho, Quantum Electron Devices: Physics and Applications
P. Solomon, D. J. Frank, S. L. Wright, and F. Canora, GaAs-Gate Semiconductor–Insulator–Semiconductor FET
M. H. Hashemi and U. K. Mishra, Unipolar InP-Based Transistors
R. Kiehl, Complementary Heterostructure FET Integrated Circuits
T. Ishibashi, GaAs-Based and InP-Based Heterostructure Bipolar Transistors
H. C. Liu and T. C. L. G. Sollner, High-Frequency-Tunneling Devices
H. Ohnishi, T. More, M. Takatsu, K. Imamura, and N. Yokoyama, Resonant-Tunneling Hot-Electron Transistors and Circuits

CONTENTS OF VOLUMES IN THIS SERIES

Volume 42 Oxygen in Silicon

F. Shimura, Introduction to Oxygen in Silicon
W. Lin, The Incorporation of Oxygen into Silicon Crystals
T. J. Shaffner and D. K. Schroder, Characterization Techniques for Oxygen in Silicon
W. M. Bullis, Oxygen Concentration Measurement
S. M. Hu, Intrinsic Point Defects in Silicon
B. Pajot, Some Atomic Configurations of Oxygen
J. Michel and L. C. Kimerling, Electrical Properties of Oxygen in Silicon
R. C. Newman and R. Jones, Diffusion of Oxygen in Silicon
T. Y. Tan and W. J. Taylor, Mechanisms of Oxygen Precipitation: Some Quantitative Aspects
M. Schrems, Simulation of Oxygen Precipitation
K. Sumino and I. Yonenaga, Oxygen Effect on Mechanical Properties
W. Bergholz, Grown-in and Process-Induced Effects
F. Shimura, Intrinsic/Internal Gettering
H. Tsuya, Oxygen Effect on Electronic Device Performance

ISBN 0-12-752140-2